D0321600

WITHDRAWN

Biological Control in IPM Systems in Africa

Biological Control in IPM Systems in Africa

Edited by
P. Neuenschwander

International Institute of Tropical Agriculture
Cotonou, Bénin

C. Borgemeister

Institute of Plant Diseases and Plant Protection
Hannover, Germany

and

J. Langewald

International Institute of Tropical Agriculture
Cotonou, Bénin

CABI *Publishing*

in association with the

ACP-EU Technical Centre for Agricultural and Rural Co-operation (CTA)

and the

Swiss Agency for Development and Cooperation (SDC)

CABI *Publishing* is a division of CAB *International*

CABI Publishing
CAB International
Wallingford
Oxon OX10 8DE
UK

Tel: +44 (0)1491 832111
Fax: +44 (0)1491 833508
E-mail: cabi@cabi.org
Website: www.cabi-publishing.org

CABI Publishing
44 Brattle Street
4th Floor
Cambridge, MA 02138
USA

Tel: +1 617 395 4056
Fax: +1 617 354 6875
E-mail: cabi-nao@cabi.org

A catalogue record for this book is available from the British Library, London, UK.

Library of Congress Cataloging-in-Publication Data
Biological control in IPM systems in Africa / edited by P. Neuenschwander, C. Borgemeister, and J. Langewald.
p. cm.
Includes bibliographical references (p.).
ISBN 0-85199-639-6 (alk. paper)
1. Pests--Biological control--Africa. I. Neuenschwander, P. (Peter)
II. Borgemeister, C. (Christian) III. Langewald, J. (Juergen)
SB975.5.A35 B56 2003
632′.9′096--dc21
2002152823

ISBN 0 85199 639 6

Published in association with:

The ACP-EU Technical Centre for Agricultural and Rural Co-operation (CTA)
Postbus 380
6700 AJ Wageningen
The Netherlands

and the

Swiss Agency for Development and Cooperation (SDC)
Freiburgstrasse 130
3003 Bern
Switzerland

Typeset in Souvenir Light by Columns Design Ltd, Reading
Printed and bound in the UK by Cromwell Press, Trowbridge

Contents

Contributors

Abraham, Yvonne J. CAB International, UK Centre, Bakeham Lane, Egham, Surrey TW20 9TY, UK.
E-mail: c/o Jerry.Bird@accenture.com

Ajuonu, Obinna International Institute of Tropical Agriculture, Biological Control Center for Africa, 08 BP 0932, Cotonou, Bénin.
E-mail: O.Ajuonu@cgiar.org

Bandyopadhyay, Ranajit International Institute of Tropical Agriculture, Oyo Road, PMB 5320, Ibadan, Nigeria.
E-mail: R.Bandyopadhyay@cgiar.org

Berner, Dana K. US Department of Agriculture, Agricultural Research Service, Foreign Disease, Weed Science Research Unit, Fort Detrick, MD 21702, USA.
E-mail: Dberner@ncifcrf.gov

Billah, Maxwell International Centre of Insect Physiology and Ecology (ICIPE), PO Box 30772, Nairobi, Kenya.
E-mail: Mbillah@icipe.org

Borgemeister, Christian Institute of Plant Diseases and Plant Protection, Herrenhäuser Str. 2, 30419 Hannover, Germany.
E-mail: Borgemeister@ipp.uni-hannover.de

Braimah, Haruna Biological Control Unit, Crops Research Institute, PO Box 3785, Kumasi, Ghana.
E-mail: Criggdp@Ghana.com

Cardwell, Kitty F. Cooperative State Research, Extension, and Education Service, US Department of Agriculture, 800 9th Street, S.W. Washington, DC 20024-2220, USA.
E-mail: KCardwell@ree.usda.gov

Cherry, Andy International Institute of Tropical Agriculture, Biological Control Center for Africa, 08 BP 0932, Cotonou, Bénin.
E-mail: A.Cherry@cgiar.org

Chilima, Clement Z. Forestry Research Institute of Malawi, PO Box 270, Zomba, Malawi.
E-mail: Frim@malawi.net

Cilliers, Carina J. Agricultural Research Council – Plant Protection Research Institute, Biological Control of Weeds, Private Bag X134, Pretoria 0001, South Africa. Present address: Box 1136, Wapadrand 0050, South Africa.
E-mail: cilliersc@envirokonsult.co.za

Cock, Matthew CABI Bioscience Switzerland, Rue des Grillons 1, CH-2800 Delémont, Switzerland.
E-mail: M.cock@cabi.org

Conlong, Des E. SA Sugar Experimental Station, Private Bag X02, Mount Edgecombe 4300, Kwazulu-Natal, South Africa.
E-mail: Xentdc@sugar.org.za

Day, Roger K. CAB International, Africa Regional Centre, PO Box 633-00621, Nairobi, Kenya.
E-mail: R.Day@cabi.org

Dimbi, Susan International Centre of Insect Physiology and Ecology (ICIPE), PO Box 30772, Nairobi, Kenya.
E-mail: Sdimbi@icipe.org

Ekesi, Sunday International Centre of Insect Physiology and Ecology (ICIPE), PO Box 30772, Nairobi, Kenya.
E-mail: Sekesi@icipe.org

Emechebe, Alphonse M. International Institute of Tropical Agriculture, Kano Station, Sabo Bakin Zuwa Road, PMB3112, Kano, Nigeria.
E-mail: A.Emechebe@cgiar.org

Everts, James CERES-LOCUSTOX Foundation, PO Box 3300, Dakar, Senegal.
E-mail:James_everts@yahoo.fr

Gerling, Dan Department of Zoology, Tel Aviv University, Ramat Aviv, 69978, Israel.
E-mail Dangr@post.tau.ac.il

Godonou, Ignace CAB International, PO Box 633-00621, Nairobi, Kenya.
E-mail: I.Godonou@cgiar.org

Gold, Clifford S. International Institute of Tropical Agriculture – Eastern and Southern Africa Regional Center (ESARC), Namulonge, PO Box 7878, Kampala, Uganda.
E-mail: C.Gold@cgiar.org

Greathead, David J. Centre for Population Biology, Imperial College, Silwood Park, Ascot, Berkshire SL5 7PY, UK.
E-mail: D.greathead@ic.ac.uk

Hanna, Rachid International Institute of Tropical Agriculture, Biological Control Center for Africa, 08 BP 0932, Cotonou, Bénin.
E-mail: R.Hanna@cgiar.org

Herren, Hans R. International Centre of Insect Physiology and Ecology (ICIPE), PO Box 30772, Nairobi, Kenya.
E-mail: Hherren@icipe.org

Hess, Dale E. Agronomy Department, Purdue University, West Lafayette, IN 47907, USA.
E-mail: Dhess@purdue.edu

Hill, Martin P. Department of Zoology and Entomology, Rhodes University, PO Box 94, Grahamstown, 6140, South Africa.
E-mail: M.p.hill@ru.ac.za

Hodges, Rick J. Natural Resources Institute, University of Greenwich, Chatham, Kent ME4 4TB, UK.
E-mail: R.J.Hodges@greenwich.ac.uk

Holst, Niels Department of Crop Protection, Danish Institute of Agricultural Sciences, Research Centre Flakkebjerg, DK-4200 Slagelse, Denmark.
E-mail: Niels.Holst@agrisci.dk

James, Braima International Institute of Tropical Agriculture, Biological Control Center for Africa, 08 BP 0932, Cotonou, Bénin.
E-mail: B.James@cgiar.org

Kairo, Moses T.K. CAB International, Caribbean and Latin America Regional Centre, Gordon Street, Curepe, Trinidad and Tobago, West Indies.
E-mail: M.Kairo@cabi.org

Kfir, Rami Agricultural Research Council – Plant Protection Research Institute, Division of Insect Ecology – Biological Control, Private Bag X134, Pretoria 0001, South Africa.
E-mail: Rietrk@plant2.agric.za

Kimenju, John University of Nairobi, Nairobi, Kenya.
E-mail: Phealth@nbnet.co.ke

Kooyman, Christiaan CAB International, Africa Regional Centre, PO Box 633, Village Market, Nairobi, Kenya.
E-mail: C.Kooyman@cabi.org

Langewald, Jürgen International Institute of Tropical Agriculture, Biological Control Center for Africa, 08 BP 0932, Cotonou, Bénin.
E-mail: J.Langewald@cgiar.org

Legg, James International Institute of Tropical Agriculture – Eastern and Southern Africa Regional Center (ESARC), Namulonge, PO Box 7878, Kampala, Uganda; and Natural Resources Institute, Greenwich University, Central Avenue, Chatham Maritime, Kent ME4 4TB, UK.
E-mail: JLegg@iitaesarc.co.ug

Lux, Slawomir A. International Centre of Insect Physiology and Ecology (ICIPE), PO Box 30772, Nairobi, Kenya.
E-mail: S.A.Lux@icipe.org

Maniania, Nguya K. International Centre of Insect Physiology and Ecology (ICIPE), PO Box 30772, Nairobi, Kenya.
E-mail: NKmaniania@icipe.org

Mitchell, Janette D.
Agricultural Research Council – Plant Protection Research Institute, PO Box X134, Pretoria 0001, South Africa.
E-mail: Rietjm@plant2.agric.za

Mohamed, Samira International Centre of Insect Physiology and Ecology (ICIPE), PO Box 30772, Nairobi, Kenya.
E-mail: Smohamed@icipe.org

Murphy, Sean T. CAB International, UK Centre, Bakeham Lane, Egham, Surrey TW20 9TY, UK.
E-mail: S.Murphy@cabi.org

Mutitu, K. Eston Kenya Forestry Research Institute, PO Box 20412, Nairobi, Kenya.
E-mail: Kefri@arcc.co.ke

Nankinga, Caroline International Institute of Tropical Agriculture – Eastern and Southern Africa Regional Center (ESARC), Namulonge, PO Box 7878, Kampala, Uganda.
E-mail: C.Nankinga@cgiar.org

Neuenschwander, Peter International Institute of Tropical Agriculture, Biological Control Center for Africa, 08 BP 0932 Cotonou, Bénin.
E-mail: P.Neuenschwander@cgiar.org

Niere, Björn International Institute of Tropical Agriculture – Eastern and Southern Africa Regional Center (ESARC), Namulonge, PO Box 7878, Kampala, Uganda.
E-mail: B.Niere@cgiar.org

Oduor, George I. CAB International – Africa Regional Centre, PO Box 633-00621, Nairobi, Kenya.
E-mail: G.Oduor@cabi.org

Ogwang, James A. Biological Control Unit, Namulonge Agricultural Research Institute, PO Box 7084, Kampala, Uganda.
E-mail: Jamesogwang@hotmail.com

Olckers, Terry Agricultural Research Council – Plant Protection Research Institute, Private Bag X6006, Hilton 3245, South Africa.
E-mail: Ntto@natal1.agric.za

Overholt, William A. Indiana River Research and Education Center, University of Florida, 2199 South Rock Road, Fort Pierce, Florida, USA.
E-mail: waoverholt@mail.ifas.ufl.edu

Peveling, Ralf Institut für Natur, Landschafts und Umweltschutz (NLU), Biogeographie, St. Johanns-Vorstadt 10, 4056 Basel, Switzerland.
E-mail: Peveling@ubaclu.unibas.ch

Sauerborn, Joachim University of Hohenheim, Institute for Plant Production and Agroecology in the Tropics and Subtropics,70593 Stuttgart, Germany.
E-mail: Sauerbn@uni-hohenheim.de

Schulthess, Fritz Postfach 112, 7004, Chur, Switzerland.
E-mail: F.Schulthess@gr-net.ch

Sétamou, Mamadou International Centre of Insect Physiology and Ecology (ICIPE), PO Box 30772, Nairobi, Kenya.
E-mail: Msetamou@icipe.org

Sikora, Richard A. Soil Ecosystem Phytopathology and Nematology, University of Bonn, Nussallee 9, D-53115 Bonn, Germany.
E-mail: Rsikora@uni-bonn.de

Simons, Sarah A. CAB International – Africa Regional Centre, PO Box 633-00621 Nairobi, Kenya.
E-mail: S.simons@cabi.org

Stolz, Ine Institut für Natur, Landschafts und Umweltschutz (NLU), Biogeographie, St. Johanns-Vorstadt 10, 4056 Basel, Switzerland.
E-mail: Muesine@yahoo.de

Tamò, Manuele International Institute of Tropical Agriculture, Biological Control Center for Africa, 08 BP 0932, Cotonou, Bénin.
E-mail: M.Tamò@cgiar.org

Timbilla, James A. Biological Control Unit, Crops Research Institute, PO Box 3785, Kumasi, Ghana.
E-mail: JTimbilla@yahoo.com

Tribe, Geoff D. Plant Protection Research Institute, Private Bag X5017, Stellenbosch 7599, South Africa.
E-mail: Vredgt@plant3.agric.za

van den Berg, Henk Laboratory of Entomology, Department of Plant Sciences, Wageningen University, PO Box 8031, 6700EH Wageningen, The Netherlands.
E-mail: Vandenberg100@wanadoo.nl

Yaninek, Steve Department of Entomology, 901 W. State Street, Purdue University, West Lafayette, IN 47907-2054, USA.
E-mail: Yaninek@purdue.edu

Zachariades, Costas Agricultural Research Council – Plant Protection Research Institute, Cedara Weeds Laboratory, Private Bag X6006, Hilton, 3245 South Africa.
E-mail: Ntczs@natal1.agric.za

Zimmermann, Helmuth G. Agricultural Research Council – Plant Protection Research Institute, Weeds Research Division, Private Bag X134, Pretoria 0001, South Africa.
E-mail: Riethgz@plant2.agric.za

Preface

In these times of instant access to new information through the various electronic media, what can a book like this contribute? We hope the reader will find out that this book goes well beyond what can be found on the Internet or in electronic databases. The main theme of this book is to show that biological control components provide a common basis for many integrated pest management (IPM) systems, although the relative importance of biological control varies in different systems. In 24 chapters, the book gives an authoritative view of clearly defined problem areas in plant protection in Africa. Some chapters describe research with preliminary results that offer promise for successful implementation in the near future. Others cover projects that have been effectively implemented in a few countries only, offering the potential for similar success in other regions. A number of the chapters describe highly successful implementation, which has in many cases resulted in huge returns to the local economies.

Each chapter tells a story without necessarily attempting complete coverage. Citations have been kept to a minimum and, at times, the reader will have to pass through the last cited review papers to find the accurate source of a statement. While we have tried to restrict authors to citations of peer-reviewed articles, citations of grey literature have been unavoidable, particularly in new and fast developing fields. Coverage of biological control in agricultural, horticultural and forestry systems is extensive but not complete, and entire fields where biological control plays an important role, such as in vector control, are not represented.

The book chapters concentrate on tropical Africa, including the islands of the Indian and Atlantic Oceans; but some of the projects described reach into regions with Mediterranean type climate. The reader will find a substantial amount of information on biological control with insects, rather less relating to the use of pathogens against insects or mites, and less still where microorganisms have been used to combat other microorganisms. We contend that this does not just show up the bias of the editors, but rather provides an accurate reflection of the scope and emphases of biological control research in the tropics.

We hope that readers from universities, technical and development agencies, national and international research centres, plant protection research institutes and extension services will find this book a useful basis from which to start their own work to strengthen environmentally friendly plant protection in Africa and beyond. If the book is successful in enthusing and inspiring its readers to play an active role in the development, promotion and implementation with farmers of biological control approaches to pest management, it will truly have been a worthwhile endeavour.

This volume is the fruit of many hundreds of cumulative years of research and implementation by 62 authors. It tells of struggles and hard won victories, but also of failures that still need redress. This book has been written by authors with a long history of work in Africa, but it would not have been possible without the help of many more colleagues, who gave their personal insights and who helped with the text or pictures. We would like to particularly acknowledge the contributions of: Matthew Cock, Georg Goergen, David Hall, Kerstin Krueger, Bernhard Löhr, Sean Murphy, Sampson Nyampong, George Oduor, Andy Polaszek, Lisbeth Riis, Di Taylor and Remi Yusuf. We would also like to thank our external reviewers from across the world, who gave their insights and helped the authors to achieve a text that was useful for general readers and not just for specialists. These reviewers were: Franz Bigler, Ted Center, Wendy Gelernter, Andy Gutierrez, Ann Hajek, Bernhard Hau, John Hoffmann, Heiki Hokkanen, Marshall Johnson, Dyno Keatinge, J.P. Michaud, Sini Olander, Hans-Michael Poehling, Jörg Romeis, Rodomiro Ortiz, Roy van Driesche, Mike Walters and Steve Wratten. We are well aware that many more were responsible for the success of this book, though they remain unnamed. We thank them all.

And last, we would like to acknowledge the continuous backing by our editors, Tim Hardwick and Zoe Gipson of CABI *Publishing*, and the generous financial contribution by the Swiss Agency for Development and Cooperation and the Technical Centre for Agricultural and Rural Co-operation (CTA) that made publication of this book possible. We would like to thank the various donors, councils and international agencies that contributed to the research for implementation reported upon in this book, for their steady support. Most of this work has resulted in public goods, providing long lasting benefits for a wide array of people, including some of world's most needy. In the name of the plant health researchers active on the African continent, we express our gratitude and hope to show with this book that investment in this type of development pays.

Peter Neuenschwander, Christian Borgemeister and Jürgen Langewald
Cotonou and Hannover

Technical Centre for Agricultural and Rural Cooperation (ACP-EU)

The ACP-EU Technical Centre for Agricultural and Rural Cooperation (CTA) was established in 1983 under the Lomé Convention between the ACP (African, Caribbean and Pacific) Group of States and the European Union Member States. Since 2000 it has operated within the framework of the ACP-EC Cotonou Agreement.

CTA's tasks are to develop and provide services that improve access to information for agricultural and rural development, and to strengthen the capacity of ACP countries to produce, acquire, exchange and utilize information in this area. CTA's programmes are designed to: provide a wide range of information products and services and enhance awareness of relevant information sources; promote the integrated use of appropriate communication channels and intensify contacts and information exchange (particularly intra-ACP); and develop ACP capacity to generate and manage agricultural information and to formulate ICM strategies, including those relevant to science and technology. CTA's work incorporates new developments in methodologies and cross-cutting issues such as gender and social capital.

CTA, Postbus 380, 6700 AJ Wageningen, The Netherlands

Dedication

We dedicate this book to Dr Chris Lomer. It was one of his last initiatives, which he started shortly before he died unexpectedly on 17 October 2001, during a visit to Australia. Dr Lomer's many contributions to the field of biological control, particularly to insect pathology, are highly regarded by colleagues in and beyond Africa. Everyone who was lucky enough to have worked with him will sadly miss his enthusiasm and optimism, his humour and his comradeship. Dr Lomer was very supportive and helped many young African scientists to start their own international careers.

Most colleagues know Dr Lomer in connection with the LUBILOSA project (biological control of grasshoppers and locusts). The use of entomopathogenic fungi, which was only one of several different options when LUBILOSA started, turned out to become the most successful approach. Today, microbial control has become popular all across Africa and within the scientific community LUBILOSA is considered to be one of the most successful projects using entomopathogens to combat outbreaks of economically important pests. Moreover, Dr Lomer was also involved in projects including classical biological control of water hyacinths and homopteran pests. Over the last few years, he lived and worked in Turkey and Denmark, but was still very much oriented towards biological control in Africa. It is an honour and a privilege for us, having been able to finalize the present book on his behalf.

Peter Neuenschwander, Christian Borgemeister and Jürgen Langewald
Cotonou and Hannover, September 2002

Foreword

Hans R. Herren

International Centre of Insect Physiology and Ecology, Nairobi, Kenya

The traditional agronomic characteristics of the African farming model, with its rich diversity of crops and crop varieties, often grown in mixed cultures and under shifting cultivation, helped to mitigate the problems of pest build-up. This agricultural system was adapted to low human population density, when space was not limited. However, with increasing population growth, modernization of agriculture, as well as the introduction of new and sometimes ill-adapted varieties, the need for new pest management tools and practices alongside other inputs has increased. Furthermore, globalization of trade and transportation of agricultural goods led to the accidental introduction of exotic pests, such as arthropods, nematodes and weeds, sometimes making emergency control actions necessary.

The different chapters in this book describe how both exotic and endemic pests are suitable objects for biological control, employed either in a single tactic approach or embedded within an integrated control strategy. The single tactic approach, or classical biological control, has been used successfully in certain cases against exotic pests, whereas the integrated pest management (IPM) approach has shown greater success in controlling endemic pests. The main feature and unique advantage of biological control over other IPM tactics is its inherent sustainability based on the co-evolution of the organisms involved. This characteristic makes (or should make) biological control the option of choice when it comes to integrated pest and vector management strategies. Biological control, however, cannot be a substitute for mismanaged plant production, in short bad farming. In order to access the full power and potential of biological control, the crop production system needs to be fully integrated in the larger agroecosystem, fulfilling the principles of agroecology. Under such a system, the powers of biological control can best be unleashed, and its synergistic effects with host plant tolerance/resistance, habitat management and agronomic practices brought to bear maximum impact.

The knowledge of the system and its functioning, which is required for the successful application of biological control, also promotes good agricultural practices. As is well documented in the following chapters, there exists a major need for increasing the knowledge about the agroecosystem and its links with pest and vector management issues, and for improving the skills of scientists, extension workers and farmers. Biological control, just as IPM, is knowledge-, skill- and wisdom-intensive and complex, unlike the energy- and capital-intensive and reductionistic chemical or genetic engineering alternatives. The many examples presented in this book exemplify the power and potential of biological control.

In addition to the need for more research, capacity and institution building in biological control and IPM, there is a strong need for better advertisement of the benefits of biological control-based IPM at farmer and government level. Better government policies and guidelines for the introduction and deployment of beneficial organisms are needed, both for exotic and endemic species. Biological control, when done by specialists and in respect of international (FAO) guidelines, is the safest and most economic method for managing pest populations, in harmony with the environment, respectful of biodiversity and safe for humans, animals and the environment.

The lack of confidence in the potential of biological control as an alternative to other pest management tools requires that greater effort be dedicated to detailed economic and environmental impact studies of successful biological control projects. Better documentation of how the farmers, the consumers and the environment can benefit from these strategies will provide the convincing arguments for financial and policy support. The programmes thus far executed in Africa, both biological-control and IPM, have often not been well evaluated, except in a few cases such as the cassava mealybug biological control project implemented in the 1980s and 1990s by the International Institute of Tropical Agriculture. There is a dire need to disseminate information on successful programmes more widely and in appropriate format to all stakeholders. Though biological control, with research funded mainly by governments or international agencies, has already contributed significantly to food security in Africa, more projects need to be executed and existing projects expanded to new areas. It is interesting to note that genetic engineering technologies, fuelled by commercial interests, are sometimes advertised as 'silver bullet' solutions to food security. It is our conviction that the solution to pest problems lies in prevention, growing crops in harmony with the environment and according to agroecological principles. If control interventions are needed, the selected tactics, and the above-mentioned silver bullets in particular, need to be tested and embedded in an IPM strategy, so that their potential long-term environmental impact will not render them the *new* problem.

1

Historical Overview of Biological Control in Africa

David J. Greathead
Centre for Population Biology, Imperial College, Ascot, UK

Introduction

The scope of this historical overview includes the area covered by the Afrotropical Zoogeographical Region, i.e. Africa south of the Sahara and the islands in the Indian and Atlantic Oceans closer to Africa than other continents. These include Madagascar, the Mascarene Islands, the Seychelles Islands and Comoros Islands in the Indian Ocean and in the Atlantic Ocean Ascension Island, St Helena, São Tome, Principe and the Cape Verde Islands on which biological control has been practised.

The history of applied entomology in this region was influenced until recently by the pattern of European colonization during the 17th to 19th centuries. Before European colonization, the Mascarene Islands, Seychelles Islands, Ascension Island and St Helena were uninhabited. Madagascar had already been colonized some 1000 years previously by settlers from Indonesia. Indonesians are also known to have reached the East African coast and traded with the inhabitants (e.g. Miller, 1969). This trade may have been responsible for the introduction of some exotic pests; the Asian cereal stem borer, *Chilo partellus* (Swinhoe) (Lepidoptera, Pyralidae), may be one of these. European colonists also brought new crops and their associated pests. These plants were brought by sea, growing in barrels or Wardian cases (specially constructed glass cabinets). Many scale insects and soil pests must have reached the region in this way. The entomological history of Mauritius is relatively well documented and provides illustrations. Both the Indonesian sugarcane stem borer, *Chilo sacchariphagus* (Bojer), and its natural enemy, *Cotesia flavipes* (Cameron) (Hymenoptera, Braconidae), had arrived on the island, presumably with planting material brought from Java, before entomologists began to study the insect fauna (Greathead, 1971 and references therein). More recently, the white grub, *Phyllophaga smithi* (Arrow) (Coleoptera, Scarabaeidae) (otherwise only known from the island of Barbados), was almost certainly introduced in tubs of soil containing sugarcane imported from Barbados (Simmonds and Greathead, 1977).

Until the advent of air travel, invasive species had to survive transport by sea, which could take several weeks and limited the number of exotic arthropod pests capable of colonizing the region. However, air travel allowed arthropods to arrive within a few hours in planting material, produce and cut flowers for sale or passengers' food, and smuggled plant material, or as adults hitchhiking in the body of the plane. Many of the new pests that reached the region after World War II and have been targeted for biological control arrived in this way. On the continent of Africa these include the cassava pests, *Mononychellus tanajoa* (Bondar) (Acari, Tetranychidae) and *Phenacoccus manihoti* Matile-Fererro (Hemiptera, Pseudococcidae), introduced from South America on illegally imported planting material (Neuenschwander, this volume), *Liriomyza trifolii* (Burgess) (Diptera, Agromyzidae), which reached Kenya on chrysanthemum cuttings from Florida (Neuenschwander *et al.*, 1987) imported for multiplication, and *Pineus boerneri* Annand (Hemiptera, Adelgidae), which is believed to have reached Africa on pine twigs imported for grafting. *Prostephanus truncatus* (Horn) (Coleoptera, Bostrychidae) arrived by sea in maize sent as famine relief (Borgemeister *et al.*, this volume). A notable example on the island of Mauritius is the south east Asian banana skipper, *Erionota thrax* (Linnaeus) (Lepidoptera, Hesperiidae), which almost certainly gained entry at the time of civil disturbances when troops were flown at night from Malaysia to help keep order (Simmonds and Greathead, 1977).

Such new arrivals are prime targets for classical biological control and are the subjects of many of the chapters in this book. Native pests have spread also and expanded their range with human assistance. The coffee mealybug, *Planococcus kenyae* Le Pelley (Hemiptera, Pseudococcidae), is an example, having spread into Kenya from Uganda (Greathead, 1971, and references therein). These too are sometimes, as in this instance, good targets. However, the majority of pests are native and many of them have a full complement of natural enemies, which leaves few opportunities for classical biological control. Here methods for conservation or augmentation may be appropriate. These are applied, consciously or not, in traditional agriculture, which, for example, makes extensive use of polyculture (mixed cropping) and have only recently been investigated by scientists developing IPM programmes. The philosophy and technology employed are largely the commonsense techniques employed by crop protection practitioners before the advent of powerful synthetic pesticides when any effective method (physical, mechanical or biological) was a welcome addition to the few relatively inefficient pesticides (botanicals, tar oils, inorganic chemicals) available. Thus, the first applied entomologists appointed by the colonial governments had to use their wits to devise effective crop protection and became enthusiastic about the opportunities offered by introducing natural enemies that offered permanent control without the need for input from farmers.

In this chapter, I trace the history of biological control and IPM by periods of time rather than by country or category of pest (the term is used to include animals, pathogens and weeds). A rigid historical sequence is not possible because many of the long-term campaigns overlap other developments. Further, it is not possible to mention every attempt against every pest in each country. Rather, I have tried to pick out programmes that have been of particular significance in the development of biological control and IPM in the region.

Many of them are treated in detail in other chapters so that only brief mention is made here with emphasis on points considered to be of historical significance. Notably, the large number of successful biological control programmes against weeds in South Africa since the end of World War II, many of them of conservation importance, is not discussed because they are reviewed in a separate chapter (Zimmermann and Olckers, this volume).

The BIOCAT database (Greathead and Greathead, 1992, and updated to end 2001) contains records of introductions of insect natural enemies made against insect pests. Figure 1.1 summarizes the data for the entire world by

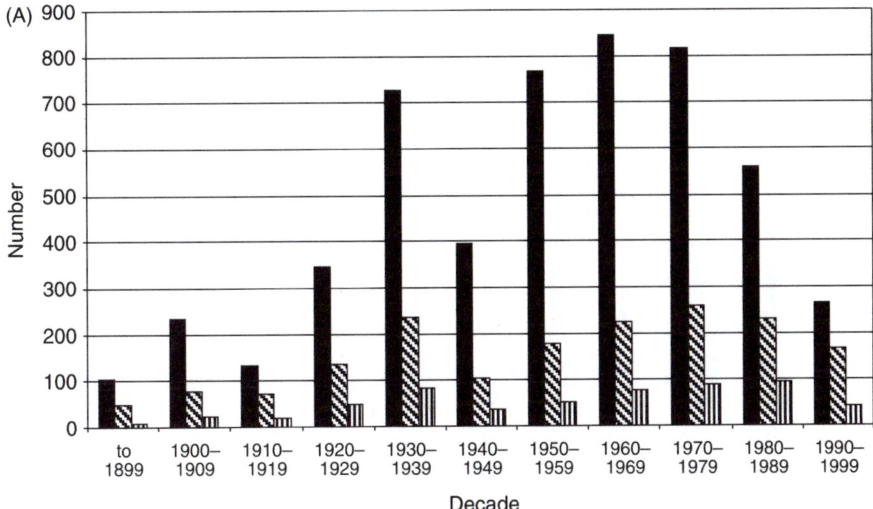

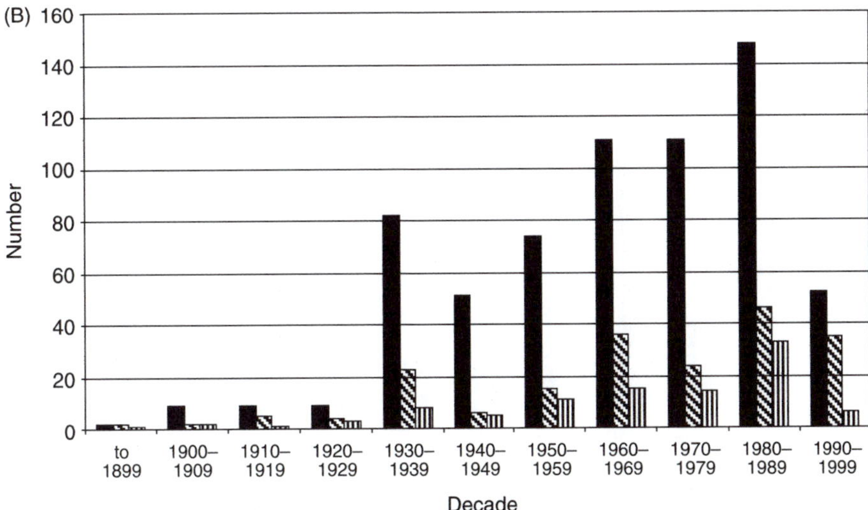

Fig. 1.1. Introductions of insect natural enemies for biological control of insect pests by decade. Numbers of introductions, establishments and numbers contributing to control (represented by the three consecutive bars) (data from the BIOCAT database). (A) World; (B) Afrotropical region.

decade. It shows that the number of introductions increased decade by decade, except for the decades including the two World Wars up to the 1960s. After the 1970s the number dropped substantially. The pattern for the Afrotropical Region is similar, except that the numbers before the 1930s are too small to exhibit the trend. The rate of successful introductions and species contributing to successful controls (Fig. 1.2) shows that, as the numbers of introductions increased, the success rates fell, especially during the 1950s to 1970s. The pattern for the Afrotropical Region is less clear and values before the 1930s are based on too few data to be reliable indicators of a trend. However, both the world figures and the Afrotropical figures show a sharp increase in the rate of

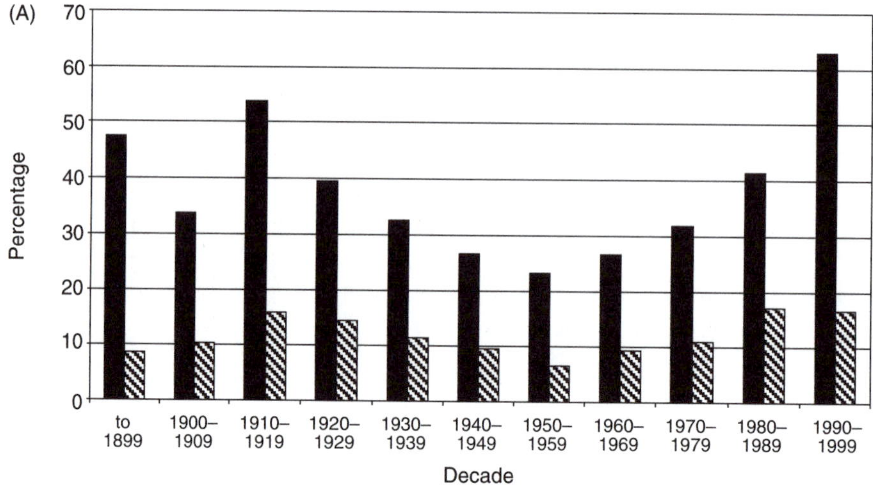

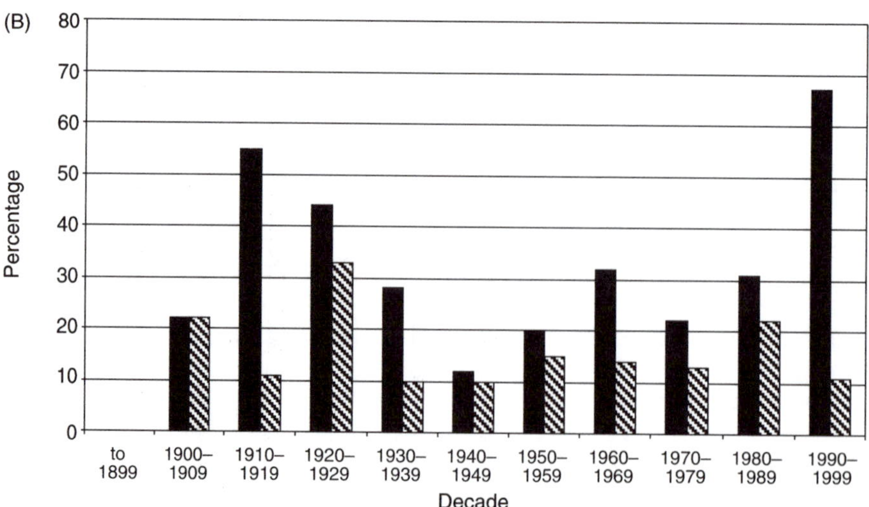

Fig. 1.2. Percentage establishments and results contributing to control by decade (represented by the two consecutive bars) (data from the BIOCAT database). (A) World; (B) Afrotropical region. Note that the 'to 1899' column in B is based on only two records and is omitted, also the columns for '1900–1909', '1910–1919' and '1920–1929' are based on only nine records.

successful controls and establishments during the 1980s. The figures for the 1990s probably show the same trend, but the final outcome of many of the successful introductions during this decade is not yet clear. Reasons for these features of the results of introductions are discussed.

Table 1.1 shows the countries of the Afrotropical Region that have made more than ten introductions and the number of insect pest species successfully controlled in each of them. As the figures for some countries, e.g. Ghana and Zambia, are affected by the large number of introductions made against a single pest species, the number of introductions made against the pest for which the largest number was made is also shown. It is of interest that those countries at the top of the table are ones that had early biological control successes. The results being obtained in Mauritius resulted in the neighbouring island countries starting biological control programmes. Similarly work in the eastern African countries was stimulated by successes in Kenya and also to some extent South Africa. It is notable that the only West African countries included in the table appear largely because of the unsuccessful campaign against *Planococcoides njalensis* (Laing) (Hemiptera, Pseudococcidae) in Ghana (Greathead, 1971) and *L. trifolii* in Senegal (Neuenschwander *et al.*, 1987). Summary information for all successful biological controls of insect pests up to 1979 is provided in Table 1.2, i.e. principally results achieved during the period before the initiation of the programmes that form the subjects of the remaining chapters of this book.

Table 1.1. Countries making more than ten introductions of insect biological control agents against arthropod pests (data from the BIOCAT database).

Country	No. of introductions and (successful controls)	No. of pests	Year started	Largest no. in any one programme
Mauritius	132 (10)	22	1913	36 *Phyllophaga smithi* (Arrow)
South Africa	106 (11)	32	1892	11 *Aonidiella aurantii* (Maskell), 11 *Chilo partellus* Swinhoe
Kenya	53 (6)	18	1911	18 *Planococcus kenyae* Le Pelley
Ghana	47 (2)	5	1948	28 *Phenacoccoides njalensis* (Laing)
Seychelles Islands	30 (6)	13	1930	5 coconut scale insects (4 spp.)
Madagascar	28 (3)	11	1948	8 *Chilo sacchariphagus* (Bojer)
Cape Verde Islands	25 (2)	10	1981	5 *Helicoverpa armigera* (Hübner), 5 *Etiella zinkenella* (Treitschke)
Uganda	24 (3)	9	1934	8 *Aulacaspis tegalensis* (Zehntner)
Réunion	22 (4)	9	1953	7 fruit fly species
Zambia	22 (2)	6	1968	12 *Phthorimaea operculella* (Zeller)
St Helena	20 (4)	6	1896	13 *P. operculella*
Sénégal	17 (1)	3	1954	9 *Liriomyza trifolii* Burgess
Tanzania	17 (3)	8	1934	5 *P. operculella*
Comoros Islands	12 (0)	2	1969	11 *C. partellus*

Table 1.2. Successful biological controls of insect pests in the Afrotropical Region during 1892–1979.

Pest	Crop	Country	Date	Agent
Hemiptera				
Aleyrodidae				
Aleurocanthus woglumi Ashby	Citrus	Seychelles	1956	*Eretmocerus serius* (Haldeman) (Aphelinidae)
		South Africa	1959	
		Kenya	1960	
Aleurothrixus floccosus (Maskell)	Citrus	Réunion	?	*Cales noaki* Howard (Aphelinidae)
Aphididae				
Eriosoma lanigerum (Hausmann)	Apple	South Africa	1920	*Aphelinus mali* (Haldeman) (Aphelinidae)
		Kenya	1927	
Asterolecaniidae				
Asterolecanium sp.	Cacao, coffee	São Tome	1965	*Trichomasthus portoricoensis* (Crawford) (Encyrtidae)
Cicadellidae				
Perkinsiella saccharicida Kirkaldy	Sugarcane	Mauritius	1956	*Tytthus mundulus* (Breddin) (Miridae)
Diaspididae				
Aspidiotus destructor Signoret	Coconut	Mauritius	1937–1939	*Chilocorus politus* Mulsant and *C. nigrita* (Fabricius) (Coccinellidae)
	Coconut, oil palm	Principe	1955	*Cryptognathus nodiceps* Marshall (Coccinellidae)
Aulacaspis tegalensis (Zehntner)	Sugarcane	Tanzania	1971	*Rhyzobius lophanthae* (Blaisdell) (Coccinellidae)
Chrysomphalus ficus Ashmead	Citrus	South Africa	1962	*Aphytis holoxanthus* DeBach (Aphelinidae)
Lepidosaphes beckii (Newman)	Citrus	South Africa	1966	*Aphytis lepidosaphes* Compére
Parlatoria blanchardii (Targioni)	Date palm	Niger (Air)	1974	*Chilocorus bipustulatus* (Linnaeus) (Coccinellidae)
Scale insects	Coconut	Seychelles	1938	*Exochomus* spp. and *Chilocorus* spp. (Coccinellidae)
Margarodidae				
Icerya purchasi Maskell	Citrus, ornamentals	South Africa	1892	*Rodolia cardinalis* (Mulsant) (Coccinellidae)
	Citrus	Ethiopia	1947	
	Citrus	Ascension	1977	
Icerya seychellarum (Westwood)	Fruit trees	Seychelles	1930	*R. cardinalis*
		Mauritius	1952	*Cryptochetum monophlebi* Skuse (Cryptochetidae)
Ortheziidae				
Orthezia insignis Browne	Ornamentals	Kenya	1945	*Hyperaspis pantherina* Fürsch (Coccinellidae)
		Tanzania	1953	
		Uganda	195?	

Pseudococcidae				
Planococcus citri (Risso)	Citrus	St Helena	1973	*Cryptolaemus montrouzieri* Mulsant (Coccinellidae)
		South Africa	1900	
Planococcus kenyae Le Pelley	Coffee	Kenya	1938	*Anagyrus* spp. (Encyrtidae)
Psyllidae				
Diaphorina citri Kuyana	Citrus	Réunion	1978	*Tamarixia radiata* (Waterston) (Eulophidae)
		Mauritius	1984	
Trioza erytreae (Del Guercio)	Citrus	Réunion	1974	*Tamarixia dryi* (Waterston) (Eulophidae)
		Mauritius	1975	
Coleoptera				
Curculionidae				
Gonipterus scutellatus Gyllenhal	Eucalyptus	South Africa	1926	*Anaphes nitens* (Girault) (Mymaridae)
		Kenya	1945	
		Mauritius	1946	
		Madagascar	1948	
		St Helena	1958	
Scarabaeidae				
Oryctes tarandus (Olivier)	Sugarcane	Mauritius	1917	*Scolia oryctophaga* Coquillett (Scoliidae)
Lepidoptera				
Gelechiidae				
Phthorimaea operculella (Zeller)	Potato	Zambia	1970	*Apanteles subandinus* Blanchard, *Bracon gelechiae* Ashmead (Braconidae); *Eriborus trochanterus* (Morley) (Ichneumonidae); *Copidosoma koehleri* Blanchard (Encyrtidae)
Hesperiidae				
Erionota thrax (Linnaeus)	Banana	Mauritius	1971–1972	*Cotesia erionotae* (Wilkinson) (Braconidae) and *Ooencyrtus erionotae* Ferrière (Encyrtidae)
Noctuidae				
Sesamia calamistis Hampson	Sugarcane	Mauritius	1951	*Cotesia sesamiae* (Cameron) (Braconidae)
		Réunion	1953	
		Madagascar	1968	*Pediobius furvus* (Gahan) (Eulophidae)

Continued

Table 1.2. *Continued.*

Pest	Crop	Country	Date	Agent
Lepidoptera				
Pyralidae				
Chilo sacchariphagus (Bojer)	Sugarcane	Madagascar	1960	*Cotesia flavipes* (Cameron) (Braconidae)
Tortricidae				
Maruca vitrata Fabricius	Pigeon pea	Mauritius	1953	*Bracon cajani* (Muesbeck) (Braconidae) and *Eiphosoma dentator* (Fabricius) (Ichneumonidae)
Diptera				
Muscidae				
Stomoxys calcitrans (Linnaeus)	Cattle	Mauritius	1966–1968	*Spalangia endius* Walker and *Spalangia nigra* Latreille (Pteromalidae)
Stomoxys niger Macquart	Cattle	Mauritius	1972	*Tachinaephagus stomoxicida* Subba Rao (Encyrtidae)

Summary information on biological control of weeds worldwide up to 1996 is contained in the fourth edition of the catalogue edited by Julien and Griffiths (1998). Successes for the Afrotropical Region from introductions made up to 1979 are included in Table 1.3. An analysis of the data in the first edition was published by Julien *et al.* (1984) and a further analysis was published after the appearance of the second edition (Julien, 1989). These analyses show an increasing number of introductions each decade, with the exception of the 1940s, and steady establishment and success rates (species contributing to control). The trend towards increasing activity in biological control of weeds has continued with both the number of new releases and the number of new weed targets increasing in each 5-year period between successive editions (Julien and Griffiths, 1998). A frequently noted and important difference between weed biological control and insect biological control is the higher establishment rate (63%) and success rate (27.9%) for weeds as compared with rates for insects: 33.5% establishments and 11.2% successes (data from BIOCAT). The weed data include all taxa of agents, principally insects, but with an increasing number of fungi. Certain fungi, principally rusts, have proved to be effective classical biological control agents with very narrow host ranges.

First Attempts at Biological Control (1892–1920)

The earliest records of the manipulation of beneficial organisms for the control of pests refer to the management of predatory ants for control of citrus pests in China in the 3rd century AD (e.g. Smith *et al.*, 1973) and the control of date palm pests in Yemen around AD 1200 (El-Haidari, 1981). No such early references to the deliberate use of beneficial organisms for pest control have been found for Africa, although it is possible that the predators were managed in traditional African agriculture.

This use of natural enemies is referred to as augmentative biological control and was practised long before introductions of exotic species were made for pest control, although this practice is now widely referred to as 'classical biological control'. The reason for this was the dramatic success of the first successful introduction of an insect for control of an insect pest, which triggered worldwide interest in the use of beneficial insects for pest control at a time when there were few reliable chemical control products available. The story of the control of the cottony cushion scale, *Icerya purchasi* Maskell (Hemiptera, Margarodidae), in Californian citrus orchards by the Australian ladybird beetle, *Rodolia cardinalis* (Mulsant) (Coleoptera, Coccinellidae), introduced in 1888, has been told many times (e.g. Caltagirone and Doutt, 1989). However, there had been earlier attempts to control pests by the introduction of generalist vertebrate predators, such as the Indian mongoose and owls, mainly on islands, e.g. Mauritius (Greathead, 1971). These introductions frequently led to the extermination of native fauna without effective control of the rats against which they had been introduced. However, one of these early introductions, the Indian mynah, *Acridotheres tristis* Linnaeus (Aves, Sturnidae), into Mauritius for control of the red locust, *Nomadacris septemfasciata* (Serville) (Orthoptera,

Table 1.3. Successful controls of weeds in the Afrotropical Region 1913–1996 (data from Julien and Griffiths, 1998).

Weed	Situation	Country	Date	Agent
Araceae				
Pistia stratiotes Linnaeus	Waterways	South Africa	1985	*Neohydronomus affinis* Hustache (Curculionidae)
		Zimbabwe	1988	
		Zambia	1991	
Asteraceae				
Ageratina riparia (Regel) R. King and H. Robinson	Uncultivated land	South Africa	1989	*Entyloma ageratinae* Barreto and Evans (Ustilaginales)
Cactaceae				
Opuntia aurantiaca Lindley	Grazing	South Africa	1935	*Dactylopius austrinus* De Lotto (Dactylopiidae)
Opuntia ficus-indica (Linnaeus) Miller, etc.	Grazing	South Africa	1933	*Cactoblastis cactorum* (Bergroth) (Pyralidae)
			1938	*Dactylopius opuntiae* (Cockerell)
Opuntia imbricata (Haworth) De Candolle	Grazing	South Africa	1970	*Dactylopius tomentosus* (Lamark)
Opuntia tuna (Linnaeus) Miller and	Grazing	Mauritius	1928	*Dactylopius ceylonicus* (Green) and *D. opuntiae*
Opuntia vulgaris Miller			1950	*C. cactorum*
O. vulgaris Miller		Kenya	1958	*D. ceylonicus*
		South Africa	1913	
		Tanzania	1957	
Opuntia spp.		Ascension	1973	*C. cactorum*
		St Helena	1971	
		Madagascar	1923	*D. opuntiae*
Clusiaceae				
Hypericum perforatum Linnaeus	Grazing	South Africa	1961	*Chrysolina quadrigemina* (Suffrian) (Chrysomelidae)
			1972	*Zeuxidiplosis giardi* (Kieffer) (Cecidomyiidae)
Ehretiaceae				
Cordia curassavica (Jacquin) Roemer and Schultes	Grazing	Mauritius	1949	*Eurytoma attiva* Burks (Eurytomidae)
			1948	*Metrogaleruca obscura* (Degeer) (Chrysomelidae)

Fabaceae				
Sesbania punicea (Cavanille) Bentham	Grazing	South Africa	1984	*Neodiplogrammus quadrivittatus* (Olivier) and *Rhyssomatus marginatus* Fåhreus (Curculionidae)
Mimosaceae				
Acacia longifolia (Andrews) Willdenow	Grazing	South Africa	1985	*Melanterius ventralis* Lea (Curculionidae)
			1982	*Trichilogaster acaciaelongifoliae* (Froggatt) (Pteromalidae)
Acacia saligna (Labillardière) Wendland	Grazing	South Africa	1987	*Uromycladium tepperianum* (Saccardo) McAlpine (Uredinales)
Prosopis spp.	Grazing	South Africa	1987	*Algarobius prosopis* (LeConte) (Bruchidae)
Pontederiaceae				
Eichhornia crassipes (Martius) Solms-Laubach	Waterways	Bénin	1991	*Neochetina eichhorniae* Warner and *N. bruchi* Hustache (Curculionidae)
		Nigeria	1993	
		South Africa	1985	
		Sudan	1978	
		Uganda	1993	
		Zimbabwe	1993	
Proteaceae				
Hakea sericea Schrader	Grazing	South Africa	1972	*Carposina autologa* Meyrick (Carposinidae)
			1975	*Erytenna consputa* Pascoe (Curculionidae)
Salviniaceae				
Salvinia molesta D.S. Mitchell	Waterways	Ghana	1996	*Cyrtobagous salviniae* Calder and Sands
		Kenya	1990	
		Namibia	1984	
		South Africa	1985	
		Zambia	1990	
		Zimbabwe	1992	
Verbenaceae				
Lantana camara Linnaeus	Grazing	Ascension	1973	*Teleonemia scrupulosa* Stål (Tingidae)
		South Africa	1971	
		St Helena	1971	

Acrididae), in 1762, is claimed a success and the first example of pest control by an introduced natural enemy (Greathead, 1971), although this widely introduced bird is usually regarded as a nuisance in the countries where it has become established.

Documented biological control on the African continent began with the independent introductions of *R. cardinalis* into the Cape Colony (Lounsbury, 1940; Greathead, 1971) and Egypt (Kamal, 1951) in 1892. Both introductions were made as a direct result of news of the outcome of its introduction into California. There followed a period of indiscriminate introduction of beneficial insects, chiefly ladybirds for aphid control, with little success. This period was referred to by C.P. Lounsbury, an American appointed as the first government entomologist at the Cape, as the 'ladybird fantasy'. Based in Cape Town, Lounsbury was well placed to benefit from the shipment of insect natural enemies round the world. Not only was he a former colleague of American government entomologists and remained in contact with them, but he also had the task of assisting entomologists tending cultures being transferred by sea when the ships called at Cape Town during the long voyage between Australia and North America (Lounsbury, 1940). Consequently, the southern African states, later to become the Union of South Africa, were highly active in the introduction of biological control agents.

In eastern Africa, the first biological control attempt was made in Kenya in 1911 against an aphid, *Schizaphis graminum* (Rondani) (Hemiptera, Aphididae), which had first appeared in 1909–1910 damaging the wheat crop. The government entomologist returned from a visit to the USA with a parasitoid, *Lysiphlebus testaceipes* (Cresson) (Hymenoptera, Braconidae), and a predator, *Hippodamia convergens* (Guérin-Méneville) (Coleoptera, Coccinellidae); both were released, but neither is known to have become established (Greathead, 1971).

In West Africa, biological control activity does not seem to have begun until after World War I, but even then was much less extensive than in other parts of the continent until the 1980s.

In spite of the largely negative results of introducing generalist vertebrate predators, biological control was the principal means for combating major pests in Mauritius, particularly in sugarcane where spraying with pesticides is both inefficient and uneconomic. On sugarcane, the first target was a white grub, *Oryctes tarandus* (Olivier) (Coleoptera, Scarabaeidae) native to Madagascar, which was readily controlled by introduction of its parasitoid, *Scolia oryctophaga* Coquillett (Hymenoptera, Scoliidae), imported from Madagascar in 1917. Less readily controlled was another white grub, *P. smithi*, which had been accidentally introduced from Barbados with sugarcane varieties shipped in tubs of infested soil. Introduction of its parasitoid, *Tiphia parallella* Smith (Hymenoptera, Tiphiidae), from Barbados in 1915 did not provide control and a campaign followed to import and release parasitoids of other white grubs, principally from Madagascar, Indonesia, the Philippines and South Africa. Of some 42 species, chiefly Scolioidea and Tachinidae, only seven became established by the time work stopped in 1951 after a misguided attempt to introduce the giant toad, *Bufo marinus* (Linnaeus) (Amphibia,

Bufonidae), from Trinidad, which fortunately failed. By then the importance of the pest had declined, probably due to a combination of the results of breeding varieties better suited to the island and improved agronomic methods as well as the establishment of parasitoids. Other sugarcane pests were more readily controlled (Table 1.2). The Seychelles and Madagascar began biological control after World War I but Réunion did not start until the 1960s.

Insects were targets for biological control of all the early efforts mentioned above. The earliest attempt to control a weed took place in South Africa when *Dactylopius ceylonicus* (Green) (Hemiptera, Dactylopiidae) was obtained from the Queensland Prickly Pear Commission in 1913 and achieved spectacular control of *Opuntia vulgaris* Miller (*Cactaceae*) within a few years. Subsequent effort to control other *Opuntia* spp. in South Africa up to the 1950s followed the lead of Queensland (Greathead, 1971; Zimmermann and Olckers, this volume).

The first attempts to use microbial agents also took place in South Africa, when, in 1896, unsuccessful attempts began to culture and distribute fungal pathogens of locusts. Then, in 1912, experiments were carried out on controlling grasshoppers with *Coccobacillus acridiorum* d'Hérelle (Bacteria), which, as in other countries, were a failure (Greathead, 1971).

These early attempts, mostly inadequately researched and unsuccessful, did not prevent the development of further biological control programmes. Activity was interrupted by World War I, but several major programmes were carried out until the availability of DDT and other synthetic pesticides after World War II caused a temporary decline in interest in biological control (Fig. 1.1). For details of all programmes, see the comprehensive review of biological control activity in the Afrotropical zoogeographical region up to 1970 by Greathead (1971). Here only a few particularly significant programmes that influenced the development of biological control activities in African countries can be mentioned, with Table 1.2 providing a summary of the results of successful introductions up to 1979.

Major Programmes and New Insights (1920–1940)

After World War I, response to the demand for biological control agents led to the setting up of the Farnham House Laboratory in 1927 under the Imperial Bureau of Entomology, to find and supply biological control agents for the British Empire. In fact, from the outset, work was also carried out for other countries. The Farnham House Laboratory was directed by W.R. Thompson, a Canadian who had worked in France for the United States Department of Agriculture laboratory, set up to find natural enemies for control of the gypsy moth, *Lymantria dispar* (Linnaeus) (Lepidoptera, Lymantriidae), in the USA. The Farnham House Laboratory was soon involved in supplying natural enemies to African countries and in assisting with several of the major biological control introduction programmes that were carried out until World War II. W.F. Jepson was employed by the laboratory to work with the Mauritius authorities on the campaign to control *P. smithi* (Greathead, 1994).

In Kenya, a landmark programme took place against a mealybug that began to devastate coffee plantations and food crops in the Kenya highlands in 1923. It was identified initially as *Planococcus lilacinus* (Cockerell) (Hemiptera, Pseudococcidae) and efforts were made to obtain natural enemies from the native home of *P. lilacinus* in South and South-East Asia. Many species were shipped to Kenya and cultures of natural enemies of other mealybugs were obtained from California, Hawaii and Japan, but all attempts made to culture them in quarantine failed. Partly as a result of these failures, it was realized that the mealybug was a new species, described as *P. kenyae* Le Pelley. Unfortunately, early efforts with natural enemies from Uganda had failed and this delayed the discovery that the mealybug had originated in Uganda, northwest Tanzania and the Congo. However, new importations from Uganda, made in 1938, included two species of *Anagyrus* (Hymenoptera, Encyrtidae), which readily bred on *P. kenyae* and rapidly established following releases in the same year. By 1949, control was good in almost all areas and incipient outbreaks were controlled by the release of parasitoids. The situation was disturbed during the early 1950s by the use of persistent chlorinated hydrocarbon insecticides to control other pests on coffee, but was re-established when non-persistent insecticides replaced the chlorinated hydrocarbons. In 1959, it was estimated that some £10 million had been saved against an outlay of a total expenditure of not more that £30,000 (Greathead, 1971 and references therein).

This programme emphasized the need for accurate identification of the pest and the need to look in its native distribution area for effective natural enemies. It also supported the concept of J.G. Myers, developed while working on biological control of sugarcane stem borers in the Caribbean using parasitoids from South America (Greathead, 1994), that ecological islands with high biodiversity exist within continental areas and are profitable places to search for natural enemies. This led the coffee research authorities in Kenya and Tanzania to fund research on biological control of coffee bugs, *Antestiopsis* spp. (Hemiptera, Pentatomidae), and leaf miners *Leucoptera* spp. (Lepidoptera, Lyonetiidae) during the 1960s (Greathead, 1971 and references therein). Unfortunately, no new and effective natural enemies of either of these two pests were found and insecticides continue to be applied for their control.

In South Africa, an Australian weevil, *Gonipterus scutellatus* Gyllenhal (Coleoptera, Curculionidae), was first discovered attacking young growth in eucalyptus plantations in 1916. It remained largely confined to coastal areas until 1925 when it began to spread rapidly into the interior. Feeding by the weevil and its larvae destroys the tender young shoots causing poor growth and distortion of trees in plantations. An entomologist was sent to Australia, where the weevil is not a pest, and he soon found an egg-parasitoid, *Anaphes nitens* (Girault) (Hymenoptera, Mymaridae). This along with other parasitoids was shipped to South Africa, but it was the only one to be successfully bred and released. By 1935, it had achieved economic control in all areas except the Highveld. Gradually the parasitoid seems, however, to have adapted to the cooler conditions at higher altitudes and control has substantially improved (Tribe, this volume). This success was achieved against predictions that egg-parasitoids are less effective than natural enemies of the later stages. It has also been repeated elsewhere, wherever the parasitoid has been released, including East Africa, Madagascar, Mauritius and St Helena (Greathead, 1971, and references therein).

In Mauritius, pest control of sugarcane white grubs dominated biological control activity during the interwar period (see above). In the Seychelles, a complex of scale insects on coconuts, principally *Eucalymnatus tessalatus* (Signoret) (Hemiptera, Coccidae), *Chrysomphalus aonidum* (Linnaeus), *Ischnaspis longirostris* (Signoret) and *Pinnaspis buxi* Bouché (Hemiptera, Diaspididae), were the most important insect pests and in 1936, investigations in the feasibility of biological control began. As there were no effective native natural enemies, coccinellid predators were introduced from East Africa and India. *Chilocorus distigma* (Klug) and two species of *Exochomus* from Africa and *Chilocorus nigrita* (Fabricius) from India became established. The results were spectacular, with control achieved in a matter of months and a substantial increase in the coconut crop from 1940 onwards. *C. nigrita* became the most abundant species and remains so to date. It was also introduced from Sri Lanka into Mauritius in 1939 for control of another diaspidid scale insect on coconuts, *Aspidiotus destructor* Signoret (Table 1.1). It has proved to be a good colonist, has reached the African mainland, and is now well established in East Africa and in southern Africa (Samways, 1989).

During this period, a major effort was made in South Africa to control prickly pear cactus (*Opuntia* spp.) (Zimmermann and Olckers, this volume). *Dactylopius* spp. were also introduced into Mauritius in 1928 and provided good control, until the establishment of the Australian coccinellid, *Cryptolaemus montrouzieri* Mulsant, in 1938 for control of the pineapple mealybug, *Dysmicoccus brevipes* (Cockerell) (Hemiptera, Pseudococcidae). No recoveries of the coccinellid were made on pineapple, but by 1950 it was affecting control of cactus, as it did in South Africa, and *Cactoblastis cactorum* (Bergroth) (Lepidoptera, Pyralidae) was introduced to maintain control of prickly pear cactus (Greathead, 1971). Otherwise there were no significant efforts to control weeds during this period.

The Response to Synthetic Pesticides (1940–1970)

At the end of World War II, new, powerful, broad-spectrum synthetic pesticides, which had successfully helped control diseases among soldiers during the war, became available for agricultural use. They revolutionized pest control and in many countries biological control was abandoned. Many of the remaining biological control practitioners responded by trying to demonstrate that biological control was cheaper and provided permanent control. At the same time, air transport was becoming universal and for the first time consignments of natural enemies could be sent across the world as eggs or pupae in a few days at most, instead of several weeks on ships, where they frequently required the attendance of an entomologist to maintain the culture. Consequently, it was tempting to economize on detailed ecological studies and the development of methods for laboratory culture by shipping large numbers of agents for direct release on arrival. In this way it was possible to send numbers of species, release them and see whether they became established, instead of sending one or very few carefully studied species for multiplication and release (Fig. 1.1). Thus, the lessons learned in the preceding period were forgotten and the success rate fell (Fig. 1.2), with the result that biological control acquired a reputation of being unlikely to succeed and was at best a last resort to be considered only if all else failed.

W.R. Thompson and some of the staff of the Farnham House Laboratory went to Canada to continue their work in 1940 and, after the war, the service became the Commonwealth Institute of Biological Control (CIBC). Work in developing countries was expanded and an East African Station opened in 1962 in Uganda and a West African Substation in Ghana in 1969 (Greathead, 1994). The purpose of these was to assist African countries and to find natural enemies for export to other regions. In francophone West Africa, Madagascar and Réunion biological control programmes started to be undertaken by staff of l'Institut de Recherches Agronomiques Tropicales (IRAT) and l'Office du Départment de Recherche Scientifique d'Outre-Mer (ORSTOM) (Jourdheuil, 1986).

One target for biological control was the potato tuber moth, *Phthorimaea operculella* (Zeller) (Lepidoptera, Gelechiidae), a native of South America, which has become a major pest of potato, tobacco and other solanaceous crops throughout the warm temperate and tropical zones of the world. Efforts to find biological control agents began as long ago as 1918 with the importation and release of North American parasitoids in Europe and South Africa, but these were ineffective. Exploratory research showed that South America was the native home of the insect and natural enemies from there appeared to have greater potential for biological control. Introduction programmes were carried out in most countries active in biological control, many of them with the assistance of CIBC, which maintained cultures at its Indian Station at Bangalore. Most anglophone southern and eastern African countries, as well as Madagascar, Mauritius and the Seychelles, participated. Only Zambia and Zimbabwe claimed spectacular results, but the practicability of relying on biological control alone is in doubt (Sankaran and Girling, 1980; Kfir, this volume).

The campaign against cereal and sugarcane lepidopterous stem borers in a number of countries, which took place during the 1950s and 1960s, is typified by the campaign in Mauritius. However, although one stem borer, *Sesamia calamistis* Hampson (Lepidoptera, Noctuidae), was controlled by introduction of its parasitoid, *Cotesia sesamiae* (Cameron) (Hymenoptera, Braconidae), from Kenya in 1951, importations of parasitoids of other genera of stem borers principally from India and Trinidad against the most damaging borer, *C. sacchariphagus*, during 1940–1965 failed to result in a single species becoming established. This despite the fact that earlier introductions of parasitoids of other *Chilo* spp. from Sri Lanka in 1939 had at least resulted in establishment, although none had exerted any impact on the stem borer. In 1961, efforts began in order to obtain parasitoids of *C. sacchariphagus* from Java. These included the breeding and release of more than 62,000 individuals of a parasitoid, *Diatraeophaga striatalis* Townsend (Diptera, Tachinidae). This parasitoid was also introduced into Réunion, where some 80,000 flies were released, but again without becoming established (Greathead, 1971 and references therein). This negative result contrasts with the success achieved in the New World tropics, where tachinid parasitoids have successfully controlled the major pest, *Diatraea saccharalis* (Fabricius) (Lepidoptera, Pyralidae) in a number of countries (Cock, 1985), which justified the effort made to establish *D. striatalis*. *S. calamistis* was also controlled in Madagascar by *Pediobius furvus* (Gahan) (Hymenoptera, Eulophidae) imported from East Africa in 1969 (Greathead, 1971). In East Africa and South Africa detailed ecological studies preceded these introductions, and in francophone West Africa, releases of parasitoids cultured in France were made, but little detail has been published. The outcomes of all these studies were comprehensively reviewed by the contributors to Polaszek (1998) (see also Overholt *et al.*, this volume).

The importation of a predatory mite, *Bdellodes lapidaria* (Acari, Bdellidae), found to be effective against the lucerne flea, *Sminthurus viridis* (L.) (Collembola, Sminthuridae), in Australia, into the Western Cape in South Africa was aimed at controlling the pest in cultivated legume-based pastures. Over 78,000 mites were released between 1963 and 1966 and successful establishment and significant impact on pest numbers were achieved (Wallace and Walters, 1974).

The Asian rhinoceros beetle, *Oryctes rhinoceros* (Linnaeus) (Coleoptera, Scarabaeidae) appeared in Mauritius in 1962 near the Port Louis docks, suggesting that it had arrived on shipped cargo. During the following decade, it spread across the island destroying coconut and ornamental palms. Introductions of insect natural enemies failed to check it, while on Pacific Islands the beetle was eventually controlled by the introduction of a host-specific virus. In 1970, this virus was also introduced into Mauritius and rapidly brought the beetle under control (Monty, 1978). This example is interesting as it is one of the few instances where an insect pathogen has proved to be an effective classical biological control agent. An African species of rhinoceros beetle (*Oryctes monoceros* (Olivier)) is a pest in the Seychelles Islands. Insect natural enemies again proved ineffective in controlling this species and in 1981–1983, an attempt was made to use the *O. rhinoceros* virus to control it. It infected *O. monoceros*, became established in the field, and caused a substantial reduction in damage levels, but the infection rate and the degree of control was less than for *O. rhinoceros* (Lomer, 1986).

In Ghana, after it had been established that the native mealybug, *P. njalensis*, was the principal vector of swollen shoot disease of cacao and that its own natural enemies did not provide adequate control, efforts were made to import and establish natural enemies of other species. These included species shipped from California, Trinidad and Kenya during 1948–1955. Since early direct releases into the field failed, parasitoids were mass reared and released during the later years of the programme. In all, some 880,000 individuals of ten species were released to no avail before the programme was abandoned (Greathead, 1971).

Another programme in which relatively large numbers of inappropriate natural enemies were released without success was the attempt to control the Karoo caterpillar, *Loxostege frustalis* Zeller (Lepidoptera, Pyralidae), a serious pest of sweet Karoo bush, *Pentzia incana* Druce (*Asteraceae*), following ecological changes resulting from overgrazing by sheep. In this instance, parasitoids of the congeneric beet web worm, *Loxostege sticticalis* (Linnaeus), were obtained from the USA and released directly into the field during 1942–1950 without any recoveries in follow-up surveys during the two seasons after releases ceased. In addition, one of the parasitoids, *Chelonus insularis* (Cresson) (Hymenoptera, Braconidae), was mass-reared on a factitious host, the pyralid *Ephestia kuehniella* Zeller. In spite of problems with diseases, just under 6 million were reared and released during 1942–1954. Initial claims of recoveries were discounted when it was discovered that they related to a similar native species, not previously recorded from the Karoo caterpillar (Greathead, 1971).

Most new initiatives for biological control of weeds during this period largely consisted of introducing agents that became available as a result of research for countries in other regions. As well as continuing efforts to control prickly pear cactus, introductions were made in East, South and West Africa and the Indian Ocean Islands for control of *Lantana camara* Linnaeus (*Verbenaceae*) and in South Africa for control of *Hypericum perforatum* Linnaeus (*Clusiaceae*) (Julien and Griffiths, 1998). However, alongside research on stem borers in cereals, studies on insects affecting witchweeds, *Striga* spp. (*Scrophulariaceae*), were carried out by the CIBC in East Africa (Greathead and Milner, 1971). New initiatives were also made to discover biological control agents for control of woody weeds, mostly of Australian origin, that were displacing native vegetation in South Africa. This work has led to the introduction of some very effective agents that are now controlling several of these plants very effectively (Julien and Griffiths, 1998; Zimmermann and Olckers, this volume).

Highly successful control resulted from the campaign in Mauritius to control *Cordia curassavica* (Jacquin) Roemer and Schultes (*Ehretiaceae*), an invader from the Caribbean, which had developed dense thickets that were displacing pasture and natural vegetation. Research in Trinidad resulted in the introduction of two leaf-feeding chrysomelid beetles in 1947. One of them, *Metrogaleruca obscura* (Degeer), became established, and by 1950 much of the scrub was dying and continued defoliation was reducing its competitive power. To combat re-colonization, seed-destroying insects were studied and one, *Eurytoma attiva* Burks (Hymenoptera, Eurytomidae), was selected for introduction and successfully established. Together these two agents have reduced the status of *C. curassavica* to that of a minor roadside weed (Greathead, 1971; Julien and Griffiths, 1998).

New Approaches to Biological Control and IPM (1970–2000)

By the 1970s, realization of the disadvantages of sole reliance on synthetic pesticides had resulted in moves towards developing IPM programmes in which biological control was a major component.

Citrus pests in southern Africa provide one of the first examples of the development of IPM in Africa. Scale insects are major pests of citrus wherever it is grown and the crop has been the subject of biological control programmes around the world. This started in California with the control of *I. purchasi* and eventually resulted in the development of IPM programmes in which biological control agents suppress all the different species of scale insects. In South Africa the success with *I. purchasi* was followed by haphazard and unsuccessful introductions of ladybirds. Interestingly, one of them, *C. montrouzieri,* only became established as an effective predator of *Planococcus citri* (Risso) in 1939, when *Dactylopius* spp., which had been established for control of *Opuntia* spp., provided alternative hosts, and annual releases were no longer required. Following the lead of California, *Aphytis* spp. (Hymenoptera, Aphelinidae) were imported and successfully controlled *C. aonidum* and *Lepidosaphes beckii* (Newman), but species introduced for control of *Aonidiella aurantii* (Maskell) (all Hemiptera, Diaspididae) failed to become established. However, pioneering work by E.C.G. Bedford showed that *A. aurantii* is suppressed by the native *Aphytis africanus* Quednau and, provided indiscriminate insecticide applications cease and steps are taken to control ants, IPM can be successful (Bedford, 1968).

Renewed confidence in biological controls also led to an end to the practice of haphazard shipment of natural enemies at minimal cost and a return to well-funded research programmes involving the selection and careful study of candidate biological control agents for control of arthropod pests prior to their introduction. This had long been done in weed control programmes, where the prevention of damage to economically important plants was a prime concern.

The establishment of the International Institute of Tropical Agriculture (IITA) at Ibadan in Nigeria in 1967, principally concerned with the breeding of improved crop varieties, eventually provided a new focus for pest management and biological control in tropical Africa, especially in West Africa, which had been the least active. The first of a new generation of international biological control programmes developed following the discovery of a mite, *M. tanajoa*, on cassava in Uganda in 1971 and a mealybug, *P. manihoti*, in 1973 in the Congo. Both new pests come from South America and are believed to have reached Africa on smuggled planting material. The CIBC soon obtained funding for research on their natural enemies in Trinidad and South America, but IITA was designated to carry out implementation of biological control. This began in 1979 with the appointment of H.R. Herren to lead the programme, which became the largest biological control programme ever undertaken. Outstanding control of *P. manihoti* was obtained with the encyrtid parasitoid *Anagyrus lopezi* (De Santis) (Hymenoptera, Encyrtidae), shipped to IITA in 1981 through a newly established CIBC quarantine facility in the UK (Neuenschwander, this volume). Progress with controlling the mite was slower and less dramatic than with the mealybug, and only began to succeed once the climates of the source area in South America and the infested areas of Africa were carefully matched and predators were obtained from areas of northwest Brazil with a similar climate. The most successful species, *Typhlodromalus aripo* DeLeon (Acari,

Phytoseiidae), is confined to shoot tips and so allows persistence of the host population and is also better able to survive on alternative sources of food when *M. tanajoa* is scarce. It is now established in some 20 countries and has reduced mite damage by more than 50% (Yaninek and Hanna, this volume). This narrow climatic dependency contrasts with *A. lopezi*, which came from Paraguay and southern Brazil, yet was rapidly successful throughout the range of climates of the infested areas in Africa.

The confidence in biological control in West Africa generated by the success with *P. manihoti* enabled rapid progress in mounting a programme for control of the mango mealybug, *Rastrococcus invadens* Williams, when it appeared in West Africa in 1982. An encyrtid parasitoid, *Gyranusoidea tebygi* Noyes, was found in its native home in India, quarantined, released and had suppressed the mealybug in Togo within 2 years (Agricola *et al.*, 1989). In conjunction with a complementary encyrtid parasitoid, *Anagyrus mangicola* Noyes, the mango mealybug has been controlled throughout the area (Neuenschwander, this volume).

There was also renewed interest in controlling cereal stem borers at the International Centre of Insect Physiology and Ecology (ICIPE) in Nairobi, which had been initiated by T.R. Odhiambo in 1970. This programme initially explored intercropping and methods of enhancing existing natural enemies, but also undertook a concerted, and eventually successful, attempt to introduce the parasitoid *C. flavipes*, for control of the major immigrant pest species *C. partellus*. Previous attempts to introduce this parasitoid by CIBC in 1968–1972 in Uganda and Kenya and by South African entomologists in 1983–1985 had failed (Polaszek, 1998; Overholt *et al.*, this volume).

Other collaborative programmes also developed, including a regional programme against forestry pests in tropical Africa, which was coordinated by the International Institute of Biological Control (formerly CIBC) from its Kenya Station, set up in 1980 to replace the former East African Station in Uganda, closed in 1979. A devastating attack on ornamental and plantation cypresses in Malawi in 1985 and later Kenya and Tanzania by an immigrant aphid, *Cinara cupressi* (Buckton) (Hemiptera, Aphididae), stimulated the development of a regional programme to find biological control for this species. Interest was also renewed in controlling *P. boerneri*, which had appeared in Kenya on exotic pine plantations in the 1960s, and after the failure of an eradication programme had been the subject of an earlier unsuccessful biological control programme. This aphid had spread in the meantime and had reached as far south as the northern provinces of South Africa (Day *et al.*, this volume).

The floating waterweed, water hyacinth, *Eichhornia crassipes* (Martius) Solms-Laubach (*Pontederiaceae*), which originated in South America and has been spread by horticulturists throughout the tropics on account of its showy flowers, has long been present on the African continent. This weed had been controlled successfully on the River Nile in the Sudan during the 1970s by introduction of insect control agents (Table 1.3; Beshir, 1984). Although present on several other rivers, it did not attract international attention until it invaded West Africa (Côte d'Ivoire, Ghana, Bénin and Nigeria) and Lake Victoria down the Kagera River from Rwanda. Its rapid spread in the lake threatened fisheries, transportation and the hydroelectric power station at Jinja in Uganda where the River Nile leaves the lake. The IIBC Kenya Station was also involved with the FAO in developing an international campaign against it, but action was delayed by disagreements among the three riparian countries (Kenya, Tanzania and Uganda) on priorities and on the safety of biological control. This has eventually been implemented with very promising initial results. Later, the Kenya station became part of a wider initiative to develop a mycoherbicide to complement the action of insect agents, the International Mycoherbicide Programme for *E. crassipes* Control in Africa (IMPECCA) also including South Africa, Malawi, Nigeria, Bénin and Egypt. Insect control agents had already been established in these countries but had not always been as successful as had been hoped (Cilliers *et al.*, this volume).

Another invasive pest, the larger grain borer, *Prostephanus truncatus*, which appeared in Tanzania in 1981 and shortly afterwards in Togo, spread into neighbouring countries causing devastating damage to stored maize and other crops. Major research programmes were initiated in West Africa in collaboration with the German Gesellschaft für Technische Zusammenarbeit (GTZ), IITA and several national programmes, and in East Africa with the British Natural Resources Institute (NRI) and the national programmes of this region. When it was realized that the beetle was breeding in natural habitats, the possibility of biological control was considered. Field studies in its native home in Mexico detected a histerid predator, *Teretrius nigrescens* (Lewis). Unexpectedly, it was attracted to *P. truncatus* pheromone traps and *P. truncatus* was shown to be, at least, a preferred host, if not its only host, and so a potential biological control agent. Releases have been made in both East and West Africa, where it is now well established. Its presence is linked to substantial reductions of *P. truncatus* in natural habitats and so colonization of grain stores has been reduced (Borgemeister *et al.*, this volume).

Biological control of pests of medical and veterinary importance has seldom been successful, but stable flies that were a serious constraint on dairy farming in Mauritius have been substantially controlled by introduced parasitoids. Puparial parasitoids of dung-breeding flies were introduced in 1966–1972 but did not solve the problem. Intensive surveys showed that they had in fact greatly reduced numbers of the dung-breeding species, *Stomoxys calcitrans* (Linnaeus) (Diptera, Muscidae), but had not affected numbers of another species, *Stomoxys niger* Macquart, which was found breeding in rotting sugarcane tops. Studies in Uganda, started as part of a worldwide survey of filth fly natural enemies, showed a substantially different parasitoid spectrum of *Stomoxys* spp. breeding in rotting vegetation from that found in dung pits. When the parasitoids from puparia in rotting vegetation were introduced during 1975–1978, a substantial drop in stable fly numbers took place and numbers remain at an acceptable level during most of the year (Greathead and Monty, 1982).

Perhaps the most innovative biological control programme was initiated in 1989 for the control of locusts and grasshoppers. The desert locust, *Schistocerca gregaria* (Forskål) (Orthoptera, Acrididae), outbreak of 1986–1988 coincided with the banning of dieldrin, which had been the mainstay of locust control since the 1960s. The FAO sought suggestions for novel environmentally benign control measures and supported the funding of work on semiochemicals at ICIPE and the development of a biopesticide by a consortium of IIBC, IITA and Département de Formation en Protection Végétaux (DFPV) of the Comité permanent Inter-Etats de Lutte contre la Sécheresse au Sahel (CILSS), which came to be known as LUBILOSA. The biopesticide programme investigated the proposition that fungi provided the best possibility of biological control using spores formulated in oil. This was based on the observation by C. Prior that oil formulations overcome the requirement for high humidity for the germination of spores of entomophagous fungi (Prior and Greathead, 1989). The concept proved to be viable and eventually resulted in the registration of a product, Green Muscle™, based on a strain of the green muscardine fungus with a narrow host range, *Metarhizium anisopliae* var. *acridum* Driver and Milner (*Deuteromycetes*), for locust control in South Africa and subsequently elsewhere (Langewald *et al.*, this volume). The discovery opens the way for the development of other biopesticides based on entomophagous fungi for the control of other arthropod pests such as termites (Langewald *et al.*, this volume).

Most biological control research in Africa has aimed at achieving classical biological control as a first objective. However, there are numerous serious pests native to Africa that do not offer obvious opportunities for this approach. For example, research on natural enemies of the boll worm *Helicoverpa armigera* (Hübner) (Lepidoptera, Noctuidae) in Africa, Asia and Australia had shown few gaps in indigenous natural enemy spectra that could be exploited. Consequently, a new initiative was launched in 1987 to look for alternatives. The CIBC Station in Kenya undertook studies on natural enemy impact on a range of important crops with the objective of exploring their potential for enhancement in IPM (van den Berg, 1993; Cherry *et al.*, this volume). Similarly, cowpea pests have been targets for IPM, exploiting natural enemies including a possibly adventive parasitoid (*Ceranisus femoratus* Gahan (Hymenoptera, Eulophidae)), which appeared in Cameroon in 1998 and has been redistributed to Bénin (Tamò *et al.*, this volume).

In Kenya, coffee is a crop where biological control has been important since biological control of the mealybug *Planococcus kenyae* was implemented. This was overlooked in the 1950s when persistent organochlorine insecticides were applied for the control of antestia bugs (*Antestiopsis* spp.). Not only did this cause resurgence of mealybugs but also outbreaks of leafminers (*Leucoptera* spp.), which had been suppressed by their native natural enemies. A change to non-persistent organophosphate insecticides, timed to coincide with peak adult leafminer numbers, allowed biological control of the mealybug to be re-established (Bess, 1964). However, spraying of copper fungicides for control of coffee berry disease was implicated in initiating outbreaks of a native species, *Icerya pattersoni* Newstead, in the early 1980s. Investigations showed that the principal natural enemy is a ladybird, *Rodolia iceryae* Janson, and efforts by growers to conserve this ladybird and other natural enemies resulted in a reduction in numbers of *I. pattersoni* by the end of the decade (Kairo and Murphy, 1992; Oduor, this volume).

During the 1980s, there was increasing concern about the impact of introduced species on natural ecosystems and, in particular, criticism of the impact of past introductions of biological control agents on non-target species, and a demand for more stringent screening of potential classical biological control agents prior to importation and release (e.g. Howarth, 1991). One response was the convening of an expert consultation by the FAO in 1991, which drafted a Code of Conduct for the import and release of exotic biological control agents, published in 1996 (FAO, 1996). Its implementation is followed by agencies involved in the introduction of biological control agents into Africa, many of whom were represented at the expert consultation, notably the Inter-African Phytosanitary Council (IAPSC), whose country members have responsibility for approval of introductions of biological control agents into African countries. Biological control in Africa has also been affected by the Agenda 21 of the Rio Earth Summit of 1992. As a result of these developments, African governments are much more aware of biological control, and biological control agents are being more thoroughly tested and evaluated before importation and release of exotic species is permitted. This will also ensure that, in the future, fewer but better researched agents are imported and will hopefully result in a higher success rate for introductions. Greater environmental awareness should also provide a spur to the development of IPM systems, minimizing the use of broad-spectrum chemicals and making greater use of indigenous biological control agents and biopesticides. However, concern for the environment and the preservation of biodiversity needs to be tempered by the realities of African agriculture, which remains predominantly the concern of resource-poor farmers. As eloquently argued by Neuenschwander and Markham (2001), the regulatory framework should not be made so prescriptive and cumbersome that biological control is replaced by more destructive alternatives, such as broad-spectrum chemical pesticides, which few farmers can afford or are equipped to use safely.

However, classical biological control is providing a benign means of limiting the damage done to natural ecosystems and endangered species by exotic pests. Progress in the control of invasive plants, principally from Australia, threatening the unique South African fynbos vegetation, is discussed elsewhere by Zimmerman and Olckers (this volume). A further example is the control of the polyphagous cosmopolitan scale insect *Orthezia insignis* Browne (Hemiptera, Ortheziidae) in St Helena, where it was threatening the survival of the national tree, the endemic gumwood, *Commidendrum robustum* (*Asteraceae*). Serendipitously, the scale had already been controlled in East Africa in the 1950s, when it was causing severe nuisance by damaging urban flowering trees, especially jacaranda, *Jacaranda mimosiflolia* G. Don (*Bignoniaceae*), by introduction of a ladybird, *Hyperaspis pantherina* Fürsch (Coleoptera, Coccinellidae) from Trinidad, since shown to be specific to the genus *Orthezia* (Booth *et al.*, 1995). Thus, it was relatively straightforward to obtain the ladybird from Kenya for quarantining and introduction into St Helena, where it has provided highly successful control (Beggs, 2001).

Conclusions

The historical development of biological control in the Afrotropical Region has followed much the same pattern as in other regions (Figs 1.1 and 1.2). It began in South Africa near the end of the 19th century during the period of enthusiasm following the successful control of *I. purchasi* in California and shortly afterwards in Kenya and Mauritius. These three countries remained committed users and the results they achieved caused neighbouring countries to undertake similar programmes. Other countries, notably those in West Africa, did not attempt biological control until after World War II and many of them did not attempt classical biological control until confronted with the ravages of the cassava mealybug during the 1970s and 1980s. The number of introductions overall was not as greatly affected by the availability of the new synthetic chemical pesticides after the end of World War II as in some other regions.

In East Africa, the concept of ecological islands, allowing movement of natural enemies from one part of a continent to another, gained support as a result of the control of the Kenya mealybug on coffee in Kenya with parasitoids from Uganda. However, subsequent investigations in Eastern Africa on other coffee pests, cereal stem borers and witchweeds did not identify many opportunities for intra-continental exchange of natural enemies. Rather, they supported the need for IPM programmes emphasizing the conservation of augmentation of existing natural enemies.

The advent of cheap and rapid air travel encouraged the importation of natural enemies on a 'try it and see' basis. Few successes were achieved and there was a return to in-depth studies, precipitated by the challenges posed by the appearance of cassava mites and mealybugs, and the devastation they caused to a vital staple food crop. The need for well-funded, in-depth studies was given further support by the worldwide concern over the conservation of biodiversity and the environment, and criticism of the damage to non-target organisms by some introduced natural enemies. It also resulted in the development of the FAO Code of Conduct for the import of exotic biological control agents, which has provided the basis for national and regional plant quarantine organizations to regulate the importation and release of exotic biological control agents.

Although there remain opportunities for classical biological control, and no doubt more will occur as a result of accidental introductions of pests and invasive species, the principal need is for IPM schemes optimizing the impact of indigenous natural enemies. This will, most likely, take the form of measures to conserve and enhance the action of arthropod natural enemies and the development of selective biopesticides for application as sprays or dusts.

References

Agricola, U., Agounké, D., Fischer, H.U. and Moore, D. (1989) The control of *Rastrococcus invadens* Williams (Hemiptera: Pseudococcidae) in Togo by the introduction of *Gyranusoidea tebygi* Noyes (Hymenoptera: Encyrtidae). *Bulletin of Entomological Research* 79, 671–678.

Bedford, E.C.G. (1968) Biological and chemical control of citrus pests in the Western Transvaal: an integrated spray programme. *South African Citrus Journal* 417, 9, 11, 13, 15, 17, 19, 21–28.

Beggs, J. (2001) Biological control – a success story. *Stowaways* 1, 15.

Beshir, M.O. (1984) The establishment and distribution of natural enemies of water hyacinth released in Sudan. *Tropical Pest Management* 30, 320–323.

Bess, H.A. (1964) Populations of the leaf-miner *Leucoptera meyricki* and its parasites in sprayed and unsprayed coffee in Kenya. *Bulletin of Entomological Research* 55, 59–82.

Booth, R.G., Cross, A.E., Fowler, S.V. and Shaw, R.H. (1995) The biology and taxonomy of *Hyperaspis pantherina* (Coleoptera: Coccinellidae) and the classical biological control of its prey, *Orthezia insignis* (Homoptera: Ortheziidae). *Bulletin of Entomological Research* 85, 307–314.

Caltagirone, L.E. and Doutt, R.L. (1989) The history of the vedalia beetle importation to California and its impact on the development of biological control. *Annual Review of Entomology* 34, 1–16.

Cock, M.J.W. (ed.) (1985) *A Review of Biological Control of Pests in the Commonwealth Caribbean and Bermuda up to 1982*. Commonwealth Institute of Biological Control, Technical Communication No. 9. Commonwealth Agricultural Bureaux, Farnham Royal, UK.

El-Haidari, H.S. (1981) The use of predator ants for control of date palm insects pests in the Yemen Arab Republic. *Date Palm Journal* 1, 129–132.

FAO (Food and Agriculture Organization of the United Nations) (1996) *Code of Conduct for the Import and Release of Exotic Biological Control Agents*. International Standards for Phytosanitary Measures, Publication No. 3. FAO, Rome.

Greathead, D.J. (1971) *A Review of Biological Control in the Ethiopian Region*. Commonwealth Institute of Biological Control, Technical Communication No. 5. Commonwealth Agricultural Bureaux, Farnham Royal, UK.

Greathead, D.J. (1994) History of biological control. *Antenna* 18, 187–199.

Greathead, D.J. and Milner, J.E.D. (1971) A survey of *Striga* spp. (Scrophulariaceae) and their insect natural enemies in East Africa with a discussion on the possibilities of biological control. *Tropical Agriculture (Trinidad)* 48, 111–124.

Greathead, D.J. and Monty, J. (1982) Biological control of stableflies (*Stomoxys* spp.): results from Mauritius in relation to fly control in dispersed breeding sites. *Biocontrol News and Information* 3, 105–109.

Greathead, D.J. and Greathead, A.H. (1992) Biological control of insect pests by insect parasitoids and predators: the BIOCAT database. *Biocontrol News and Information* 12, 61N-68N.

Howarth, F.G. (1991) Environmental impacts of classical biological control. *Annual Review of Entomology* 36, 485–509.

Jourdheuil, P. (1986) La lutte biologique à l'aide d'arthropodes entomophages. Bilan des activités des services français de recherche et de développement. *Cahiers de Liason O.P.I.E.*, 20, 3–48.

Julien, M.H. (1989) Biological control of weeds worldwide: trends, rates of success and the future. *Biocontrol News and Information* 10, 299–306.

Julien, M.H. and Griffiths, M.W. (1998) *Biological Control of Weeds: a World Catalogue of Agents and their Target Weeds*. 4th edn. CAB International, Wallingford, UK.

Julien, M.H., Kerr, J.D. and Chan, R.R. (1984) Biological control of weeds: an evaluation. *Protection Ecology* 7, 3–25.

Kamal, M. (1951) Biological control projects in Egypt, with a list of introduced parasites and predators. *Bulletin de la Société Fouad 1er d'Entomologie* 35, 205–220.

Kairo, M.T.K. and Murphy, S.V. (1992) An analysis of insecticide use in Kenya coffee IPM and outbreaks of *Icerya pattersoni*. *Proceedings of the Brighton Crop Protection Conference – Pests and Diseases*, 1992, pp. 1027–1032.

Lomer, C.J. (1986) Release of *Baculovirus oryctes* into *Oryctes monoceros* populations in the Seychelles. *Journal of Invertebrate Pathology* 47, 237–246.

Lounsbury, C.P. (1940) The pioneer period of economic entomology in South Africa. *Journal of the Entomological Society of Southern Africa* 3, 9–29.

Miller, J.I. (1969) *The Spice Trade of the Roman Empire 29 B.C. to A.D. 641*. Clarendon Press, Oxford, UK.

Monty, J. (1978) The coconut palm rhinoceros beetle *Oryctes rhinoceros* (L.) in Mauritius and its control. *Revue Agricole et Sucrière de l'Ile Maurice* 57, 60–76.

Neuenschwander, P. and Markham, R. (2001) Biological control in Africa and its possible effects on biodiversity. In: Wajnberg, E., Scott, J.K. and Quimby, P.C. (eds) *Evaluating Indirect Effects of Biological Control*. CAB International, Wallingford, UK, pp. 127–146.

Neuenschwander, P., Murphy, S.T. and Coly, E.V. (1987) Introduction of exotic parasitic wasps for the control of *Liriomyza trifolii* (Dipt., Agromyzidae) in Senegal. *Tropical Pest Management* 33, 290–297.

Polaszek, A. (1998) *African Stem Borers: Economic Importance, Taxonomy, Natural Enemies and Control*. CAB International, Wallingford, UK.

Prior, C. and Greathead, D.J. (1989) Biological control of locusts: the potential for the exploitation of pathogens. *FAO Plant Protection Bulletin* 37, 37–48.

Samways, M.J. (1989) Climate diagrams and biological control: an example from the areography of the ladybird *Chilocorus nigritus* (Fabricius, 1798) (Insecta, Coleoptera, Coccinellidae). *Journal of Biogeography* 16, 345–351.

Sankaran, T. and Girling, D.J. (1980) The current status of biological control of the potato tuber moth. *Biocontrol News and Information* 1, 207–211.

Simmonds, F.J. and Greathead, D.J. (1977) Introductions and pest and weed problems. In: Cherrett, J.M. and Sagar, G.R. (eds) *Origins of Pest, Parasite, Disease and Weed Problems*. Blackwell Scientific Publications, Oxford, UK, pp. 109–124.

Smith, R.R., Mittler, T.E. and Smith, C.N. (1973) *History of Entomology*. Annual Reviews Inc., Palo Alto, California, USA.

van den Berg, H. (1993) Natural control of *Helicoverpa armigera* in smallholder crops in East Africa. PhD thesis, Agricultural University, Wageningen, The Netherlands.

Wallace, M.M.H. and Walters, M.C. (1974) The introduction of *Bdellodes lapidaria* (Acari: Bdellidae) from Australia into South Africa for the biological control of *Sminthurus viridis* (Collembola). *Australian Journal of Zoology* 22, 505–517.

2

Biological Control of Alien Plant Invaders in Southern Africa

Helmuth G. Zimmermann[1] and Terry Olckers[2]

[1]*Agricultural Research Council – Plant Protection Research Institute, Weeds Research Division, Pretoria, South Africa;* [2]*Agricultural Research Council – Plant Protection Research Institute, Hilton, South Africa*

Introduction

The considerable diversity of floristic biomes in southern Africa, including succulent deserts, mediterranean shrubland, alpine highland, grasslands, savannah and subtropical forests (Low and Rebelo, 1996), provide many niches for the more than 1000 introduced plant species that have become established in South Africa (Wells *et al.*, 1986). Of these, some 500 species have naturalized, i.e. they reproduce consistently and sustain populations without human interventions. Furthermore, 198 species are listed as problematic and are either prohibited or regulated in terms of the Conservation of Agricultural Resources Act, 1983 (Act No. 43 of 1983) (CARA) of South Africa (Henderson, 2001). According to Richardson *et al.* (2000), an 'alien invasive plant' represents a naturalized introduced species that reproduces prolifically and has the potential to spread extensively and invade relatively undisturbed ecosystems (Henderson, 2001). Many are also 'transformer species', which represent invasive plants that change the character, condition, form or nature of ecosystems over substantial areas relative to the extent of the ecosystems (Richardson *et al.*, 2000). Alien invasive plants have invaded some 10.1 million ha, or 6.8% of the total landscape, in South Africa and are responsible for the loss of 7% (3300 million m^3) of the mean annual runoff through transpiration by woody species in catchments, rivers, wetlands and other water resources (Versfeld *et al.*, 1998).

Despite concerted efforts in South Africa to control many of these species using conventional methods, limited success has been achieved (Moran and Annecke, 1979; Fenn, 1980; Wassermann *et al.*, 1988). However, the launch of the 'Working for Water' (WFW) Programme in South Africa in 1996 heralded the start of the largest programme ever initiated in Africa to address the threat of alien plant invaders in a holistic manner, with biological control fully integrated into the national strategy (Olckers *et al.*, 1998). This signified a turning

(eds P. Neuenschwander, C. Borgemeister and J. Langewald)

point in alien weed control in South Africa because the integration of biological, chemical and mechanical control at this level had not previously been attempted. New initiatives by the Global Invasive Species Programme (GISP) and International Union for the Conservation of Nature (IUCN), with regional support from the New Partnership for Africa's Development (NEPAD) may see similar programmes launched within other Southern African Development Community (SADC) countries. It is expected that biological control will feature significantly in any such new ventures. In this chapter, we review the status of biological weed control in southern Africa, highlighting some unique features and addressing key issues such as the value of biological control in integrated weed management (IWM) and the prospects for its further development and exploitation in the region.

History of Biological Weed Control in Southern Africa

Biological control of weeds in southern Africa was first practised in 1913 when the Brazilian cactus, *Opuntia monocantha* Haw. (*Opuntia vulgaris* Mill. misapplied) (*Cactaceae*), was controlled in South Africa by the introduced cochineal insect, *Dactylopius ceylonicus* (Green) (Homoptera, Dactylopiidae), after equally dramatic success in India (Tryon, 1910). In the ensuing 89 years, more than 95 species of biological control agents have been introduced against 47 weed species in South Africa alone (Olckers and Hill, 1999). Table 2.1 provides a summary of the agents released and established in South Africa and the overall degree of control achieved for each weed species targeted. In South Africa, 11 of the targeted species (23%) are considered to be under 'complete biological control' (i.e. no other control measures are required to maintain the weed populations at acceptable densities). A further 15 species (32%) are under 'substantial control' in that conventional control measures are still needed, but at reduced rates. Seven species (15%) are under 'negligible control' because there has been virtually no reduction in the need for conventional control methods, despite damage inflicted by the agents. One species (2%) is under 'no control' because the agents failed to establish, and on the remaining 13 species (28%), releases of agents have been too recent to allow any meaningful assessment. Overall, this translates into a success rate of 76% for all evaluated projects.

South Africa has supplied 15 species of biological control agents to other African countries and Australia since 1995 for the control of mainly water weeds, but information on the degree of success of these introductions is limited. The outcome of earlier introductions of biological control agents into other African countries, mainly through the efforts of the Commonwealth Institute of Biological Control (CIBC), are discussed by Greathead (1971). The international catalogue of biological control agents released against weeds worldwide (Julien and Griffiths, 1999) also lists all releases carried out in Africa until 1996.

During the early stages of weed biological control in southern Africa, programmes were largely reliant on successes achieved elsewhere in the world. For example, the entire programme targeting invasive *Cactaceae* in East Africa and southern Africa relied initially on earlier work done in Australia (Annecke and

Table 2.1. Success ratings of biological control programmes undertaken against 47 alien plant species in South Africa. Details on the biological control agents involved are provided in Julien and Griffiths (1999).

Weed species	Resource affected[a]	Agents released/ established[b]	Control status[c]
Araceae			
Pistia stratiotes	Wat, Con	1/1 (insect)	Complete
Asteraceae			
Ageratina adenophora	For	2/2 (insect and pathogen)	Negligible
Ageratina riparia	For	1/1 (pathogen)	Unknown
Chromolaena odorata[e]	Pas, Con, For, Wat	3/1? (insect)	Unknown
Cirsium vulgare	Pas	2/1 (insect)	Negligible
Silybum marianum	Pas	1/1[d] (insect)	Unknown
Azollaceae			
Azolla filiculoides[e]	Wat, Con	1/1 (insect)	Complete
Bignoniaceae			
Macfadyena unguis-cati[e]	Con, For	1/1 (insect)	Negligible
Cactaceae			
Cereus jamacaru	Pas	2/2[d] (insects)	Substantial
Harrisia martinii	Pas	2/2 (insects)	Complete
Opuntia aurantiaca	Pas, Con	5/2 (insects)	Substantial
Opuntia exaltata	Pas	1/1[d] (insect)	Unknown
Opuntia ficus-indica	Pas, Con	4/4 (insects)	Substantial
Opuntia fulgida (= *O. rosea*)	Pas	1/1[d] (insect)	Negligible
Opuntia imbricata	Pas	2/1 (insect)	Substantial
Opuntia leptocaulis	Pas	1/1[d] (insect)	Complete
Opuntia lindheimeri	Pas	2/2[d] (insects)	Substantial
Opuntia monocantha (= *O. vulgaris*)	Pas	3/3[d] (insects)	Complete
Opuntia salmiana	Pas	1/1[d] (insect)	Substantial
Opuntia spinulifera	Pas	1/1[d] (insect)	Unknown
Opuntia stricta	Pas, Con	2/2[d] (insects)	Complete
Pereskia aculeata	Con, For	1/1 (insect)	Negligible
Clusiaceae			
Hypericum perforatum	Pas	6/2 (insects)	Complete
Convolvulaceae			
Convolvulus arvensis	Pas	1/0	None
Fabaceae			
Acacia cyclops[e]	Pas, Con, Wat	2/2 (insects)	Substantial
Acacia dealbata[e]	Pas, Con, Wat	1/1 (insect)	Unknown
Acacia decurrens[e]	Pas, Con, Wat	1/1? (insect)	Unknown
Acacia longifolia[e]	Pas, Con, Wat	2/2 (insects)	Substantial
Acacia mearnsii[e]	Pas, Con, Wat	1/1 (insect)	Unknown
Acacia melanoxylon[e]	Pas, Con, Wat	1/1 (insect)	Substantial
Acacia pycnantha[e]	Pas, Con	1/1 (insect)	Complete
Acacia saligna[e]	Pas, Con, Wat	2/2 (pathogen and insect)	Complete
Caesalpinia decapetala[e]	Pas, Con, For, Wat	1/1 (insect)	Unknown
Leucaena leucocephala[e]	Con	1/1 (insect)	Unknown

Continued

Table 2.1. *Continued.*

Weed species	Resource affected[a]	Agents released/ established[b]	Control status[c]
Fabaceae (*Continued*)			
Paraserianthes lophanta[e]	Pas, Con	1/1 (insect)	Unknown
Prosopis species[e]	Pas, Wat	3/2 (insects)	Negligible
Sesbania puniceae	Con, Wat	3/3 (insects)	Complete
Haloragaceae			
Myriophyllum aquaticum[e]	Wat	1/1 (insect)	Substantial
Myrtaceae			
Leptospermum laevigatum[e]	Pas, Con	2/2 (insects)	Unknown
Pontaderiaceae			
Eichhornia crassipes	Wat, Con	6/6 (4 insects, 1 mite and 1 pathogen)	Substantial
Proteaceae			
Hakea gibbosa[e]	Con, Wat	2/2[d] (insects)	Negligible
Hakea sericea[e]	Con, Wat	4/4 (insects)	Substantial
Salviniaceae			
Salvinia molesta	Wat, Con	1/1 (insect)	Complete
Solanaceae			
Solanum elaeagnifolium[e]	Cro, Pas	4/2 (insects)	Substantial
Solanum mauritianum[e]	Pas, Con, For, Wat	1/1 (insect)	Unknown
Solanum sisymbriifolium[e]	Pas, For	1/1 (insect)	Substantial
Verbenaceae			
Lantana camara	Pas, Con, For, Wat	20/8 (insects)	Substantial

[a]Resources affected include croplands (Cro), pastures and rangelands (Pas), conservation areas (Con), forestry plantations (For) and water resources including catchments (Wat).
[b]Number of agent species released versus the number that established or that extended their host ranges following releases on closely related weed species. Information in brackets refers to established agents.
[c]Complete, no other control methods needed; Substantial, other methods still needed, but at reduced rates; Negligible, other methods still needed at the same rates; None, no control; Unknown, recent or unevaluated projects.
[d]Involves or includes agents that extended their host ranges.
[e]Weeds targeted without assistance from programmes in other countries.

Moran, 1978; Moran and Zimmermann, 1991; Zimmermann and Moran, 1991). The first use of a biological control agent that was funded and initiated by South Africa occurred in 1958, with the introduction and release of *Metamasius spinolae* Gyllenhal (Coleoptera, Curculionidae) from Mexico to improve the biological control of *Opuntia ficus-indica* (L.) Mill. (Annecke and Moran, 1978). Increasingly, local initiatives followed and the first uniquely South African project started in 1962, with the targeting of *Hakea sericea* Schrad. and J.C. Wendl. (*Proteaceae*) for biological control (Kluge and Neser, 1991). The introduction and release of the grasshopper *Paulinia acuminata* (Degeer) (Orthoptera, Pauliniidae), against *Salvinia molesta* D.S. Mitch. (*Salviniaceae*) in Zimbabwe in 1969 (Julien and Griffiths, 1999) was probably the first African attempt at weed biological control outside of South Africa. The benefits of weed biological control in African countries other than South Africa have been underexploited but, given South Africa's excellent record of successes, biological control still presents excellent opportunities for many countries.

Conflicts of Interest with Weed Biological Control in Southern Africa

In South Africa, 17 of the 40 most aggressive alien plant invaders were deliberately introduced for forestry and agroforestry purposes. These include well-known species such as *Prosopis* species, *Acacia saligna* (Labill.) H.L.Wendl., *Acacia melanoxylon* R.Br., *Acacia cyclops* A. Cunn. ex G. Don, *Acacia mearnsii* De Wild., *Acacia dealbata* Link, *Acacia decurrens* (Wendl.) Willd., *Leucaena leucocephala* (Lam.) de Wit (all *Fabaceae*), *Casuarina equisetifolia* L. (*Casuarinaceae*), *Pinus pinaster* Aiton and *Pinus radiata* D. Don (both *Pinaceae*). These species all have commercial value and are utilized by some parties, but are viewed as troublesome and as a threat by others. The ensuing conflicts of interest between their benefits and threats limit the extent to which biological control can be implemented. The problem is compounded where biological control is the only feasible way of controlling the spread of some populations, but their commercial value limits the type of agents that can be used. Consequently, only seed-destroying agents were considered for reducing the invasive potential of species such as *A. mearnsii*, *A. cyclops*, *A. melanoxylon*, *L. leucocephala*, *Prosopis* species and *P. pinaster* (Olckers *et al.*, 1998; Dennill *et al.*, 1999; Impson *et al.*, 1999). The merits of using seed control as an efficient means of reducing invasive potential is debatable, but evidence emerging from the successful projects on *Acacia longifolia* (Andr.) Willd. (*Fabaceae*) (Dennill *et al.*, 1999) and *Sesbania punicea* (Cav.) Benth. (*Fabaceae*) (Hoffmann and Moran, 1999) has confirmed the feasibility of this approach. Because of the seriousness of invasions by *A. saligna*, a widely used agroforestry species (Henderson, 1998), a full-scale biological control programme was launched in 1988 with the release of the Australian rust fungus, *Uromycladium tepperianum* (Sacc.) McAlp. (Basidiomycetes, Uredinales), followed by the release of the seed weevil, *Melanterius compactus* Lea (Coleoptera, Curculionidae). This programme resulted in the weed being brought under complete biological control (Table 2.1). Reports of similar invasions by *A. saligna* and other agroforestry species in other regions of Africa and the Mediterranean are emerging, but no initiatives aimed at curbing seed pollution have ensued. Host-specific seed-feeding agents are best utilized during the early stages of a weed's invasion, when the effect of seed control will be more efficient in reducing the levels of invasion (Neser and Kluge, 1986; Hughes, 1995).

A different conflict of interest emerged in Madagascar following the effective control of large infestations of introduced *Opuntia* species (*Cactaceae*) by the cochineal bug *Dactylopius opuntiae* (Cockerell) (Homoptera, Dactylopiidae) in 1923 (Julien and Griffiths, 1999). Although the insect cleared large areas of the invasive *O. monocantha* (*O. vulgaris*) (Greathead, 1971), it similarly removed a valuable human food source, a spineless cactus variety, causing large-scale starvation (Middleton, 1999). Large parts of southern Madagascar have since been re-invaded by the related *Opuntia stricta* (Haw.) Haw. var. *dillenii* (Ker Gawl) L.Benson and *O. ficus-indica*, which suggests that *D. ceylonicus* and not *D. opuntiae* was the agent that was used in the 1920s. The taxonomic confusion surrounding the biological control agents and their hosts in Madagascar needs to be clarified before the events of the 1920s can be fully interpreted. However, the serious conflict between useful and invasive *Opuntia* species in Madagascar today continues to constrain the implementation of biological control of invasive cacti.

Despite very serious invasions of *O. ficus-indica* in the lowlands of Tigray, Ethiopia, it was decided not to resort to any biological control in view of the agricultural importance of this plant in the highlands where it is less invasive (Behailu and Tegegne, 1997). Instead, an attempt is under way to improve the utilization of the weed as a method of reducing its density and further invasion, but it remains to be seen whether this approach will be sufficient to reduce infestations to acceptable and manageable levels. A similar situation exists in Eritrea, but the problem is compounded because of serious invasions by *O. stricta*. However, the use of host-specific biotypes of *D. opuntiae* that will control *O. stricta* without attacking *O. ficus-indica* (Volchansky *et al.*, 1999) presents a solution for African countries that are concerned about protecting *O. ficus-indica* populations.

Similar concerns emerged when biological control was envisaged as a means of curbing the increasing invasions of *Prosopis* species in the Sudan, Eritrea, Ethiopia, Kenya, Malawi, Namibia, Botswana and South Africa. So far, two seed-feeding beetles, *Algarobius prosopis* (LeConte) and *Neltumius arizonensis* Schaeffer (both Coleoptera, Bruchidae), have been released in South Africa and Namibia (Impson *et al.*, 1999). However, their ability to prevent further spread and densification relies on labour-intensive management practices (Impson and Hoffmann, 1998) and additional biological control agents that would further reduce seed production are under consideration. This strategy is being pursued in parallel with increased efforts to utilize the weed and thereby limit further spread. The early establishment of seed-feeding biological control agents in many other African countries where *Prosopis* species are increasing presents a pre-emptive option that, in combination with improved utilization programmes, could prevent future disasters.

Biological Control of Water Weeds

The waterways and lakes of Africa have been extensively invaded by several species of floating alien water weeds which have seriously impeded access to water, affected water quality and reduced aquatic biodiversity (see Cilliers *et al.*, in this volume). The worst of these species is undoubtedly water hyacinth, *Eichhornia crassipes* (Mart.) Solms Laubach (*Pontederiaceae*). Biological control of this and other species, including *S. molesta* Mitchell (*Salviniaceae*) and *Pistia stratiotes* L. (*Araceae*), has been widely implemented in Africa with considerable success (Julien and Griffiths, 1999), but only after unnecessarily long delays caused by scepticism and prolonged duplication of host-specificity tests.

Variable levels of success with the water hyacinth programme, notably in South Africa, have been caused by the effect of cooler climates in high altitude areas, injudicious use of herbicides, high nutrient levels in the water and periodic flooding of open water systems (Hill and Cilliers, 1999; Hill and Olckers, 2001). The time span needed to achieve significant population declines can vary considerably, often creating the perception of failures amongst politicians and decision makers. To overcome some of these problems and improve the levels of biological control, South Africa has screened additional natural enemies to supplement the two weevils, *Neochetina eichhorniae* Warner and *Neochetina bruchi* Hustache (both Coleoptera, Curculionidae), which are widely used in Africa. The newly released agents include *Eccritotarsus catarinensis* (Carvalho) (Heteroptera, Miridae), which is now widely established in South Africa (Hill and Cilliers, 1999) and has also been released in Bénin, Malawi, Zambia and Zimbabwe. Other new agents scheduled for release include *Cornops aquaticum* (Bruner) (Orthoptera, Acrididae) and *Thrypticus* sp. (Diptera, Dolichopodidae) (Hill and Cilliers, 1999).

Unique to South Africa are the highly successful programmes against red water fern (*Azolla filiculoides* Lam.) (*Azollaceae*) (Hill, 1999) and parrot's feather (*Myriophyllum aquaticum* (Vell.) Verdc.) (*Haloragaceae*) (Cilliers, 1999). These programmes should also be further extended to other neighbouring SADC countries where these weeds are also problematic. Indeed, the highly effective weevil, *Stenopelmus rufinasus* Gyllenhal (Coleoptera, Curculionidae), has already been released in Zimbabwe to combat *A. filiculoides* and has spread naturally to Mozambique.

Pathogens for Control of Alien Plant Invaders

Introduced pathogens for classical biological control have been used in South Africa against *A. saligna*, *Ageratina adenophora* (Spreng.) R.M. King and H. Rob., *Ageratina riparia* (Regel) R.M. King and H. Rob. (*both Asteraceae*), *E. crassipes* and *Lantana camara* L. (*Verbenaceae*). Native pathogens in the form of mycoherbicides have also been used in both classical and inundative biological control of *A. mearnsii* and *H. sericea* (Morris *et al.*, 1999).

Early projects in South Africa include the introduction and release of the leaf-spot pathogen *Phaeoramularia eupatorii-odorati* (Yen) Liu and Guo. (*Hyphomycetes, Moniliales*) for the control of *A. adenophora* (Morris, 1991). The fungus, which has a close association with the gall-forming fly *Procecidochares utilis* Stone (Diptera, Tephritidae), released in 1984 (Bennett and Van Staden, 1986), occasionally defoliates plants in the more subtropical regions of KwaZulu-Natal Province, but has had a negligible impact overall. Severe defoliation of *A. riparia*, caused by the fungus *Entyloma compositarum* Farlow (*Basidiomycetes, Ustilaginales*), which was released in South Africa in 1989 (Morris, 1991), has also been reported but the impact of this has not been evaluated.

Since 1986, several pathogens of unknown origin have been recorded on water hyacinth in South Africa (Morris *et al.*, 1999). *Cercospora rodmanii* Conway (*Hyphomycetes, Moniliales*) was introduced from Florida in 1988 and tested against water hyacinth, although Morris (1990) regarded it as synonymous with *Cercospora piaropi* Tharp, which had already become widely established in South Africa. This pathogen now occurs extensively on *E. crassipes* in the Western Cape Province, although there have not been any significant declines in weed populations due to fungal attack. Further work on other naturally occurring pathogens on water hyacinth included studies on pathogen–insect interactions (Den Breeÿen, 1999) and the development of a mycoherbicide.

One of the most outstanding successes involving fungal agents is the contribution of the gall-rust fungus, *U. tepperianum*, to the control of *A. saligna* in South Africa's Western Cape Province (Morris, 1999). Releases carried out from 1988 to 1996 ensured that the rust now occurs throughout the areas of infestation, while field evaluations showed a high incidence of galling with up to 500 galls per large tree. Heavily infected larger trees became defoliated and died, while only between one and five galls were needed to kill young saplings. After 5 years, the density of living trees at the evaluation sites was reduced to 5–10% of the original tree density. However, several native parasitic fungi have recently become established on the galls and are reducing the efficacy of this highly effective agent (Morris, 1999). Because many infected trees are still able to produce seeds, the release and establishment of the seed-feeding weevil, *M. compactus*, in 2001 should reduce the annual accumulation of seed into the seed bank and improve the already impressive levels of biological control.

Two native fungal species have been formulated as mycoherbicides and used for inundative biological control. The first of these involved dried preparations of wheat bran pieces infected with a native strain of *Colletotrichum gloeosporioides* (Penz.) Sacc. (*Coelomycetes*) that were used to treat *H. sericea* seedlings (Morris, 1989, 1999). A granular product has since been developed and granted provisional registration as Hakatak® in 1990 for use in South Africa. New formulations of this product are being developed for improved aerial applications. The second mycoherbicide involved local isolates of *Cylindrobasidium laeve* (Pers.: Fr.) Chamuris (*Basidiomycetes*), which were developed as a cut-stump treatment containing basidiospores in a mineral oil suspension, which is then further diluted with vegetable oil before treatment. This product is now registered as Stumpout® to prevent the resprouting of cut stumps of *A. mearnsii* and *Acacia pycnantha* Benth. (*Fabaceae*) (Morris *et al.*, 1999). This product offers a cheaper and more environmentally friendly alternative to certain herbicides, often used in a diesel carrier, that are used to treat cut stumps of invading Australian *Acacia* species along watercourses or in other sensitive areas.

One of the more recent programmes in South Africa involved the release in 2002 of a isolate from Florida (USA) of the fungus *Mycovelosiella lantanae* (Chupp) Deighton against *L. camara*. It is still too early to determine the outcome of these releases and to predict what the agent's impact on the weed will be. In addition, the use of fungal agents is being considered for other important species that are also invasive in other African countries, such as *Chromolaena odorata* (L.) R.M. King and H. Rob. (*Asteraceae*) and *Prosopis* species. Efforts to improve existing mycoherbicides may also yield considerable benefits for these countries.

Biological Control of Crop Weeds

Few crop weeds have been subjected to biological control in Africa. The most significant project so far has involved the North American weed, *Solanum elaeagnifolium* Cav. (*Solanaceae*), which invades both croplands and rangelands, seriously affecting crop production in South Africa (Wassermann *et al.*, 1988) and in Morocco (Tanji *et al.*, 1985). Chemical and mechanical control of *S. elaeagnifolium* is largely ineffective in South Africa because of the weed's extensive and resistant root system, which permits rapid recovery. The release in South Africa of two leaf-feeding beetles, *Leptinotarsa texana* Schaeffer and *Leptinotarsa defecta* Stål (both Coleoptera, Chrysomelidae), has given good control in certain situations (Olckers *et al.*, 1999). Repeated defoliation of the weed, by *L. texana* in particular, gradually diminishes the root reserves and the plants become smaller and stunted. Further suppression of the weed is achieved by competition when infested lands are temporarily converted to perennial pastures before being used again for annual crop production.

Convolvulus arvensis L. (*Convolvulaceae*) is a similar perennial crop weed that was targeted. However, the leaf-galling mite *Aceria malherbae* Nuzzaci (Acari, Eriophyidae), which was released in 1994 (Craemer, 1995), failed to establish in South Africa. The innovative use of the native fungus, *Fusarium nygamai* Burgess and Trimboli (*Hyphomycetes*), which is applied as a protective seed coat treatment on sorghum, against the parasitic weed *Striga hermonthica* (Del.) Benth. (*Scrophulariaceae*) in the Sudan, was only partially successful. Considerable research has also been done on native weevils in the genus *Smicronyx* (Coleoptera, Curculionidae), whose larvae develop in fruit galls of *Striga* species and prevent seed production (Anderson and Cox, 1997). In general, the use of native natural enemy species to control a native crop weed is unlikely to be very successful.

Biological Control of Cactus Weeds

Several introduced species in the family *Cactaceae* have become naturalized in Africa and many have become serious invaders (Moran and Zimmermann, 1984; Brutsch and Zimmermann, 1993). Aspects of cactus invasions in southern Africa have already been discussed in this chapter. At least two species were deliberately introduced as agroforestry species or as a source of fruit and fodder, for example, the many cultivars of *O. ficus-indica* and *Opuntia robusta* H.L. Wendl. and, to a lesser extent, *Pereskia aculeata* Mill. (Brutsch and Zimmermann, 1995). Most species were, however, introduced as ornamentals or as barrier plants.

Brutsch and Zimmermann (1995) list the cactus invaders in Africa and the introduction of biological control agents for their control. These were mainly early projects, which were highly successful (e.g. several species of *Opuntia*), having relied on the outstanding successes achieved in other countries. The biological control of a species of *Opuntia* (probably *O. monocantha*) in Madagascar around 1923 (Greathead, 1971) was initially well received because of the extent of pre-biological control invasions, but later reports and analyses showed that the demise of this species also resulted in severe famine (Middleton, 1999). New invasions of *O. stricta* var. *dillenii* in southern Madagascar are now severely impacting on agriculture. Although it was shown that a host-specific biotype of *D. opuntiae* is now available that would attack this weed but not the beneficial *O. ficus-indica*, fears of non-target effects on *O. ficus-indica* may well preclude its implementation.

Recent research in South Africa on host-adapted biotypes of cochineal (*Dactylopius* species) has provided exciting new opportunities for more effective biological control of *Opuntia* species. Indeed, both *O. ficus-indica* and *O. stricta* (Haw.) Haw. var. *stricta* (Haw.) Haw. have been successfully controlled by distinct host-adapted biotypes of *D. opuntiae* (Plates 1 and 2), while biotypes of *Dactylopius tomentosus* (Lamarck) provide a similar possibility for *Opuntia imbricata* (Haw.) DC. and *Opuntia fulgida* Engelm. (previously misidentified as *Opuntia rosea* DC. in South Africa) (Githure *et al.*, 1999; Volchansky *et al.*, 1999; Hoffmann *et al.*, 2003). The use of host-specific biotypes of an agent species provides the opportunity to control problem species (e.g. *O. stricta*) without harming congeneric cultivated or beneficial species (e.g. *O. ficus-indica*).

Biological Control and IWM

The aim of IWM is to combine all relevant control options in such a way that the most cost-effective control is achieved. Herbicides used to control alien plant invaders are often antagonistic to biological control (Moran and Zimmermann, 1991; Olckers *et al.*, 1998; Ueckermann and Hill, 2001) and several modifications to conventional control methods have been devised to increase the overall efficiency of biological control (Kluge *et al.*, 1986). The establishment of 'insect reserves' in areas where clear-felling is practised ensures that the natural enemies can persist and migrate to regrowth of the weed at a very early stage of reinvasion (Olckers *et al.*, 1998). The careful selection of herbicides and their additives can also improve the performance of natural enemies, as occurred in the chemical control of *Opuntia aurantiaca* Lindl. (*Cactaceae*) in South Africa when researchers adopted a less insect-toxic herbicide that was more compatible with the cochineal agent (Zimmermann, 1979). The negative impacts of chemical control operations on natural enemies of water hyacinth in South Africa could also be reduced if the additives, which were shown to have an insecticidal effect, were changed or removed from the herbicide formulations (Ueckermann and Hill, 2001).

The modification of control operations can influence the efficacy of biological control significantly (Zimmermann and Malan, 1981; Olckers *et al.*, 1998). When herbicidal follow-up operations against *O. stricta* are delayed until the regrowth starts to fruit, it allows the cactus moth, *Cactoblastis cactorum* (Bergroth) (Lepidoptera, Phycitidae), sufficient time to infest the regrowth, which then becomes stunted and consequently takes considerably longer to mature, thus saving time and money. At the stage when chemical control is applied again, a concerted effort is made to destroy only the large cactus plants, so that a reasonable stock of small plants is left for the continuous development of *C. cactorum* (Hoffmann *et al.*, 1998; Lotter and Hoffmann, 1998).

The modification of rangeland management practices can increase the efficacy of natural enemies considerably (Olckers *et al.*, 1998; Campbell and Kluge, 1999). To improve the efficacy of the two seed-feeding beetle species, *A. prosopis* and *N. arizonensis*, for the control of *Prosopis* species, livestock are excluded from infestations for a period of several months, allowing the beetles unrestricted access to the seeds. Very high levels of seed destruction can be achieved, depending on the time that the seeds can remain exposed to the beetles (Moran *et al.*, 1993; Impson *et al.*, 1999).

Since the inception of the WFW Programme in South Africa, which is clearing large tracts of land of alien invaders, a special task team has been appointed to ensure that maximum benefit is derived from biological control. This implementation programme includes several mass-rearing facilities that provide important biological control agents to wherever they are needed as well as personnel that are actively involved in distributing agents and monitoring their impact (Olckers, 2000). The programme also encourages increased exploitation of seed-destroying agents, as opposed to vegetative feeders, that can prevent rapid reinvasion from soil seed-banks and thus reduce the intensity of follow-up control operations (Zimmermann and Neser, 1999). The control of invaders that are not specifically targeted by the WFW Programme will continue to rely more on biological control.

Intensive utilization of invasive plants can have commercial benefits and can also complement biological control, provided the populations of the target weed are reduced in the process. Some woody invaders are used extensively to produce charcoal and firewood (E.J. Azorin, 1992 unpublished report of the National Energy Council, Cape Town), timber products (Geldenhuys, 1986), as well as human food and stockfeed (Harding, 1991; Brutsch and Zimmermann, 1995). Other novel uses include the synthesis of biopesticides ((e.g. *Melia azedarach* L. (*Meliaceae*)) and food or cosmetic colourants (e.g. rearing the carmine-dye cochineal *Dactylopius coccus* O. Costa on *O. ficus-indica* infestations; Zimmermann, 1990).

The Value of Biological Weed Control

The overall cost of alien plant invaders to South Africa's water and natural resources is difficult to determine. However, the cost to clear South Africa of alien plant invaders was estimated at US$1.39 billion in 1998 (Versfeld *et al.*,

1998; Le Maitre, 2001). The latter studies also estimated that existing biological control programmes have already reduced the amount needed for the control of alien plant invaders by 19.8%, which represents a saving of US$276 million, while the potential savings, if biological control were to be fully exploited, would be around US$816 million.

Only a few detailed cost–benefit studies have been made to demonstrate the economic benefits of biological weed control. McConnachie *et al.* (2003) indicated that the benefit:cost ratio of the highly successful project on *A. filiculoides* was 74:1. Benefit:cost ratios for six other projects that were recently evaluated amounted to 8:1 for *L. camara*, 104:1 for *A. longifolia*, 251:1 for *H. sericea* and 709:1 for *O. aurantiaca*, all of which are deemed to be under 'substantial control', while the ratios for species under 'complete control' amounted to 22:1 for *S. punicea* and 665:1 for *A. pycnantha* (Van Wilgen *et al.*, 2003). These studies have indicated that investing in research into biological control will, under most scenarios, deliver positive and even considerable returns on investment.

Trends and Future Prospects in Southern Africa

The impacts and threats of alien plant invaders to agricultural and natural resources and to biodiversity are increasingly recognized by conservationists. The research that demonstrated that more than 10 million ha are already invaded, and that invasions are increasing at a rate of 5% annually, has been the main driving force behind the creation of South Africa's WFW Programme. This and various other partnership organizations and specialist groups are now dedicated to preventing, controlling and limiting alien plant invasions in South Africa. Recent benefit:cost studies have highlighted the value and opportunities for biological control for playing a more positive role in achieving these objectives on a regional basis (e.g. amongst other SADC countries). However, recent reports of non-target effects of biological control agents in other countries (Follett and Duan, 1999; Wajnberg *et al.*, 2001) have raised negative perceptions and increased safety standards to unnecessary levels, although the safety record for South African projects is impeccable. This can prolong the research time needed to get agents cleared for release and make projects considerably more expensive. However, advances in biological weed control technology worldwide have made it possible to select 'winners' more accurately and to waste less time on unsuitable agents.

International and regional cooperation and partnerships are of considerable benefit for countries that are starting to invest in weed biological control projects. Indeed, South Africa's considerable experience and successes can be extremely valuable to other countries in Africa that share the same or similar weed problems. International organizations, including the Food and Agriculture Organisation (FAO) of the United Nations, IUCN, GISP and others, serve as 'watchdogs' and provide much-needed guidance to ensure minimum risks and maximum benefits from biological control (Wittenberg and Cock, 2001).

However, an unfortunate outcome of implementing the Convention on Biodiversity (CBD), aimed primarily at protecting biodiversities and their exploitation, is the limitations placed on accessing the same biodiversities for natural enemies to facilitate biological control. Indeed, despite the CBD's stipulation that countries should make their biodiversities accessible to other countries, it has become almost impossible to legally import natural enemies from some countries where the invaders are native.

Many tree species that were introduced for forestry and agroforestry purposes in southern Africa have become invasive and there is a general realization that future introductions of similar species will aggravate the problems caused by alien invasions. The problem of new invasions is particularly acute in Africa where the demand for new crops, fuel wood and fodder is very high. Adherence to strict protocols for new introductions (Hughes, 1995) provides a defence against new invasions, but this is unlikely to occur in the short term. In the interim, the early establishment, before or during the early stages of invasion, of effective seed-destroying agents for species that are known to be aggressive, could prevent or delay new disasters and retain the usefulness of potentially invasive species.

Conclusions

Despite the impressive track record of biological weed control in southern Africa, many countries and state departments remain sceptical and suspicious of biological control and insist that further exhaustive safety tests are done on agents that have been safely implemented elsewhere for years. Projects driven on a regional basis will hopefully prevent this problem. Recent negative publicity from a few isolated cases of non-target effects in biological weed control in other countries has caused decision makers to become overcautious and to impose unrealistic constraints on the practice of biological control (Zimmermann *et al.*, 2000). It is, however, also encouraging to notice the acceptance of biological control using seed-feeders in commercial forestry and agroforestry in South Africa as a safety mechanism for preventing seed pollution. Regrettably, this is not the case in the rest of Africa.

Several favourable circumstances in South Africa have coincided to create the WFW Programme, which has had a major impact on the control and management of alien plant invasions. These circumstances include important information regarding the impact of alien plant invasions on biodiversity and water loss, the need for poverty alleviation projects that include elements for uplift of the unemployed, good scientific and technical capacity and political buy-in. Improved public awareness of the benefits of control programmes and the dangers of alien invasions has also contributed to general public support of the WFW Programme. The programme has accepted the value of biological control in seeking a long-term solution to the problem of alien invading plants. Favourable returns on investment, following recent successes, have contributed to these sentiments.

International involvement and cooperation have offered new opportunities, but have also required greater commitments and added responsibilities. Alien weed control is rapidly becoming a regional issue (e.g. the threat of *C. odorata* to SADC countries) and will need greater involvement in the sharing of scarce scientific capacity and facilities, firm agreements on minimum standards and protocols for the introduction and safety of biological control organisms and the provision of specialized training. The sharing of capacity between leading biological control organizations and international facilities, and the co-funding of projects targeting common weeds, can also improve success rates and reduce the costs of such projects in the future. In this regard, South Africa with its considerable weed biological control expertise, impeccable safety record and high success rate is poised to have a far greater influence on weed biological control in southern Africa and elsewhere on the continent.

References

Anderson, D.M. and Cox, M.L. (1997) *Smicronyx* species (Coleoptera: Curculionidae), economically important seed predators of witchweeds (*Striga* spp.) (Scrophulariaceae) in sub-Saharan Africa. *Bulletin of Entomological Research* 87, 3–17.

Annecke, D.P. and Moran, V.C. (1978) Critical reviews of biological pest control in South Africa. 2. The prickly pear, *Opuntia ficus-indica* (L) Miller. *Journal of the Entomological Society of Southern Africa* 41, 161–188.

Behailu, M. and Tegegne, F. (eds) (1997) *Opuntia* in Ethiopia: state of knowledge in *Opuntia* research. *Proceedings of the International Workshop*, Mekelle University (Ethiopia) and Wiesbaden Polytechnic (Germany).

Bennett, P.H. and Van Staden, J. (1986) Gall formation in crofton weed, *Eupatorium adenophorum* Spreng (syn. *Ageratina adenophora*), by the *Eupatorium* gall fly *Procecidochares utilis* Stone (Diptera: Trypetidae). *Australian Journal of Botany* 34, 473–480.

Brutsch, M.O. and Zimmermann, H.G. (1993) The prickly pear (*Opuntia ficus-indica*) (Cactaceae) in South Africa: utilization of the naturalized weed, and of cultivated plants. *Economic Botany* 47, 154–162.

Brutsch, M.O. and Zimmermann, H.G. (1995) Control and utilization of wild opuntias. In: Barbera, G., Inglese, P. and Pimienta-Barrios, E. (eds) *Agro-ecology, Cultivation and Uses of Cactus Pear.* FAO Plant Production and Protection Paper 132, Rome.

Campbell, P.L. and Kluge, R.L. (1999) Development of integrated control strategies for wattle. 1. Utilization of wattle, control of stumps and rehabilitation with pastures. *South African Journal of Plant and Soil* 16, 24–30.

Cilliers, C.J. (1999) Biological control of parrot's feather, *Myriophyllum aquaticum* (Vell.) Verdc. (Haloragacaeae), in South Africa. In: Olckers, T. and Hill, M.P. (eds) *Biological Control of Weeds in South Africa (1990–1998). African Entomology Memoir* 1, 113–118.

Craemer, C. (1995) Host specificity, and release in South Africa, of *Aceria malherbae* Nuzzaci (Acari: Eriophyoidea), a natural enemy of *Convolvulus arvensis* L. Convolvulaceae). *African Entomology* 3, 213–215.

Den Breeÿen, A. (1999) Biological control of water hyacinth using plant pathogens: dual pathogenicity and insect interaction. In: Hill, M.P., Julien, M.H. and Center, T.D. (eds) *Proceedings of the First IOBC Global Working Group Meeting for the*

Biological and Integrated Control of Water Hyacinth, 16–19 November 1999, Harare, Zimbabwe, pp. 75–79.

Dennill, G.B., Donnelly, D., Stewart, K. and Impson, F.A.C. (1999) Insect agents used for the biological control of Australian *Acacia* species and *Paraserianthes lophanta* (Willd.) Nielsen (Fabaceae) in South Africa. In: Olckers, T. and Hill, M.P. (eds) *Biological Control of Weeds in South Africa (1990–1998). African Entomology Memoir* 1, 45–54.

Fenn, J.A. (1980) Control of hakea in the western Cape. In: Neser, S. and Cairns, A.L.P. (eds) *Proceedings of the Third National Weeds Conference of South Africa, 1979, Pretoria, South Africa*, pp. 167–173.

Follett, P.A. and Duan, J.J. (eds) (1999). *Nontarget Effects of Biological Control*. Kluwer Academic Publishers, Boston.

Geldenhuys, C.J. (1986) Costs and benefits of the Australian blackwood, *Acacia melanoxylon*, in South African Forestry. In: Macdonald, I.A.W., Kruger, F.J. and Ferrar, A.A. (eds) *The Ecology and Management of Biological Invasions in Southern Africa.* Oxford University Press, Cape Town, South Africa, pp. 275–284.

Githure, C.W., Zimmermann, H.G. and Hoffmann, J.H. (1999) Host specificity of biotypes of *Dactylopius opuntiae* (Cockerell) (Homoptera: Dactylopiidae): prospects for biological control of *Opuntia stricta* (Haworth) Haworth (Cactaceae) in Africa. *African Entomology* 7, 43–48.

Greathead, D.J. (1971) A review of biological control in the Ethiopian Region. *Technical Communication, CIBC* 5.

Harding, G.B. (1991) Sheep can reduce recruitment of invasive *Prosopis* species. *Applied Plant Science* 5, 25–28.

Henderson, L. (1998) Invasive alien woody plants of the southern and southwestern Cape region, in South Africa. *Bothalia* 28, 91–112.

Henderson, L. (2001) *Alien Weeds and Invasive Plants.* Plant Protection Research Institute Handbook, Pretoria, South Africa, 12.

Hill, M.P. (1999) Biological control of red water fern, *Azolla filiculoides* Lamarck (Pteridophyta: Azollaceae), in South Africa. In: Olckers, T. and Hill, M.P. (eds) *Biological Control of Weeds in South Africa (1990–1998). African Entomology Memoir* 1, 119–124.

Hill, M.P. and Cilliers, C.J. (1999) A review of the arthropod natural enemies, and factors that influence their efficacy, in the biological control of water hyacinth, *Eichhornia crassipes* (Mart.) Solms-Laubach (Pontederiaceae), in South Africa. In: Olckers, T. and Hill, M.P. (eds) *Biological Control of Weeds in South Africa (1990–1998). African Entomology Memoir* 1, 103–112.

Hill, M.P. and Olckers, T. (2001) Biological control initiatives against water hyacinth in South Africa: constraining factors, success and new courses of action. In: Julien, M.H., Hill, M.P., Center, T.D. and Jianqing, D. (eds) *Proceedings of the Second Meeting of the Global Working Group for the Biological and Integrated Control of Water Hyacinth.* 9–12 October 2000, Beijing, China. *ACIAR Proceedings* 102, 33–38.

Hoffmann, J.H. and Moran, V.C. (1999) A review of the agents and factors that have contributed to the successful biological control of *Sesbania punicea* (Cav.) Benth. (Papilionaceae) in South Africa. In: Olckers, T. and Hill, M.P. (eds) *Biological Control of Weeds in South Africa (1990–1998). African Entomology Memoir* 1, 75–88.

Hoffmann, J.H., Moran, V.C. and Zeller, D.A. (1998) Long-term population studies and the development of an integrated management programme for control of *Opuntia stricta* in Kruger National Park, South Africa. *Journal of Applied Ecology* 35, 156–160.

Hoffmann, J.H., Impson, F.A.C. and Volchansky, C.R. (2003) Biological control of cactus weeds: implications of hybridisation between control agent biotypes. *Journal of Applied Ecology* (in press).

Hughes, C.E. (1995) Protocols for plant introductions with particular reference to forestry: changing perspectives on risks to biodiversity and economic development. In: Stirton, C.H. (ed.) *British Crop Protection Council Symposium Proceedings 64: Weeds in a Changing World.* British Crop Protection Council, Farnham, pp. 15–32.

Impson, F.A.C. and Hoffmann, J.H. (1998) Competitive interactions between larvae of three bruchid species (Coleoptera) in mesquite seeds (*Prosopis* spp.) under laboratory conditions. *African Entomology* 6, 376–378.

Impson, F.A.C., Moran, V.C. and Hoffmann, J.H. (1999) A review of the effectiveness of seed-feeding bruchid beetles in biological control of mesquite, *Prosopis* species (Fabaceae), in South Africa. In: Olckers, T. and Hill, M.P. (eds) *Biological Control of Weeds in South Africa (1990–1998). African Entomology Memoir* 1, 81–88.

Julien, M.H. and Griffiths, M.W. (1999) *Biological Control of Weeds: a World Catalogue of Agents and their Target Weeds,* 4th edn. CAB International, Wallingford, UK.

Kluge, R.L. and Neser, S. (1991) Biological control of *Hakea sericea* (Proteaceae) in South Africa. *Agriculture, Ecosystems and Environment* 37, 91–113.

Kluge, R.L., Zimmermann, H.G., Cilliers, C.J. and Harding, G.B. (1986) Integrated control of invasive alien weeds. In: Macdonald, I.A.W., Kruger, F.J. and Ferrar, A.A. (eds) *The Ecology and Management of Biological Invasions in Southern Africa.* Oxford University Press, Cape Town, pp. 295–304.

Le Maitre, D.C., Van Wilgen, B.W., Gelderblom, C.M., Baily, C., Chapman, R.A. and Nel, J.A. (2001) Invasive alien trees and water resources in South Africa: case studies of the costs and benefits for management. *Forest Ecology and Management* 160, 143–159.

Lotter, W.D. and Hoffmann, J.H. (1998) An integrated management plan for the control of *Opuntia stricta* (Cactaceae) in the Kruger National Park, South Africa. *Koedoe* 48, 241–255.

Low, A.B. and Rebelo, A.G. (eds) (1996) *Vegetation of South Africa, Lesotho and Swaziland.* Department of Environmental Affairs and Tourism, Pretoria, South Africa.

McConnachie, A.J., De Wit, M.P., Hill, M.P. and Byrne, M.J. (2003) Economic evaluation of the successful biological control of *Azolla filiculoides* in South Africa. *Biological Control* (in press).

Middleton, K. (1999) Who killed 'malagasy cactus'? Science, environment and colonialism in southern Madagascar (1924–1930). *Journal of Southern African Studies* 25, 215–248.

Moran, V.C. and Annecke, D.P. (1979) Critical reviews of biological control in South Africa. 3. The jointed cactus, *Opuntia aurantiaca* Lindley. *Journal of the Entomological Society of Southern Africa* 42, 299–329.

Moran, V.C. and Zimmermann, H.G. (1984) The biological control of cactus weeds: achievements and prospects. *Biocontrol News and Information* 5, 297–320.

Moran, V.C. and Zimmermann, H.G. (1991) Biological control of cactus weeds of minor importance in South Africa. *Agriculture, Ecosystems and Environment* 37, 37–55.

Moran, V.C., Hoffmann, J.H. and Zimmermann, H.G. (1993) Objectives, constraints and tactics in the biological control of mesquite weeds (*Prosopis*) in South Africa. *Biological Control* 3, 80–83.

Morris, M.J. (1989) Host specificity studies of a leaf spot fungus, *Phaeoramularia* sp., for the biological control of crofton weed (*Ageratina adenophora*) in South Africa. *Phytophylactica* 21, 281–283.

Morris, M.J. (1990) *Cercospora piaropi* recorded on the aquatic weed, *Eichhornia crassipes*, in South Africa. *Phytophylactica* 22, 255–256.

Morris, M.J. (1991) The use of plant pathogens for biological weed control in South Africa. *Agriculture, Ecosystems and Environment* 37, 230–255.

Morris, M.J. (1999) The contribution of the gall-forming rust fungus *Uromycladium tepperianum* (Sacc.) McAlp. to the biological control of *Acacia saligna* (Labill.) Wendl. (Fabaceae) in South Africa. In: Olckers, T. and Hill, M.P. (eds) *Biological Control of Weeds in South Africa (1990–1998). African Entomology Memoir* 1, 125–128.

Morris, M.J., Wood, A.R. and Den Breeÿen, A. (1999) Plant pathogens and biological control of weeds in South Africa: a review of projects and progress during the last decade. In: Olckers, T. and Hill. M.P. (eds) *Biological Control of Weeds in South Africa (1990–1998). African Entomology Memoir* 1, 129–138.

Neser, S. and Kluge, R.L. (1986) The importance of seed-attacking agents in the biological control of invasive alien plants. In: Macdonald, I.A.W., Kruger, F.J. and Ferrar, A.A. (eds) *The Ecology and Management of Biological Invasions in Southern Africa*. Oxford University Press, Cape Town, South Africa, pp. 285–294.

Olckers, T. (2000) Implementing biological control technology into the management of alien invasive weeds: South African experiences and challenges. In: Copping, L.G. (ed.) *British Crop Protection Proceedings 74: Predicting Field Performance in Crop Protection*. British Crop Protection Council, Farnham, UK, pp. 111–122.

Olckers, T. and Hill, M.P. (eds) (1999) Biological Control of Weeds in South Africa (1990–1998). *African Entomology Memoir No. 1.*

Olckers, T., Zimmermann, H.G. and Hoffmann, J.H. (1998) Integrating biological control into the management of alien invasive weeds in South Africa. *Pesticide Outlook* 9, 9–16.

Olckers, T., Hoffmann, J.H., Moran, V.C., Impson, F.A.C. and Hill, M.P. (1999) The initiation of biological control programmes against *Solanum elaeagnifolium* Cavanilles and *S. sisymbriifolium* Lamarck (Solanaceae) in South Africa. In: Olckers, T. and Hill, M.P. (eds) *Biological Control of Weeds in South Africa (1990–1998). African Entomology Memoir* 1, 55–64.

Richardson, D.M., Pyšek, P., Rejmánek, M., Barbour, M.J., Panetta, F.D. and West, C.J. (2000) Naturalization and invasion of alien plants: concepts and definitions. *Diversity and Distribution* 6, 93–107.

Tanji, A., Boulet, C. and Hammoumi, M. (1985) Etat actuel de l'infestation par *Solanum elaeagnifolium* Cav. pour les différentes cultures du perimètre du Tadla (Maroc). *Weed Research* 25, 1–9.

Tryon, H. (1910) The 'wild cochineal insect', with reference to its injurious action on prickly pear (*Opuntia* spp.) in India. *Queensland Agricultural Journal* 25, 188–197.

Ueckermann, C. and Hill, M.P. (2001) Impact of herbicides used in water hyacinth control on natural enemies released against the weed for biological control. *Water Research Commission Report* 915/1/01, Pretoria, South Africa.

Van Wilgen, B.W., De Wit, M.P., Anderson, H.J., Le Maitre, D.C., Kotze, I.M., Ndala, S., Brown, B. and Rapholo, M.B. (2003) Costs and benefits of biological control of invasive alien plants: case studies from South Africa. *Ecological Economics* (in press).

Versfeld, D.B., Le Maitre, D.C. and Chapman, R.A. (1998) Alien invading plants and water resources in South Africa: a preliminary assessment. *Water Research Commission Report*. TT 99/98, Pretoria, South Africa.

Volchansky, C.R., Hoffmann, J.H. and Zimmermann, H.G. (1999) Host-plant affinities of two biotypes of *Dactylopius opuntiae* (Homoptera: Dactylopiidae): enhanced prospects for biological control of *Opuntia stricta* (Cactaceae) in South Africa. *Journal of Applied Ecology* 36, 85–91.

Wajnberg, E., Scott, J.K. and Quimby, P.C. (eds) (2001) *Evaluating Indirect Ecological Effects of Biological Control*. CAB International, Wallingford, UK.

Wassermann, V.D., Zimmermann, H.G. and Neser, S. (1988) The weed silverleaf bitter apple (satansbos) (*Solanum elaeagnifolium* Cav.) with special reference to its status in South Africa. *Technical Communication 214*. Department of Agriculture and Water Supply, Pretoria, South Africa.

Wells, M.J., Balsinhas, A.A., Joffe, H., Engelbrecht, V.M., Harding, G. and Stirton, C.H. (1986) A catalogue of problem plants in southern Africa. *Memoirs of the Botanical Survey of South Africa 53*. Botanical Research Institute, Pretoria, South Africa.

Wittenberg, R. and Cock, M.J.W. (eds) (2001) *Invasive Alien Species: a Toolkit of Best Prevention and Management Practices*. CAB International, Wallingford, UK.

Zimmermann, H.G. (1979) Herbicidal control in relation to distribution of *Opuntia aurantiaca* Lindley and effects on cochineal populations. *Weeds Research* 19, 89–93.

Zimmermann, H.G. (1990) The utilisation of an invader cactus weed as part of an integrated control approach. In: Delfosse, E.S. (ed.) *Proceedings of the Seventh International Symposium on Biological Control of Weeds*. Istituto Sperimentale per la Patalogia Vegetale, Ministero dell' Agricoltura e delle Foreste, Rome, Italy, pp. 429–432.

Zimmermann, H.G. and Malan, D.E. (1981) The role of imported natural enemies in suppressing re-growth of prickly pear, *Opuntia ficus-indica*, in South Africa. In: Delfosse, E.S. (ed.) *Proceedings of the Fifth International Symposium on Biological Control of Weeds*. CSIRO, Canberra, pp. 375–381.

Zimmermann, H.G. and Moran, V.C. (1991) Biological control of prickly pear, *Opuntia ficus-indica* (Cactaceae), in South Africa. *Agriculture, Ecosystems and Environment* 37, 29–35.

Zimmermann, H.G. and Neser, S. (1999) Trends and prospects for biological control of weeds in South Africa. In: Olckers, T. and Hill. M.P. (eds) *Biological Control of Weeds in South Africa (1990–1998)*. *African Entomology Memoir* 1, 165–173.

Zimmermann, H.G., Moran, V.C. and Hoffmann, J.H. (2000) The renowned cactus moth *Cactoblastis cactorum:* its natural history and threat to native *Opuntia* floras in Mexico and the United States of America. *Diversity and Distributions* 6, 259–269.

3

Biological Control of Cassava and Mango Mealybugs in Africa

Peter Neuenschwander
International Institute of Tropical Agriculture, Cotonou, Bénin

Introduction

In the 1970s, the exotic mealybug *Phenacoccus manihoti* Matile-Ferrero (Homoptera, Pseudococcidae) (Plate 3) was inadvertently introduced into Africa. It quickly spread across almost all the continent and, in the absence of efficient adapted natural enemies, became the prime pest on cassava. Fifteen years later, *P. manihoti* was under complete biological control thanks to an international collaborative effort across three continents, but mainly in Africa. The story of this well-documented success in classical biological control illustrates the different stages of such a project, from desperate, though inefficient, attempts by farmers to control the pest, through collaborative research to implement classical biological control on a large scale. It is also the story of winning over the minds of the various people in the research establishment and government to allow introductions of exotic species, of monitoring the ups and downs of the pest populations, and of communicating these results back to the client farmers, local collaborators and government officials. Finally, the international community and the donor agencies had to be convinced that despite the slowness of the process, classical biological control was a good investment, deserving long-term external support. More recently, dangers of these interventions to the environment through possible effects on non-target organisms had to be evaluated and the results used in the effort to codify the conduct of biological control practitioners.

On the strength of this model of biological control in Africa, several other projects with the same international organizations and national collaborators were executed. These projects profited from established procedures and the trust already achieved among colleagues. One of these projects, the fight against the mango mealybug, *Rastrococcus invadens* Williams (Homoptera, Pseudococcidae) (Plate 6), which invaded West Africa in the early 1980s, shows striking parallels, but also a few highly interesting differences.

Biological control of cassava mealybug has been reviewed on several occasions (for complete bibliographies, see Herren and Neuenschwander, 1991; Neuenschwander, 2001). The mango mealybug story, by contrast, is much less well known (Neuenschwander *et al.*, 1994; Neuenschwander, 1996). Here, we review the parallels and explore the impact in a larger context of IPM, citing only a few recent papers.

Origin of Host Plants and Invading Mealybug Species

Cassava, *Manihot esculenta* Crantz (*Euphorbiaceae*), was introduced to Africa from South America by the Portuguese in the 16th century, but penetrated into the interior of the continent only in the 20th century. Today, this hardy plant is the main staple for about 200 million Africans, with a particular importance in poor countries.

Because of its plant defences, namely a high cyanide and latex content in leaves, stems and tubers, it had few pest insects attacking it in Africa. By contrast, the list of co-evolved arthropods in its native South America comprises some 200 species. In Africa, several diseases of African origin also affected yields before breeding provided resistant or tolerant varieties. The situation changed dramatically when, in the early 1970s, the cassava mealybug was accidentally introduced into the Republic of Congo and D.R. Congo (then Zaire). *P. manihoti* was described as a new species of presumably Neotropical origin, but it took several years of search before this mealybug was discovered in its original home land, Paraguay, and later in Brazil and Bolivia. In most instances, the densities of the few infestations that were found in South America were low.

As *P. manihoti* spread across most of Africa, its high infestation levels immediately made it the number one pest of cassava on the continent, severely affecting the livelihood, particularly of poor people. New foci of infestation were discovered in Senegal/Gambia in 1976, Nigeria/Bénin in 1979, and in Sierra Leone in 1985. In West Africa, the mealybug spread at 300 km year^{-1}; but in East Africa barriers like the Rift Valley or vast areas with little cassava led to a slower spread and an initially sporadic distribution. New pest outbreaks, sometimes hundreds of kilometres away from the next infestation, were found in Malawi in 1985, Mozambique in 1986, Tanzania in 1987 and Kenya in 1989. By 2000, the entire cassava belt south to the Lowveld of the Republic of South Africa, with the main exceptions of Madagascar and the Indian Ocean islands, was infested.

Where infestations were high, crawlers of cassava mealybug also settled on the surrounding vegetation, where they sometimes even reproduced. Without a high population on cassava, these infestations on other host plants ceased, however, within a short time. Two *Manihot* spp. and their hybrid, all introduced from South America, proved to be the only lasting host plants. Recent studies shed light on the adaptation of cassava mealybug to its host plant. *P. manihoti* is capable of metabolizing cyanogenic glycosides and three flavonoid glycosides, among them rutin, which are translocated in the phloem sap with seasonal fluctuations (Catalayud *et al.*, 1994a,b). These compounds are responsible for the antibiosis that cassava mealybug, unlike other insects, is capable of counteracting to a large extent.

In the 1980s, another exotic mealybug, the mango mealybug, was observed for the first time in massive infestations in Lomé, Togo, and Cotonou, Bénin. This new plague then spread rapidly along the coast west to Ghana and Côte d'Ivoire and east to Nigeria. By the mid-1990s, it had invaded most of West and Central Africa, from Senegal to D.R. Congo. In the urban environment, this mealybug attacked indiscriminately mango, citrus, shade trees (mainly *Terminalia* spp.), but also plants of many other families; over 100 host plant species were registered. In Bénin, for instance, *R. invadens* was observed most often in and around large cities, being less abundant in commercial orchards and even less so on local mango varieties in farmers' fields.

After its introduction into Africa, a renewed search led to the identification of this species within a complex of mealybug species on mango on the Indian sub-continent. *R. invadens* is now known from India to Indonesia, where it seems to be of minor economic importance. Interestingly, it is still absent from East Africa. Similarly to the situation with cassava mealybug, once the populations of mango mealybug had crashed due to biological control, it also became restricted to mango as its main host.

Impact of Mealybugs on Farming Communities and First Control Measures

Damage by cassava mealybug is mainly due to the distortion of shoot tips, which are colonized preferentially. Because leaf production stops, carbohydrate accumulation in the tubers ceases and, in the following season, early mobilization of sugars causes severe quality decline of tubers. Early tuber losses were estimated at 80%. In addition, stems became contorted and unfit for use as planting material, so that cassava production collapsed over vast areas. In countries where leaves are consumed as vegetable, this food source also disappeared. This combined effect sometimes led to famine conditions requiring food aid from abroad.

Due to the adaptation of local predators to this new food source and due to farmers' selection of less susceptible cassava varieties, these losses declined within 5 years so that losses were about 40% of pre-invasion conditions in the savannah and highlands and an estimated 20% in the rainforest zone. In most countries, farmers and extension services experimented with insecticide applications. These invariably did not give satisfactory results because of the habit of the insect to hide and find protection in the mass of contorted leaves at the shoot tips.

By contrast, when the mango mealybug first appeared, the first victims were urban home owners and the damage concerned not only the loss of a widely consumed fruit, but also the spoilage of shade trees used as meeting places, by black sooty mould and dripping honeydew. The initial panic reaction of the homeowners consisted of cutting down affected trees. Insecticide spraying on these, sometimes huge, shade trees proved difficult and inefficient, and a quick move to less susceptible varieties was impossible because of the slow growth of these trees. As a result, mango production in the southern part of Bénin, for instance, stopped almost completely in the early 1990s. Mango mealybug infestation then moved quickly north into the commercial production zones.

Both mealybugs invaded a production system based on exotic crops that had been mostly free of important pests and therefore without plant protection measures. In both cases, early local interventions did not offer any relief and losses were catastrophic, though mostly non-quantified.

Initiating a Biological Control Project: From Discovery to Release

As both species were evidently of exotic origin, classical biological control was the method of choice. Lengthy foreign exploration was undertaken by three international institutions in Central and South America, but was not immediately successful. The first recovery of natural enemies from northern South America from a purported cassava mealybug consisted in parasitoids that, when tested, did not accept *P. manihoti* as host. Later, this mealybug was described as a new species, *Phenacoccus herreni* Williams, and the search had to be continued. Eventually, in 1981, *P. manihoti* was found serendipitously in Paraguay and later in neighbouring provinces of Brazil and Bolivia. The first recovered parasitoid, (*Anagyrus*, *Epidinocarsis*) *lopezi* De Santis (Hymenoptera, Encyrtidae) (Plate 4), which had already been described earlier from an unknown host from northern Argentina, i.e. from the same general area of the La Plata valley, later proved to be the key species for successful biological control in Africa.

Following rearing and host specificity tests in quarantine in England by what is now CABI *Bioscience*, wasps from several different sources were sent to the IITA in Ibadan, Nigeria, and, later, to IITA in Cotonou, Bénin, for further study and mass-rearing. This transfer was sanctioned by permits from the Nigerian (and later Bénin) Plant Quarantine and executed under the umbrella of the IAPSC. A national biological control committee was established to guide activities, with the IITA representative as the only non-Nigerian member.

At the same time, the non-glamorous task of producing enough *A. lopezi* for release was pursued. Because, according to quarantine regulations, it was not possible to transport cassava plants infested with parasitized mealybugs across borders, enough adult wasps had to be produced and delivered within a short time to assure survival and establishment. After some experience with specially designed high-technology equipment for rearing, eventually a special, decidedly low-tech, rearing cage consisting of a central plastic column filled with coconut husks and watered by drip irrigation equipment, into which 150 cassava sticks could be placed, the whole frame being covered with gauze, gave a simple and efficient rearing unit (Plate 5). Over 1.6 million wasps were produced for releases all over Africa.

First releases in 1981 led to immediate establishment and rapid spread (up to 20 km per generation). Subsequently, the same procedures were used for another 125 releases of *A. lopezi* across Africa (Fig. 3.1). With a few exceptions, these releases were done from the ground. Though a novel release system by airplane had been developed, it was eventually used only rarely in the operational phase of the project because *A. lopezi* proved to be such a good disperser. Within about 10 years, *A. lopezi* was established across all ecological zones from Senegal to the Republic of South Africa.

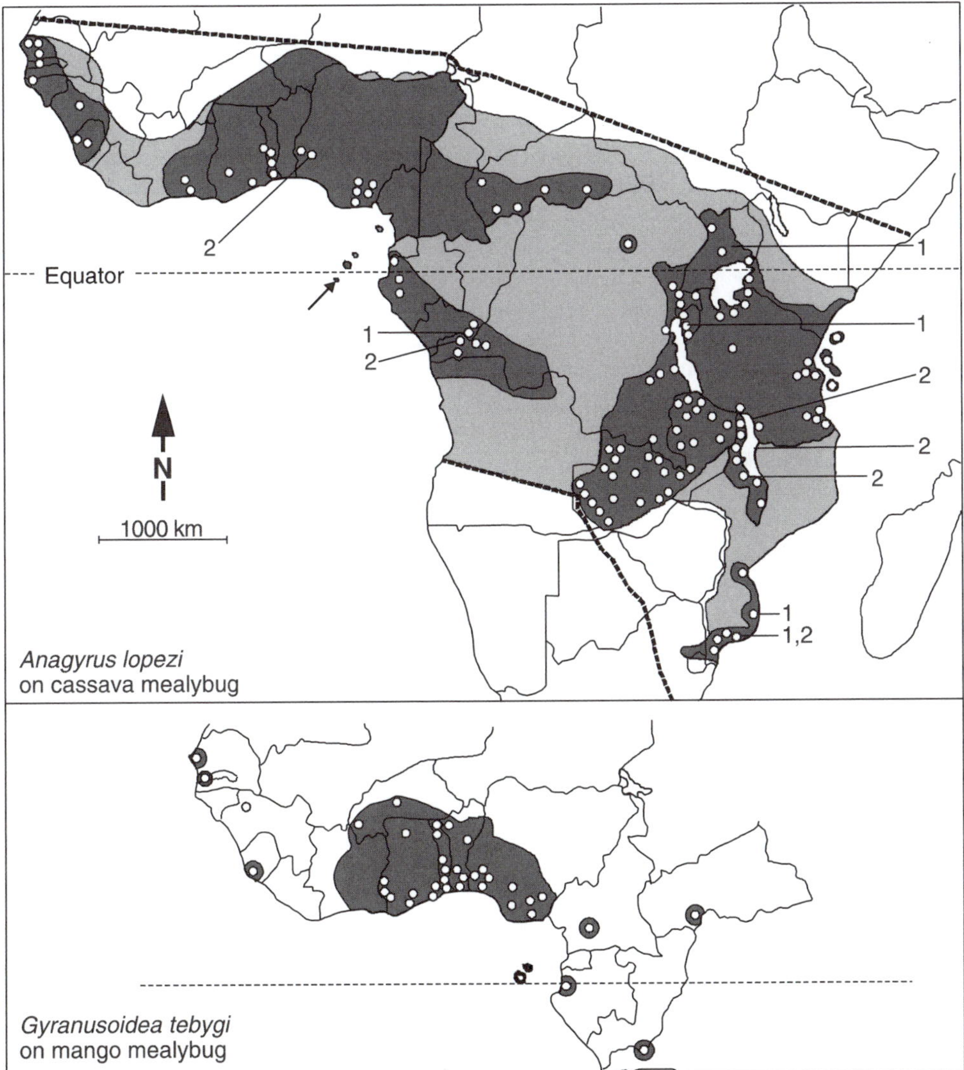

Fig. 3.1. Distribution of two hymenopterous parasitoids and their respective hosts in Africa. White dots: release sites (sometimes one dot covers two sites); dark grey: establishment confirmed; light grey: establishment expected; 1: establishment of *Hyperaspis notata*; 2: establishment of *Diomus hennesseyi*.

A few years later, a sister species, *Anagyrus diversicornis* (Howard), was recovered from mealybug infestations set out on potted plants in Brazil. It was brought to Africa, but despite releases in 14 countries, no long-term recoveries were ever made. Several exotic coccinellids and one hemerobiid predator were also imported from Paraguay and/or Brazil. They were released in various African countries (see list in Neuenschwander, 2001), but mostly without establishing permanently (i.e. with recoveries after more than 1 year). Only *Hyperaspis notata* Mulsant and *Diomus hennesseyi* Fürsch (both Coleoptera, Coccinellidae) persisted locally in central and eastern Africa. In most places, indigenous coccinellids and other predators became more numerous on cassava mealybug, which constituted a new and abundant food source. These indigenous natural enemies were capable of reducing peak populations, but they could not stabilize mealybug populations at acceptable levels.

On *R. invadens* two new parasitoids were discovered in India. *Gyranusoidea tebygi* Noyes (Hymenoptera, Encyrtidae) (Plate 7) was passed through quarantine in England and released through a project supported by the Gesellschaft für technische Zusammenarbeit and the FAO. The project was led by the national programme in Lomé, Togo, and in 1987 the parasitoid was released in six localities in Togo. The following year, *G. tebygi* was also sent to IITA Cotonou, to be complemented later by *Anagyrus mangicola* Noyes (Hymenoptera, Encyrtidae) (Plate 8). Both were reared in normal rearing cages and released from the ground. Of *G. tebygi*, 59,000 adults were released on 48 occasions, and of *A. mangicola*, 15,000 wasps were released on 57 occasions, in another ten countries. Both species are now established on all *R. invadens* infestations in West and Central Africa (Fig. 3.1). While *G. tebygi* established immediately in all situations, *A. mangicola* was more difficult because its adults proved to be highly fragile and short-lived. When parasitized mealybugs on potted plants from local rearing units were brought to the field, the species finally established on the remaining high mealybug populations, mainly in urban centres.

Impact Assessment: Effect of Biological Control on Host Populations

The impact of the exotic parasitoids on their hosts was assessed by IITA and its collaborating national programmes in Africa and has been reviewed in detail (Neuenschwander, 1996).

As a first step, it was necessary to determine whether *A. lopezi* was capable of suppressing its host and to prove its efficiency to the scientists involved and the donors. This was done through a series of exclusion experiments, whereby the parasitoids were excluded from host colonies by sleeve cages, by ants or by chemical treatment. In each case, the result was compared with the situation where the parasitoid had free access to its host. Chemical exclusion gave the most striking and immediate results with population explosions of 40–80-fold over the control plots where parasitoids were active. In three physical exclusion experiments using sleeve cages, which had the advantage of repetitions, parasitoids that had entered through the open base of the sleeve reduced mealybug populations to a mean of 43.5% and 2.3% for *P. manihoti* and 37.0% for *R.*

invadens. Where ants were prevented from protecting the mealybugs, whose sweet honeydew they crave, parasitoids lowered population levels to 66.7% of the mealybug population found in the closed sleeves.

Long-term population dynamics studies were started immediately following the release (Fig. 3.2). Of course, such studies should start a long time before the releases in order to develop baseline data, but in the heat of implementing a promising biological control project, we succeeded in capturing an early high host population in only four long-term studies, two on cassava and two on mango. Sampling procedures initially were developed to the best knowledge of the scientists involved, but were later revisited by scientifically developed sampling plans, which gave confidence intervals for different sample sizes. In retrospect, the previously chosen sampling procedure proved to be acceptable. Larger sample sizes would certainly have given better resolution, but were unacceptable because of time limits. In all study sites, samples were taken from physiologically different host plants: susceptible versus relatively unsusceptible cassava varieties in the case of *P. manihoti*, young, susceptible versus old and relative unsusceptible mango leaves in the case of *R. invadens*.

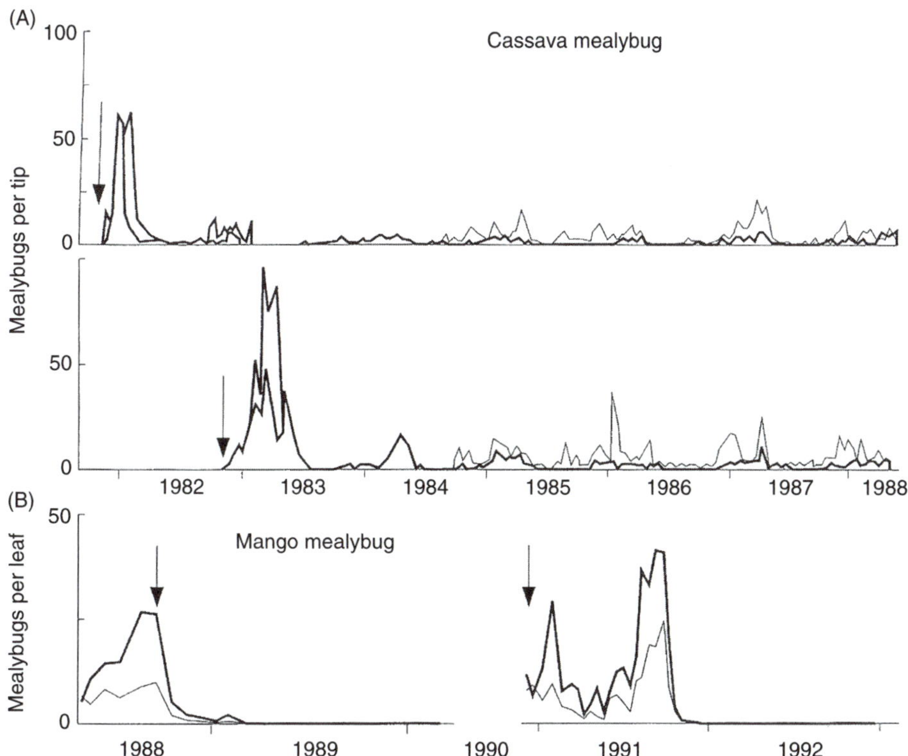

Fig. 3.2. Population dynamics of two mealybug species following the release of exotic parasitoids (arrow). (A) Cassava mealybug in Ibadan (top) and Abeokuta (below), both Nigeria; thick line: on IITA cassava variety; thin line: on farmers' variety. (B) Mango mealybug in Ouega (left) and Calavi (right), both Bénin; thick line on young mango leaves, thin line: on old mango leaves.

Overall, populations of *P. manihoti* crashed after the last peak following the release of *A. lopezi* and within 2 years reached low levels with irregular peaks that were about one-tenth as high as the former population peaks just following the releases (Fig. 3.2).

On mango mealybug, which eventually spread across all of West and some parts of Central Africa (Fig. 3.1), the newly established *G. tebygi* immediately reduced host numbers. *A. mangicola* was more difficult to establish and became mostly restricted to mealybug infestations in towns. Population densities of *R. invadens* crashed within 1–2 years, but in contrast to the punctuated equilibrium shown during 7 years of monthly samples of two cassava varieties each, in two locations, mango mealybug populations collapsed totally, leading to local extinction. This was indicated by no recovery of mealybugs during all sampling dates throughout a full year (Fig. 3.2).

Between 1983 and 1995, repeated surveys on cassava were conducted in 11 African countries, and from 1989 to 1991, a survey on mango and other host plants of *R. invadens* was repeated yearly in Bénin. At fixed intervals, regular samples were taken and evaluated in order to document establishment of the exotic species, quantify mealybug populations together with those of their natural enemies and hyperparasitoids, and assess host plant parameters, including yield. The results of the surveys were analysed by multiple regression analysis, published and reviewed (Neuenschwander, 1996, 2001). The following points emerged from these surveys:

1. Establishment of *A. lopezi* was highly successful (close to 100%), even under difficult conditions and sometimes with small numbers (a few hundred) of long-travelled and already weakened wasps. The same success was observed with *G. tebygi*. All other species were far less successful, as already mentioned.
2. Both *A. lopezi* and *G. tebygi*, once established, immediately spread. The detailed studies of *A. lopezi* indicate that this spread occurred long before the local host population had been fully exploited.
3. Mealybug populations collapsed only after 2 (in highlands in East Africa, 4) years. With a generation time of the parasitoid of about 2 (to 3) weeks, this is a strikingly slow effect. We hypothesize that *A. lopezi* starts having an impact on host populations only when emigration rates are compensated for by immigration rates.
4. Astonishingly, establishment and good impact by *A. lopezi* were observed across all ecological zones from the Sahel to the rainforest to the East African highlands, despite the rather restricted ecological zone of collection in South America, which did not always match the climate in the establishment zones in Africa. Similarly, *G. tebygi* and *A. mangicola* cover the entire range of their host in West Africa from the forest to the Sudan savannah, except for some dry areas in the north, where invasion of the pest has occurred only recently.
5. Most importantly, in surveys that had been repeated many times (Malawi: five yearly surveys; Zambia: ten half-yearly surveys), the duration of *A. lopezi*'s presence in an area proved to be the single most important factor predicting suppression of cassava mealybug populations. This factor was more important than rainfall or other ecological or plant factors. For mango mealybug popula-

tion densities, the duration of the presence of *G. tebygi* (Bénin: three yearly surveys) was the second most important factor, after 'human population density'. That mango mealybug is particularly abundant on mangos near taxi stands and is a more important pest in urban conditions than in villages or farmers' fields might in fact be attributable to air pollution (Bell *et al.*, 1993).

While sampling both mealybug systems, care was taken to also quantify populations of indigenous natural enemies and competitors. On cassava, 135 species were linked with *P. manihoti*. On mango, the food web was much more restricted, but covered essentially the same species. Though other Homoptera were found, neither exotic mealybug species seemed to have important competitors and, before biological control had been achieved, both mealybugs were the dominant insect species on their host plants across the entire range. Other, indigenous, mealybug species were not attacked by the released exotic wasps, which indicates that both biological control projects against mealybugs in Africa did not endanger biodiversity through unwanted non-target effects. Indigenous parasitoids of mealybugs, mostly *Anagyrus* spp. (Hymenoptera, Encyrtidae), were recovered only occasionally from *P. manihoti* or *R. invadens*; they evidently could not adapt to these new hosts either. Their hyperparasitoids, however, transferred successfully to the three introduced primary parasitoids. The same dozen or so African hyperparasitoid species, though in different proportions, were recovered from both mealybugs, sometimes already on the first sampling date. Exotic hyperparasitoids had been excluded by quarantine, and none were ever recorded in Africa. On *P. manihoti*, the strict density dependence of these hyperparasitoids could be demonstrated. This was highly evident when, in each new establishment area of *A. lopezi*, hyperparasitism rates were in the order of 50%, which worried the local communities. As mealybug populations collapsed and parasitism rates fell, hyperparasitism rates dropped to around 10%. It is worth mentioning that all the above estimates of impact of *A. lopezi* were achieved in the presence of these hyperparasitism rates.

With the arrival of these exotic mealybugs, which both immediately built up huge population densities, indigenous predators of mealybugs, like coccinellids, neuropterans, and the predatory caterpillar of the lycaenid butterfly *Spalgis lemolea* Druce, immediately became abundant. A total of 32 species of coccinellids were recovered from *P. manihoti*, and many of them later on *R. invadens*. They were reducing peak populations of the mealybugs, but could not stabilize host populations at an acceptable level. A simulation model credited them with an overall reduction of *P. manihoti* of about 25%. Once the encyrtid parasitoids exerted control and kept mealybug populations low, these predators (and also the few indigenous *Anagyrus* spp. parasitoids) lost these abundant hosts and their populations became again inconspicuous on cassava and mango. Only two exotic coccinellid species could be established. While the large *H. notata* had to compete with other, abundant *Hyperaspis* spp., *D. hennesseyi* found a particular niche, as indigenous small coccinellids of the tribe Scymnini are relatively rare on cassava. Both released coccinellids remained, however, rather local and never seemed to play an important role in the reduction of cassava mealybug populations.

Understanding Impact: From Observational Studies to Simulation Models

Field studies across the entire range of the cassava mealybug in Africa unequivocally demonstrated the impact of *A. lopezi* and this project was correctly classified as a major success in biological control. From the beginning, there was, however, skepticism about this claim coming from two angles.

First, *A. lopezi*, though present on all mealybug infestations, was not equally effective in all conditions. On several occasions, our own countrywide surveys indicated that about 5% of all randomly chosen fields had unacceptably high mealybug infestations even after *A. lopezi* had reduced populations. Such fields, invariably, were associated with extremely poor cassava because the soil was pure sand and had no mulch cover whatsoever. In experiments, mulching of such poor soils improved water retention capacity of the soil and the nutrient status of cassava and had a measurable effect on parasitism.

Second, performance measurements of *A. lopezi* in the laboratory and the field initially did not reveal a promising biological control agent. In fact, throughout the indicated field studies, observed parasitism rates remained relatively low (below 50%). Detailed analysis showed that mean parasitism rates slightly increased up to about 25% at a host density of 10 mealybugs per tip, and that the percentage of tips with at least one parasitoid steeply increased to reach a plateau of 70% at the same density of 10 mealybugs per tip. This indicates a density dependent reaction of the parasitoid in response to its host population in this population range, a condition that is considered an important attribute of efficient parasitoids. Above this population density, parasitism rates dropped, however, sharply. In fact, it was observed that highly infested leafless tips with hosts exhibiting strong jerking defence reactions became unattractive to searching parasitoid females. Similarly, life-table parameters were not impressive.

Detailed observations, particularly of the host selection behaviour of the female, gave important clues as to the biological mechanisms that would explain *A. lopezi*'s success. As common in many encyrtids, *A. lopezi* was shown to be a frequent host feeder, particularly on young host stages for which host mortality due to host feeding equalled and was additional to the one through egg laying. In addition, observation of the behaviour of released females in relation to different (manipulated) host colony sizes demonstrated the high host-finding capacity of this species, in relation to other species of natural enemies that were also released in the same experiment. Anecdotal evidence of this high host-finding capacity had already been accumulated from observations following releases. On several occasions, the initial isolated release field had been harvested soon after the release; yet the few survivors of the first generation had still managed to find new fields, sometimes tens of km away. Several laboratories have now reported that *A. lopezi* responds to the odours of the host plant (Nadel and van Alphen, 1987; Souissi and Le Rü, 1999), which is not the case for its coccinellid competitors.

The attributes that made *A. lopezi* special were placed into relief in a parallel study with a sister species, *A. diversicornis*, which was released widely, but could not establish. That *A. diversicornis* is a poor competitor to *A. lopezi* was explained by the small range of host stages on which it could produce female offspring compared with *A. lopezi*, which can produce females even from small hosts. Also, *A. diversicornis* showed a comparatively low host-finding capacity and, in the larval stage, a low intrinsic capacity to compete in multiparasitized hosts.

These observations were integrated into a weather-driven simulation model that predicted plant and tuber growth as well as *P. manihoti* populations. The model has recently been expanded to include a spatial component, accounting for the irregular distribution of mealybug colonies (Gutierrez *et al.*, 1999). Adding *A. lopezi* in this model lowered mealybug populations to 10%, which corresponded to observed field data. The same model also indicated that *A. diversicornis* was competitively displaced by *A. lopezi* in all circumstances. Prior release of this weaker competitor would allow temporary establishment and some impact on *P. manihoti*; but the immigration of *A. lopezi* into *A. diversicornis* territory would still lead to *A. diversicornis'* extinction. These predictions confirm the lack of establishment of *A. diversicornis* in Africa. As for South America, it is predicted that this species, which had only been recovered from *P. manihoti*-infested potted plants, probably has other mealybug hosts of larger size.

By contrast, on mango mealybug, the two parasitoids *G. tebygi* and *A. mangicola* co-exist, both in the field and in cage studies. This is possible because their niches do not totally overlap: *A. mangicola* is far larger than *G. tebygi,* which allows it to overpower large hosts (fourth instars). *G. tebygi* prefers third instar hosts, but uses much time in handling them and grooming after oviposition. As a result, its fitness return is greatest on first instars. This means, that in a new host colony, *G. tebygi* can pre-empt an attack by *A. mangicola.* Both species accept hosts already parasitized by the other species, but inside the same host, *A. mangicola* larvae win more often. As a result of these subtle differences between the two species, where the advantage does not rest on the same side in all conditions, both species co-exist in the field. Thus an early prediction by another simulation model, based on preliminary data, which warned against the release of *A. mangicola,* was refuted.

Economic Impact: What Does it Mean for the Farmer?

An early economic study with preliminary data predicted a high return on investment (146 to 1). This study was taken up again (Zeddies *et al.*, 2001) with new data on yield losses in different conditions, which also included the initial reduction of yield loss due to indigenous predators and the farmers' choice of relatively tolerant varieties. The impact of *A. lopezi* in different ecological zones, with the now known slow impact of the parasitoid, was superimposed. While yield loss in the first year can be expressed in tons of cassava, the action of the farmer in the following year cannot. Different scenarios were discussed, like the farmer growing a larger acreage of cassava to compensate for loss, or growing maize, or buying maize or cassava, or even receiving food aid. For all scenarios, a dollar value was attributed and added up with depreciation taken into account. Interestingly, the result from this much larger study essentially confirmed the previous one, with returns, depending on the scenario, estimated at between 200 and over 500 for each dollar from donors invested.

For mango mealybug biological control, the returns from southern Bénin alone gave a return on investment in Bénin and Togo, where this project had originated, of 145 to 1. The survey also demonstrated the good knowledge about biological control and the appreciation of its impact by mango producers (Bokonon-Ganta *et al.*, 2002).

In both cases, this is, however, not the whole story. Both projects had brought a sustainable and environmentally friendly solution without any health hazards, at no costs to the farmer and with benefits accruing directly to the farmers, without additional administrative expenditures. As is customary with most biological control projects, these ecological and social benefits had not been taken into account in the above calculations. In addition, the training and awareness building that accompanied particularly the cassava mealybug project had a beneficial effect on the acceptance of other biological control projects in Africa. Today, most African entomologists working in plant protection have at one time or the other profited from courses in biological control given by IITA staff working on the biological control of cassava mealybug.

The Future: IPM Considerations

Both exotic mealybugs were and are sustainably controlled by their specific exotic parasitoids, introduced in the framework of a classical biological control programme. Interactions with African hyperparasitoids and competition with indigenous coccinellids and other predators were described; they are principally outside the scope of farmers' interventions.

In the case of cassava, there was a high degree of experimenting with new varieties at the height of the cassava mealybug infestation. As *A. lopezi* started to exert control, switchover to other varieties slowed down considerably. Of course, IITA and national institutes tested their varieties for resistance. Some varieties that were less attacked by *P. manihoti* and suffered less damage, were distributed by institutes and extension services, but their popular acceptance was based more on other favourable attributes than on their tolerance towards *P. manihoti*. It is interesting to note that varieties with some antibiosis against *P. manihoti* had a negative effect also on life-table parameters of *A. lopezi* (Souissi and Le Rü, 1998) and negatively influenced life-table parameters of coccinellids feeding on them (Le Rü and Mitsipa, 2000).

Pest population levels, even after biological control had stabilized, were much higher under bad agronomic conditions, for instance on extremely poor sandy and unmulched soils, than where plants were healthy. This opens up the opportunity for advocating IPM measures, like mulching, which – while reducing mealybug infestations – also strengthens plant growth and improves yields.

It has now been shown in the laboratory that mineral fertilization increases resistance of cassava to *P. manihoti* (Le Rü *et al.*, 1994). In the field, it was shown that increasing soil fertility led to larger mealybugs, which in turn resulted in a higher proportion of female *A. lopezi* and an enhanced biological control effect (Schulthess *et al.*, 1997).

While no insecticides are applied in Africa on cassava, insecticide drift from neighbouring fields sometimes affected *A. lopezi* and led to mealybug explosions. Such situations were observed where the pyrgomorphid grasshopper *Zonocerus variegatus* L. was treated, sometimes from the air, or where neighbouring cotton fields were heavily treated with insecticides.

As African agriculture is intensified and might in the future use more insecticides, such IPM situations that require taking into account environmental effects of insecticide use will become more frequent. Similarly, with soil nutrient and soil organic matter being steadily depleted under shortened fallow systems, the proportion of cassava fields susceptible to a mealybug attack despite the presence of *A. lopezi* will probably increase from its present level of about 5%. On mango, it was striking that highest infestations occurred in towns, and mainly at bus and taxi stops, where air pollution might have affected *G. tebygi*. Again, despite government efforts, air pollution is likely to increase in the near future, and with it the situations where high mango mealybug populations make shade trees unattractive.

Conclusion

The inadvertent introduction of the cassava mealybug into Africa led to a unique alliance of institutions across three continents. The search for the original home proved long and arduous and, as in many other projects, the first introductions of natural enemies failed because of taxonomic confusion. Due to the huge damage done to a subsistence crop, donor support was maintained and eventually led to a great success. The prime natural enemy, *A. lopezi*, on first glance did not look like the best candidate and came from different ecological conditions from those encountered in Africa. It nevertheless adapted to the new conditions, competed against indigenous predators of mealybugs and permanently lowered *P. manihoti* to erratic small peak populations. In many areas, this re-established cassava cultivation and led to a good economic return. Along the way, new rearing methods, monitoring and impact assessment techniques, as well as evaluation by means of computer simulation models were developed so that the project became one of the best-researched biological control endeavours. The early inclusion of numerous African collaborators, their institutions and government agencies, extensive awareness building and training at all levels and material and technical support to African institutions constituted a new approach and had a beneficial effect on the acceptance of other biological control projects in Africa.

The next project done with the same collaborators concerned the equally successful, though less widespread, control of mango mealybug. Though the two mealybugs came from different continents, the scenarios of their invasion and biological control were highly similar. The largest biological difference consisted in the complementarity of the two parasitoids established on mango mealybug. By contrast, on cassava mealybug, a sole parasitoid, *A. lopezi*, kept the host populations low, thereby excluding another similar parasitoid with a different choice of host stages, as well as the indigenous predators, which needed higher host populations to stay in the field.

Both biological control projects had an institution building effect and served as a beacon for a new orientation in plant protection in Africa.

References

Bell, J.N.B., McNeill, S., Houldon, G., Brown, V.C. and Mansfield, P.J. (1993) Atmospheric change: effect on plant pests and diseases. *Parasitology* 106, S11–S24.

Bokonon-Ganta, A.H., de Groote, H. and Neuenschwander, P. (2002) Socio-economic impact of biological control of mango mealybug in Bénin. *Agriculture, Ecosystems and Environment* 93, 367–378.

Catalayud, P.A., Rahbé, Y., Delobel, B., Khuong-Huu, F., Tertuliano, M. and Le Rü, B. (1994a) Influence of secondary compounds in the phloem sap of cassava on expression of antibiosis towards the mealybug *Phenacoccus manihoti*. *Entomologia Experimentalis et Applicata* 72, 47–57.

Catalayud, P.A., Tertuliano, M. and Le Rü, B. (1994b) Seasonal changes in secondary compounds in the phloem sap of cassava in relation to plant genotype and infestation by *Phenacoccus manihoti* (Homoptera: Pseudococcidae). *Bulletin of Entomological Research* 84, 453–459.

Gutierrez, A.P., Yaninek, J.S., Neuenschwander, P. and Ellis, C.K. (1999) A physiological-based tritrophic metapopulation model of the African cassava food web. *Ecological Modelling* 123, 225–242.

Herren, H.R. and Neuenschwander, P. (1991) Biological control of cassava pests in Africa. *Annual Review of Entomology* 36, 257–283.

Le Rü, B. and Mitsipa, A. (2000) Influence of the host plant of cassava mealybug *Phenacoccus manihoti* on life-history parameters of the predator *Exochomus flaviventris*. *Entomologia Experimentalis et Applicata* 95, 209–212.

Le Rü, B., Diangana, J.P. and Beringar, N. (1994) Effects of nitrogen and calcium on the level of resistance of cassava to the mealybug *P. manihoti*. *Insect Science and its Application* 15, 87–96.

Nadel, H. and van Alphen, J.J.M. (1987) The role of host and host-plant odours in the attraction of a parasitoid, *Epidinocarsis lopezi*, to the habitat of its host, the cassava mealybug, *Phenacoccus manihoti*. *Entomologia Experimentalis et Applicata* 45, 181–186.

Neuenschwander, P. (1996) Evaluating the efficacy of biological control of three exotic Homopteran pests in tropical Africa. *Entomophaga* 41, 405–424.

Neuenschwander, P. (2001) Biological control of the cassava mealybug in Africa: a review. *Biological Control* 21, 214–229.

Neuenschwander, P., Boavida, C., Bokonon-Ganta, A., Gado, A. and Herren, H.R. (1994) Establishment and spread of *Gyranusoidea tebygi* Noyes and *Anagyrus mangicola* Noyes (Hymenoptera: Encyrtidae), two biological control agents released against the mango mealybug *Rastrococcus invadens* William (Homoptera: Pseudococcidae) in Africa. *Biocontrol Science and Technology* 4, 61–69.

Schulthess, F., Neuenschwander, P. and Gounou, S. (1997) Multi-trophic interactions in cassava, *Manihot esculenta*, cropping systems in the subhumid tropics of West Africa. *Agriculture, Ecosystems and Environment* 66, 211–222.

Souissi, R. and Le Rü, B. (1998) Influence of the host plant of the cassava mealybug *Phenacoccus manihoti* (Hemiptera: Pseudococcidae) on biological characteristics of its parasitoid *Apoanagyrus lopezi* (Hymenoptera: Encyrtidae). *Bulletin of Entomological Research* 88, 75–82.

Souissi, R. and Le Rü, B. (1999) Behavioural responses of the endoparasitoid *Apoanagyrus lopezi* to odours of the host and hosts' cassava plants. *Entomologia Experimentalis et Applicata* 90, 215–220.

Zeddies, J., Schaab, R.P., Neuenschwander, P. and Herren, H.R. (2001) Economics of biological control of cassava mealybug in Africa. *Agricultural Economics* 24, 209–219.

4

Cassava Green Mite in Africa – a Unique Example of Successful Classical Biological Control of a Mite Pest on a Continental Scale

Steve Yaninek[1] and Rachid Hanna[2]

[1]*Department of Entomology, Purdue University, West Lafayette, Indiana, USA;* [2]*International Institute of Tropical Agriculture, Cotonou, Bénin*

Introduction

Cassava was introduced into Africa from the New World more than 400 years ago. It took more than 300 years before cassava became an important food staple and a commodity of choice for the rapidly expanding rural and urban populations during the 20th century. Then the rapidly growing human population quickly adopted a crop that was easy to grow under a wide range of agronomic conditions. Today, cassava is the primary source of carbohydrates for more than 200 million people, including the poorest on the continent, and provides food security and economic means to a majority of subsistence farmers and an increasing number of commodity entrepreneurs interested in commercializing new cassava products. Cassava production systems continue to change as increasing production demands degrade existing ecosystems, reduce traditional fallow periods and cultivation extends into new and, often, marginal areas. Increasing production demands coupled with finite agricultural resources pose a range of threats to the sustainability of cassava agroecosystems in Africa (Nweke *et al.*, 2002).

An important constraint to production is the combined effect of pests, diseases, and weeds that together can substantially reduce cassava yields; the effect of cassava green mite alone can be as large as 45% reduction in cassava yield (Yaninek *et al.*, 1990). This impact of cassava green mite *Mononychellus tanajoa* (Bondar) (Acari, Tetranychidae) (Plate 9) in combination with diseases such as the cassava mosaic virus disease (Legg and Thresh, 2000) and pests such as the Africa root and tuber scale *Stictococcus vayssierei* Richard (Homoptera, Stictococcidae) can cause near total loss of the crop. The evolution of cassava pest problems in Africa can be best understood in terms of the crop/pest equilibrium dynamics found in the Neotropics where cassava originally evolved. Cassava in Central and South America, particularly locally adapted varieties used in traditional farming systems, are relatively tolerant of

 Biological Control in IPM Systems in Africa (eds P. Neuenschwander, C. Borgemeister and J. Langewald)

indigenous pests (Bellotti *et al.*, 1987). Cultivars adapted locally over relatively long periods help create stable and productive cassava agroecosystems, and although numerous pest species can be found in these traditional systems, they usually do not cause serious problems. Changes in cultivar, farming practices or the environment can disrupt this equilibrium and create new pest problems.

The cassava agroecosystem in Africa changed profoundly during the 20th century, a relatively short adaptive period for any cropping system, particularly one with an average crop cycle of 12–18 months. Cassava germplasm in Africa is limited in relation to the crop's vast area of origin. This situation, however, is rapidly changing as hundreds of new lines with good agronomic characteristics and resistance/tolerance to multiple pests and diseases have been recently developed and are presently being evaluated by national research institutions for agroecological adaptations under local conditions (Fokunang *et al.*, 2000; Dixon *et al.*, 2002). It is true, however, that the combination of desirable agronomic characters observed in farmers' varieties of a region tend to be homogeneous when cassava has been cultivated for many years, which tends to accelerate the rapid spread of pests and diseases. In many newly cultivated areas, especially drier regions, the germplasm is probably inadequate for a wide range of agronomic and ecological conditions, which narrows the genetic response available to farmers cultivating cassava in new or changing ecosystems. The exotic nature of the crop in Africa makes cassava especially susceptible to exotic pests. Jones (1959) noted relatively few serious arthropod pests of cassava during the first half of the last century. However, waves of exotic pests, including plant pathogens, and weeds invaded the continent during the second half of the last century. All of the major pests were foliar in nature, and swept the continent without well-adapted local natural enemies.

The introduction of the cassava mealybug *Phenacoccus manihoti* Mat.-Ferr. (Homoptera, Pseudococcidae) (Neuenschwander, this volume) and cassava green mite into Africa prompted the IITA to initiate a campaign in the 1980s to control these exotic pests using classical biological control. Since little tradition of biological control existed in cooperating national programmes, a comprehensive training and development programme was developed and implemented to create this new capacity. Purpose-built facilities and equipment were designed and constructed for the state-of-the-art programme, while human resources were developed by creating a practical biological control curriculum, and delivering in-service, specialized and post-graduate training to hundreds of collaborators from 25 countries in the continent over the last 18 years. The expertise, experiences and infrastructure created by the cassava biological control projects provided the basis for other biological control activities in sub-Saharan Africa today (Yaninek and Schulthess, 1993).

Cassava Pest Management in Africa

Cassava and most other subsistence food crops were neglected by agricultural researchers early in the 20th century, as cassava was considered a low-value food crop with little prestige as a subject for research. Work on cassava eventually began with resistance breeding on East African and African cassava mosaic virus in the 1930s (Bock and Guthrie, 1978). In the 1960s, research on cassava in Africa expanded beyond disease resistance and yield improvement to include agronomy and early farming systems research. Research on cassava became multi-disciplinary in the 1970s. A farming systems approach was widely adopted to measure the production constraints confronted by peasant farmers. This research model was also used to test technologies being developed by research stations in farmers' fields. The approach was 'top-down', in the sense that the client farmers were not involved at any stage in the technology development and testing. Most technologies were too difficult to transfer, and many pest constraints were overlooked.

Research to develop, test and adopt 'appropriate' technologies on-farm expanded during the 1980s. Increasing concerns over the widespread devastation caused by several exotic pests and the need for an urgent, sustainable solution provided the attention needed to attract support for pest management research. Entomologists, plant pathologists and weed scientists began working together to address cassava pests as subjects of pest management research, and not simply as objects of resistance breeding. By the 1990s, a systems approach with biological control as the centre of sustainable pest management became the basis for developing environmentally sound and economically feasible plant protection for basic food crops.

Traditionally, plant protection components were developed by research teams isolated within their disciplines. Such narrow perspectives often resulted in control measures that were inadequate, impractical, unacceptable or unnecessary. Multi-disciplinary teams are much more likely to be well suited to resolve the often complex and interrelated ecological, agronomic and socioeconomic issues that confront a farmer. Diagnosis with input from several disciplines is more likely to capture the range of conditions experienced by farmers. Likewise, such an approach provides a basis for selecting and developing intervention technologies that are more likely to consider the critical needs of farmers. Early and continuous interactions with farmers and extension agents familiar with their local agroecology enhance the appropriateness of plant protection strategies being developed.

Integrating the management of pests of cassava into a strategy that met the needs of local farmers required a conceptual framework for development and implementation. Concerns for pests, diseases and weeds without regard for the availability of appropriate interventions, local preferences, and labour and market constraints were of little interest to client farmers. A number of promising technologies were already available for a range of cassava production constraints. However, many had never been tested on farmers' fields, and even fewer were available to extension agents and their client farmers. In theory, interventions developed in an interdisciplinary manner had a better chance of success. A systems analysis that included multi-trophic interactions for a given agroecosystem and the socioeconomic concerns of the farmer provided therefore the paradigm for linking the ecological, agronomic and socioeconomic milieu in which constraints were evaluated and informed decisions were made.

Cassava Green Mite in Africa

The cassava green mite was among several introduced exotic pests that caused major ecological perturbations in the cassava agroecosystem late in the 20th century. This neotropical spider mite was discovered attacking cassava in East Africa in the early 1970s. It quickly spread throughout the cassava belt, causing locally up to 80% reduction in cassava root yield. Cassava green mite threatened production in many marginal areas where cassava was often the last crop available for harvest when all other crops had failed. Prior to its discovery in Africa, virtually nothing was known of cassava green mite. This mite was originally described in 1938 from cassava in Bahia, Brazil (Bondar, 1938), but only achieved notoriety 33 years later when it was found in Uganda (Z.M. Nyiira, Kampala, Uganda, 1972, unpublished data). Historically, there had been no serious mite problems in small-scale food crop production in tropical Africa, thus no need for local expertise in acarology. This left national programmes at a loss for the next step. Early efforts to control the mite included the use of chemicals, but the futility of this approach quickly became clear given the low income of most cassava farmers (Yaninek and Herren, 1988). Control efforts eventually shifted to resistance breeding, the mainstay of most commodity improvement programmes, and the dominant pest management practice in many national programmes (Hahn *et al.*, 1989).

Cassava in the Neotropics is relatively tolerant of indigenous pests, particularly locally adapted varieties used in traditional farming systems (Bellotti *et al.*, 1987). In the Neotropics, natural enemies were shown to cause substantial reductions in cassava green mite populations leading to the preservation of at least 30% of potential cassava root production (Braun *et al.*, 1989). This prompted the development of a sustainable pest management initiative to complement the ongoing efforts in resistance breeding. The project began with a series of baseline studies that quantified the dispersal and agronomic damage caused by a mite similar in bionomic attributes to other spider mites (Yaninek *et al.*, 1989a,b, 1990; Toko *et al.*, 1996). This included the predictable effects of temperature and humidity, the impact of rainfall on its geographical distribution and seasonal phenology, host plant effects on growth rates and population densities, and how agronomic practices enhance cassava plant vigour and boost pest populations. This package of agroecological data exposed windows of intervention opportunities, which formed the basis of the formulation of the classical biological control strategy including the direction of foreign exploration, development of systems for high production of quality host mites, and generation of quantitative sampling protocols for cassava green mite researchers.

The problems posed by cassava green mite are noteworthy at several levels. Ecologically, the mite lives and feeds on young leaves and green stems, and increases in population density during the transition period between wet and dry seasons (Yaninek *et al.*, 1989a,b). When populations reach and sustain high levels, the mite becomes an agronomic pest that reduces yields by damaging the photosynthetically active leaf surface area of the plant (Yaninek *et al.*, 1990). From a socioeconomic perspective, the significance of this pest depends on the importance of the crop to the farmer. Without adequate and appropriate intervention technologies, mite problems will persist, and intensify, as cassava production collapses under the strain of poorly adapted natural enemies, limited germplasm and ineffective agronomic practices. Therefore, cassava green mite as a pest posed multi-dimensional problems and required multi-dimensional solutions.

The model used for cassava green mite management was realized in the Ecologically Sustainable Cassava Plant Protection (ESCaPP) project, implemented in four countries in West Africa. ESCaPP was a regional project to develop, test and adapt sustainable technologies for protecting cassava against the most important arthropod, pathogen and weed pests found in West Africa (Yaninek *et al.*, 1994). Multi-disciplinary teams of national plant protection experts joined regionally with international experts to share expertise and pool efforts across agroecological zones. Project activities were divided into three interrelated and overlapping phases. First, the major cassava pests were identified in targeted agroecologies through diagnostic surveys. Second, farmers' participation highlighted the development and testing of appropriate intervention technologies. The third component was an evaluation of the training objectives and technology implementation. Training, including in-service and postgraduate, was an important feature of all phases of the project. ESCaPP undoubtedly accelerated the eventual impact of cassava green mite control throughout the continent because of technologies and knowledge developed in part by this regional project. A significantly enhanced cassava plant production and pest management capacity in participating national programmes was an additional benefit of the project.

Cassava Green Mite Biological Control Campaign

The exotic nature of the pest and its host plant in Africa prompted scientists in 1984 to initiate a classical biological control programme to complement the efforts in ongoing resistance breeding. The aim was to fill gaps in the knowledge of the cassava agroecosystem via empirical evaluation and to develop successful release and follow-up experiences needed to facilitate continent-wide implementation by national programmes. Selective aspects of the pest's biology and key interactions in the surrounding agroecosystem were characterized to facilitate the appropriate development and testing of ecologically sustainable corrective measures. The initial task was to understand the biotic potential of cassava green mite in its area of origin and in the newly affected farming systems in Africa. Once it was clear that local natural enemies in the pest's area of origin were potential key factors in the control of this introduced mite, the control campaign refocused on identifying and selecting phytoseiid predators of neotropical origin.

The taxonomic challenges encountered in this project where typical of any classical biological control programme entering uncharted taxonomic waters. First, very little was known about the indigenous mite fauna in sub-Saharan Africa at the start of the project, and most known pest and natural enemy species on the continent were based on a handful of regional surveys (Pritchard and Baker, 1962; Matthysse and Denmark, 1981; Meyer, 1974, 1981). Second, identifying mites in the area of origin was equally challenging because of disputes over the taxonomy of the pest found in Africa, and the paucity of information on tropical phytoseiids. A dispute arose when *M. tanajoa* (Bondar) in Africa was later identified as *Mononychellus progresivus* Doreste (Gutierrez, 1987). The new identification has been questioned because the determination was based exclusively on samples from Africa. It was also discovered that *M. progresivus* may be invalid because the original description was based on specimens preserved in methanol, which distorted the integument of the mites causing the dorsal setae to appear relatively larger than those normally found on *M. tanajoa*. The phytoseiids associated with cassava green mite were a poorly known group of predators; little was known about their composition in Africa and almost nothing was known about their biology and ecology from anywhere in the world. Surveys for a unique continent-wide inventory of indigenous phytoseiids and their host plant relationships in cassava agroecosystems were undertaken to collect background information on indigenous phytoseiids in preparation for the introduction of exotic phytoseiid species for the control of cassava green mite. More than 160,500 mite specimens representing 79 genera and 170 species in 33 families were collected from 496 host plant species in 26 countries (S. Yaninek and R. Hanna, Cotonou, 2002, unpublished data). By design, the largest group of specimens belonged to the Phytoseiidae, which were represented by 18 genera and 103 known and 26 undescribed species from 402 host plant species. This has led to descriptions of 21 new species, and re-descriptions of 45 species in collaboration with Gilberto De Moraes initially with Empresa Brasileira de Pesquisa Agropecuaria (EMBRAPA) and more recently with the University of Sao Paulo in Piracicaba in Brazil (Moraes and McMurtry, 1988; Moraes *et al.*, 1989a,b, 2001a,b; Zannou *et al.*, 2002).

In addition to the taxonomic survey, a series of studies were conducted on the development and reproductive biology, temperature and humidity tolerances, feeding ecology, and population dynamics of both indigenous and exotic phytoseiids (e.g. Bruce-Oliver *et al.*, 1996; Mebelo, 1999). These studies contributed to the general knowledge of phytoseiids, particularly tropical species, helped identify gaps in the phytoseiid composition in Africa that were exploited during foreign exploration for natural enemies, and provided a quantitative basis for evaluating and measuring the impact of phytoseiids introduced as natural enemies of cassava green mite.

Working with collaborators in Africa and South America on cassava green mite and its natural enemies required quick and reliable monitoring protocols, as well as effective and efficient field procedures. Procedures were developed for large-scale mass production of natural enemies, pre-release surveys of cassava green mite incidence/abundance/severity, novel and efficient methods to release natural enemies, post-release follow-up monitoring, pest and natural enemy identification, and impact assessment (Yaninek *et al.*, 1989c). The repeatable methodologies developed for sampling and handling cassava green mite and its natural enemies are now used by collaborators and practitioners on two continents, and provide the basis for monitoring progress and measuring impact in the biological control campaign.

As phytoseiid predators were identified as the most promising predators of cassava green mite, extensive foreign explorations were undertaken in collaboration with EMBRAPA and the Centro Internacional de Agricultura Tropical in Colombia. Initially, phytoseiid predators were selected and shipped through quarantine in Europe (initially at Imperial College in England, then later at the University of Amsterdam in The Netherlands) en route to Africa for experimental releases based on their abundance and frequency on cassava. Between 1984 and 1988, more than 5.2 million phytoseiids belonging to seven species of Colombian origin were imported into Africa and released in 348 sites in ten countries (Table 4.1). None of these species and populations ever became established in the wide range of agronomic and ecological conditions tested, apparently because of extended periods of low relative humidity or inadequate alternative food sources when cassava green mite densities were low (Yaninek *et al.*, 1993).

Foreign exploration was adjusted in 1988 to focus on neotropical regions that were agrometeorologically homologous to areas in Africa where the potential for severe cassava green mite damage existed (Yaninek and Bellotti, 1987). Natural enemies associated temporally and spatially with cassava green mite and capable of surviving periods of low cassava green mite densities on alternative food sources in the new exploration sites were given selection priority. Several natural enemy candidates were immediately identified in northeast Brazil and shipped to Africa. Approximately 6.1 million phytoseiids of the species *Neoseiulus idaeus* (Denmark and Muma), *Typhlodromalus manihoti* Moraes (Plate 10), and *Typhlodromalus aripo* DeLeon (Plate 11) of Brazilian origin were released in more than 400 sites in 20 countries between 1989 and 2000 (Table 4.1).

Table 4.1. Release and recovery of phytoseiids from Colombia and Brazil into Africa 1984–2001.

Phytoseiid species and origin	No. countries	No. shipments	Predators shipped	Release sites	Rate of recovery[a]
Colombia					
Amblyseius aerialis	7	80	141,335	26	0.00
Euseius concordis	8	77	170,262	44	0.00
Galendromus annectens	10	124	516,728	41	0.13
Neoseiulus anonymus	10	169	1,921,580	68	0.30
Neoseiulus californicus	8	83	443,512	30	0.28
Neoseiulus idaeus	10	187	1,819,779	81	0.47
Phytoseiulus mexicanus	1	1	726	1	0.00
Typhlodromalus manihoti	6	56	86,077	24	0.19
Typhlodromalus tenuiscutus	1	8	33,452	4	0.00
Brazil					
A. aerialis	1	1	731	1	0.00
N. anonymus	1	6	108,403	3	0.00
N. idaeus	12	189	3,737,067	160	0.43
Typhlodromalus aripo	16	120	400,587	220	0.90
Typhlodromalus limonicus	3	31	144,000	20	0.00
T. manihoti	9	137	2,044,442	124	0.54

[a]Proportion of releases with subsequent recoveries.

The campaign significantly benefited from the ESCaPP project during most of the 1990s. Foreign exploration, natural enemy shipments, quarantine, training, development of new technologies, and implementation of release and follow-up activities were all integral components of ESCaPP during this period. Systematic post-release monitoring eventually revealed that *N. idaeus* was established in Bénin and Kenya (Yaninek *et al.*, 1992), *T. manihoti* was established in Bénin, Burundi, Ghana and Nigeria (Yaninek *et al.*, 1998), and *T. aripo* was established and spread to 20 countries spanning the cassava belt of Africa in less than 10 years (Hanna and Toko, 2001).

To complement cassava green mite biological control by phytoseiid predators, a parallel effort was initiated to introduce fungal pathogens of cassava green mite, which were reported to cause spectacular epizootics in cassava green mite populations in northeastern Brazil (Delalibera *et al.*, 1992). An international workshop was organized in 1994 to identify issues critical to the development of *Neozygites* as a fungal biological pesticide or classical biological control agent for cassava green mite. A survey in West Africa revealed the presence of several previously unknown but ubiquitous fungal pathogens on cassava green mite and other spider mites, including the first record of *Neozygites floridana* (Weiser and Muma) (*Entomophthorales*, *Neozygitaceae*), later renamed *N. tanajoae* n. sp. (I. Delalibera, Cornell University, 2002, personal communication) (Plate 12), in Africa (Yaninek *et al.*, 1996). At the same time, this fungus was found attacking cassava green mite during foreign exploration in humid valleys of northeastern Brazil. Subsequent studies were initiated with collaborators in Brazil to evaluate the specificity and effects of temperature, humidity and photoperiod on the development and virulence of *N. tanajoae* isolates found attacking cassava green mite (Oduor *et al.*, 1995, 1996). This work showed the importance of fungal pathogens as a source of mortality for tropical spider mite pests generally, and the potential of exotic isolates of *N. tanajoae* as a natural enemy of cassava green mite in particular. Virulent strains of *N. tanajoae* have been shipped to Africa for culturing and experimental releases. Studies in Africa have been completed on specificity (Hountondji *et al.*, 2002a) and experimental field releases have been evaluated in Bénin, where the fungus is apparently established and is causing an average of 25% infection level in cassava green mite populations (Hountondji *et al.*, 2002b). We continue to monitor the spread of the fungus in Bénin (R. Hanna, A. Cherry, F. Hountondji, IITA Cotonou, Bénin, 2002, unpublished data), and a parallel effort is continuing to develop molecular tools for distinguishing fungal isolates of *N. tanajoae* (I. Delalibera, A. Hajek, A. Cherry and R. Hanna, Cornell University and IITA Cotonou, Bénin, 2002).

Biological Control Impact and Prospects for Success

Prospects for control of cassava green mite were initially inferred from the impact of natural enemies in their area of origin, but eventually dramatic field results told the real story. *N. idaeus* never spread beyond the two original release and 'establishment' sites, and has probably become extinct because of insufficient mite prey on cassava and associated host plant species. *T. manihoti* became established and spread slowly around the original release sites. However, its impact on cassava green mite has been difficult to quantify at the low predator densities found in the field. In contrast, *T. aripo*, a predator that rests in the growing tip of cassava plants during the day and forages on cassava leaves at night (Onzo *et al.*, 2003), was the big surprise.

T. aripo became established in 20 countries and rapidly spread beyond most original release sites. This predator was introduced into Ikpinlé, Bénin, in October 1993, and through further releases and natural spread it was reported established by the year 2000 in the following 20 countries in sub-Saharan Africa and covering more than 3.8 million km^2: Bénin, Burundi, Cameroon, Central African Republic, Democratic Republic of Congo, Gabon, Ghana, Guinea Conakry, Ivory Coast, Kenya, Liberia, Malawi, Mozambique, Nigeria, Republic of Congo, Sierra Leone, Tanzania, Togo, Uganda and Zambia (Hanna and Toko, 2001).

Initially, the impact of *T. aripo* on cassava green mite populations in the field was measured by evaluating mite populations in 468 farmers' fields distributed from Cameroon to Ghana over three seasons as the exotic predator spread across West Africa. Cassava green mite populations in cassava fields where *T. aripo* was present were 43% lower on average (ranging from 16% to 60%) compared with nearby fields where *T. aripo* had yet to arrive (S. Yaninek, IITA Cotonou, 1996, unpublished data). Similarly, in the southern African countries of Malawi and Mozambique, cassava green mite densities were reduced by an average of 40% 2 years after the establishment of *T. aripo* (R. Hanna and I. Zannou, IITA Cotonou, 2002, unpublished data). Continuous longer-term population dynamics monitoring provided a picture of a textbook example of successful classical biological control – initially high pest densities are brought down to more or less stable levels by a corresponding increase in the introduced natural enemies, as shown in the example from Ikplinlé, Bénin, over a period of 30 months (S. Yaninek and R. Hanna, 2002, IITA Cotonou, unpublished data; Fig. 4.1).

The impact of cassava green mite biological control on cassava production was equally impressive. In yield trials in several countries over different seasons, cassava from which *T. aripo* was eliminated from the shoot tips with low rate of permethrin (0.0625 g a.i. of Imperator 25 EC l^{-1} of water) during the period when cassava green mites were most damaging resulted in an average root yield loss of 35% compared with untreated control fields. The economic return for this predator has been estimated to be equivalent to hundreds of millions of dollars in food aid each growing season if yield losses had to be replaced (O. Coulibaly, S. Yaninek, R. Hanna and V. Manyong, IITA Cotonou, Bénin, 2002, unpublished data). Studies have also shown that the introduction of *T. aripo* has had no negative non-target effects, as this predator is restricted to cassava and its impact is restricted to the control of cassava green mite (I. Zannou and R. Hanna, IITA, Cotonou, Bénin, 2002, unpublished data).

T. aripo has been more successful than the other introduced phytoseiid predators because it does not overexploit its prey as measured by predation

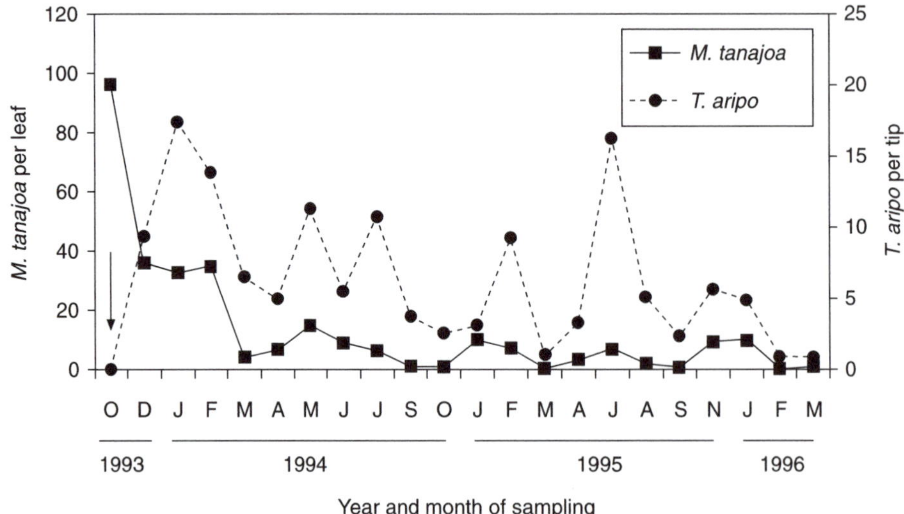

Fig. 4.1. Population size of *Mononychellus tanajoa* on leaves (filled squares and solid line) and *Typhlodromalus aripo* in cassava tips (filled circles and dashed line) in Ikpinlé, Bénin, over a period of 30 months, from October 1993, the date of *T. aripo* release, to March 1996. Arrow indicates date of *T. aripo* release.

rates, and is able to survive, develop and oviposit on the widest range of diets tested. This promotes the persistence of *T. aripo* populations over a wide range of *M. tanajoa* densities and facilitates the predator–prey encounters important for effective biological control. Similar conclusions were drawn from a tri-trophic meta-population model that compared interacting pest and natural enemy species in the African cassava food web (Gutierrez *et al.*, 1999). In addition, *T. aripo* inhabits the apex of the plant where it is well protected from the climatic extremes confronting the more common leaf dwelling phytoseiids. The rapid spread of *T. aripo* across the continent probably reflects its ability to inhabit the shoot tip, in which it is transported in the vibrant commercial trade in cassava stem cuttings as field propagation seed stock.

Constraints Overcome and Lessons Learnt

Several important constraints were overcome to implement successfully this campaign against a mite pest. The exotic phytoseiids had to be available in sufficient numbers and at the right time to assure worthwhile experimental releases. This was partly resolved by new mass production technology that was developed to facilitate decentralized rearing of phytoseiid predators by a cadre of highly trained and dedicated project staff. Release fields were judiciously selected and faithfully monitored to understand the conditions needed to achieve widespread establishment. Post-release follow-up monitoring was one of the most difficult tasks for national programme staff to accomplish because of many constraints. Continued training and close mentoring by experienced collaborators helped to resolve the problem. Ultimately, close and continuous contact between national agricultural research programmes and extension services with specific local capacities and needs, international programmes with unique expertise and implementation resources, and a donor community committed to long-term support set the stage for the success achieved in this campaign.

The ecological risks associated with biological control, particularly classical biological control, have been the topic of considerable discussion in recent years. Evidence of non-target effects by a few natural enemies introduced before stringent taxonomic specificity standards were adopted, and increasing concern about these introductions as potential invasive species, has put classical biological control practices under considerable scrutiny. The concern is the ecological impact these introductions have on non-target species, associated habitats and surrounding ecosystems. In this project, candidate natural enemies were systematically evaluated for potential multi-trophic, intra-guild, community and ecosystem impact as part of the selection and pre-release procedures. There was no evidence that any released natural enemy has had or would have an unanticipated impact or pose an ecological risk anywhere in the African cassava belt. Biological control will continue to be a desirable and appropriate intervention tactic in the future. However, success and impact will be measured in part by the risks that are mitigated in the process.

Human resources are among the most precious commodity available to a research programme. Most research centres realized this and put training as one of their principle objectives which accounts for a significant proportion of every scientist's time and effort. Training programmes were developed that catered to the needs of in-service, specialized and graduate training every year. Similarly, bench training was executed for 25 trainees with specialized needs and implemented through a period of attachment. Fifteen PhD, nine MSc and nine BSc students were trained in this manner. This training created a basis for all subsequent research for development activities in collaborating national programmes and accounts to a significant degree for many of the accomplishments outlined above.

Ultimately, the success of the classical biological control campaign of cassava green mite rested on finding and establishing effective natural enemies. This effort provided a unique example in which the least likely predator, *T. aripo*, turned out to be the best natural enemy for classical biological control on a continental scale. This predator, although less voracious and increasing at a slower rate than *T. manihoti* and *N. idaeus*, was best at establishment, dispersal and persistence on cassava plants owing to its use of cassava tips as refuge (Onzo *et al.*, 2003), its efficient ability to locate its prey (Gnanvossou *et al.*, 2001), and last but not least its ability to persist at low prey densities because of its use of multiple resources, including cassava green mite, pollen and cassava extrafloral exudates (S. Yaninek, M. Sabelis, G.J. De Moraes, R. Hanna, A.P. Gutierrez, IITA Cotonou and University of Amsterdam, 2002, unpublished data).

Conclusions

The introduction of major exotic pests of cassava in the 1970s brought much-needed pest management attention to a 'low-value' food crop in Africa and helped galvanize the development of the sustainable multi-disciplinary pest management strategy being promoted today. The introduced cassava green mite was one of these catalysts. A major campaign with multi-institutional and multi-disciplinary collaboration was put in place. The concept was to implement comprehensive cassava plant production and protection practices to meet the needs identified by farmers, local cassava specialists and regional scientists with unique disciplinary expertise working together as a team. This strategy was a success and elements of this model are now being promoted throughout the cassava belt of Africa.

Prior to this work on cassava green mite, the role of biological control of mites in agriculture had been largely limited to a few ubiquitous phytophagous species found in intensively cultivated orchard and greenhouse agroecosystems worldwide. Contrary to conventional wisdom, this project with collaborators on four continents is now successfully controlling cassava green mite in Africa using classical biological control in a field crop on a continent-wide scale – both firsts for a mite pest anywhere in the world. Generous donors supported the substantial research and development effort over 15 years, while innovations successfully targeted the pest and its integrated biological control as part of a sustainable plant protection system in Africa. An analytical systems approach was adopted to characterize the conditions that prompt pest outbreaks and to identify natural enemies for introduction. The project developed field-sampling procedures, implementation protocols and experimental designs used today by most researchers working on cassava green mites in Africa and Latin America. The project was a rare comprehensive study of a pest and a solution to a problem that included the pest's biology, yield loss assessment, phytoseiid predator biology, fungal pathogens, natural enemy production systems, release and follow-up monitoring, implementation through a network of trained national collaborators, and documented in a number of unique databases, in addition to the scientific literature. The effort resulted in the establishment of three exotic phytoseiid predators, two of which are spreading, and one of which reduces pest populations by half and increases yields by a third – the economic impact in four countries in West Africa alone is estimated at more than $200 million per season.

The campaign to control cassava green mite in Africa through classical biological control far exceeded most expectations. Because there were no comparable mite biological control programmes to model this campaign after, there was scepticism in the scientific community about the likelihood of its success. The real surprise was not only the degree and magnitude of the biological control campaign, but also the rate at which selected introduced natural enemies became established and spread covering most important cassava-producing countries in less than 10 years. Blending the biological control campaign with a multi-disciplinary crop protection programme in West Africa certainly accelerated the implementation and post-release evaluation of the control campaign. It also gave participating national programme staff a sense of ownership and accomplishment by being part of an action programme that yielded tangible field results.

References

Bellotti, A.C., Mesa, N., Serrano, M., Guerrero, J.M. and Herrera, C.J. (1987) Taxonomic inventory and survey activity for natural enemies of cassava green mites in the Americas. *Insect Science and Its Application* 8, 845–849.

Bock, K.R. and Guthrie, E.J. (1978) African mosaic disease in Kenya. In: Brekelbaum, T., Bellotti, A. and Lozano, J.C. (eds) *Proceedings of the Cassava Protection Workshop, CIAT, Cali, Colombia, 7–12 November, 1977*. International Agriculture in the Tropics (Series CE-14), pp. 41–44.

Bondar, G. (1938) Notas entomologicas da Bahia III. *Revista Entomologica Rio de Janeiro* 9, 441–445.

Braun, A.R., Bellotti, A.C., Guerrero, J.M. and Wilson, L.T. (1989) Effect of predator exclusion on cassava infested with tetranychid mites (Acari: Tetranychidae). *Environmental Entomology* 18, 711–714.

Bruce-Oliver, S.J., Hoy, M. and Yaninek, J.S. (1996) Effect of some food sources associated with cassava in Africa on development success, fecundity, and longevity of *Euseius fustis* (Pritchard and Baker) (Acari: Phytoseiidae). *Experimental and Applied Acarology* 20, 73–85.

Delalibera, I., Jr, Sosa-Gomes, D.R., De Moraes, G.J., De Alencar, J.A. and Farias-Araujo, W. (1992) Infection of *Mononychellus tanajoa* (Acari: Tetranychidae) by the fungus *Neozygites* sp. (Entomophthorales) in Northeastern Brazil. *Florida Entomologist* 75, 145–147.

Dixon, A.G.O., Ngeve, J.M. and Nukenine, E.N. (2002) Response of cassava genotypes to four biotic constraints in three agroecologies of Nigeria. *African Crop Science Journal* 10, 11–21.

Fokunang, C.N., Akem, C.N., Ikotun, T. and Dixon, A.G.O. (2000) Evaluation of a cassava germplasm collection for reaction to three major diseases and the effect on yield. *Genetic Resources and Crop Evolution* 47, 63–71.

Gnanvossou, D., Hanna, R., Dicke, M. and Yaninek, J.S. (2001) Response of the predatory mites *Typhlodromalus manihoti* Moraes and *Typhlodromalus aripo* DeLeon (Acari: Phytoseiidae) to volatiles from cassava plants infested by cassava green mite. *Entomologia Experimentalis et Applicata* 101, 291–298.

Gutierrez, J. (1987) The cassava green mite in Africa: one or two species? (Acari: Tetranychidae). *Experimental and Applied Acarology* 3, 163–168.

Gutierrez, A.P., Yaninek, J.S., Neuenschwander, P. and Ellis, K.C. (1999) A physiologically-based tritrophic metapopulation model of the African cassava food web. *Ecological Modelling* 123, 225–242.

Hahn, S.K., Isoba, J.C.G. and Ikotun, T. (1989) Resistance breeding in root and tuber crops at the International Institute of Tropical Agriculture (IITA), Ibadan, Nigeria. *Crop Protection* 8, 147–168.

Hanna, R. and Toko, M. (2001) Cassava green mite biological control in Africa: project overview and summary of progress. In: Hanna, R. and Toko, M. (eds) *Proceedings of the Regional Meeting of the Cassava Green Mite Biocontrol Project, Dar es Salaam, Tanzania, 15–17 November 2000*. IITA, Cotonou, Bénin, pp. 4–22.

Hountondji, F.C.C., Lomer, C., Hanna, R. and Cherry, A. (2002a) Release and establishment of *Neozygites floridana* (Entomophtherales: Neozygitacea) for the microbial control of cassava green mite in Bénin. *Biocontrol Science and Technology* 12, 361–370.

Hountondji, F.C.C., Yaninek, J.S., De Moraes, G.J. and Odour, G.I. (2002b) Host specificity of the cassava green mite pathogen *Neozygites floridana*. *BioControl* 47, 61–66.

Jones, W.O. (1959) *Manioc in Africa*. Stanford University Press, Stanford, California, USA.

Legg, J.P. and Thresh, J.M. (2000) Cassava mosaic virus disease in East Africa: a dynamic disease in a changing environment. *Virus Research* 71, 135–150.

Matthysse, J.G. and Denmark, H.A. (1981) Some phytoseiids of Nigeria (Acarine: Mesostigmata). *Florida Entomologist* 64, 340–357.

Mebelo, M. (1999) Selection of phytoseiids for introduction into Zambia to control cassava green mite, *Mononychellus tanajoa* (Acari: Tetranychidae). PhD dissertation, Imperial College, Ascot, UK.

Meyer, M.K.P.S. (1974) *A Revision of the Tetranychidae of Africa (Acari) with a Key to the Genera of the World.* Entomology Memoirs, No. 36, Department of Agriculture Technical Service, South Africa.

Meyer, M.K.P. (1981) *Mite Pests of Crops in Southern Africa.* Science Bulletin 397, Plant Protection Research Institute, South Africa, pp. 1–92.

Moraes, G.J. De and McMurtry, J.A. (1988) Some phytoseiid mites from Kenya, with description of three new species. *Acarologia* 29, 13–18.

Moraes, G.J. De, McMurtry, J.A. and Yaninek, J.S. (1989a) Some phytoseiid mites (Acari: Phytoseiidae) from tropical Africa with descriptions of one new species. *International Journal of Acarology* 15, 95–102.

Moraes, G.J. De, van den Berg, H. and Yaninek, J.S. (1989b) Phytoseiid mites (Acari: Phytoseiidae) of Kenya, with descriptions of five new species and redescriptions of eight species. *International Journal of Acarology* 15, 79–93.

Moraes, G.J. De, Oliveira, A.R. and Zannou, I.D. (2001a) New phytoseiid species (Acari: Phytoseiidae) from tropical Africa. *Zootaxa* 8, 1–10.

Moraes, G.J. De, Ueckermann, E.A., Oliveira, A.R. and Yaninek, J.S. (2001b) Phytoseiid mites of the genus *Euseius* (Acari: Phytoseiidae) from Sub-Saharan Africa. *Zootaxa* 3, 1–70.

Nweke, F.I., Spencer, D.S.C. and Lynam, J.K. (2002) *The Cassava Transformation: Africa's Best-Kept Secret.* Michigan State University Press, East Lansing, Michigan, USA.

Oduor, G.I., De Moraes, G.J., Yaninek, J.S. and van der Geest, L.P. (1995) Effect of temperature, humidity and photoperiod on mortality of *Mononychellus tanajoa* (Acari: Tetranychidae) infected by *Neozygites* cf. *floridana* (Zygomycetes: Entomophthorales). *Experimental and Applied Acarology* 19, 571–579.

Oduor, G.I., Yaninek, J.S., van der Geest, L.P.S. and De Moraes, G.J. (1996) Germination and viability of capilliconidia of *Neozygites floridana* (Zygomycetes: Entomophthorales) under constant temperature, humidity and light conditions. *Journal of Invertebrate Pathology* 67, 267–278.

Onzo, A., Hanna, R., Sabelis, M.W. and Yaninek, J.S. (2003) Dynamics of refuse use: diurnal vertical migration of predatory and herbivorous mites within cassava plants. *Oikos* (in press).

Pritchard, A.E. and Baker, E.W. (1962) Mites of the family Phytoseiidae from Central Africa, with remarks on genera of the world. *Hilgardia* 33, 205–309.

Toko, M., Yaninek, J.S. and O'Neil, R.J. (1996) Response of *Mononychellus tanajoa* (Acari: Tetranychidae) to cropping systems, cultivars and pest interventions. *Environmental Entomology* 25, 237–249.

Yaninek, J.S. and. Bellotti, A.C. (1987) Exploration for natural enemies of cassava green mites based on agrometeorological criteria. In: Rijks, D. and Mathys G. (eds) *Proceedings of the Seminar on Agrometeorology and Crop Protection in the Lowland Humid and Sub-Humid Tropics, 7–11 July 1986, Cotonou, Bénin*. World Meteorological Organization, Geneva, Switzerland, pp. 69–75.

Yaninek, J.S. and Herren, H.R. (1988) Introduction and spread of the cassava green mite, *Mononychellus tanajoa* (Bondar) (Acari: Tetranychidae), an exotic pest in Africa and the search for appropriate control methods: a review. *Bulletin of Entomological Research* 78, 1–13.

Yaninek, J.S. and Schulthess, F. (1993) Developing environmentally sound plant protection for cassava in Africa. *Agriculture, Ecosystems and the Environment* 46, 305–324.

Yaninek, J.S., Gutierrez, A.P. and Herren, H.R. (1989a) Dynamics of *Mononychellus tanajoa* (Acari: Tetranychidae) in Africa: experimental evidence of temperature and host plant effects on population growth rates. *Environmental Entomology* 18, 633–640.

Yaninek, J.S., Herren, H.R. and Gutierrez, A.P. (1989b) Dynamics of *Mononychellus tanajoa* (Acari: Tetranychidae) in Africa: seasonal factors affecting phenology and abundance. *Environmental Entomology* 18, 625–632.

Yaninek, J.S., De Moraes, G.J. and Markham, R.H. (1989c) *Handbook on the Cassava Green Mite* Mononychellus tanajoa *in Africa: a Guide to their Biology and Procedures for Implementing Classical Biological Control.* IITA Publication Series, Ibadan, Nigeria.

Yaninek, J.S., Gutierrez, A.P. and Herren, H.R. (1990) Dynamics of *Mononychellus tanajoa* (Acari: Tetranychidae) in Africa: impact on dry matter production and allocation in cassava, *Manihot esculenta. Environmental Entomology* 19, 1767–1772.

Yaninek, J.S., Mégevand, B., De Moraes, G.J., Bakker, F., Braun, A. and Herren, H.R. (1992) Establishment of the neotropical predator *Amblyseius idaeus* (Acari: Phytoseiidae) in Bénin, West Africa. *Biocontrol Science and Technology* 1, 323–330.

Yaninek, J.S., Onzo, A. and Ojo, B. (1993) Continent-wide experiences releasing neotropical phytoseiids against the exotic cassava green mite in Africa. *Experimental and Applied Acarology* 16, 145–160.

Yaninek, J.S., James, B.D. and Bieler, P. (1994) Ecologically sustainable cassava plant protection (ESCaPP): a model for environmentally sound pest management in Africa. *African Crop Science* 2, 553–562.

Yaninek, J.S., Saizounou, Z., Onzo, A., Zannou, I. and Gnanvossou, D. (1996) Seasonal and habitat variability in the fungal pathogens, *Neozygites* cf. *floridana* and *Hirsutella thompsonii*, associated with cassava mites in Bénin, West Africa. *Biocontrol Science and Technology* 6, 23–33.

Yaninek, J.S., Mégevand, B., Ojo, B., Cudjoe, A.R., Abole, E.A., Onzo, A. and Gnanvossou, D. (1998) Establishment and spread of *Typhlodromalus manihoti*, an introduced phytoseiid predator of *Mononychellus tanajoa*, in Africa. *Environmental Entomology* 27, 1496–1505.

Zannou, I., De Moraes, G.J. and Hanna, R. (2002) New species of phytoseiid mites from Mozambique and Malawi. *Zootaxa* 79, 1–6.

5

Biological Control of the Potato Tuber Moth *Phthorimaea operculella* in Africa

Rami Kfir
Agricultural Research Council – Plant Protection Research Institute, Division of Insect Ecology – Biological Control, Pretoria, South Africa

Introduction

The potato tuber moth, *Phthorimaea operculella* (Zeller) (Lepidoptera, Gelechiidae) (Plate 13) is a widespread pest of potatoes in all the warmer parts of the world. It occurs in Africa wherever potatoes are grown and also attacks tobacco, aubergines and tomatoes. The potato tuber moth probably originated in the Bolivian Andes in South America, where the potato plant itself most probably also came from. The pest, which is the most serious pest of potatoes in the region, has been known in South Africa since the late 19th century. Damage is caused by the larvae, which attack leaves, stems and the tubers of potato plants (Plate 14a) and, under optimum conditions, may defoliate and eventually destroy the epigeal plant. Most larvae mine between the upper and lower surfaces of the leaves, forming typical 'windows', but some also tunnel into the stems, where their presence goes unseen unless the stems break. When the foliage of the plant dies back at the end of the growing season, the larvae tend to move down the plant towards the tubers in the soil. Damage to tubers may then exceed 50%, and may also continue into the storage period (Whiteside, 1981a; Annecke and Moran, 1982).

Indigenous natural enemies

A granulosis virus is the only disease found attacking potato tuber moth in South Africa. Minor outbreaks of the disease, in which up to 10% of tuber moth larvae become infected, occur sporadically and are fairly common, but they do not usually have much impact on tuber moth larval populations. However, severe outbreaks sometimes occur, in which up to 80% of larvae may become infected (Broodryk and Pretorius, 1974; Whiteside, 1981a).

(eds P. Neuenschwander, C. Borgemeister and J. Langewald)

Several species of ladybirds (Coleoptera, Coccinellidae) such as *Lioadalia flavomaculata* De Geer and *Hippodamia variegata* (Goeze) prey on eggs and young larvae of tuber moth in potato fields (Findlay, 1975). In addition, larvae of lacewings (Neuroptera, Chrysopidae), *Chrysopa* spp. and big-eyed bugs (Hemiptera, Geocoridae), *Geocoris* spp., ground beetles (Coleoptera, Carabidae), such as *Harpalus lugubris* Boheman, *Harpalus natalensis* Boheman and *Harpalus nanniscus* Péringuey together with earwigs (Dermaptera) and staphylinid beetles also prey on all stages of the pest. Also a wasp, *Synagris abysinica* Geursp. (Hymenoptera, Eumenidae), preys on larvae after cutting them out of the leaves.

Five indigenous hymenopteran parasitoids attack potato tuber moth in South Africa. These are three ichneumonids (*Diadegma mollipla* Holmgren (Plate 14b), *Diadegma* sp., *Temelucha picta* Holmgren) and two braconids (*Orgilus parcus* Turner, *Chelonus curvimaculatus* Cameron) (Watmough *et al.*, 1973). Detailed studies of the biology of these parasitoids revealed that *C. curvimaculatus* attacked the egg stage of potato tuber moth (Broodryk, 1969a), whereas the others attacked young larvae (Broodryk, 1969b, 1971). All killed the host just before pupation. Until then, the larvae develop normally and show no visible external signs of parasitism (Whiteside, 1980). Since potato tuber moth is not native to South Africa, all these parasitoids have additional hosts as well. The ichneumonid *D. mollipla*, which is native to South Africa, is a common parasitoid of the diamondback moth (*Plutella xylostella* (Linnaeus)) in South Africa (Kfir, 1997) and in other countries in sub-Saharan Africa, including some Indian Ocean and South Atlantic islands (Azidah *et al.*, 2000). While the braconid *O. parcus* was reared from the lesser armyworm (*Spodoptera exigua* (Boisduval)) in potato fields (Broodryk, 1969b), the other braconid, *C. curvimaculatus,* has been recorded in South Africa from *S. exigua,* the armyworm (*Spodoptera exempta* (Walker)), cotton leaf worm (*Spodoptera littoralis* Hübner), African bollworm (*Helicoverpa armigera* Hübner), Mediterranean flour moth (*Ephestia kuehniella* Zeller), dried-fruit moth (*Ephestia cautella* (Walker)), false codling moth (*Cryptophlebia leucotreta* (Meyrick)) (Broodryk, 1969a), spotted stem borer (*Chilo partellus* (Swinhoe)) (Kfir, 1990), African stem borer (*Busseola fusca* (Fuller)) (Kfir, 1995), and the diamondback moth (Kfir, 1997).

These parasitoids frequently achieve high levels of parasitism of potato tuber moth larvae. Field sampling indicated that *D. mollipla* was always the dominant parasitoid, frequently parasitizing over 70% of tuber moth larvae. The remaining species were always less important, seldom achieving >10% parasitism individually (Broodryk, 1967). Despite this parasitism, losses from potato tuber moth damage nevertheless remained high (Annecke and Moran, 1982).

In areas such as the Highveld region of South Africa where winters are cold and generally dry, potatoes are grown as a summer crop with wide spacing between rows, which allows for effective ridging. In these areas and in the presence of the indigenous parasitoid complex, potato production was successful, but efficient ridging was essential. Evidently the indigenous parasitoids, thriving in the absence of insecticides, exerted sufficient natural control to keep the pest below its economic threshold. However, growers have tended to increasingly rely on insecticides for tuber moth control.

In other regions of South Africa, such as the Northern Province, potatoes are grown as an irrigated winter crop. Production costs are high due to costly irrigation, and yields are low. In order to obtain maximum return per unit area, potatoes are therefore planted at higher densities with narrow interrow spacings. Ridging is also not practised because of the closely spaced plant rows. Under these conditions the indigenous parasitoids have proven ineffective in controlling potato tuber moth and total destruction of foliage by tuber moth sometimes occurs. An intensive chemical control regime is therefore applied throughout most of the growing season (Watmough *et al.*, 1973).

Biological Control

Because ridging was not always possible and the insecticides applied often failed to provide sufficient crop protection, additional measures became necessary. Exotic parasitoids from the country of origin of the pest were therefore imported to reinforce the existing indigenous parasitoid complex (Watmough *et al.*, 1973). The first attempt at biological control of the potato tuber moth in South Africa was made in 1944 when the braconids *Chelonus phthorimaeae* Gahan and *Bracon gelechiae* Ashmead were imported from the USA. Although the parasitoids were released, no recoveries were made (Bedford, 1954). A second attempt was made in 1965 after several parasitoids, resulting from exploratory work initiated by the Commonwealth Institute of Biological Control (CIBC) in South America were shipped to South Africa. The following parasitoids came by way of the CIBC Indian Station after becoming established there: the polyembryonic encyrtid parasitoid *Copidosoma koehleri* Blanchard, also referred to as *Copidosoma uruguayensis* Tachikawa by Annecke and Mynhardt (1974), the braconids *Apanteles subandinus* Blanchard and *Orgilus lepidus* Muesebeck, and the ichneumonid *Campoplex haywardi* Blanchard (Greathead, 1971). Of these species, *O. lepidus* and *C. koehleri* were successfully bred and released. An additional consignment of *A. subandinus*, received from Australia in 1967 and reared in the laboratory, was also successfully released. The braconid *O. lepidus*, which proved so successful in the coastal areas of Australia (Callan, 1974), did not establish itself in South Africa and releases were consequently discontinued (Findlay, 1975).

C. koehleri and *A. subandinus* were released in potato fields in different parts of South Africa until July 1969, and both became established over wide areas. Recoveries many kilometres from the original release sites were common (Watmough *et al.*, 1973). The two parasitoids have now become widely dispersed and have been recorded from all potato- and tobacco-growing areas of South Africa (Findlay, 1975). Their spread was obviously accelerated by the transportation of potatoes, some of which contained parasitized tuber moth larvae, to all parts of the country. Both species also pass the winter within their hosts in unharvested potatoes in the soil.

The mean maximum parasitism at several sites in the Highveld Region of South Africa, for which appropriate data from 1965 to 1980 exist, rose from about 70% to 90–95%, once the two exotic parasitoids became fully established (Fig. 5.1) (Broodryk, 1967; Watmough *et al.*, 1973; Findlay, 1975; Whiteside, 1980). The introduced parasitoids largely displaced the less effective indigenous ones, which accounted for only about 10% of the maximum parasitism from 1975 to 1979 (Whiteside, 1980).

In a trial conducted in 1988 in South Africa, tuber moth larvae were collected from potato leaves and tubers towards the end of the season from a field sprayed weekly with various pesticides. Between 71% and 82% of the tuber moth larvae were nevertheless found to be parasitized by *C. koehleri*, which indicated that the parasitoid could tolerate relatively severe pesticide spray regimes (Kfir, 1989). This corroborated observations that insecticide applications against tuber moth have a negligible effect on the maximum parasitism by its natural enemies (Whiteside, 1980). More recently, in 1994, where no insecticides were applied, *C. koehleri* and *A. subandinus* were responsible for up to 86% parasitism, while all other parasitoids together attained only 4.5% parasitism (Charleston *et al.*, 2003).

Each year since their establishment, *C. koehleri* and *A. subandinus* complemented each other, providing a seasonal succession. The latter species predominated in the first part of the potato-growing season as tuber moth numbers built up after winter, but was replaced by *C. koehleri* towards the end of the season, when parasitism reached its maximum levels (Whiteside, 1981b). The introduction of *A. subandinus* has markedly improved biological control of tuber moth by reinforcing parasitism at a time when *C. koehleri* populations were still low. Before *A. subandinus* had been introduced, *D. mollipla* occupied this same niche (Watmough *et al.*, 1973).

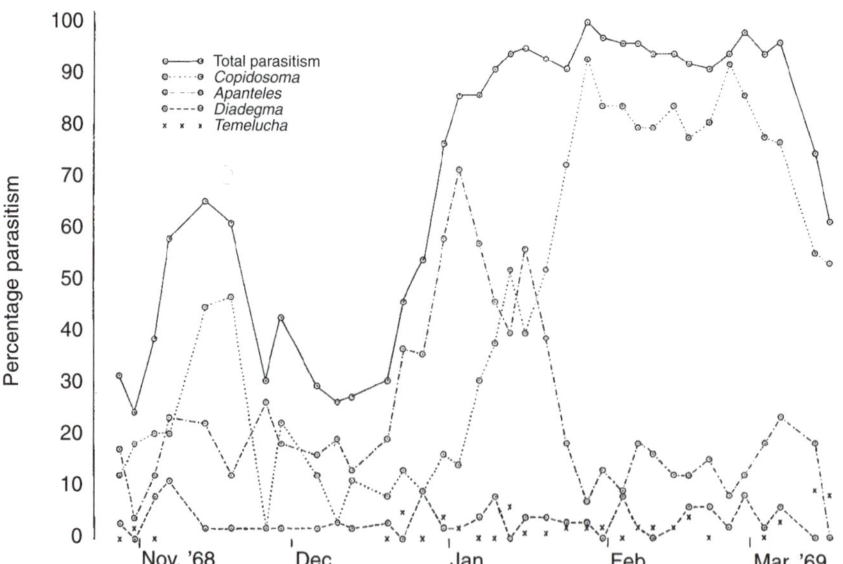

Fig. 5.1. Parasitism of potato tuber moth larvae on a farm in the Highveld region of South Africa in the 1968/69 potato-growing season after *Apanteles subandinus* and *Copidosoma koehleri* had become established and had almost completely replaced the indigenous parasitoids (from Watmough *et al.*, 1973).

The change in the relative abundance of the two parasitoids during the potato-growing season may be due to different relative humidities requirements. *C. koehleri* is extremely susceptible to low relative humidities (Kfir, 1981), whereas *A. subandinus* prefers low relative humidities (Cardona and Oatman, 1975). At the beginning of the season when potato plants are still small and more bare ground is exposed, the air within the growing crop is warmer and drier than at 1.2 m above ground level. However, as the canopy develops, conditions within the crop cause an increase in the relative humidity (Hirst and Stedman, 1960). It was observed that in volunteer potatoes, which develop dense foliage earlier than the crop, *C. koehleri* became more abundant than *A. subandinus* earlier in the season (Kfir, 1981).

C. koehleri and *A. subandinus* were introduced from South Africa to Zimbabwe and Zambia, where they have become successfully established. In Zimbabwe, 1.5 million *C. koehleri* were released from 1967 to 1969 and 25,000 *A. subandinus* again from 1969 to 1970. Total parasitism was thus significantly raised during the critical period, June to October, when indigenous local parasitoids proved ineffective (Mitchell, 1978).

In Zambia, unspecified numbers of *C. koehleri* were released from 1969 to 1971 and *A. subandinus* from 1970 to 1971. According to Cruickshank and Ahmed (1973), parasitism by *C. koehleri* attained 90% with a substantial decrease in crop damage in the main commercial potato areas.

Between February and April 1965, seven consignments totalling about 59,000 *C. koehleri* were sent from the CIBC Indian Station to Tanzania. Six field releases of 3000–9000 individuals were made on unsprayed farms, but the parasitoid was not recovered in follow-up surveys during 1965 and 1967. Further introductions and releases of *C. koehleri* were made again in 1968. Additional shipments of *C. haywardi* Blanchard and *Nytobia* sp. (Hymenoptera, Ichneumonidae), *A. subandinus*, and *O. lepidus* were received and released in West Kilimanjaro and Moshi Districts in July 1969 (Greathead, 1971).

A survey of parasitoids of the potato tuber moth in Mauritius revealed the presence of *C. curvimaculatus* and *D. millipla*. However, these could not prevent heavy infestations of tuber moth from developing on tobacco and potato crops. A biological control programme started on the island in 1964 and *C. koehleri, C. haywardi* and *A. subandinus* were imported from the CIBC Station in Colton, California. During 1965, about 3 million parasitoids were therefore released, which by the next year resulted in a marked reduction in the tuber moth infestation, which was mainly attributed to *C. koehleri* (Greathead, 1971).

Shipments of *C. koehleri* were also sent from Mauritius to Seychelles in 1966 and released on Praslin Island, but no recoveries were made (Greathead, 1971). During 1967 and 1968, shipments of *C. koehleri* from Mauritius, *D. mollipla* from South Africa, *Apanteles scutellaris* Muesebeck, *A. subandinus* and *C. haywardi* from the Indian Station were all sent for release in Madagascar, where potato tuber moth is the principal pest of tobacco (Appert *et al.*, 1969). The results of this project are unknown.

Cultures of the indigenous parasitoids of the potato tuber moth *O. parcus, D. mollipla* and *C. curvimaculatus* were also shipped from South Africa to the Indian Station in 1964–1965, for distribution to other countries attempting biological control of the pest (Greathead, 1971). *C. koehleri* is also known to have been distributed from South Africa to California (during 1969), Greece (1968 and 1969), Madagascar (1967 and 1968) (Greathead, 1971), Italy (1991) and Israel (1992 and 2003) (R. Kfir, unpublished).

Control Recommendations

Following the introduction and establishment of the parasitoids, it has been repeatedly demonstrated that little benefit is derived from applying insecticides against potato tuber moth in table potato fields in the Highveld region of South Africa. If farmers conserve and exploit the action of these beneficials and maintain good cultivation practices, there should be little need for insecticides. It is very important that initial populations of potato tuber moth be kept as low as possible by preventing an overlap from tuber moth infested volunteer plants with the new plantings. All volunteer potato plants should therefore be destroyed beforehand. It is also recommended that ridging be correctly implemented at least three times during the growing season, since this covers the tubers and prevents tuber moth larvae from reaching them. A layer of soil will also kill moths, which experience difficulties in reaching the surface, following their emergence from pupae (Annecke and Moran, 1982; Anon., 1992).

Economic Benefits

An indication of the economic benefits resulting from this biological control programme in South Africa are reflected in the records kept by the South African Potato Marketing Board in Pretoria. After the introduction and establishment of *C. koehleri* and *A. subandinus* in 1966, there was a sharp decline, from about 14% to less than 5% in the number of 15-kg bags of spoilt potatoes, either downgraded or rejected, at the municipal markets due to tuber moth damage (Anon., 1992). In 1978, about 1.8 million fewer bags of potatoes were rejected than would have normally been the case prior to the introduction of these parasitoids. This saving to the South African producer still recurs every year and represents a sum of money far in excess of the total investment into the biological control programme of the potato tuber moth in South Africa (Annecke and Moran, 1982).

In Zambia, the financial benefits from a similar biological control programme were calculated on the basis of an increased yield of 22% (Cruickshank and Ahmed, 1973), while in Zimbabwe, the outcome of a similar project was that, within 2 years of the first *C. koehleri* release, the potato tuber moth, which was once regarded as the most injurious pest of potatoes, has been relegated to an insect of minor economic importance (Mitchell, 1978).

Conclusion

The potato tuber moth, *P. operculella*, probably originated together with the potato in the Bolivian Andes in South America and is currently a widespread pest in all the warmer parts of the world. It occurs throughout Africa wherever potatoes are grown. Damage is caused by the larvae, which attack the leaves, stems and tubers.

In South Africa, prior to 1965, a granulosis virus and several insect predators were recorded as attacking potato tuber moth, but they did not sufficiently reduced the pest to prevent severe crop loss. Five indigenous parasitoids, of which *D. mollipla* was the most abundant species, also attacked the potato tuber moth. These parasitoids often parasitized up to 70% of tuber moth larvae, but despite this, losses from potato tuber moth were still too high.

Between 1965 and 1967, four parasitoid species of South American origin were introduced into South Africa and two parasitoids, *C. koehleri* and *A. subandinus*, become established. As these parasitoids spread they displaced the less effective indigenous parasitoid complex and increased parasitism up to 90–95%. Since their establishment, *C. koehleri* and *A. subandinus* have complemented each other, providing a seasonal succession each year, *A. subandinus* predominating at the start of the potato-growing season, as tuber moth numbers built-up, with *C. koehleri* assuming dominance towards the end of the season.

The introduction and establishment of *C. koehleri* and *A. subandinus* in South Africa marked a turning point in the fight against the pest. By 1978, about 1.8 million fewer bags were either downgraded or rejected at municipal markets due to tuber moth damage than would have normally been the case prior to the establishment of these beneficial parasitoids. These recurring savings for the South African producer are far in excess of the total investment made into the biological control programme for potato tuber moth in South Africa.

Both parasitoids were also successfully introduced from South Africa into Zambia and Zimbabwe, where they became established. In Zambia, the financial returns from this biological control programme gave an increased yield of 22%, while in Zimbabwe, where potato tuber moth was once regarded as the most injurious potato pest, it was relegated within a matter of 2 years to an insect of minor importance.

References

Annecke, D.P. and Mynhardt, M.J. (1974) On the identity of *Copidosoma koehleri* Blanchard, 1940 (Hymenoptera: Encyrtidae). *Journal of the Entomological Society of Southern Africa* 37, 31–33.

Annecke, D.P. and Moran, V.C. (1982) *Insects and Mites of Cultivated Plants in South Africa*. Butterworth, Pretoria, South Africa.

Anonymous (1992) Potato tuber moth: biological control of economic benefit. *Plant Protection News*. Bulletin of the Plant Protection Research Institute, Pretoria 29.

Appert, J., Betbeder-Matibet, M. and Ranaivosoa, H. (1969) Vingt années de lutte biologique à Madagascar. *Agronomie Tropicale* 24, 555–572.

Azidah, A.A., Fitton, M.G. and Quicke, D.L.J. (2000) Identification of the *Diadegma* species (Hymenoptera: Ichneumonidae, Campopleginae), attacking the diamond-back moth, *Plutella xylostella* (Lepidoptera: Plutellidae). *Bulletin of Entomological Research* 90, 375–389.

Bedford, E.C.G. (1954) Summary of the biological control of insect pests in South Africa 1895–1953. Department of Agriculture Technical Services, Pretoria, Special Report 2, 53–54.

Broodryk, S.W. (1967) Bioecological studies on the potato tuber moth *Gnorimoschema operculella* (Zeller) and its hymenopterous parasitoids in South Africa. PhD thesis, University of Pretoria, Pretoria, South Africa.

Broodryk, S.W. (1969a) The biology of *Chelonus* (*Microchelonus*) *curvimaculatus* Cameron (Hymenoptera; Braconidae). *Journal of the Entomological Society of Southern Africa* 32, 169–189.

Broodryk, S.W. (1969b) The biology of *Orgilus parcus* (Turner) (Hymenoptera; Braconidae). *Journal of the Entomological Society of Southern Africa* 32, 243–257.

Broodryk, S.W. (1971) The biology of *Diadegma stellenboschense* (Cameron) (Hymenoptera: Ichneumonidae) a parasitoid of potato tuber moth. *Journal of the Entomological Society of Southern Africa* 34, 413–423.

Broodryk, S.W. and Pretorius, L.M. (1974) Occurrence in South Africa of a granulosis virus attacking potato tuber moth, *Phthorimaea operculella* (Zeller). *Journal of the Entomological Society of Southern Africa* 37, 125–128.

Callan, E.M. (1974) Changing status of the parasites of potato tuber moth, *Phthorimaea operculella* (Lepidoptera: Gelechiidae) in Australia. *Entomophaga* 19, 97–101.

Cardona, C. and Oatman, E.R. (1975) Biology and physical ecology of *Apanteles subandinus* Blanchard (Hymenoptera: Braconidae), with notes on temperature responses of *Apanteles scutellaris* Muesebeck and its host, the potato tuberworm. *Hilgardia* 43, 1–51.

Charleston, D.S., Kfir, R., van Rensburg, N.J., Barnes, B.N., Hattingh, V., Conlong, D.E., Visser, D. and Prinsloo, G.J. (2003) Integrated pest management in South Africa. In: Maredia, K., Dakouo D. and Mota-Sanchez, D. (eds) *Integrated Pest Management in the Global Arena*. CAB International, Wallingford, UK (in press).

Cruickshank, S. and Ahmed, F. (1973) Biological control of the potato tuber moth *Phthorimaea operculella* (Zell.) (Lep.: Gelechiidae) in Zambia. *Technical Bulletin 16, Commonwealth Institute of Biological Control*, pp. 147–162.

Findlay, J.B.R. (1975) Biological and cultural control of the potato tuber moth in table potatoes. *Proceedings, First Congress of the Entomological Society of Southern Africa*, pp. 147–150.

Greathead, D.J. (1971) A review of biological control in the Ethiopian Region. *Technical Communication 5, Commonwealth Institute of Biological Control.* CAB International, Wallingford, UK.

Hirst, J.M. and Stedman, O.J. (1960) The epidemiology of *Phytophthora infestans*. I. Climate, ecoclimate and the phenology of disease outbreak. *Annals of Applied Biology* 48, 471–488.

Kfir, R. (1981) Fertility of the polyembryonic parasite *Copidosoma koehleri*, effect of humidities on life length and relative abundance as compared with that of *Apanteles subandinus* in potato tuber moth. *Annals of Applied Biology* 99, 225–230.

Kfir, R. (1989) Effect of pesticides on *Copidosoma koehleri* Blanchard (Hymenoptera: Encyrtidae), a parasite introduced into South Africa for the biological control of the potato tuber moth. *Journal of the Entomological Society of Southern Africa* 52, 180–181.

Kfir, R. (1990) Parasites of the spotted stalk borer, *Chilo partellus* (Lepidoptera: Pyralidae) in South Africa. *Entomophaga* 35, 403–410.

Kfir, R. (1995) Parasitoids of the African maize stem borer, *Busseola fusca* (Lepidoptera: Noctuidae), in South Africa. *Bulletin of Entomological Research* 85, 369–377.

Kfir, R. (1997) Parasitoids of diamondback moth, *Plutella xylostella* (L.) (Lepidoptera: Yponomeutidae), in South Africa: an annotated list. *Entomophaga* 42, 517–523.

Mitchell, B.L. (1978) The biological control of the potato tuber moth *Phthorimaea operculella* (Zeller) in Rhodesia. *Rhodesia Agricultural Journal* 75, 55–58.

Watmough, R.H., Broodryk, S.W. and Annecke, D.P. (1973) The establishment of two imported parasitoids of potato tuber moth (*Phthorimaea operculella*) in South Africa. *Entomophaga* 18, 237–249.

Whiteside, E.F. (1980) Biological control of the potato tuber moth *(Phthorimaea operculella*) in South Africa by two introduced parasitoids (*Copidosome koehlerii* and *Apanteles subandinus*). *Journal of the Entomological Society of Southern Africa* 43, 239–255.

Whiteside, E.F. (1981a) The potato tuber moth. *Farming in South Africa* G8, 1–6.

Whiteside, E.F. (1981b) The results of competition between two parasites of the potato tuber moth, *Phthorimaea operculella* (Zeller). *Journal of the Entomological Society of Southern Africa* 44, 359–365.

6

Biological Control of Whiteflies in Sub-Saharan Africa

James Legg,[1,2] Dan Gerling[3] and Peter Neuenschwander[4]
[1]*International Institute of Tropical Agriculture – Eastern and Southern Africa Regional Center, Namulonge, Kampala, Uganda;*
[2]*Natural Resources Institute, Greenwich University, Kent, UK;*
[3]*Department of Zoology, Tel Aviv University, Ramat Aviv, Israel;*
[4]*International Institute of Tropical Agriculture, Cotonou, Bénin*

Introduction

Whiteflies (Hemiptera, Aleyrodidae) occur as a diverse assemblage of about 1500 species throughout the tropics, and to a lesser degree in warmer temperate environments, roughly a quarter of the total number being reported from Africa (Mound and Halsey, 1978; Martin *et al.*, 2000). In Africa, only six species are outstanding as causing economic damage (Table 6.1). In this article, we will focus on three of these for which biological control approaches are either currently being developed to complement other control tactics (*Bemisia tabaci* (Gennadius)) or for which measures have already been developed and implemented (*Aleurodicus dispersus* Russell (Plate 16) and *Aleurothrixus floccosus* (Maskell) (Plate 15)).

Managing an Indigenous Annual Pest and Virus Vector – the Case of *Bemisia tabaci*

Background

B. tabaci belongs to a small group of insects that could reasonably be considered the most globally destructive pests. It occurs worldwide, out of doors throughout the tropical and subtropical regions of the world and in greenhouses within the temperate and cold countries. *B. tabaci* causes physical damage through sucking plant cells and contamination by honeydew on a wide range of annual crops (Gerling *et al.*, 2001) and transmits more than 70 disease-causing viruses. Recent molecular evidence suggests that the genus *Bemisia* originated in Africa (Campbell *et al.*, 1996). From there, it probably spread in recent times, throughout the tropics, subtropics and finally into temperate zone greenhouses. Two species of *Bemisia* occur frequently on agricul-

 Biological Control in IPM Systems in Africa (eds P. Neuenschwander, C. Borgemeister and J. Langewald)

Table 6.1. Principal whitefly species of agricultural importance in sub-Saharan Africa.

Species	Economic host plant	Location	Severity	Control	Comments
Aleurocanthus spiniferus	Citrus	S. Africa, Tanzania, Swaziland, Zanzibar, Nigeria, Uganda	++	Biological	
Aleurocanthus woglumi	Citrus	East and southern Africa	++	Biological	
Aleurodicus dispersus	Many hosts	West Africa, D.R. Congo,	+++[a]	Biological	Recent introduction to Africa
Aleurothrixus floccosus	Citrus	East and southern Africa	+++[a]	Biological	Recent introduction to Africa
Bemisia tabaci	Many spp. mainly annuals	Worldwide	++++	Mainly chemical	A major pest worldwide
Trialeurodes vaporariorum	Many annuals	Highlands of East and southern Africa	+++	Biological and chemical	A major pest worldwide

[a]Before biological control.

tural crops in Africa, *B. tabaci* and *Bemisia afer* Priesner and Hosny. Of the two, the latter has often been recorded on cassava, but is not considered a severe pest. *B. tabaci*, by contrast, was recorded as a pest of cotton in the Sudan as early as the 1920s and later caused extreme damage to that crop (Dittrich *et al.*, 1985). *B. tabaci* was first reported as the likely vector of cassava mosaic geminiviruses in the 1930s (Kufferath and Ghesquière, 1932) and in more recent years it has been demonstrated to be the vector of a range of leaf curl disease-inducing geminiviruses of tomatoes, particularly in eastern and southern Africa (Pietersen *et al.*, 2000). In sweet potato, *B. tabaci* transmits a closterovirus, *Sweet potato chlorotic stunt virus*, which interacts synergistically with the aphid-borne *Sweet potato feathery mottle virus*. Together they cause the 'sweet potato virus disease' (SPVD) that constitutes the most important disease constraint to sweet potato production in sub-Saharan Africa.

A *B. tabaci* biotype occurring on cassava throughout sub-Saharan Africa has been shown to be largely host-restricted (Legg, 1996). It exists sympatrically, but probably does not interbreed with a second polyphagous biotype that colonizes non-cassava crops such as cotton, okra and sweetpotato. More recent molecular evidence suggests that another cassava-colonizing genotype is associated with an epidemic of severe cassava mosaic disease in Uganda (Legg *et al.*, 2002). Major whitefly outbreaks in both the new world (Perring *et al.*, 1991) and more recently in the Indian subcontinent (J. Colvin *et al.*, unpublished results) have been associated with the spread of the highly fecund and polyphagous 'B' biotype of *B. tabaci*, also described as a separate species, *B. argentifolii* Bellows and Perring (Bellows *et al.*, 1994). Damage attributed to this 'B' biotype amounted to over US$500 million in the USA alone (Perring *et al.*, 1991), and major losses are presently being recorded from all areas of

North and South America to which this biotype had spread. The 'B' biotype has been recorded from Egypt and Sudan in North Africa and from South Africa (Pietersen *et al.*, 2000) but does not yet seem to have become established in the main tropical zones in the eastern, central and western parts of the continent. It clearly does, however, pose an important threat to crop production in these zones, and this threat is relevant to our discussion here, given the breadth of experience with biological control of the 'B' biotype in the USA and elsewhere.

B. tabaci *and cassava mosaic virus disease (CMD)*

CMD emerged as a major constraint to cassava production in the 1920s and 1930s and was subsequently recorded spreading throughout the cassava-growing zones of tropical sub-Saharan Africa. A recent review of the status of CMD in Africa estimated losses of fresh tuberous roots at between 12 and 23 million t (Thresh *et al.*, 1997). In the final decade of the 20th century, an unusually severe epidemic of CMD was recorded from north-central Uganda (Otim-Nape *et al.*, 1997), which subsequently spread to cover much of Uganda and expanded into neighbouring Kenya, Tanzania and Rwanda (Legg, 1999). Losses in many affected areas were almost total (Otim-Nape *et al.*, 1997), and in many situations farmers abandoned growing cassava altogether. Detailed characterization of the virus/viruses occurring in severely diseased plants in affected areas revealed the presence of a novel recombinant hybrid form of East African cassava mosaic virus (EACMV), referred to as the Uganda variant of EACMV (EACMV-Ug) (Deng *et al.*, 1997). This has now been shown to be consistently associated with the 'pandemic' of severe CMD, and has most recently been detected in severely CMD-diseased cassava plants in both the western portion of the Democratic Republic of Congo and the Republic of Congo (Neuenschwander *et al.*, 2002). An important feature of the CMD pandemic has been the elevated whitefly populations (Legg and Ogwal, 1998) combined with the occurrence of a pandemic-associated genotype (Legg *et al.*, 2002). In addition, however, there is preliminary evidence to show that *B. tabaci* is also unusually abundant on the virus resistant varieties that have been introduced to and multiplied in pandemic-affected areas of East Africa. Furthermore, these increased *B. tabaci* populations have, for the first time, begun to cause physical damage to cassava plants characterized by chlorosis, reduction in leaf size, sooty mould and leaf abscission.

Relation of B. tabaci *to other plant virus epidemics in Africa*

B. tabaci populations in East and Southern Africa are economically important vectors of geminiviruses affecting tomatoes and are becoming an increasingly severe constraint to tomato production (Pietersen *et al.*, 2000). However, there are few data describing the incidence of disease, the viruses associated with the epidemics or the scale of losses. Research elsewhere has shown that the geminiviruses affecting tomato are tremendously diverse, and that sources of infection are often weed species occurring in uncultivated grounds surrounding areas of tomato cultivation (J. Colvin *et al.*, 2002, unpublished results).

SPVD is most damaging in the Lake Victoria zone of East Africa, and occurs at incidences of up to 80% in parts of western Kenya, southern Uganda and northwestern Tanzania (Aritua *et al.*, 1998). The amount of inoculum of the viruses causing SPVD in surrounding fields and *B. tabaci* abundance are the principal factors determining the rate of spread of SPVD into initially SPVD-free fields (Aritua *et al.*, 1999).

The potential for biological control of **B. tabaci**

The principal approach used in the management of diseases caused by viruses transmitted by *B. tabaci* involves the development and deployment of host plant resistance. High levels of CMD resistance have been developed through classical breeding approaches based initially on inter-specific crosses between cultivated cassava and Ceara rubber (*Manihot glaziovii* Muell.-Arg.) (Jennings, 1957). The IITA has been the centre for this breeding work for more than 30 years, and the most recently developed germplasm combines a newly identified source of resistance obtained from Nigerian cassava landraces with the *M. glaziovii* source (Mahungu *et al.*, 1994). New clones bred in this way have the advantage of combining high levels of CMD resistance with farmer-preferred agronomic and postharvest qualities. The deployment of virus-resistant germplasm has been the frontline approach used in the effort to tackle the CMD pandemic (Legg, 1999) and considerable success has been realized with more than a quarter of the total cassava area now cropped with these new varieties in Uganda alone. However, the use of a single strategy in managing a disease is a high-risk approach, particularly in view of the known potential for geminiviruses to recombine. The probability of these recombination events is also enhanced where there is a high frequency of whitefly-mediated virus/virus and virus/plant contact events. This is clearly enhanced by increased whitefly abundance. Efforts are therefore under way to investigate approaches to preserve virus resistance, to minimize virus spread and to prevent the possibility of physical damage due to whitefly feeding by exploring whitefly control approaches. Two potential approaches for cassava, which is primarily a subsistence crop, are the development of resistance to whitefly infestations and the use of biological control. Since *B. tabaci* has been present in Africa for many years, and has a diverse natural enemy fauna, the most promising biological control approaches will involve conservation and augmentation of existing natural enemies and the introduction of supplementary natural enemies from other locations with similar environmental characteristics to those of the target areas for control (Naranjo, 2001). Conservation is particularly appropriate where the pest–natural enemy balance has been disrupted, commonly as a result of pesticide use. Various approaches to manipulating the cropping pattern, such as the use of intercrops or banker plants, may enhance natural control (Barbosa, 1998) and should be considered. Augmentation, which is a more resource-demanding approach, particularly where exotic natural enemies are used, is also a viable option, as demonstrated in the successful utilization of *B. tabaci* parasitoids in the southwestern USA (Goolsby *et al.*, 2000; Hoelmer *et al.*, 2000).

Biological control of B. tabaci *in Africa*

Distribution of natural enemies

Surveys of natural enemies of *B. tabaci* in Africa have covered only a small part of the continent, have often been very localized, and have only considered a few of the crop hosts on which *B. tabaci* causes damage, either directly or through virus transmission. Fishpool and Burban (1994) drew attention to the relative dearth of information on natural enemies of *B. tabaci* in Africa. The most recent attempt to collect a large quantity of material from a wide geographical range was the survey conducted during the diagnostic phase of the Whitefly IPM Project, under the umbrella of the Consultative Group for International Agricultural Research's (CGIAR) Systemwide Programme for IPM. Surveys in Africa yielded primarily parasitoids and a few predators from ten sub-Saharan African countries (Ghana, Bénin, Nigeria, Cameroon, Uganda, Kenya, Sudan, Tanzania, Malawi and Madagascar). Prior to this, area-specific studies had been conducted in Malawi, parts of Ethiopia (for cotton only), Zimbabwe, Kenya and, more recently, in Uganda. Hymenopterous wasps parasitizing whiteflies are known from four genera, the large majority being aphelinids from the genera *Encarsia* (34 species) and *Eretmocerus* (14 species) (Gerling *et al.*, 2001) (Plate 15). Sub-Saharan Africa differs markedly in its *Encarsia* species complex from North Africa, exemplified by Egypt. Typically, two species common to the Mediterranean and Mid-Eastern countries, *Encarsia lutea* (Masi) and *Encarsia inaron* (Walker), predominate in Egypt and occur as far south as the Sudan, with isolated occurrences of the former in Nigeria and South Africa (Table 6.2). Results from surveys in sub-Saharan Africa (Gerling, 1985; W. Dirorimwe *et al.*, Zimbabwe, 1999; G. Terefe *et al.*, Ethiopia, 1999, unpublished results) (Table 6.2) indicate that *Encarsia sophia* is the most frequently found species in all countries, with *Eretmocerus* nr. *mundus* often being equally abundant. The Sudan has both Mediterranean (*E. lutea* and *Encarsia mineoi* Viggiani) and Africo-tropical species (*E. sophia*) (D. Gerling, unpublished information). As a consequence of the dominance of *E. sophia* in sub-Saharan Africa, coupled with its apparent absence from North Africa, it was introduced into Egypt in 1998 in a bid to control *B. tabaci* populations there (Abd Rabou, personal communication). *Eretmocerus diversiciliatus* Silvestri was described by Silvestri in 1915 from Nigeria, and has since been recorded also in Egypt (Abd Rabou, 1998). *Eretmocerus* nr. *emiratus* Zolnerowich and Rose was collected in Ethiopia and released against *B. tabaci* in Texas (Goolsby *et al.*, 2000). The only introduction so far made into sub-Saharan Africa from outside the continent was *Encarsia luteola* (Howard) introduced from the USA through Israel by D. Gerling into Malawi in 1983–1985, but no information on its establishment exists.

Table 6.2. Records of *Bemisia* parasitoids in sub-Saharan Africa.

	Species							
	Encarsia						*Eretmocerus*	
Country	*sophia*	nr. *nigricephala*	*lutea*[a]	*mineoi*	*formosa*[b]	*hispida*	nr. *mundus* complex	*diversiciliatus*
Bénin	+						+	
Cameroon	+						+	
Ethiopia	+	+					+	
Ghana	+							
Kenya	+				+		+	
Malawi	+						+	
Nigeria		+	+	+				+
S. Africa			+		+	+		
Sudan	+		+	+			+	
Tanzania	+	+					+	
Uganda	+	+					+	
Zimbabwe	+		+				+	

[a]May be a species complex '*lutea* group' rather than one species.
[b]This typical parasitoid of the greenhouse whitefly has been found attacking *B. tabaci* in many countries including Kenya, but no record from *B. tabaci* is available for South Africa.

Gerling *et al.* (2001) listed 117 predator species attacking *B. tabaci* worldwide, of which many are known from Africa. However, there are only a few records of predators attacking *B. tabaci* in Africa. Predatory mites of the family Phytoseiidae occur widely on agricultural crops throughout Africa. In Sudan, species from the genera *Typhlodromus* and *Amblyseius* as well as *Euseius scutalis* (Athias-Henriot) were noted as predators of *B. tabaci* on cotton. Mites have also been recorded predating on *B. tabaci* in Ivory Coast and coastal Kenya and some other generalist predators were observed feeding on *B. tabaci* in Ivory Coast (Fishpool and Burban, 1994). These generalists included the coccinellid *Stethorus jejunus* Casey, the staphylinid *Holoborus pallidicornis* (Cameron) and the thrips *Scolothrips latipennis* Priesner. The coniopterygid, *Conwentzia africana* Meinander, is considered to be one of the most important predators of *B. tabaci* in East and southern Africa, and has been observed feeding directly on nymphs in Malawi and Kenya (D. Gerling, personal records) as well as in Uganda (Legg, 1995).

The impact of natural enemies on B. tabaci *populations in Africa*

Little information is available on the impact of either parasitoids or predators on *B. tabaci* in Africa. Legg (1995) demonstrated from a partial life table analysis for *B. tabaci* on cassava in Uganda that parasitoids accounted for 34% of the mortality in the fourth nymphal instar and the two dominant species were shown to be *E. sophia* and *Eretmocerus* sp. Provisional data suggest that under certain conditions, low *B. tabaci* populations may be sustained by high levels of parasitism, but more detailed studies assessing the effect of environmental conditions on tritrophic interactions will be required before more general conclusions can be drawn on what conditions favour parasitism. Moreover, success with the introduction of parasitoids against whitefly in other countries and the extensive fauna of *B. tabaci* parasitoids worldwide indicate the utility of examining the possibilities of introducing additional species of parasitoids.

For predators, there are to date no published studies that describe their effects in controlling *B. tabaci* in sub-Saharan Africa. There is therefore a clear need for a concerted effort to characterize the natural enemy fauna, examine pest/natural enemy/host interactions, and implement some of the approaches of conservation, augmentation and introduction so successfully deployed in managing outbreak populations of the 'B' biotype in the USA (Hoelmer *et al.*, 1999, 2000; Goolsby *et al.*, 2000).

Managing a Polyphagous Exotic Pest through Biological Control – the Case of the Spiralling Whitefly, *Alenrodicus dispersus*

Background

The spiralling whitefly, *A. dispersus*, is a polyphagous whitefly species probably of neotropical origin, which was first noticed when it invaded Hawaii in 1978. It has since been reported from Australia, India, the Pacific Region and South-East Asia, but also from the Canary Islands. In 1990, *A. dispersus* was reported from 38 genera in 27 plant families, on crops such as cocoa, guava, papaya and mango, as well as on many ornamentals. Following its accidental introduction into southwestern Nigeria around 1992, it was observed inflicting heavy damage on many food crops, including cassava, soybean, pigeon pea, citrus and others (Akinlosotu *et al.*, 1993).

Biological control

Following its successful control by the coccinelid *Nephaspis oculata* (Blatchley) (= *N. amnicola* Wingo) and the aphelinid *Encarsia ?haitiensis* Dozier on Hawaii and some Pacific islands and recently India (Waterhouse and Norris, 1989; Ramani *et al.*, 2002), attempts were made to introduce these natural enemies into Bénin, where *A. dispersus* appeared for the first time in 1993. By the time preparations had been made for their introduction, it was discovered that two parasitoid species, *E. ?haitiensis* and *Encarsia guadeloupae* Viggiani (Hymenoptera, Aphelinidae) had already colonized *A. dispersus* in Bénin. On regularly sampled guava trees, host populations fluctuated, but there was a general decline of 80% between 1993 and 1996, while parasitism rates increased and stabilized, as indicated by the number of pupae (Fig. 6.1). The impact of these parasitoids was also quantified in four surveys covering the entire country. *A. dispersus* was more abundant in towns and spread into farmland only later. At least initially, *E. ?haitiensis* was the more abundant parasitoid. Population densities of *A. dispersus*, the pro-

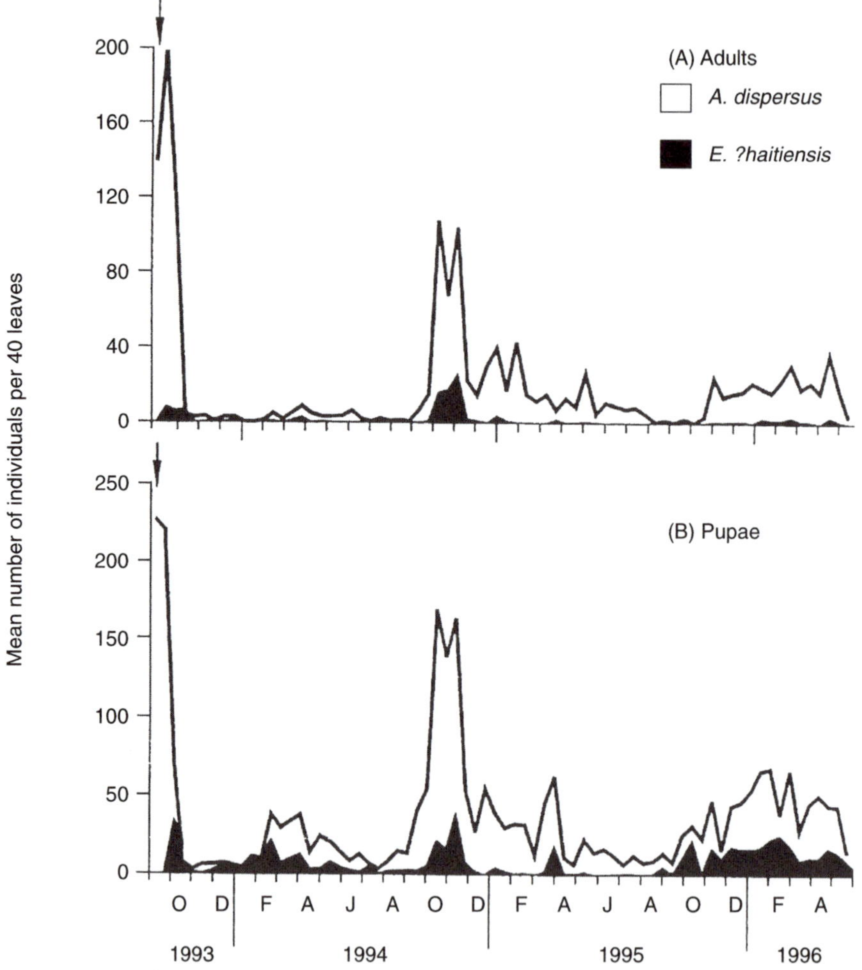

Fig. 6.1. Impact of the aphelinid parasitoid, *Encarsia ?haitiensis*, on populations of the spiralling whitefly, *Aleurodicus dispersus*, on guava trees in Bénin.

portion of infested trees and damage scores all declined significantly with increasing duration of parasitoid presence and these changes were attributed to the impact of biological control (d'Almeida *et al.*, 1998). *A. dispersus* has been recorded as spreading further in Nigeria and Bénin, and additionally to Togo, Ghana, Sierra Leone and Senegal in West Africa, and to D.R. Congo and possibly also Angola in Central Africa. In all countries in West Africa, *E. ?haitiensis* and sometimes also *E. guadeloupae* have been reported shortly after the establishment of the pest. Within West Africa, therefore, natural spread of these parasitoids seems able to follow that of the pest and ensure that any losses sustained are only temporary. The status in Central Africa remains to be clarified and it has been reported that on the Cape Verde Islands *A. dispersus* occurs without the parasitoids. IITA is therefore preparing to supply parasitoids to countries where they do not spread fortuitously.

Since these studies, *A. dispersus* populations in the entire region have stayed low, while another, rather similar, whitefly, *Paraleurodes minei* (Iaccarino), has sometimes become locally abundant on the same trees as *A. dispersus* (G. Goergen and A. Ajuonu, Cotonou, 2000, unpublished results).

Prior to the establishment of the two *Encarsia* species and their subsequent successful control of *A. dispersus*, the introduction of *N. oculata* had been considered as an additional natural enemy to control the newly introduced pest. Host specificity tests, following the FAO Code of Conduct for the Import and Release of Biological Control Agents, were planned for this predator in view of the increased concern over the safety of introduced natural enemies. Following confusion over the identity of the most appropriate predator species, these studies were eventually conducted with *N. bicolor* Gordon, which proved to be specific to aleyrodid prey, a feature that is believed to be common to other *Nephaspis* species (Lopez and Kairo, 2003). To date, however, no situation is currently known in Africa where *A. dispersus* cannot be completely brought under control by the two parasitoid species, thus no further introductions of predators are envisaged.

Managing an Exotic Pest of a Tree Crop with Biological Control – the Case of the Citrus Woolly Whitefly, *Aleurothrixus floccosus*

Background

The citrus woolly whitefly, *A. floccosus*, was first recorded from Jamaica and appears to be a native of tropical and subtropical America. It has a wide host range, having been reported from more than 50 plant species in 31 families from Hawaii alone (Paulson and Beardsley, 1986). Heaviest infestations are commonly found on *Citrus* spp., where nymphs secrete copious quantities of honeydew on which sooty moulds develop. Feeding and sooty mould build-up leads to reduction in leaf size and sometimes leaf loss in heavy infestations and these in turn result in reduced fruit yields. The first record from mainland USA was from the early part of the 20th century when damage was reported from eastern citrus groves. Spread across the USA was relatively slow, and the pest only arrived on the west coast in the 1960s (DeBach and Rose, 1976). The pest spread accidentally to Europe in the late 1960s and was recorded for the first time from Morocco in North Africa in 1973 (Abbassi, 1975). The pattern of spread of *A. floccosus* within the African continent was not well documented, but records were made of its occurrence in Egypt in the mid-1970s, Réunion in 1984, São Tome and Principe in the Gulf of Guinea in the same year, and informal reports were made from mainland West Africa in the mid-1980s. In East Africa, the Kenya Agricultural Research Institute (KARI) began working on the pest in the late 1980s and subsequent reports were made of its presence in a number of East and southern African countries, including Uganda, Tanzania, Zambia, Rwanda, Burundi and, finally, Malawi (in 1994). In Kenya it rapidly became the country's most important citrus production constraint.

Biological control

A diverse range of natural enemies obtained primarily from southern and central America was introduced to Europe and California in the 1970s (Onillon, 1972; DeBach and Rose, 1976). The two most successful species were *Cales noacki* Howard (Hymenoptera, Aphelinidae) obtained from Chile, and *Amitus spiniferus* (Brèthes) (Hymenoptera, Proctotrupidae) from Mexico (Miklasiewicz and Walker, 1990).

Surveys in Kenya, Malawi, Tanzania, Uganda and Zambia revealed a pattern of generally higher incidence and severity of infestation in Uganda, Kenya and Tanzania and at higher altitudes of up to 1800 m above sea level, while no woolly whiteflies were found on the Kenyan coast. It seemed that indigenous natural enemies, all coccinellids like *Cheilomenes* sp., *Chilocorus* sp., *Platynaspis* sp. and others, were inadequate to control this whitefly, and insecticide-based control measures applied in Kenya and Zambia were not successful either. It was concluded that the most appropriate management approach would be classical biological control, as it had been so successfully implemented in previous years in Europe and the USA.

In 1995, both *C. noacki* and *A. spiniferus* were therefore sent through the Switzerland-based quarantine station of the International Institute of Biological Control (IIBC) to the GTZ-financed Horticulture IPM Project, in collaboration with the Biological Control Programme of Uganda's National Agricultural Research Organization (NARO). Eventually, more than 1500 live *C. noacki* adults were received by NARO at their Namulonge station towards the end of 1996. *C. noacki* was mass-multiplied and more than 80,000 parasitoids were released on to infested citrus trees in the three most important citrus-growing districts of Uganda from 1996 to 1998. *C. noacki* became established and its populations increased with a concomitant decrease in the abundance and infestation levels of woolly whitefly (Fig. 6.2), but only limited parasitoid spread was recorded (Molo, 1998).

Following the success of the Ugandan biological control programme, *C. noacki* was imported into Kenya from Uganda and released in 1998 at eight sites. The parasitoids established at all release sites, and within 6 months had reduced woolly whitefly by more than 95% at two of the sites. A total of 400 and 1000 *C. noacki* adults from the insectary at the National Biological Control Unit, Muguga, were subsequently sent to Tanzania and Malawi, respectively. Although no data are currently available from either of these two countries, the classical biological programme for *A. floccosus* has had a major and sustainable impact on citrus production in Uganda and Kenya. Given the diversity of environments in these two countries, prospects seem good for the replication of these successes with *C. noacki* throughout the citrus growing zones of sub-Saharan Africa. In view of the successes realized in biocontrol of *A. floccosus* using *C. noacki*, there are currently no plans to introduce further natural enemies, although this may be reviewed in future, and the first candidate for such an introduction would be *A. spiniferus*.

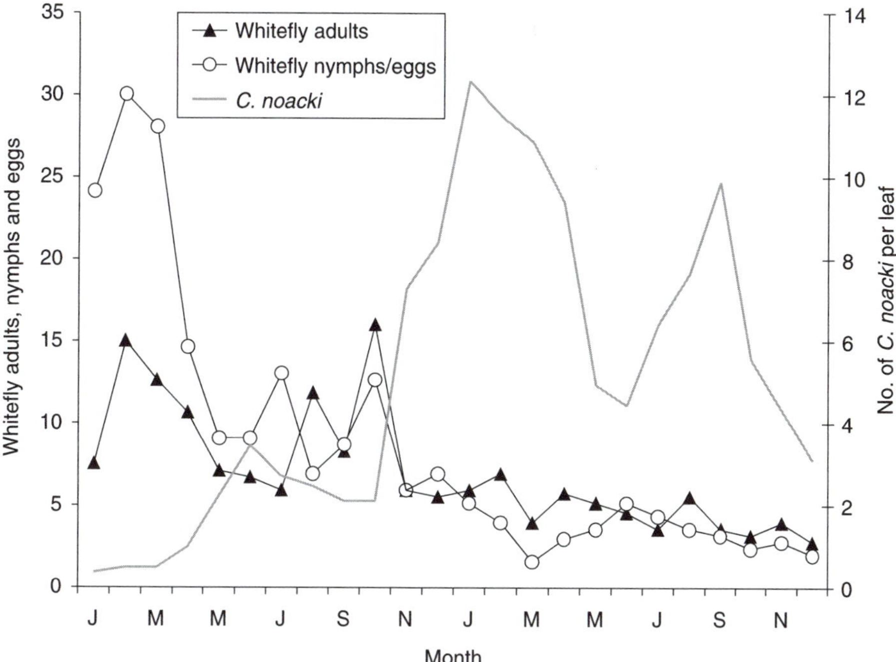

Fig. 6.2. Impact of introduced *Cales noacki* parasitoids on abundance of adult and immature stages of *Aleurothrixus floccosus* at Lwangosa, Uganda (from Molo, 1998).

Conclusions

The three whitefly pests discussed in this chapter highlight the diversity of challenges for the biocontrol researcher and practitioner posed by insects with contrasting continental origins and life histories. *B. tabaci* is orders of magnitude more important than either *A. dispersus* or *A. floccosus* in terms of its economic importance in sub-Saharan Africa and worldwide. Its genetic diversity, ability to transmit numerous disease-causing viruses, extreme polyphagy and environmental adaptability make this pest species a unique challenge to manage. By contrast, startling successes that have yielded major positive economic impact have been achieved with the classical biological control programmes of the two exotic introductions, *A. dispersus* and *A. floccosus*, albeit fortuitously through the joint pest/parasitoid introduction in the case of the former. These two whitefly species will undoubtedly spread to new areas, demanding continued attention in the future and the extension of the existing parasitoid-based biological control programmes.

These programmes are just two examples of successful whitefly management in Africa through biocontrol. Others, such as the highly successful biological control of *Aleurocanthus spiniferus* (Quaintance) in South Africa (van den Berg and Greenland, 1997) through the introduction of the exotic *Encarsia* nr. *smithi* (Silvestri), attest to the usefulness and economic sense of the classical biological control approach. There may also be a future requirement for the execution of similar programmes to address any further exotic whitefly introductions, as increased international travel increases the likelihood of their introduction.

In many respects, however, the most exciting future challenge will be the development of biologically based approaches to manage the most destructive of Africa's whitefly pests, *B. tabaci*. Great potential lies in the approaches of conservation and augmentation, although these require an intimate knowledge of the local extant fauna of natural enemies, and their interactions with pest and host plant. Whilst such measures may provide less immediate impact than the classical programmes used for managing *A. dispersus* and *A. floccosus*, the huge economic importance of *B. tabaci* and the viruses it transmits in sub-Saharan Africa demand that all such methods be incorporated as components within IPM control programmes. Only through the application of a multi-component IPM approach, including biological control, can we hope to achieve effective and sustainable management of existing and likely future threats to African agriculture posed by *B. tabaci*.

References

Abbassi, M. (1975) Présence au Maroc d'une nouvelle espèce d'aleurode *Aleurothrixus floccosus* Maskell (Homoptera, Aleurodidae). *Fruits* 30, 173–176.

Abd-Rabou, S. (1998) A revision of the parasitoids of whiteflies from Egypt. *Acta Phytopathologica et Entomologica Hungarica* 33, 193–215.

Akinlosotu, T.A., Jackai, L.E.N., Ntonifor, N.N., Hassan, A.T., Agyakwa, G.W., Odebiyi, J.A., Akingbohungbe, A.E. and Rossel, H.W. (1993) Spiralling whitefly, *Aleurodicus dispersus,* in Nigeria. *FAO Plant Protection Bulletin* 41, 127–129.

d'Almeida, Y.A., Lys, J.A., Neuenschwander, P. and Ajuonu, O. (1998) Impact of two accidentally introduced *Encarsia* species (Hymenoptera: Aphelinidae) and other biotic and abiotic factors on the spiralling whitefly *Aleurodicus dispersus* (Russell) (Homoptera: Aleyrodidae), in Bénin, West Africa. *Biocontrol Science and Technology* 8, 163–173.

Aritua, V., Adipala, E., Carey, E.E. and Gibson, R.W. (1998) The incidence of sweet potato virus disease and virus resistance of sweet potato grown in Uganda. *Annals of Applied Biology* 132, 399–411.

Aritua, V., Legg, J.P., Smit, N.E.J.M. and Gibson, R.W. (1999) Effect of local inoculum on the spread of sweet potato virus disease: limited infection of susceptible cultivars following widespread cultivation of a resistant sweet potato cultivar. *Plant Pathology* 48, 655–661.

Barbosa, P. (ed.) (1998) *Conservation Biological Control*. Academic Press, San Diego, California, USA.

Bellows, T.S., Perring, T.M., Gill, R.J. and Headrick, D.H. (1994) Description of a species of *Bemisia* (Homoptera: Aleyrodidae). *Annals of the Entomological Society of America* 87, 195–206.

Campbell, B.C., Steffen-Campbell, J.D. and Gill, R. (1996) Origin and radiation of whiteflies: an initial molecular phylogenetic assessment. In: Gerling, D. and Mayer, R.T. (eds) Bemisia *1995. Taxonomy, Biology, Damage, Control and Management.* Intercept, Andover, UK, pp. 29–51.

DeBach, P. and Rose, M. (1976) Biological control status of woolly whitefly. *California Agriculture* 30, 4–7.

Deng, D., Otim-Nape, G.W., Sangare, A., Ogwal, S., Beachy, R.N. and Fauquet, C.M. (1997) Presence of a new virus closely associated with cassava mosaic outbreak in Uganda. *African Journal of Root and Tuber Crops* 2, 23–28.

Dittrich, V., Hassan, S.O. and Ernst, G.H. (1985) Sudanese cotton and the whitefly: a case study of the emergence of a new primary pest. *Crop Protection* 4, 161–176.

Fishpool, L.D.C. and Burban, C. (1994) *Bemisia tabaci*: the whitefly vector of African cassava mosaic geminivirus. *Tropical Science* 34, 55–72.

Gerling, D. (1985) Parasitoids attacking *Bemisia tabaci* (Hom.: Aleyrodidae) in Eastern Africa. *Entomophaga* 30, 163–165.

Gerling, D., Alomar, O. and Arno, J. (2001) Biological control of *Bemisia tabaci* (s.l.) using predators and parasitoids. *Crop Protection* 20, 779–799.

Goolsby, J.A., Ciomperlik, M.A., Kirk, A.A., Jones, W.A., Legaspi, B.C., Legaspi, J.C., Ruiz, R.A., Vacek, D.C. and Wendel, L.E. (2000) Predictive and empirical evaluation for parasitoids of *Bemisia tabaci* (Biotype “B”), based on morphological and molecular systematics. In: Austin, A. and Dowton, M. (eds) *Hymenoptera: Evolution, Biodiversity, and Biological Control.* 4th International Hymenopterists Conference, 1999, Canberra. CSIRO, Collingwood, Victoria, Australia, pp. 347–358.

Hoelmer, K., Kirk, A.A. and Simmons, G. (1999) An overview of natural enemy explorations and evaluations for *Bemisia* in the U.S. *5ème Conférence Internationale sur les Ravageurs en Agriculture*. Montpellier, France, 3, 689–697.

Hoelmer, K., Roltsch, W., Simmonds, G., Gould, J., Pickett, C., Goolsby, J. and Andress, E. (2000) Reviewing a multi-agency biological control programme for *Bemisia argentifolii* in the southwest United States: establishing, conserving and augmenting new parasitoids and evaluating their impact. *Proceedings of the 11th International Congress of Entomology*, Fos do Iguassu, Brazil, 2, 999.

Jennings, D.L. (1957) Further studies in breeding cassava for virus resistance. *The East African Agricultural Journal* 22, 213–219.

Kufferath, H. and Ghesquière, J. (1932) La mosaïque du manioc. *Compte-Rendu de la Société de Biologie Belge* 109, 1146–1148.

Legg, J.P. (1995) The ecology of *Bemisia tabaci* (Gennadius) (Homoptera), vector of African cassava mosaic geminivirus in Uganda. Doctoral thesis, University of Reading, UK.

Legg, J.P. (1996) Host-associated strains within Ugandan populations of the whitefly *Bemisia tabaci* (Genn.), (Hom., Aleyrodidae). *Journal of Applied Entomology* 120, 523–527.

Legg, J.P. (1999) Emergence, spread and strategies for controlling the pandemic of cassava mosaic virus disease in east and central Africa. *Crop Protection* 18, 627–637.

Legg, J.P. and Ogwal, S. (1998) Changes in the incidence of African cassava mosaic geminivirus and the abundance of its whitefly vector along south–north transects in Uganda. *Journal of Applied Entomology* 122, 169–178.

Legg, J.P., French, R., Rogan, D., Okao-Okuja, G. and Brown, J.K. (2002) A distinct, invasive *Bemisia tabaci* (Gennadius) (Hemiptera: Sternorrhyncha: Aleyrodidae) genotype is associated with the epidemic of severe cassava mosaic virus disease in Uganda. *Molecular Ecology* 11, 1219–1229.

Lopez, V.F. and Kairo, M.T.K. (2003) Prey range of *Nephaspis bicolor* Gordon (Coleoptera: Coccinellidae), a potential biological control agent of *Aleurodicus dispersus* and other *Aleurodicus* spp. (Homoptera: Aleyrodidae). *International Journal of Pest Management* 49, 75–86.

Mahungu, N.M., Dixon, A.G.O. and Mkumbira, J. (1994) Breeding cassava for multiple resistance in Africa. *African Crop Science Journal* 2, 539–552.

Martin, J.H., Mifsud, D. and Rapisarda, C. (2000) The whiteflies (Hemiptera: Aleyrodidae) of Europe and the Mediterranean Basin. *Bulletin of Entomological Research* 90, 407–448.

Miklasiewicz, T.J. and Walker, G.P. (1990) Population dynamics and biological control of the woolly whitefly (Homoptera: Aleyrodidae) on citrus. *Environmental Entomology* 19, 1485–1490.

Molo, R. (1998) Impact of *Cales noacki* How. on citrus woolly whitefly in Uganda. Uganda NARO, Biological Control Unit Technical Report 3 (December 1997).

Mound, L.A. and Halsey, S.H. (1978) *Whitefly of the World. A Systematic Catalogue of the Aleyrodidae (Homoptera) with Host Plant and Natural Enemy Data*. John Wiley & Sons, Chichester, UK.

Naranjo, S.E. (2001) Conservation and evaluation of natural enemies in IPM systems for *Bemisia tabaci*. *Crop Protection* 20, 835–852.

Neuenschwander, P., Hughes, J. d'A., Ogbe, F., Ngatse, J.M. and Legg, J.P. (2002) Occurrence of the Uganda variant of *East African cassava mosaic virus* (EACMV-Ug) in western Democratic Republic of Congo and the Congo Republic defines the westernmost extent of the CMD pandemic in East/Central Africa. *Plant Pathology* 51, 385.

Onillon, J.C. (1972) Possibilités de régulation des populations d'*Aleurothrixus floccosus* Mask. (Homopt., Aleurodidae) sur agrumes par *Cales noacki* How. (Hymenopt., Aphelinidae). *Bulletin Organisation Européenne et Méditerranéenne pour la Protection des Plantes* 3, 17–26.

Otim-Nape, G.W., Bua, A., Thresh, J.M., Baguma, Y., Ogwal, S., Semakula, G.N., Acola, G., Byabakama, B. and Martin, A. (1997) *Cassava Mosaic Virus Disease in Uganda: the Current Pandemic and Approaches to Control*. Natural Resources Institute, Chatham, UK.

Paulson, G.S. and Beardsley, J.W. (1986) Development, oviposition and longevity of *Aleurothrixus floccosus* (Maskell) (Homoptera: Aleyrodidae). *Proceedings of the Hawaiian Entomological Society* 26, 97–99.

Perring, T.M., Cooper, A., Kazmer, D.J., Shields, C. and Shields, J. (1991) New strain of sweetpotato whitefly invades California vegetables. *California Agriculture* 45, 10–12.

Pietersen, G., Idris, A.M., Krüger, K. and Brown, J.K. (2000) Tomato curly stunt virus a new begomovirus within the TYLCV-Is cluster, causing a severe disease of tomato in South Africa. *Plant Disease* 84, 810.

Ramani, S., Poorani, J. and Bhumannavar, B.S. (2002) Spiralling whitefly, *Aleurodicus dispersus,* in India. *Biocontrol News and Information* 23, 55N–62N.

Thresh, J.M., Otim-Nape, G.W., Legg, J.P. and Fargette, D. (1997) African cassava mosaic disease: the magnitude of the problem? *African Journal of Root and Tuber Crops* 2, 13–19.

van den Berg, M.A. and Greenland, J. (1997) Classical biological control of the spiny blackfly, *Aleurocanthus spiniferus* (Hem.: Aleyrodidae) on citrus in southern Africa. *Entomophaga* 42, 459–465.

Waterhouse, D.F. and Norris, K.R. (1989) Aleurodicus dispersus *Russell. Biological Control: Pacific Prospects – Supplement 1*. ACIAR monograph 12. Australian Centre for International Agricultural Research, Canberra, Australia, pp. 13–22.

7

Biological Control of Homopteran Pests of Conifers in Africa

Roger K. Day,[1] Moses T.K. Kairo,[2] Yvonne J. Abraham,[3] Rami Kfir,[4] Sean T. Murphy,[3] K. Eston Mutitu,[5] and Clement Z. Chilima[6]

[1]CAB International, Africa Regional Centre, Nairobi, Kenya; [2]CAB International, Caribbean and Latin America Regional Centre, Curepe, Trinidad and Tobago, West Indies; [3]CAB International, UK Centre, Egham, Surrey, UK; [4]Agricultural Research Council – Plant Protection Research Institute, Division of Insect Ecology – Biological Control, Pretoria, South Africa; [5]Kenya Forestry Research Institute, Nairobi, Kenya; [6]Forestry Research Institute of Malawi, Zomba, Malawi

Introduction

Planted forests account for less than 4% of total forest area globally, but their importance is increasing along with concerns about the sustainability of harvesting natural forests. In Africa, planted forests covered about 5.7 million ha in 1995, around 5% of the world total. The African countries with the largest areas are South Africa, Morocco, Tunisia and Libya, together accounting for over half the total, but another 16 countries have 0.1 million ha or more. As in other tropical and subtropical regions, *Pinus* and *Eucalyptus* are the most planted genera, but *Acacia* is also important.

Monocultures of exotic trees are prone to pest attack, and increased global travel and trade increases the risk of introduction of exotic pests. This chapter describes biological control of three such pests of conifers in Africa, the cypress aphid, *Cinara cupressivora* Watson and Voegtlin (Homoptera, Aphididae), the woolly adelgid, *Pineus boerneri* Annand (Homoptera, Adelgidae), and the black pine aphid, *Cinara cronartii* Tissot and Pepper (Homoptera, Aphididae).

Plantation forestry has been considered a particularly appropriate target system for 'classical' or 'introduction' biological control for a number of reasons. First, the value increment per year is often relatively low, making regular interventions such as chemical control economically unviable, while biological control also offers the prospect of a permanent solution. Second, the long rotation time means that changing tree species or provenance is expensive and/or slow (although there are a number of examples of this approach in Africa).

(eds P. Neuenschwander, C. Borgemeister and J. Langewald)

Third, it is reasoned that the environmental stability of a forest is likely to be more conducive to biological control, although the evidence suggests that this applies more to the establishment of agents than to the magnitude of their impact on target pests (Hall *et al.*, 1980).

Murphy and Day (1998) re-analysed the data from Greathead and Greathead (1992) focusing on biological control of tree pests. At that time, 43 agents had been introduced against 17 pest species, with 15 out of 35 projects judged to be successful. Greathead and Greathead (1992) had shown that Homoptera is the order most frequently and most successfully targeted with introduced natural enemies.

Cinara cupressivora

The cypress aphid, *C. cupressivora,* was first reported in Malawi in 1986 but, within a relatively short period, spread to many countries in eastern and southern Africa (Mills, 1990; FAO, 1991). Initially the pest was identified as *Cinara cupressi* (Buckton). However, recent studies have shown that aphids previously identified as *C. cupressi* actually belong to a species complex (Watson *et al.*, 1999). These authors conducted morphometric analyses of material from different regions around the world, studied host plant range, parasitoid records and host associations and revealed that the species causing problems in Africa was *C. cupressivora*. Watson *et al.* (1999) also hypothesized that the native range of this species is most likely the region from eastern Greece to just south of the Caspian Sea. *C. cupressivora* is now widely distributed throughout eastern, central and southern Africa, the margins of western Europe, countries bordering the Mediterranean Sea, the Middle East, Yemen, Mauritius and Colombia. The taxonomic complexity of the group indicates that caution is required when interpreting earlier literature, particularly with respect to western Europe and the Mediterranean region.

In Africa, *C. cupressivora* develops throughout the year producing both apterous and alate populations largely in response to changes in density and host plant quality. Previous studies have elucidated the basic developmental and reproductive biology of the aphid (Kairo and Murphy, 1999a). The aphid is highly aggregative and exploits a wide range of feeding sites ranging from green branches to woody stems. Damage is characterized by dieback, and heavy infestations can cause the death of mature trees. In Africa, *C. cupressivora* attacks a wide range of Cupressaceae, including indigenous species such as Mulanje cedar, *Widdringtonia cupressoides* (L.) Endl. (the national tree of Malawi) and *Juniperus procera* Hochst. ex Endl., an important tree in many water catchment areas in Kenya. In most countries, however, the greatest damage is on exotic plantation or ornamental species, particularly the Mexican cypress, *Cupressus lusitanica* Mill, an important plantation tree grown in large monocultures.

By 1991 it was estimated that the aphid had killed US$41 million worth of trees in Africa, and was causing an annual loss of growth increment worth US$13.5 million (Murphy, 1996). Orondo and Day (1994) reported 12% mortality to an old seed stand of *C. lusitanica* in Kenya, but also noted that some severely damaged trees were recovering.

A range of options for managing *C. cupressivora* was initially considered that included chemical control, the use of resistant varieties and silvicultural techniques. However, it was clear from the outset that biological control was the most promising option (Mills, 1990). Discussions in several regional workshops held during the early 1990s arrived at the same conclusions (FAO, 1991).

A biological control programme commenced in 1991, spearheaded by CAB *International* in collaboration with several national and regional agencies. At that time the pest's native range was not certain, and there was uncertainty with regard to the taxonomic status of two closely related North American species. Against this background, exploratory surveys covered a broad geographical range including Mexico, USA, Europe, North Africa and Pakistan. Aphid specimens were also collected for more detailed biogeographical studies. Exploratory surveys focused on parasitoids since these were more likely to have narrow host ranges and thus a greater chance of satisfying requirements for safety with respect to non-target impact (Murphy *et al.*, 1994). Subsequent to clarification of the taxonomy of the group, additional surveys were undertaken in Syria.

Parasitoids were collected from several species in the sub-genus *Cinara Cupressobium* feeding mainly on *Cupressaceae* including, *Cinara fresai* Blanchard, *Cinara juniperi* de Geer and material that is now known to belong to the *C. cupressi* complex. Four species of parasitoids, all in the genus *Pauesia* (Hymenoptera, Braconidae) were assessed. Of these, *Pauesia juniperorum* (Starý) (Plate 17) showed the greatest promise, and it was introduced into Africa following quarantine assessment and screening at CAB *International*, UK. Like other aphidiines, *P. juniperorum* is a solitary endoparasitoid. It attacks all stages of *C. cupressivora,* but development is best in older hosts (>9 days old) (Kairo and Murphy, 1999b). Several other *Pauesia* spp. collected on *Cinara* spp. on pine were also assessed, but none parasitised *C. cupressivora.*

Pauesia juniperorum was first described by Starý (1960) from specimens reared out of *C. juniperi.* The species is widely distributed in mountain and sub-mountain habitats of Europe, mainly on *Juniperus* spp. and *Cupressus macrocarpa* Hartw. ex Gordon. Field host records include, *C. juniperi, C. fresai, C. ?fresai* and *C. cupressivora.* Populations of *P. juniperorum* were collected from *C. cupressivora* in England and *C. fresai* in France and England. While the parasitoid was recovered from *C. cupressivora*, it was more commonly found attacking *C. fresai.* Based on more recent knowledge on the distribution of *C. cupressivora*, it seems likely that the preferred host is *C. fresai* and the association with *C. cupressivora* is recent.

Prior to the introduction of *P. juniperorum* into Africa, a dossier was prepared to support applications to the national regulatory authorities, along the lines prescribed by the Code of Conduct for the Import and Release of Exotic Biological Control Agents (FAO, 1996), although at that time the code had not yet been published. The first shipments were made to Kenya in 1993, and the import permit was granted under the conditions that the insects be held in quarantine, and host-specificity tests conducted to demonstrate that it would not attack coccinellids, predatory mites or other local conifer aphids. Permission for field releases was granted in 1994 following presentation of a report on quarantine and host-specificity testing, during which the insect was cultured for nine generations. In Malawi and Uganda, permission for direct releases was granted on the basis of the same dossier. The first releases were made in Malawi and Kenya in 1994, and in Uganda in 1995. Repeated releases of small numbers (<500 wasps) were made over a period of 1–2 years in all three countries. Following more than a year of releases in Malawi, there was no evidence of establishment, but initial results in field cages were encouraging (Chilima *et al.*, 1995). Further releases were made in 1995, and the following year establishment was confirmed on both *C. lusitanica* and *W. cupressoides*. Establishment was slower in Kenya and Uganda. The last releases were made in late 1996 with no sign of establishment, but in early 1999 the parasitoid was found in both countries, over 2 years after it had last been seen in the field.

P. juniperorum is now widespread in Malawi and Kenya, and impact assessment has commenced in Kenya. The severity of damage by the aphid has declined in both countries, but this cannot definitely be attributed solely to the action of *P. juniperorum,* as some reduction in the aphid population had been noted in Kenya even before the agent was released.

Several studies have shown potential for the use of plant resistance and/or tolerance as part of an IPM strategy for cypress aphid. In a study on the variation and inheritance of resistance in *C. lusitanica* against *C. cupressivora*, Kamunya *et al.* (1997) found a marked variation in aphid survival both between and within host tree families. They demonstrated an individual-tree narrow-sense heritability of 0.76 ± 0.30, indicating strong additive genetic control and suggesting that potential exists for selection and breeding for resistance. These conclusions have been corroborated by other studies on *C. lusitanica* in Kenya and Tanzania (Mugasha *et al.*, 1997; Kamunya *et al.*, 1999).

Chemical control has been used, but in most circumstances it is prohibitively expensive and logistically difficult in plantations. Ornamental trees and hedges have been sprayed in towns; for example; in Mauritius methamidophos has been applied with motorized mistblowers.

Pineus boerneri

Pineus spp. are native to the temperate zones of the northern hemisphere. All species feed on conifers and produce a characteristic white, woolly covering. The most acute pest problems have been caused by the introduction of two species, *Pineus pini* (Macquart), which is indigenous to Europe, and *P. boerneri* into countries of the southern hemisphere. *P. boerneri* has been reported as a pest of pines in East, Central and southern Africa. It was described from *Pinus radiata* D. Don in California (Annand, 1928), but possibly originates in East Asia, where it has been recorded under the name *Pineus laevis* (Maskell) (McClure, 1984). Their complex polymorphic life cycles, a shortage of distinctive taxonomic features and the use of various synonyms have caused great confusion over the true identities and origins of these two species. For example, *P. boerneri* has been recorded under the name of *P. laevis* Gmelin (Australia, New Zealand and Hawaii), *Pineus havrylenkoi* Blanchard (South America) and

P. pini (East Africa). A study by Blackman *et al.* (1995) clarified the taxonomic status of pest *Pineus* spp. They showed that the African species, first reported as *P. pini* from Kenya and Zimbabwe in 1968 (Mills, 1990), shows closer affinity to *P. boerneri*. These findings substantiate the suspicions first raised by Barnes *et al.* (1976) that *P. boerneri* was probably accidentally introduced into Zimbabwe on *Pinus taeda* L. scions from Australia in 1962.

Heavily infested trees develop yellowing needles prior to needle drop, shoot death and dieback of growing tips. Infested needles are reduced in length, and this can lead to a loss of up to 50% of growth increment under warm dry conditions (Mailu *et al.*, 1978). Young and stressed trees are more prone to infestation and, under some circumstances, heavy infestations can lead to tree death. The impact of *P. boerneri* on wood production has been investigated in several countries. Up to 20% tree mortality was reported from *Pinus* spp. research plots in Kenya (Odera, 1974), and studies in Tanzania found that the shoots and stems of infested *Pinus patula* Schiede ex Schltdl. & Cham. seedlings lost 20.9% of their dry weight after 24 weeks (Madoffe and Austara, 1990). Pine cones are also damaged; in one study 31.8% of *Pinus pinaster* Aiton cones were infested and the seed yield from affected cones was reduced by 71.7% (Zwolinski *et al.*, 1989).

Classical biological control programmes have been implemented against *Pineus* spp. in a number of countries worldwide. Research carried out for these programmes has identified a range of insect predators of *Pineus* spp. but there are no known parasitoids. Several of these predators have been introduced as biological control agents. The most successfully utilized have been the specialized *Leucopis* spp. (Diptera, Chamaemyiidae) which have been credited with the control of outbreaks of *Pineus* spp. in Hawaii, New Zealand and Chile. During the 1970s, various predators (Table 7.1) were introduced into Kenya and Tanzania to control *P. boerneri*. These included *Scymnus* spp. (Coleoptera, Coccinellidae) and *Leucopis* spp., but reports indicate that all failed to establish (EAAFRO, 1970–1976; KARI, 1977–1980). In 1975, *Tetraphleps raoi* Ghauri (Hemiptera, Anthocoridae) was introduced to Kenya (Mailu *et al.*, 1980) and Tanzania (FAO, 1991) and has established in both countries.

In 1991, further biological control activities against *P. boerneri* were initiated with the aim of redistributing *T. raoi* and/or introducing *Leucopis tapiae* Blanchard. However, before any introductions of *Leucopis tapiae* could be made, it was found attacking *P. boerneri* in Malawi. It appears to have established either as a result of the previous introductions in East Africa in the 1970s, or by accidental introduction on imported pines. As there is no evidence for its presence in Kenya where it had been released, the latter explanation is perhaps more likely. *T. raoi* was introduced to Uganda (1996–1997) and Zambia (1996) from Kenya; in both cases direct releases were permitted on the basis of a dossier.

Predator exclusion studies conducted by Mailu *et al.* (1980) on pest populations of *P. boerneri* in Kenya concluded that eight species of indigenous predators (mainly coccinellids) killed about 12% of *P. boerneri* in the field. Unfortunately, any additional mortality due to *T. raoi* was not established. A later field experiment in Kenya failed to demonstrate the impact of *T. raoi*, perhaps because high

Table 7.1. Predators introduced into Kenya and Tanzania in the 1970s as potential biological control agents of *Pineus boerneri* Annand.

Predator species	Origin	Imported	Released	Established
Chamaemyiidae				
Leucopis argenticollis Zetterstedt	India	Kenya 1975	No	No
Leucopis manii Tanasijtshuk	India	Kenya 1975 and 1977	No	No
		Tanzania 1970s	?	?
Leucopis nigraluna McAlpine	Pakistan	Kenya 1974–1977	Yes	No
		Tanzania 1970s	?	?
Leucopis tapiae Blanchard	Europe	Kenya 1972	Yes	? No
		Tanzania 1970s	?	?
Leucopis spp.	Austria	Kenya 1971	Yes	? No
	India	Kenya 1978–1979	?	?
Coccinellidae				
Scymnus suturalis Thunberg	Austria/ Germany	Kenya 1971–1972	Yes	?
Scymnus nigrinus Kugelann	Austria	Kenya 1971–1972	No	No
Anthocoridae				
Tetraphleps raoi Ghauri	Pakistan	Kenya 1975	Yes	Yes
		Tanzania 1975	Yes	Yes

?, unknown.
Sources: EAAFRO (1970–1976); KARI (1977–1980); Mills (1990); Blackman *et al.* (1995); Greathead (1995).

levels of between-tree variability in *Pineus* populations caused experimental difficulties. The *T. raoi* released in Uganda are thought to have established, but there is no information on the outcome of the releases in Zambia.

There are two principal options for control of *Pineus* spp. in Africa apart from biological control: the application of insecticides and silvicultural techniques. In Kenya, experimental control of *P. boerneri* was achieved with both 0.075% benzene hexachloride and 0.05% dimethoate (FAO, 1991), and in Tanzania Thiodan (endosulfan) and Teepol (a detergent) 1% solution, and propoxur (Baygon E.C.) were effective (FAO, 1991). However, aside from being environmentally undesirable, chemical control of *P. boerneri* is costly and not feasible on a large scale.

Silvicultural options centre on the replacement of existing trees with different species or provenances of pines that are less susceptible to attack. However, re-planting is a long-term and possibly expensive solution that may incur hidden costs in reduced yield or quality. For example, *P. radiata* has been more or less abandoned in East Africa due to needle blight caused by *Dothistroma pini* Rostrup (*Mycosphaerellaceae*) and has been largely replaced with *P. patula*, a species less susceptible to *P. boerneri* (Barnes *et al.*, 1976), but inferior in fibre quality. Other measures include devoting more attention to the selection of suitable planting sites and to avoid placing trees under stress, a factor that can make them more vulnerable to attack.

Cinara cronartii

The black pine aphid, *C. cronartii*, is widely distributed across eastern North America, from Quebec in the north, southwards to Florida and then westwards to Texas and Arkansas (Pepper and Tissot, 1973). Its distribution closely follows that of fusiform rust disease caused by the fungus *Cronartium fusiforme* Hedge & N.R. Hunt ex Cummins (*Cronartiaceae*), a destructive disease that causes stem cankers on pine trees. The black pine aphid is not considered to be a pest in North America and lives almost exclusively in the cankers formed by this disease where it is guarded by various ant species (Pepper and Tissot, 1973).

The black pine aphid was first recorded in South Africa during 1974 (van Rensburg, 1979). Populations peak in winter (May–August), and following its appearance, it occurred annually in large numbers in pine plantations in the summer rainfall region of South Africa and Swaziland. In the summer of 1979, drought-stressed pine plantations of *P. taeda, Pinus elliottii* Engelm. (originally from the USA) and *P. patula* (from Mexico), the principal commercial timber species in the country, suffered severe damage from heavy aphid infestations as the tops of trees died back (van Rensburg, 1981). Since the costs of chemical control by aerial spraying in pine plantations are prohibitive, a biological control programme was initiated. Natural enemies of *C. cronartii* were absent in South Africa (van Rensburg, 1981), whereas at least one aphidiid parasitoid was known to attack it in the USA (Pepper and Tissot, 1973).

A survey conducted in the eastern USA in 1983 revealed a *Pauesia* sp., subsequently described as *Pauesia cinaravora* Marsh, (1991), parasitising *C. cronartii* colonies. Its biology was studied by Kirsten and Kfir (1991). A laboratory culture of this parasitoid was established in Athens, Georgia, USA, and five consignments of newly mummified aphids were shipped to South Africa (Kfir *et al.*, 1985, 2003). To prevent the possible introduction of fusiform rust into South Africa, rust-free logs were used in the aphid and parasitoid rearing facility in Athens. Before dispatching consignments to South Africa, mummies were surface-sterilized by dipping them in a 1% sodium hypochloride solution to destroy any fungal spores (Kfir *et al.*, 2003).

An import and release permit was granted on the basis of documentation submitted to a committee of experts, similar to the procedure in the present FAO Code of Conduct. The authorities required only one quarantine generation and no host-specificity tests since it was known from the literature that *Pauesia* spp. are specific to lachnids, and there are no lachnids indigenous to South Africa.

Following quarantine, logs bearing mummified aphids were transported to release sites in various pine plantations where they were suspended about 2m above the ground in pine trees heavily infested with *C. cronartii* (Kfir *et al.*, 1985) (Plate 18). Additional releases used open-sided shelters specially erected in two pine plantations in Mpumalanga Province. The parasitoids were released directly on to heavily infested branches that had been sawn off pine trees and piled up in the shelters. There, parasitoids reproduced in large numbers before dispersing into the adjacent plantations (Kfir *et al.*, 2003).

During the winter months of 1983, a total of 210,000 parasitoids were released at eight sites in commercial pine plantations in Mpumalanga, and 16,000 were released at one site in the Kwazulu-Natal midlands. Three releases, totalling 6000 parasitoids, were also made in June 1983 in one isolated pine stand (about 1 ha in size) at the Plant Quarantine Research Station at Buffelspoort near Rustenburg. This site was approximately 250 km from the nearest commercial pine plantation, and was therefore used for intensive observations following parasitoid release (Kfir *et al.*, 1985).

In 1983, the parasitoid became established in the isolated pine stand at Buffelspoort and brought the aphid population under control (Kfir and Kirsten, 1991). Two months after the initial release, the parasitoids dispersed to another pine stand about 1 km distant that had served as a control plot (Kfir *et al.*, 1985).

In the commercial pine plantations, living aphids had completely disappeared from some trees within 4 weeks of parasitoid release, leaving only mummies. Eight weeks after release, mummies were recorded up to 500 m away from the release points, indicating successful parasitoid dispersal (Kfir *et al.*, 1985). Towards the end of winter in August 1983, it became apparent that *P. cinaravora* had successfully established in all release sites in Mpumalanga and Kwazulu-Natal Provinces.

The question whether or not the parasitoid would be able to survive the summer in South Africa was of great concern to the project. In the USA, *C. cronartii* is commonly found in the fusiform rust galls during the summer, whereas in South Africa, where this disease does not occur, the black pine aphid becomes extremely rare during the rainy season and occurs singly under the bark or subterraneously on the roots of host trees (van Rensburg, 1979). However, in April 1984, dissected aphids collected at Tweefontein plantation (Mpumalanga) were found to contain parasitoid larvae. Later dissections of field collected aphids from numerous commercial pine plantations in Mpumalanga and Kwazulu-Natal as well as from the Northern province and Swaziland revealed >80% parasitism. By 1984, maximum dispersal was up to 154 km from the nearest release site (Kfir *et al.*, 2003). By the end of July 1984, *P. cinaravora* had extended its range to the northern part of the Eastern Cape and southern parts of the Northern Province and Swaziland, all of which are in the summer rainfall region (Kfir *et al.*, 2003), and had invaded all sites.

Since the introduction of *P. cinaravora* into South Africa, substantial reduction in *C. cronartii* populations has occurred throughout all pine plantations in the summer rainfall regions of the country. Surveys to determine levels of infestation by *C. cronartii* were conducted both before and after the introduction of parasitoids. The percentage infested trees declined drastically from about 99% to 2% following the release. Infestation levels per tree also decreased markedly (Kfir *et al.*, 2003). In addition to this sharp decline in pest populations, no treetop die-off has been reported since the establishment of *P. cinaravora*. A few minor outbreaks of black pine aphid occurred annually until 1994, but were confined to relatively small areas of 1–2 ha in size. Such outbreaks were normally of short duration (1–2 months) and collapsed as parasitoid numbers increased. No outbreaks of *C. cronartii* have been reported since 1994, indicating that *P. cinaravora* has successfully controlled the black pine aphid throughout the summer rainfall region of South Africa and Swaziland. The pest is now of no economic importance and the South African Forestry Council, which funded the work, concluded that the savings in any one year far exceeded the total cost of the project (Kfir *et al.*, 2003).

Discussion and Conclusion

A homopteran pest in a perennial habitat is statistically one of the most successful scenarios for classical biological control. Of the cases described here, one was a total success, while in the other two cases agents were established, but success to date has been partial at best, although impact has not been quantified. Impact assessment is probably more difficult in partial successes and, in the case of the cypress aphid, the project ended some time before the agent had even established. If *P. cinaravora* had been slower to establish, or the impact on the pest less dramatic, would further resources have been spent on searching for additional agents or on quantifying the partial success? Similarly, if funding for further work on cypress aphid was to become available, would it be spent on impact assessment of *P. juniperorum*, or on evaluation and introduction of the potentially more effective, but unidentified, *Pauesia* species found in Syria in 1997? More well-documented impact studies are required in biological control in Africa, but with limited resources, improving control is usually the first priority.

The establishment of *P. juniperorum* was slow in comparison with that of *P. cinaravora*, and its impact much less obvious. A major difference between the two situations is the number of insects released; for *P. juniperorum* it was in the order of thousands over a period of 2 years, while over 200,000 *P. cinaravora* were released over a much shorter period of time. This may account for the success of initial establishment of *P. cinaravora*, but it is not clear why *P. juniperorum* required over 2 years to build up to detectable levels in Kenya and Uganda. The possibility that *C. cupressivora* is not the primary host of *P. juniperorum* may also have contributed to the slow establishment. A genetic comparison between the *P. juniperorum* populations now established in Africa and those in Europe could be illuminating.

The importance of taxonomy in biological control is well illustrated by the examples in this chapter. *C. cronartii* was identified soon after it was found in South Africa and its area of origin was well known, so surveys for agents could be undertaken without delay in a well-defined region. In contrast, *C. cupressivora* was initially thought to be *C. cupressi*, and only after extensive taxonomic research were its correct identity and probable area of origin determined (Watson *et al.*, 1999). This meant that the search for biological control agents extended over many regions, reducing the likelihood that an effective agent would be found. While *P. juniperorum* has now established in three countries in Africa, it is likely that its preferred host is *C. fresai* or *C. juniperi*. Had the taxonomy of *C. cupressivora* been known at the outset, the *Pauesia* species in Syria would have been found much earlier, and a different course of events might have occurred. The taxonomic confusion surrounding the identity of the woolly adelgid in Africa did not have such a marked effect on the progress of the biological control programme as candidate biological control agents were already known from elsewhere, and the pest was not of such high priority.

A key point in any biological control programme is the decision whether or not to import an agent, and if so, under what conditions. Historically, the decision focused on whether the agent was likely to be successful, and thus often large numbers of species were released against a single pest. At least eight species of *Pineus* predators were introduced into Kenya and Tanzania in the 1970s. Since that time, the potential negative impact of introduced agents has become an issue of much greater concern, and in 1996 the Code of Conduct for the Import and Release of Exotic Biological Control Agents was published (FAO, 1996), stipulating consideration of possible non-target effects.

While the general principles of the Code of Conduct were followed in the later introductions described in this chapter, it is doubtful that the large numbers of agents for *Pineus* could now be introduced, even if it was thought to be biologically desirable. The cost of preparing the necessary dossiers necessitates greater selectivity before applications for the importation of agents are made. Nevertheless, considerable differences were experienced in the conditions attached to importation of the agents used in the biological control programmes reported here. In South Africa, a single quarantine generation was required for *P. cinaravora*, primarily to eliminate potential contaminant organisms. In Kenya, with similar evidence available as for *P. cinaravora,* importation of *P. juniperorum* had much stricter conditions attached, delaying field releases by nearly a year. In contrast, Uganda and Malawi allowed direct releases on the basis that the insects originated in a quarantine facility in Europe. Thus adoption of the Code of Conduct might speed introductions of biological control agents in some cases, and slow them in others, but overall the application of more harmonized procedures should be beneficial.

In none of the three cases described here have other control methods played a significant role, so there were no issues of integrating biological control with other control tactics. For both the cypress aphid and woolly adelgid, it is clear that different species and provenances of host trees vary in susceptibility to the pests. In the long term, these approaches may provide an important complement to biological control.

References

Annand, P.N. (1928) *A Contribution Toward a Monograph of the Adelgidae (Phylloxeridae) of North America.* Stanford University Press, California, USA.

Barnes, R.D., Jarvis, R.F., Schweppenhauser, M.A. and Mullin, L.J. (1976) Introduction, spread and control of the pine woolly aphid, *Pineus pini* (L.), in Rhodesia. *South African Forestry Journal* 96, 1–11.

Blackman, R.L., Watson, G.W. and Ready, P.D. (1995) The identity of the African pine woolly aphid: a multidisciplinary approach. *OEPP/EPPO Bulletin* 25, 337–341.

Chilima, C.Z., Day, R.K. and Meke, G. (1995) *The 1994 Release and Establishment of* Pauesia juniperorum, *a Natural Biological Control Agent for Cupress Aphids in Malawi.* Forestry Research Institute of Malawi Report No. 950007, Zomba, Malawi.

EAAFRO (1970–1976) *Records of Research*. Annual Reports 1970–1976. East African Agriculture and Forestry Research Organisation, Nairobi, Kenya.

FAO (1991) *Exotic Aphid Pests of Conifers: a Crisis in African Forestry. Proceedings of FAO Workshop, 3–6 June 1991, Muguga, Kenya*. FAO, Rome, Italy.

FAO (1996) *Code of Conduct for the Import and Release of Exotic Biological Control Agents*. FAO, Rome, Italy.

Greathead, D.J. (1995) The *Leucopis* spp. (Diptera: Chamaemyiidae) introduced for biological control of *Pineus* sp. (Homoptera: Adelgidae) in Hawaii: implications for biological control of *Pineus ?boerneri* in Africa. *Entomologist* 114, 83–90.

Greathead, D.J. and Greathead, A.H. (1992) Biological control of insect pests by insect parasitoids and predators: the BIOCAT database. *Biocontrol News and Information* 13, 61–68.

Hall, R.W., Ehler, L.E. and Bisabri-Ershadi, B. (1980) Rate of success in classical biological control of arthropods. *Bulletin of Entomological Society of America* 26, 111–114.

Kairo, M.T.K. and Murphy, S.T. (1999a) Temperature and plant nutrient effects on development survival and reproduction of *Cinara* sp., an invasive pest of cypress trees in Africa. *Entomologia Experimentalis et Applicata* 92, 147–156.

Kairo, M.T.K. and Murphy, S.T. (1999b) Host age choice for oviposition in *Pauesia juniperorum* (Hymenoptera: Braconidae: Aphidinae) and its effect on the parasitoids's biology and host population growth. *Biocontrol, Science and Technology* 9, 475–486.

Kamunya, S.M., Day, R.K., Chagala, E. and Olng'-otie, P.S. (1997) Variation and inheritance of resistance to cypress aphid, *Cinara cupressi* Buckton in *Cupressus lusitanica* Miller. *Annals of Applied Biology* 130, 27–36.

Kamunya, S.M., Olng'-otie, P.S., Chagala, E., Day, R.K. and Kipkore, W.K. (1999) Genetic variation and heritability of resistance to *Cinara cupressi* in *Cupressus lusitanica*. *Journal of Tropical Forest Science* 11, 587–598.

KARI (1977–1980) *Records of Research*. Annual Reports 1977–1980. Kenya Agricultural Research Institute, Nairobi, Kenya.

Kfir, R. and Kirsten, F. (1991) Seasonal abundance of *Cinara cronartii* (Homoptera: Aphididae) and the effect of an introduced parasite, *Pauesia* sp. (Hymenoptera: Aphidiidae). *Journal of Economic Entomology* 84, 76–82.

Kfir, R., Kirsten, F. and van Rensburg, N.J. (1985) *Pauesia* sp. (Hymenoptera: Aphidiidae): a parasitoid introduced into South Africa for biological control of the black pine aphid, *Cinara cronartii* (Homoptera: Aphididae). *Environmental Entomology* 14, 597–601.

Kfir, R., van Rensburg, N.J. and Kirsten, F. (2003) Biological control of the black pine aphid, *Cinara cronartii* (Homoptera: Aphididae), in South Africa. *African Entomology* 11, 117–121.

Kirsten, F. and Kfir, R. (1991) Rate of development, host instar preference and progeny distribution by *Pauesia* sp. (Hymenoptera: Aphidiidae), a parasitoid of *Cinara cronartii* Tissot & Pepper (Homoptera: Aphididae). *Journal of the Entomological Society of Southern Africa* 54, 75–80.

Madoffe, S.S. and Austara, O. (1990) Impact of pine woolly aphid, *Pineus pini* (Macquart) (Hom., Adelgidae), on growth of *Pinus patula* seedlings in Tanzania. *Journal of Applied Entomology* 110, 421–424.

Mailu, A.M., Khamala, C.P.M. and Rose, D.J.W. (1978) Evaluation of pine woolly aphid damage to *Pinus patula* and its effect on yield in Kenya. *East African Agricultural and Forestry Journal* 43, 259–265.

Mailu, A.M., Khamala, C.P.M. and Rose, D.J.W. (1980) Population dynamics of pine woolly aphid, *Pineus pini* (Gmelin) (Hemiptera: Adelgidae), in Kenya. *Bulletin of Entomological Research* 70, 483–490.

Marsh, P.M. (1991) New species of *Pauesia* (Hymenoptera, Braconidae, Aphidiinae) from Georgia and introduced into South Africa against the black pine aphid (Homoptera: Aphididae). *Journal of Entomological Science* 26, 81–84.

McClure, M.S. (1984) *Pineus boerneri* Annand (Homoptera: Adelgidae): a new or another record from the Peoples Republic of China? *Proceedings of the Entomological Society of Washington* 82, 460–461.

Mills, N.J. (1990) Biological control of forest aphid pests in Africa. *Bulletin of Entomological Research* 80, 31–36.

Mugasha, A.G., Chamshama, S.A.O., Nshubemuki, L., Iddi, S. and Kindo, A.I. (1997) Performance of thirty two families of *Cupressus lusitanica* at Hambalawei, Lushoto, Tanzania. *Silvae Genetica* 46, 185–192.

Murphy, S.T. (1996) Status and impact of invasive conifer aphid pests in Africa. In: Nair, K.S.S., Sharma, J.K. and Varma, R.V. (eds) *Impact of Diseases and Insect Pests in Tropical Forests. Proceedings of the IUFRO Symposium, 23–26 November 1993, Peechi, India.* Kerala Forest Research Institute, Peechi, India, pp. 289–297.

Murphy, S.T. and Day, R.K. (1998) Biological control in tropical plantation forest pest management: challenges and opportunities. In: Saini, R.K. (ed.) *Tropical Entomology, Proceedings of the 3rd International Conference on Tropical Entomology, 30 October–4 November 1994, Nairobi.* ICIPE Press, Nairobi, Kenya, pp. 9–26.

Murphy, S.T., Chilima, C.Z., Cross, A.E., Abraham, Y.J., Kairo, M.T.K., Allard, G.B. and Day, R.K. (1994) Exotic conifer aphids in Africa: ecology and biological control. In: Leather, S.R., Watt, A.D., Mills, N.J. and Walters, K.F.A. (eds) *Individuals, Populations and Patterns in Ecology*. Intercept, Andover, UK, pp. 233–242.

Odera, J.A. (1974) The incidence and host trees of the pine woolly aphid, *Pineus pini* (L.), in East Africa. *Commonwealth Forestry Journal* 53, 128–136.

Orondo, S.B.O. and Day, R.K. (1994) Cypress aphid (*Cinara cupressi*) damage to a cypress seed stand in Kenya. *International Journal of Pest Management* 40, 141–144.

Pepper, J.O. and Tissot, A.N. (1973) Pine feeding species of *Cinara* in the Eastern United States (Homoptera: Aphididae). *Florida Agricultural Experiment Station Monograph* series 3.

Starý, P. (1960) A taxonomic revision of the European species of the genus *Paraphidius* Starý, (Hymenoptera: Braconidae, Aphidiinae). *Acta Faunistica Entomologica Musei Nationalis Prague* 6, 5–44.

van Rensburg, N.J. (1979) *Cinara cronartii* on the roots of pine trees (Homoptera: Aphididae). *Journal of the Entomological Society of Southern Africa* 42, 151–152.

van Rensburg, N.J. (1981) A technique for rearing the black pine aphid, *Cinara cronartii* T & P, and some features of its biology (Homoptera: Aphididae). *Journal of the Entomological Society of Southern Africa* 44, 367–379.

Watson, G.W., Voegtlin, D.J., Murphy, S.T. and Foottit, R.G. (1999) Biogeography of the *Cinara cupressi* complex (Hemiptera: Aphididae) on Cupressaceae, with description of a pest species introduced into Africa. *Bulletin of Entomological Research* 89, 271–283.

Zwolinski, J.B., Grey, D.C. and Mather, J.A. (1989) Impact of pine woolly aphid, *Pineus pini* (Homoptera: Adelgidae) on cone development and seed production of *Pinus pinaster* in the southern Cape. *South African Forestry Journal* 148, 1–6.

8

Biological Control of Defoliating, and Phloem- or Wood-feeding Insects in Commercial Forestry in Southern Africa

Geoff D. Tribe
Plant Protection Research Institute, Stellenbosch, South Africa

Introduction

Exploitation of Africa's indigenous forests began with European settlement at the Cape of Good Hope in 1652 and gradually extended eastwards as local resources became exhausted. The first settlers found a land poorly endowed with natural forests where only 0.3% of South Africa's land area was afforested (Van der Zel, 1994). Timber was required for shipbuilding, housing, ox-wagons, furniture and fuel. The early Cape governors realized the necessity for the importation of fast-growing exotic tree species and the first introductions were the common oak (*Quercus robur* L., *Fagaceae*) followed by cluster and stone pines (*Pinus pinaster* Aiton and *Pinus pinea* L., *Pinaceae*) (Department of Forestry, 1980). By 1847, attempts were made to halt the exploitation and conserve what remained of the indigenous forests by restricting access. Increased demand for timber arose in the 1870s for railway sleepers, telegraph poles and props for mines. The first commercial plantation consisting of eucalypts was established at Worcester in 1876 to provide firewood for railway locomotives and the first tan wattle plantations were established in Natal shortly after this (Department of Forestry, 1980). However, it was only during World War II when timber imports were disrupted and prices escalated that a boom in afforestation began and continued after the war with a total area of commercial plantations in South Africa now exceeding 1 million ha.

Various pine and eucalypt (*Myrtaceae*) species were selected and established in those parts of southern Africa where they were most suited (Poynton, 1979) and this has had a fundamental influence on the distribution of forest pests. A major reason why exotic trees are often able to grow rapidly and outperform those in their country of origin is that they are not subjected to the depredations of the many pests and diseases that afflict them in their natural environment. Such contiguous and monospecific trees are, however, highly vulnerable should a pest become established on them. Most southern African plan-

(eds P. Neuenschwander, C. Borgemeister and J. Langewald)

tations occur in areas where forests previously never existed, with the consequence that there are no reservoirs of generalized predators of forest insects that could immediately adapt to and exploit a newly arrived pest. Although about 800 insect species have been recorded from indigenous trees and shrubs in South Africa, and thus represent a potential threat to plantations, only a small percentage occasionally become pests (Geertsema, 1982). Nearly all of these are defoliators of *Pinus* species and are mostly lepidopteran larvae (e.g. Grobler, 1957; Geertsema and Van den Berg, 1973; Van den Berg, 1973, 1974, 1975; Geertsema, 1975) or adult coleopterans (Tribe, 1991a). Wattles (*Acacia mearnsii* De Wild., *Mimosaceae*) are defoliated by the wattle bagworm (Hepburn, 1973) and while many insects are recorded as feeding on eucalypts (Poynton, 1979) they are of little consequence except for the termites (Atkinson *et al.*, 1992).

The earliest use of biological control in South African plantations was the introduction of domesticated pigs in 1911 against the indigenous willow-tree emperor moth, *Gonimbrasia tyrrhea* (Cramer), defoliating pines in the Transvaal (Hardenberg, 1912). Pigs were similarly used against the pine tree emperor moth, *Imbrasia cytherea* (Fabricius) (both Lepidoptera, Saturniidae) (Plate 19), in the Western Cape since 1925 (Geertsema, 1975) and in Natal (Van den Berg, 1969). The pigs were trained to unearth and feed on the pupae of these moths.

Biological Control of Insect Pests of *Eucalyptus* Species

The arrival of the eucalyptus snout beetle, *Gonipterus scutellatus* Gyllenhal (Coleoptera, Curculionidae), in South Africa in 1916 in a consignment of apples from Tasmania, demonstrated just how damaging a pest unrestrained by its natural enemies could be in a monospecific plantation (Annecke and Moran, 1982). This defoliator changed the face of the hardwood industry because the eucalypt species then grown were also those most susceptible to the beetle (Tooke, 1926; Poynton, 1979; Richardson and Meakins, 1986). It was the introduction of the egg parasitoid *Anaphes nitens* (Girault) (Hymenoptera, Mymaridae) in 1926, combined with the growing of less-susceptible eucalypt species, that brought the beetle under successful control (Tooke, 1955). A parasitism rate of up to 96% was achieved in coastal regions, but control in highland regions above 1200 m remained erratic (Tooke, 1942). The problem occurs in winter when *Gonipterus* beetles enter a hibernation period of 5 months during which they lay no eggs, thus breaking the cycle because *A. nitens* wasps normally live for only a few days. As late as January, only 0.5–5% parasitism is obtained (Tooke, 1942). This erratic control has persisted over the years, but has recently been exacerbated by the re-introduction of susceptible eucalypt species, which grow best at higher altitudes. Similarly, hybridization with susceptible species has also led to increased defoliation.

Two methods were considered in the 1980s to remedy the erratic biological control of *G. scutellatus* in highland areas: the augmentation of *A. nitens* with wasps collected from coastal regions in spring, and the introduction of additional cold-hardy parasitoid species or strains of *A. nitens* from Australia. To determine the present level of parasitism, 20 host egg-capsules were collected every fortnight from George (33° 58′S, 22° 28′E), Grabouw (34° 12′S, 19° 07′E) and Cape Town (33° 55′S, 18° 25′E) for 5 years between 1984 and 1989. This confirmed that the mean parasitism rate had remained unchanged over the last 65 years but varied between Cape Town (76%), George (82%) and Grabouw (89%). Parasitism peaks occurred in spring (October) at Cape Town (96%) and Grabouw (September: 92%) and in autumn (May) at George (78%), which coincided with peaks in host egg production. Thus it is possible to recruit parasitoids from coastal regions for release in highland areas, although no trials have been conducted to ascertain what numbers are needed or whether it is cost effective. This monitoring also revealed that it is the lack of fresh foliage due to a lack of rain in winter, and not exclusively cold temperatures, which is responsible for the failure of biological control at higher altitudes.

A search for additional parasitoids was made in Tasmania, whence the beetle originated (Annecke and Moran, 1982), the established *A. nitens* having been imported from Penola in South Australia (Tooke, 1955). Two new *Anaphes* species, *A. tasmaniae* Huber and Prinsloo and *A. inexpectatus* Huber and Prinsloo, 1990, were reared from *G. scutellatus* eggs and imported into South Africa where they were mass-produced for release in Lesotho. However, they succumbed on the return journey to Pretoria after entry to Lesotho was denied despite having the correct documentation. Future searches for egg parasitoids should perhaps be made in the highland summer rainfall region of New South Wales where there is also no flush in winter, unlike that of Tasmania with its all-year rainfall. Other egg and larval parasitoids of *G. scutellatus* have been recorded (Malloch, 1934; Tooke, 1942, 1955; Forestry Commission Tasmania, 1981, unpublished results). Furthermore, three parasitoid species sampled at Hobart, namely a tachinid species, the proctotrupid *Proctotrupes turneri* Dodd and the braconid *Apanteles* sp., accounted for 54% larval parasitism, with the latter species being the most important (R. Bashford, Tasmania, 1987 personal communication).

Eucalyptus trees in Australia are subjected to attacks from a wide array of insect species and consequently are likely to tolerate a high degree of defoliation without losing incremental growth. The threshold point at which the level of defoliation affects incremental growth has not been determined. Thus partial defoliation may not mean that biological control by *A. nitens* has failed.

The eucalypt borers, *Phoracantha semipunctata* (Fabricius) and *Phoracantha recurva* Newman (Coleoptera: Cerambycidae), were first discovered in the Western Cape in 1906 (Lounsbury, 1917) and probably arrived in South Africa in railway sleepers imported from Australia at the turn of last century (Tooke, 1935). The beetles occur throughout southern Africa wherever eucalypts are grown and are regarded as sporadic pests, being particularly damaging to drought-stressed eucalypts. Debarking of felled trees is mandatory to prevent the beetles from ovipositing within the bark. Not only are stressed trees killed, but the timber becomes riddled with tunnels where the larvae enter the wood to pupate. Unless otherwise specified, the unpublished results of J.A. Moore (PPRI, Stellenbosch) have been used to compile the following section on the biological control of *Phoracantha* spp.

Parasites of the family Megalyridae were introduced to Cape Town from New South Wales in 1910 as biological control agents, but all attempts to rear them failed (Webb, 1974). This was presumably because they were originally believed to be larval parasitoids, but they have since been successfully reared on pupae (Moore, 1993). It must be assumed that the parasites either escaped or were discarded because in 1962 *Megalyra fasciipennis* Westwood was recovered from *Phoracantha*-infested logs in the southwestern Cape (Gess, 1964). The ovipositor is drilled through the plug blocking the entrance to the pupal chamber of the beetle and an egg is laid on the incapacitated pupa. During the past 37 years, and perhaps for as long as 91 years, *M. fasciipennis* has not dispersed beyond the southwestern Cape. However, the impact of *M. fasciipennis* has remained unnoticed, until a survey in 1993 recorded a parasitism rate of 52.5%, permitting the recruitment of 60 wasps for release in the Tzaneen (23° 50′S, 30° 09′E) district (Moore, 1993). It is unknown whether they have become established there.

Although activity is restricted to the summer months (between October and March) when daily temperatures are greater than 16°C, *M. fasciipennis* is an effective parasitoid largely in synchrony with the pupal stage of its host. *Phoracantha* spp. have up to three generations per year (Mendel *et al.*, 1984; Scriven *et al.*, 1986) with oviposition occurring during 8 months of the year between September and April (Tooke, 1935) when night temperatures rise above 15.5°C (Duffy, 1963). Although larval development varies from 70 to 180 days (Scriven *et al.*, 1986) and the period from the completion of the pupal cell to pupation is highly variable because many larvae overwinter before pupation, the pupal stage occurs mainly in summer (Duffy, 1963). The long ovipositor of *M. fasciipennis* (42.71 ± 5.33 mm) reaches most host pupal chambers except those embedded deeply in trunks of large trees with thick bark. The host pupal stage is about 20 days (Scriven *et al.*, 1986).

A second attempt at biological control was made in 1969 with the importation of braconid larval parasitoids *Syngaster lepidus* Brullé, *Bracon capitator* F. and *Doryctes* spp. (Webb, 1974). Of these, only *S. lepidus* was eventually released but failed to become established (Drinkwater, 1975). The eggs of *Phoracantha*, of which there are 40 species in Australia (Wang, 1995), are laid in batches in crevices with gaps less than 0.65 mm wide (Way *et al.*, 1992), in the bark of stressed or felled trees. This represents a large resource, which theoretically should be attractive to parasitoids, and consequently a simple trap was devised to locate and monitor such parasitoids (Cillié and Tribe, 1991). By pinning sloughed bark to felled eucalypts, the encyrtid egg parasitoid *Avetianella longoi* Siscaro (1992) was discovered in Australia and introduced into California (Hanks *et al.*, 1995), from where it was imported into South Africa and released in 1993 (Hanks *et al.*, 1996). From release sites near Cape Town and Tzaneen, *A. longoi* dispersed rapidly, covering 130 km along the southern Cape coast in 2 years (Hanks *et al.*, 1995). Similarly, *A. longoi* dispersed approximately 70 km from its release site at Chipata in Zambia within 10 months.

Although *A. longoi* has proved to be an effective biological control agent in Italy (Longo *et al.*, 1993), California (Hanks *et al.*, 1996) and South Africa where parasitism rates of over 90% have been recorded, there are certain limitations. Each female wasp lays up to 250 eggs during her 20-day adult life (Paine *et al.*, 1995), but in South Africa the activity period is restricted to only a few months of the year (January–March) with a peak (54%) in February. Both *Phoracantha* species are active throughout the year in Africa, although oviposition peaks in summer. *P. recurva* tends to colonize the within-canopy trunk and larger branches and emerge in early summer, whereas *P. semipunctata* colonizes the main stump and emerges mostly in late summer. Because the activity of *A. longoi* is partly synchronous with the peak oviposition period of *P. semipunctata*, most of the eggs laid earlier by *P. recurva* are not parasitized. This has led to a progressive skewing of the population ratio in favour of *P. recurva*, which has become more numerous – a situation that also pertains in California (Millar, 1998, personal communication). The origin of *A. longoi* accidentally introduced into the Mediterranean (Hanks *et al.*, 1996) is unknown, but this population may possess beneficial differences, such as a wider activity period, to consider import to complement the existing gene pool within South Africa.

Because egg parasitoids prevent the damaging larval stage from developing, they are a key factor in the control of *Phoracantha*. Although *A. longoi* from the South African culture has been established in Zambia (1995), Uruguay (1998) and Chile (2000), the parasitoid has yet to be established in KwaZulu-Natal and other African countries. Further collections of *A. longoi* in Australia would not only enlarge the gene pool and so, perhaps, confer an extended activity period, but may also uncover additional egg-parasitoid species able to complement the behaviour of *A. longoi*.

In 1995 a second attempt was made to introduce larval parasitoids against the *Phoracantha* spp. Three braconid species from Australia were reared in California and dispatched as pupae to South Africa where *S. lepidus* (Hanks *et al.*, 2001) and *Jarra phoracantha* Marsh & Austin became established in the Tzaneen district, but *Jarra maculipennis* Marsh & Austin was not recovered (Kirsten, 2001). Minor releases of all three species were also made in the Cape Peninsula but did not lead to establishment.

An unidentified *Cleonymus* sp. (Hymenoptera, Pteromalidae) was discovered parasitizing *Phoracantha* spp. larvae in the Cape Peninsula in 1993, and was released in the Tzaneen district in 1996, where it became established. The origin of this parasitoid is unknown, but it is believed to be an undescribed indigenous species (G.L. Prinsloo, Pretoria, 1996, personal communication), which is rare in the field, but can destroy a *Phoracantha* culture if inadvertently introduced. The host larva is paralysed by the wasp through the bark and eggs are dropped into the chamber. About 20 parasite larvae feed externally on the host larva and pupate in the chamber. Because of its unknown origin no attempts have been made to mass-rear this parasitoid or to distribute it to other countries.

The complex of biological control agents arrayed against *Phoracantha* has not reduced damage to acceptable levels. Intraspecific larval competition (Hanks *et al.*, 1993), which accounts for an average of 30% natural mortality (Mendel *et al.*, 1984), has been reduced by the actions of the parasitoids, but this has been compensated for by increased fecundity in the surviving females. However, the full extent of biological control cannot be accurately gauged until all parasitoid species are uniformly distributed throughout the country. An assessment would then determine whether additional species are required to augment the present parasitoid complex. Such natural enemies originating in Australia (Moore, 1972; Scriven *et al.*, 1986; Austin *et al.*, 1994; Paine *et al.*, 1995) and their associated hyperparasites (Mendel *et al.*, 1984) have been recorded, and there may exist other undiscovered species that are specific to this host/habitat complex. Local parasitoids that have become established on *Phoracantha* include the European platygasterid egg parasitoid *Platystasius transversus* (Thomson) in Morocco (Fraval and Haddan, 1988) and two Californian parasitoids (Scriven *et al.*, 1986), which are of course not host specific.

There followed a period of several decades in which no new pests of eucalypts arrived in southern Africa, until in 1982 an Australian tortoise beetle, *Trachymela tincticollis* (Blackburn) (Coleoptera, Chrysomelidae), was discovered near Cape Town on severely defoliated *Eucalyptus gomphocephala* DC trees. By 1985, it had spread 800 km to Port Elizabeth. Both adults and larvae fed on the new leaves of 13 *Eucalyptus* species cultivated in South Africa, including the important commercial species *E. grandis* Hill ex Maiden (Tribe and Cillié, 1997). Females lay an average of 11 eggs day^{-1} from mid-August to the end of December. Eggs are deposited at dusk in fissures 0.5–1.0 mm wide in the bark of the host tree and hatching larvae hide in fissures near their feeding sites during the day. They have a bimodal activity pattern, feeding just before sunrise (93%) or after sunset (7%). Various traps were used to monitor the phenology of the different life stages, namely Velcro® (eggs), flattened-cork wrapped around tree trunks (larvae) and sticky traps (adults) (Tribe and Cillié, 1997).

Several natural enemies of *T. tincticollis* were discovered in southwestern Australia, and four egg-parasitoid species were selected as biological control agents and imported into South Africa (Tribe, 2000). Two of these species, *Enoggera reticulata* Naumann (Hymenoptera, Pteromalidae) and *Procheiloneurus* sp. nr. *triguttatipennis* Girault (Hymenoptera, Encyrtidae), were specific to *T. tincticollis* eggs, and two, *Enoggera nassaui* (Girault) and *Neopolycystus insectifurax* Girault (Hymenoptera, Pteromalidae), occasionally emerged from parasitized eggs. Four hyperparasitoids were associated with the first three parasitoids. *E. reticulata* parasitized 42% of *T. tincticollis* egg batches placed in Ludlow Tuart Forest, of which 27% were hyperparasitized by *Neblatticida* sp. nr. *lotae* (Girault) (Hymenoptera, Encyrtidae).

Only *E. reticulata* became established, achieving a consistently high parasitism rate with a maximum of 96% over the 4-month oviposition period of the host (Tribe and Cillié, 2000). Within 2 years, *E. reticulata* had dispersed throughout the distribution range of its host, and up to 1330 km from the release sites. With its high reproductive rate, high powers of dispersal and host specificity, *E. reticulata* had been identified as the key species in limiting the numbers of *T. tincticollis*. In Australia, *E. reticulata* had been severely restricted by the hyperparasites *Neblatticida* sp. (64%) and *Signiphora* sp. (24%) (Hymenoptera, Signiphoridae), whose exclusion was largely responsible for the level of success of the parasitoid in South Africa. *Procheiloneurus* sp., which was only present near the end of the oviposition period of its host and only in small numbers on late flushing trees in Australia, failed to become established in South Africa. The two non-specific parasitoid species, *E. nassaui* and *N.*

insectifurax, which normally search the leaves of eucalypts, where their major hosts deposit their eggs, also failed to become established. In the absence of their major host species, they were unable to survive on *T. tincticollis* eggs hidden in crevices in the bark.

E. reticulata orientate to a habitat/host complex and should have no detrimental effect on indigenous insect species. This was determined by placing *T. tincticollis* egg batches at set intervals through contiguous eucalypt and pine plantations, and recording weekly which batches had been parasitized. *E. reticulata* readily parasitized (50.3%) eggs attached to *E. gomphocephala* trees, but only rarely parasitized (2.6%) those placed on *P. pinaster* (Aiton) trees (Tribe and Cillié, 2000).

Honey production and, indirectly, pollination services in South Africa depend on eucalypts, of which *E. grandis* is the most important nectar plant (Johannsmeier, 1976). A decline in honey production from an average 28 kg per hive to 15 kg per hive has been attributed to the presence of *Drosophila flavohirta* Malloch (Diptera, Drosophilidae) larvae in the flower cups where they feed on the nectar (Du Toit, 1987). A search for natural enemies in Yuraygir National Park in Australia led to the discovery of two parasitoid species from *D. flavohirta* pupae. The eulophid *Aprostocetus* sp. accounted for 97% of the parasitized pupae and an eucoilid for the remainder. The average parasitism was 15% (range 5–28%) (Tribe, 1991b). The egg and larval stages of *D. flavohirta* are submerged in nectar and it is only the exposed pupae that become accessible to parasitoids when they are attached to the styles for 4–6 days before adult emergence.

Aprostocetus sp. was reared in Pretoria either in field-collected *D. flavohirta* or laboratory-reared *D. melanogaster* Meigen pupae, but only in the presence of eucalypt flowers (Tribe *et al.*, 1989). Two small field releases were made near Tzaneen in 1988, but the parasitoid failed to become established. These releases were made too late in the season and the culture was abruptly terminated. The ease with which *Aprostocetus* sp. was reared indicates that the lack of success was not due to some intrinsic flaw within the parasitoid. The high mortality experienced in Australia when adult *D. flavohirta* disperse in search of scattered flowering sources is largely absent in South Africa because of the contiguous nature of these plantations. Thus, although the percentage parasitism is rather low, the conditions that favour the rapid increase of *D. flavohirta* in South African plantations will also favour the fly's parasitoids. Because *D. flavohirta* breeds in an ephemeral resource which is widely scattered, the finding of a localized, mass-flowering host within Australia during the summer months will be crucial to successful reintroduction of *Aprostocetus* sp.

Biological Control of Insect Pests of *Pinus* Species

Three bark beetle species from Europe cause considerable loss to *Pinus* species grown in southern Africa. The beetles colonize stressed, moribund and felled trees where the larvae feed on the inner bark. All three species are often found in the same tree and the associated blue-stain fungi further degrade the timber.

The pine bark beetle *Hylastes angustatus* (Herbst), which feeds on the green bark of roots and root-collars of pine seedlings, is the most damaging. During its maturation-feeding phase, it may kill over 50% of 1-year-old seedlings by girdling of the stems when feeding under the bark (Tribe, 1990a). The European bark beetle *Orthotomicus erosus* (Wollaston) was discovered in Stellenbosch in 1968 and is now found throughout the country wherever pines are grown. The males release an aggregation pheromone, which allows the species to utilize temporary habitats (such as wind-blown trees) to their fullest extent. It is essentially a Mediterranean species, and although a secondary pest, it does cause substantial losses of trees stressed by adverse climatic conditions, other pests or scorched by fire (Tribe, 1990b). The red-haired pine bark beetle, *Hylurgus ligniperda* (Fabricius) is another secondary pest (Tribe, 1991c).

The three bark beetle species have different phenologies in the Western Cape with *O. erosus* active in summer, *H. angustatus* in spring and *H. ligniperda* more evenly distributed throughout the year with a peak in autumn (Tribe, 1991d). In summer rainfall regions, *H. angustatus* is active throughout the summer due to the high moisture, with consequent loss of seedlings. Each species colonizes a different site on the tree, although there is some overlap. In experimental trials, *O. erosus* occupied the above-ground parts of the logs (98%), *H. angustatus* the roots (64%) and root-collars (36%), while *H. ligniperda* was found mostly below soil level (86%). The latter two species dug independently to logs buried horizontally in the soil at a depth of 40 cm (Tribe, 1992). More adult *O. erosus* were present in the plantation (88%) than the other two species combined.

Six biological control agents of *O. erosus* were located in Israel in 1984, introduced to South Africa and released in a plantation near Sabie (Kfir, 1986). None of these became established, except for a consignment of the braconid *Dendrosoter caenopachoides* Ruscha, which was released in the southwestern Cape (Tribe and Kfir, 2001) and which had been the single most important parasitoid species of *O. erosus* in Israel (51.8%) (Mendel, 1986). Because of its small size, it only parasitises *O. erosus* in thin-barked trees and although parasitism rates ranged between 30% and 70% within individual stumps, this rose to 100% when the bark was exceptionally thin. Within 20 years of its discovery, *O. erosus* had dispersed throughout the country (over 2200 km) whereas *D. caenopachoides* dispersed over a distance of 455 km to Wilderness in 15 years. This slower dispersal of the parasitoid in comparison with its host is attributed to the fact that *D. caenopachoides* is limited by the availability of thin-barked trees, whereas its host is not. In 1980, an undescribed *Dendrosoter* sp. nr. *labdacus* Nixon of unknown origin was discovered parasitizing *O. erosus* in the southwestern Cape, to which it is confined at present.

Additional biological control agents that enter the tunnels of their hosts are required for the root-colonizing bark beetle species, which are immune to parasitism by *Dendrosoter* spp. Such natural enemies are recorded from Europe and the Mediterranean Basin (Mendel and Halperin, 1981; Mills, 1983; Mendel, 1986). Consideration should also be given to importing natural enemies of scolytid species that are attracted to the same pheromones as *O. erosus* (Giesen *et al.*, 1984; Klimitzek and Vité, 1986; Mendel, 1988), to which they will respond as kairomones. The presence of an established reservoir of biological control agents would be an insurance in the event of further, and more destructive, scolytid species entering the country. For predators that enter host tunnels, the staggered activity peaks of the three bark beetle species would ensure that hosts were available throughout the year.

The woodwasp *Sirex noctilio* Fabricius (Hymenoptera, Siricidae), which is endemic to Eurasia and North Africa with almost 76% of its distribution range in the Mediterranean bioclimatic zone, was discovered in Cape Town in 1994 (Tribe, 1995). Adults of *S. noctilio* are attracted to stressed trees, which they kill by injecting a phytotoxic mucus and the symbiotic fungus *Amylostereum areolatum* (Fr.) Boidin into the wood during oviposition (Bedding, 1993). Woodwasp larvae feed on the fungus within the wood and pupate just under the bark of the tree. Immediate steps were taken to introduce those biological control agents that had been successful in Australasia, where previously losses of up to 70% of a compartment had been recorded.

The nematode *Deladenus siricidicola* Bedding (Nematoda, Neotylenchidae) was first established, because the later-introduced hymenopterous parasitoid species do not themselves become parasitized by the nematode, yet may use such parasitized larvae as hosts. The free-living form of *D. siricidicola* occurs in the tracheids, where they feed on the *A. areolatum* fungus, but when juvenile nematodes encounter *Sirex* larvae, the high CO_2 levels and low pH micro-environment around them trigger the nematodes to develop into the parasitic form with stylets, with which they penetrate the cuticle of their hosts (Bedding, 1993). When the host pupates, the nematodes migrate into the reproductive organs, sterilizing the females, which lay eggs filled with nematodes that then infect the progeny of all the wasps that oviposit in the same tree. The males carry the nematodes but do not transfer them to the female during copulation. The virulent Kamona strain, originating in Hungary, was inoculated into trees infested with *S. noctilio* larvae in 1995 within a 90-km radius of Cape Town. In the third year, a parasitism rate of 94.5% was recorded in the Tokai Plantation.

Two hymenopterous parasitoid species were introduced to augment *D. siricidicola*. The key agent *Ibalia leucospoides* (Hochenwarth) (Hymenoptera, Ibaliidae) was collected in Uruguay, which it had entered with its host, and was introduced into South Africa in 1998. A total of 456 wasps of both sexes were released within the Western Cape over the following 3 years, until its establishment in Stellenbosch was confirmed in 2002. *Ibalia* oviposits down the oviposition channel bored by the *Sirex* wasp, which it detects from the odour of the symbiotic fungus, and lays an egg in the hatching larva. The parasite larva feeds internally in the host larva initially, but emerges to feed externally before pupating (Neumann *et al.*, 1987).

The ichneumonid, *Megarhyssa nortoni nortoni* (Cresson), originally from California, was obtained from Tasmania in October 1998. From the 42 females imported, 41 males and 38 females were reared and released in the southwestern Cape a year later. Because the numbers released were so small, no follow-up has yet been conducted to determine whether they have become established. *Megarhyssa nortoni* effectively parasitizes late-instar larvae in thick tree trunks. *Rhyssa persuasoria* (Linnaeus), a common parasitoid from Europe, parasitizes late-instar larvae in thinner trunks and branches and thus would complement the action of *M. nortoni*. However, the effectiveness of the present control measures has resulted in its importation being indefinitely postponed.

Sirex continues to expand its range at a rate of about 50 km annum^{-1} and has dispersed 255 km along the west coast (Vanrhynsdorp) and 400 km along the southern coast (Sedgefield). It is expected eventually to occur throughout southern Africa wherever pines are grown. The hymenopterous parasitoid species disperse readily with their host, but *D. siricidicola* disperses slowly between plantations and must therefore be continually brought to the dispersal front to minimize losses. In the absence of prolonged droughts or extensive fires, the biological control presently in force has proved effective in keeping *Sirex* losses at very low levels. Not reported from other countries is an association in South Africa between *Sirex* and the borer *Arhopalus syriacus* (Reitter) (Coleoptera, Cerambycidae). This Mediterranean beetle was previously found only in debris on the plantation floor, but since the arrival of *Sirex* it had increased rapidly in numbers and is found in the lower 1–2 m of trees infested with *Sirex* larvae. Effective control of *Sirex* will probably result in *A. syriacus* resuming its non-pest status.

The Deodar weevil, *Pissodes nemorensis* Germar (Coleoptera, Curculionidae) is mostly a secondary pest in its native eastern North America, but in South Africa it causes considerable losses. In the winter rainfall region, between 42% and 45% of the terminal leaders within a *P. radiata* compartment may be killed. This results in a loss in height increment although the girth increases, multiple or forked stems, and an eventual loss in timber. Attacks usually begin when trees are 7 years old, but this is reduced to 3 years if seedlings are fertilized at planting, or to 1 year if they are grown from cuttings. *P. nemorensis* feeds and breeds in the phloem of leaders of dominant trees in winter. Accumulative feeding weakens the trees' defences and is followed by the laying of one, but sometimes up to four, eggs in the feeding pits, which are then sealed with a faecal plug. Hatching larvae girdle the leader by under-bark feeding and kill seedlings.

By monitoring 700 trees in Tokai plantation, it was determined that dispersal within a compartment is low but that there is an exponential increase in *P. nemorensis* numbers with time, and that although dispersal appears to be random, the beetles select the fast-growing, succulent leaders. Based on this knowledge, a trial was conducted at Ruitersbos plantation in the southern Cape where all infested leaders in a severely attacked 7-year-old compartment were pruned and burnt. The trees fully recovered and the infestation was brought to a halt. Although widespread within South Africa, *P. nemorensis* damage is restricted mainly to the winter rainfall region, where its active period is in synchrony with the growth of its host, and to *P. radiata* with its bare (sparse needles) elongated leader. This region is regarded as marginal for commercial plantations and the implementation of control measures against *P. nemorensis* has consequently been of low priority. The importation of biological control agents is also on hold because of a possible clash of interests. Seed-feeding biological control agents, in which *Pissodes* spp. feature prominently, are being considered for importation to curb the invasiveness of *P. pinaster* in the winter rainfall biome.

Pissodes Germar has a holarctic distribution and a wide range of natural enemies (Taylor, 1929; Harman and Kulman, 1967; Kenis, 1994, 1997; Kenis and Mills, 1994) and both European and North American parasitoids should be considered against *P. nemorensis* on the African continent. Four candidates (all Hymenoptera) have been selected for importation: the larval parasitoid *Coeloides pissodes* (Ashmead) (Braconidae) from North America and the European species *Coeloides sordidator* (Ratzeburg), the larval parasitoid *Eubazus semirugosus* (Nees) (Braconidae) and the pupal parasitoid *Dolichomitus terebrans* (Ratzeburg) (Ichneumonidae).

Discussion and Conclusion

The major insect pests of *Eucalyptus* spp. grown in southern Africa all originate from Australia and comprise both defoliators and stem borers. *Pinus* spp. have more exotic pests consisting of bark beetles, stem borers and sap-suckers originating from Europe (seven) or North America (two). Various indigenous insect species, mainly Lepidopteran (seven) or Coleopteran (two) defoliators, have adapted to pines. These belong mostly to the Saturniidae (four) or Lasiocampidae (two), which become sporadic pests when conditions temporarily favour their increase and before they are again brought under control by their natural enemies.

Over the last 117 years, since pines and eucalypts were first planted in South Africa, an average of one major forestry pest has arrived every 7.8 years. Over the last 30 years, this average time lapse between arrivals has dropped to 5 years, probably due to the opening up of markets, containers and the increased mobility of people. Early insect problems were those of indigenous defoliators, such as the pine tree emperor moth *I. cytherea*, and *Euproctis terminalis* Walker (Lepidoptera, Lymantriidae) (Plate 20), adapting to introduced *Pinus* spp. It was only in 1924 that biological control was initiated against a forestry pest, the Eucalyptus snout beetle *G. scutellatus*, using the egg parasitoid *A. nitens*. The value of biological control as a cost effective, long-term solution was clearly demonstrated by the 96% parasitism rate achieved.

An analysis of biological control programmes enacted against forestry pests in South Africa reveals that egg parasitoids were exceptionally successful against defoliators such as *G. scutellatus* and *T. tincticollis*, which lay their eggs in clumps on the outside of their hosts, on leaves or in crevices in the bark. Although the *Phoracantha* borers also lay eggs in clumps in crevices and *A. longoi* achieves a parasitism rate of over 90%, the required level of control has not been achieved. The narrow active period of *A. longoi* between January and February is not fully synchronized with that of its hosts, especially *P. recurva*, which are most active between September and April. Neither does the distribution of *A. longoi*, which has been recorded only from the Melbourne/Ballarat area of southeastern Australia, concur with the extensive distribution of their hosts (Wang, 1995). This suggests that its true host may be a *Phoracantha* or *Coptocerus* species other than *P. semipunctata* and *P. recurva*. A wider gene pool may result in a longer active period in synchrony with its hosts, but if the distribution of *A. longoi* is indeed so restricted, it is probable that this niche is occupied by another species in the rest of Australia. The failure of *A. longoi* to give the required protection therefore indicates that a complex of biological control agents should be imported.

Insect species that lay single eggs in the host tree, within which their larvae develop, appear to need a complex of natural enemies to achieve sufficient control. This applies to *S. noctilio* and the scolytid species, and is predicted also for *P. nemorensis* and *A. syriacus*.

The biological control of the three aphid species, which are pests of conifers, is discussed by Day *et al.* (this volume).

Many forest pests, especially bark beetles and woodwasps, attack stressed trees and are therefore regarded as secondary pests. IPM programmes therefore combine the introduction of biological control agents with correct site selection for the tree and with forest hygiene. The simultaneous replacement of susceptible eucalypts with resistant species was instrumental in the highly successful control of *G. scutellatus* with the *A. nitens* egg parasitoid. However, other economic considerations may override the damage inflicted by pests and prevent the implementation of control measures. Though *P. pinaster* is resistant to defoliation by *I. cytherea*, *P. radiata* is still grown for its superior wood qualities despite occasional defoliation and loss of growth. Defoliators may be divided into flush or senescent feeders, with both groups utilizing the free nitrogen as it is being transported either from the senescent needles (e.g. the lachnid *Eulachnus rileyi* (Williams)) or to the growing tips (e.g. *T. tincticollis*). During the quiescent period, the older leaves either do not retain sufficient nitrogen to allow completion of larval development, or the leaves become too tough. This restricts the activity period of the pests and their parasitoids to certain times of the year. Unlike most crops, trees require 15 years or more before harvesting and may be subjected to repeated insect attacks that have a cumulative impact on the growth and survival of the trees. Under such circumstances, biological control remains the most effective control measure.

Insect pests have also been responsible for the discontinuation of trials to plant new forestry species. For example, the planting of pure-stand African mahogany, *Khaya (nyassica) anthotheca* (Welw.) C. DC. (*Meliaceae*) in the northern regions of South Africa was abandoned due to the damage caused by the shoot borer *Mussidia nigrivenella* Ragonot (Lepidoptera, Pyralidae). This moth lays its eggs in the leaf axils and the hatching larvae bore into the stem, killing the growing shoot, which results in a deformed tree. Prior to these trials, isolated *K. anthotheca* trees grown in botanical gardens in South Africa had reached maturity without hindrance from this moth. In its native central and eastern Africa, the polyphagous *M. nigrivenella* is commonly known as a maize cob borer and a pest of pre-harvest and stored maize (Sétamou *et al.*, 1999). No attempt has been made to isolate possible biological control agents of *M. nigrivenella* on its natural hosts, although efficient parasitoids have been recorded.

New pests arriving on the African continent will disperse across national borders to wherever pines and eucalypts are grown. Thus, quarantine and inspection services remain the first line of defence against unwanted importations. There is also the danger that hyperparasites of established biological control agents are inadvertently introduced, which could severely disrupt or even negate established control measures. Similarly, a secondary pest could become a serious problem should it become a vector of an exotic fungal pathogen such as pitch canker, *Fusarium subglutinans* f. sp. *pini*. Most exotic pests are theoretically amenable to biological control, but it is those species that are already a problem in their native range that pose the most serious threat to the industry. The recent designation of various forestry species as plant invaders may eventually lead to opposing biological control strategies being applied against the same tree species, namely that of simultaneously protecting the tree from exotic pests while curbing its invasiveness with imported agents such as seed feeders.

References

Annecke, D.P. and Moran, V.C. (1982) *Insects and Mites of Cultivated Plants in South Africa*. Butterworth, Durban, South Africa.

Atkinson, P.R., Nixon, K.M. and Shaw, M.J.P. (1992) On the susceptibility of *Eucalyptus* species and clones to attack by *Macrotermes natalensis* Haviland (Isoptera: Termitidae). *Forest Ecology and Management* 48, 15–30.

Austin, A.D., Quicke, D.L.J. and Marsh, P.M. (1994) The hymenopterous parasitoids of eucalypt longicorn beetles, *Phoracantha* spp. (Coleoptera: Cerambycidae) in Australia. *Bulletin of Entomological Research* 84, 145–174.

Bedding, R.A. (1993) Biological control of *Sirex noctilio* using the nematode *Deladenus siricidicola*. In: Bedding, R.A., Akhurst, R.J. and Kaya, H. (eds) *Nematodes and the Biological Control of Insect Pests*. CSIRO, Canberra, Australia, pp. 11–20.

Cillié, J.J. and Tribe, G.D. (1991) A method for monitoring egg production by the *Eucalyptus* borers *Phoracantha* spp. (Cerambycidae). *South African Forestry Journal* 157, 24–26.

Department of Forestry (1980) Forestry in South Africa. Green Heritage Committee of the Forestry Council, Pretoria, South Africa.

Drinkwater, T.W. (1975) The present status of Eucalyptus borers *Phoracantha* spp. in South Africa. *Proceedings of the 1st Congress of the Entomological Society of Southern Africa, Stellenbosch 1974*, Entomological Society of Southern Africa, Pretoria, South Africa, pp. 119–129.

Duffy, E.A.J. (1963) *A Monograph of the Immature Stages of Australasian Timber Beetles (Cerambycidae)*. British Museum (Natural History), London, UK.

Du Toit, A.P. (1987) Nectar flies, *Drosophila flavohirta* Malloch (Diptera: Drosophilidae) breeding in *Eucalyptus* flowers. *South African Forestry Journal* 143, 53–54.

Fraval, A. and Haddan, M. (1988) *Platystasius transversus* (Hymenoptera: Platygasteridae) parasitoïde oophage de *Phoracantha semipunctata* (Col.: Cerambycidae), au Maroc. *Entomophaga* 33, 381–382.

Geertsema, H. (1975) Studies on the biology, ecology and control of the pine tree emperor moth, *Nudaurelia cytherea cytherea* (Fabr.) (Lepidoptera: Saturniidae). *Annals of the University of Stellenbosch* 50A(1), 1–170.

Geertsema, H. (1982) A historical review of forest entomology in South Africa. *Saasveld 50*, Directorate of Forestry, Department of Environment Affairs, 199–207.

Geertsema, H. and Van den Berg, M.A. (1973) A review of the more important forest pests of South Africa. *South African Forestry Journal* 85, 29–34.

Gess, F.W. (1964) The discovery of a parasite of the *Phoracantha* beetle (Coleoptera, Cerambycidae) in the Western Cape. *Journal of the Entomological Society of Southern Africa* 27, 152.

Giesen, H., Kohnle, U., Vité, J.P., Pan, M.-L. and Franke, W. (1984) Das Aggregation-spheromon des mediterranen Kiefernborkenkäfers *Ips* (*Orthotomicus*) *erosus*. *Zeitschrift für angewandte Entomologie* 98, 95–97.

Grobler, J.H. (1957) *Some Aspects of the Biology, Ecology and Control of the Pine Brown Tail Moth* Euproctis terminalis *Walk*. South African Department of Agriculture Bulletin, Government Printer, Pretoria, South Africa.

Hanks, L.M., Paine, T.D. and Millar, J.G. (1993) Host species preference and larval performance in the wood-boring beetle *Phoracantha semipunctata* F. *Oecologia* 95, 22–29.

Hanks, L.M., Gould, J.R., Paine, T.D., Millar, J.G. and Wang, Q. (1995) Biology and host relations of *Avetianella longoi* (Hymenoptera: Encyrtidae), an egg parasitoid of the Eucalyptus longhorned borer (Coleoptera: Cerambycidae). *Annals of the Entomological Society of America* 88, 666–671.

Hanks, L.M., Paine, T.D. and Millar, J.G. (1996) Tiny wasp helps protect eucalypts from eucalyptus longhorned borer. *California Agriculture* 50, 14–16.

Hanks, L.M., Millar, J.G., Paine, T.D., Wang, Q. and Paine, E.O. (2001) Patterns of host utilization by two parasitoids (Hymenoptera: Braconidae) of the eucalyptus longhorned borer (Coleoptera: Cerambycidae). *Biological Control* 21, 152–159.

Harman, D.M. and Kulman, H.M. (1967) *Parasites and predators of the white pine weevil*, Pissodes strobi *(Peck)*. University of Maryland, Natural Resources Institute, Lavale, Maryland, USA, Contribution No. 323, 35pp.

Hardenberg, C.B. (1912) The willow tree caterpillar (*Angelica tyrrhea* Cramer, Order Lepidoptera: family, Saturniidae). A destructive pest in forest plantations. *Agricultural Journal of the Union of South Africa* 55, 3–24.

Hepburn, G.A. (1973) The wattle bagworm. A review of investigations conducted from 1899 to 1970. *Report of the Wattle Research Institute* 1972–1973, 75–93.

Huber, J.T. and Prinsloo, G.L. (1990) Redescription of *Anaphes nitens* (Girault) and description of two new species of *Anaphes* Haliday (Hymenoptera: Mymaridae), parasites of *Gonipterus scutellatus* Gyllenhal (Coleoptera: Curaulionidae) in Tasmania. *Journal of the Australian Entomological Society* 29, 333–341.

Johannsmeier, M.F. (1976) Factors affecting the nectar flow in *Eucalyptus grandis* (Hill) Maiden. In: Fletcher, D.J.C. (ed.) *African Bees: Taxonomy, Biology and Economic Use*. Apimondia, Pretoria, South Africa, pp. 184–194.

Kenis, M. (1994) Variations in diapause among populations of *Eubazus semirugosus* (Nees) (Hym.: Braconidae), a parasitoid of *Pissodes* spp. (Col.: Curculionidae). *Norwegian Journal of Agricultural Sciences*, Supplement 16, 77–82.

Kenis, M. (1997) Biology of *Coeloides sordidator* (Hymenoptera: Braconidae), a possible candidate for introduction against *Pissodes strobi* (Coleoptera: Curculionidae) in North America. *Biocontrol Science and Technology* 7, 153–164.

Kenis, M. and Mills, N.J. (1994) Parasitoids of European species of the genus *Pissodes* (Col: Curculionidae) and their potential for the biological control of *Pissodes strobi* (Peck) in Canada. *Biological Control* 4, 14–21.

Kfir, R. (1986) Releases of natural enemies against the pine bark beetle *Orthotomicus erosus* (Wollaston) in South Africa. *Journal of the Entomological Society of Southern Africa* 49, 391–392.

Kirsten, J.F. (2001) Biological control of the eucalyptus borers, *Phoracantha semipunctata* and *P. recurva*, in the summer-rainfall forestry regions of South Africa: establishment of two larval parasitoids, *Syngaster lepidus* and *Jarra phoracantha. Proceedings of the 13th Entomological Congress, Entomological Society of Southern Africa, 2–5 July 2001*. Pietermaritzburg, Entomological Society of Southern Africa, Pretoria, South Africa, p. 34.

Klimetzek, D. and Vité, J.P. (1986) Die Wirkung insektenbürtiger Duftstoffe auf das Aggregationsverhalten des mediterranean Kieferenborkenkäfers *Orthotomicus erosus*. *Journal of Applied Entomology* 101, 239–243.

Longo, S., Palmeri, V. and Sommariva, D. (1993) Sull'attività di *Avetianella longoi* ooparassitoide di *Phoracantha semipunctata* nell'Italia meridionale. *Redia* 76, 223–239.

Lounsbury, C.P. (1917) *The* Phoracantha *beetle. A borer pest of* Eucalyptus *trees*. Local Series, Division of Entomology, Department of Agriculture, Pretoria, South Africa, 24, 1–10.

Malloch, J.R. (1934) Notes on Australian Diptera *34*. *Proceedings of the Linnaean Society of New South Wales* 59, 1–8.

Mendel, Z. (1986) Hymenopterous parasitoids of bark beetles (Scolytidae) in Israel: host relation, host plant, abundance and seasonal history. *Entomophaga* 31, 113–125.

Mendel, Z. (1988) Attraction of *Orthotomicus erosus* and *Pityogenes calcaratus* to a synthetic aggregation pheromone of *Ips typographus*. *Phytoparasitica* 16, 109–117.

Mendel, Z. and Halperin, J. (1981) Parasites of bark beetles (Coleoptera: Scolytidae) on pine and cypress in Israel. *Entomophaga* 26, 375–379.

Mendel, Z., Golan, Y. and Madar, Z. (1984) Studies on the phenology and some mortality factors of the eucalypt borer *Phoracantha semipunctata* in Israel. *La Ya'aran* 34, 41–44.

Mills, N.J. (1983) The natural enemies of scolytids infesting conifer bark in Europe in relation to the biological control of *Dendroctonus* spp. in Canada. *Biocontrol News and Information* 4, 305–328.

Moore, J.A. (1993) Rediscovery and releases of parasitoid of Eucalyptus borer. *Plant Protection News* 33, 4 and 7.

Moore, K.M. (1972) Observations on some Australian forest insects. 25. Additional information on some parasites and predators of longicorns (Cerambycidae: Phoracanthini). *Australian Zoologist* 17, 26–29.

Neumann, F.G., Morey, J.L. and McKimm, R.J. (1987) The *Sirex* wasp in Victoria. *Bulletin No. 29*, Lands and Forests Division, Department of Conservation, Forests and Lands, Melbourne, Australia, pp. 1–41.

Paine, T.D., Millar, J.G. and Hanks, L.M. (1995) Integrated program protects trees from eucalyptus longhorned borer. *California Agriculture* 49, 34–38, 40.

Poynton, R.J. (1979) *Tree Planting in Southern Africa*. Vol. 1. *The Pines*, Vol. 2. *The Eucalypts*. Department of Forestry, Pretoria, South Africa.

Richardson, K.F. and Meakins, R.H. (1986) Inter- and intra-specific variation in the susceptibility of eucalypts to the snout beetle *Gonipterus scutellatus* Gyll. (Coleoptera: Curculionidae). *South African Forestry Journal* 139, 21–31.

Scriven, G.T., Reeves, E.L. and Luck, R.F. (1986) Beetle from Australia threatens eucalyptus. *California Agriculture* 40, 4–6.

Sétamou, M., Schulthess, F., Bosque-Pérez, N.A., Poehling, H.-M. and Borgemeister, C. (1999) Bionomics of *Mussidia nigrivenella* (Lepidoptera: Pyralidae) on three host plants. *Bulletin of Entomological Research* 89, 465–471.

Siscaro, G. (1992) *Avetianella longoi* sp. n. (Hymenoptera Encyrtidae) egg parasitoid of *Phoracantha semipunctata* F. (Coleoptera Cerambycidae). *Bolletino di Zoologia Agraria e di Bachicoltura* 24, 205–212.

Taylor, R.L. (1929) The biology of the white pine weevil *Pissodes strobi* (Peck) and a study of its insect parasites from an economic view point. *Entomologica Americana* 9, 166–246; 10, 1–86.

Tooke, F.G.C. (1926) Eucalyptus snout beetle. Extent to which different kinds of *Eucalyptus* are attacked. *Entomology Notes Series Number 50*. Department of Agriculture, Pretoria, South Africa.

Tooke, F.G.C. (1935) The *Phoracantha* beetle. Insects injurious to forest and shade trees. *Bulletin of the Department of Agriculture of South Africa* 142, 33–39.

Tooke, F.G.C. (1942) The biological control of the Eucalyptus snout beetle, *Gonipterus scutellatus*, Gyll. A summarized report. *Science Bulletin* No. 235, Department of Agriculture and Forestry, Pretoria, South Africa.

Tooke, F.G.C. (1955) The Eucalyptus snout beetle, *Gonipterus scutellatus* Gyll.: a study of its ecology and control by biological means. *Entomology Memoirs 3*. Division of Entomology, Department of Agriculture, Pretoria, South Africa, Pamphlet 142.

Tribe, G.D. (1990a) Phenology of *Pinus radiata* log colonization by the pine bark beetle *Hylastes angustatus* (Herbst) (Coleoptera: Scolytidae) in the south-western Cape Province. *Journal of the Entomological Society of Southern Africa* 53, 93–100.

Tribe, G.D. (1990b) Phenology of *Pinus radiata* log colonization and reproduction by the European bark beetle *Orthotomicus erosus* (Wollaston) (Coleoptera: Scolytidae) in the south-western Cape Province. *Journal of the Entomological Society of Southern Africa* 53, 117–126.

Tribe, G.D. (1991a) Phenology of *Oosomus varius* (Curculionidae) and *Prasoidea sericea* (Chrysomelidae) on *Pinus radiata* in the south-western Cape Province. *South African Forestry Journal* 157, 47–51.

Tribe, G.D. (1991b) *Drosophila flavohirta* Malloch (Diptera: Drosophilidae) in *Eucalyptus* flowers: occurrence and parasites in eastern Australia and potential for biological control on *Eucalyptus grandis* in South Africa. *Journal of the Australian Entomological Society* 30, 257–262.

Tribe, G.D. (1991c) Phenology of *Pinus radiata* log colonization by the red-haired pine bark beetle *Hylurgus ligniperda* (Fabricius) (Coleoptera: Scolytidae) in the south-western Cape Province. *Journal of the Entomological Society of Southern Africa* 54, 1–7.

Tribe, G.D. (1991d) Phenology of three exotic pine bark beetle species (Coleoptera: Scolytidae) colonising *Pinus radiata* logs in the south-western Cape Province. *South African Forestry Journal* 157, 27–31.

Tribe, G.D. (1992) Colonisation sites on *Pinus radiata* logs of the bark beetles, *Orthotomicus erosus*, *Hylastes angustatus* and *Hylurgus ligniperda* (Coleoptera: Scolytidae). *Journal of the Entomological Society of Southern Africa* 55, 77–84.

Tribe, G.D. (1995) The woodwasp *Sirex noctilio* Fabricius (Hymenoptera: Siricidae), a pest of *Pinus* species, now established in South Africa. *African Entomology* 3, 215–217.

Tribe, G.D. (2000) Ecology, distribution and natural enemies of the *Eucalyptus*-defoliating tortoise beetle *Trachymela tincticollis* (Blackburn) (Chrysomelidae: Chrysomelini: Paropsina) in southwestern Australia, with reference to its biological control in South Africa. *African Entomology* 8, 23–45.

Tribe, G.D. and Cillié, J.J. (1997) Biology of the Australian tortoise beetle *Trachymela tincticollis* (Blackburn) (Chrysomelidae: Chrysomelini: Paropsina), a defoliator of *Eucalyptus* (Myrtaceae), in South Africa. *African Entomology* 5, 109–123.

Tribe, G.D. and Cillié, J.J. (2000) Biological control of the *Eucalyptus*-defoliating Australian tortoise beetle *Trachymela tincticollis* (Blackburn) (Chrysomelidae: Chrysomelini: Paropsina) in South Africa by the egg parasitoid *Enoggera reticulata* Naumann (Hymenoptera: Pteromalidae: Asaphinae). *African Entomology* 8, 15–22.

Tribe, G.D. and Kfir, R. (2001) The establishment of *Dendrosoter caenopachoides* (Hymenoptera: Braconidae) introduced into South Africa for the biological control of *Orthotomicus erosus* (Coleoptera: Scolytidae), with additional notes on *D.* sp. nr. *labdacus*. *African Entomology* 9, 5–198.

Tribe, G.D., Du Toit, A.P., Van Rensburg, N.J. and Johannsmeier, M.F. (1989) The collection and release of *Tetrastichus* sp. (Eulophidae) as a biological control agent for *Drosophila flavohirta* Malloch (Drosophilidae) in South Africa. *Journal of the Entomological Society of Southern Africa* 52, 181–182.

Van den Berg, M.A. (1969) Die bestryding van *Nudaurelia cytherea cytherea* (Fab.) met varke. *South African Forestry Journal* 69, 14–15.

Van den Berg, M.A. (1973) Host plants of, and degree of defoliation by three saturniids injurious to forest trees. *Phytophylactica* 5, 65–70.

Van den Berg, M.A. (1974) Bio-ecological studies on forest pests. 1. *Pseudobunaea irius* (F.) (Lepidoptera: Saturniidae). *Forestry in South Africa* 15, 1–16.

Van den Berg, M.A. (1975) Bio-ecological studies on forest pests. 2. *Holocerina smilax* (Westw.) (Lepidoptera: Saturniidae). *Forestry in South Africa* 16, 1–11.

Van der Zel, D.W. (1994) Indigenous forests in South Africa. *Arbor* 2, 10.

Wang, Q. (1995) A taxonomic revision of the Australian genus *Phoracantha* Newman (Coleoptera: Cerambycidae). *Invertebrate Taxonomy* 9, 865–958.

Way, M.J., Cammell, M.E. and Paiva, M.R. (1992) Studies on egg predation by ants (Hymenoptera: Formicidae) especially on the eucalyptus borer *Phoracantha semipunctata* (Coleoptera: Cerambycidae) in Portugal. *Bulletin of Entomological Research* 82, 425–432.

Webb, D.V.V. (1974) Forest and timber entomology in the Republic of South Africa. *Entomology Memoir*, Department of Agricultural Technical Services, South Africa 34, 1–21.

9

Biological Control of Gramineous Lepidopteran Stem Borers in Sub-Saharan Africa

William A. Overholt,[1] Des E. Conlong,[2] Rami Kfir,[3] Fritz Schulthess[4] and Mamadou Sétamou[5]

[1]Indian River Research and Education Center, University of Florida, Florida, USA; [2]SA Sugar Experimental Station, KwaZulu-Natal, South Africa; [3]Agricultural Research Council – Plant Protection Research Institute, Division of Insect Ecology – Biological Control, Pretoria, South Africa; [4]Postfach 112, Chur, Switzerland; [5]International Centre of Insect Physiology and Ecology, Nairobi, Kenya

Introduction

Lepidopteran cereal stem and cob borers are generally considered to be the most injurious insect pests of maize, sorghum, millet, rice and sugarcane in sub-Saharan Africa (Kfir *et al.*, 2002). Maes (1998) lists 21 species considered to be of economic importance; however, within any region/crop combination, only a small subset are damaging (Kfir *et al.*, 2002). Among all the stem borers associated with cereal crops and sugarcane, two species are exotic to Africa, and thus present traditional targets for classical biological control; *Chilo partellus* (Swinhoe) (Plate 23), which attacks maize and sorghum in eastern and southern Africa, and the sugarcane borer, *Chilo sacchariphagus* (Bojer) (Lepidoptera, Crambidae), which was recently confirmed to be present in Mozambique. New association biological control (Hokkanen and Pimentel, 1989) and redistribution of natural enemies within Africa (Schulthess *et al.*, 1997) have been attempted against some native and exotic borers, thus far with limited success. This chapter summarizes work on biological control of cereal stem borers using parasitoids in eastern and southern Africa, West Africa and South Africa and work on sugarcane stem borer biological control in mainland Africa and the Mascarene Islands. A list of the parasitoids introduced into Africa for stem borer biological control can be found in Overholt (1998).

(eds P. Neuenschwander, C. Borgemeister and J. Langewald)

Cereal Stem Borers

Eastern and southern Africa

Maize and sorghum are the most important cereals for human consumption in eastern and southern Africa (Seshu Reddy, 1998). Two stem borers are dominant in these crops: *Busseola fusca* Fuller (Lepidoptera, Noctuidae) and *C. partellus* (Kfir *et al.*, 2002). *B. fusca* is found at elevations >600 m, whereas *C. partellus* occurs from sea level to 2300 m, but is most common at lower elevations (Zhou *et al.*, 2001b). In addition to these two species, *Eldana saccharina* (Walker) (Lepidoptera, Pyralidae) (Plate 22) is important in Uganda (Girling, 1980), and *Chilo orichalcociliellus* (Strand) (Crambidae) is abundant in the low elevation coastal areas of Kenya and Tanzania (Nye, 1960). *Sesamia cretica* Lederer (Noctuidae) is considered to be a major pest of sorghum in Sudan (Tams and Bowden, 1953).

Introduction biological control

Due to its status as an introduced pest, *C. partellus* has been the primary target of two classical biological control programmes in eastern and southern Africa. The first attempt was conducted by the CIBC from 1968 to 1972 (CIBC, 1968–1972) in Kenya, Tanzania and Uganda. Eight parasitoids, mostly old association natural enemies of *C. partellus* from India, were released in Uganda, Tanzania and Kenya, but none of these established, and there is little information available on the reasons for the failures.

In 1991, the ICIPE initiated a second biological control programme against *C. partellus*. The gregarious koinobiont larval parasitoid, *Cotesia flavipes* Cameron (Hymeoptera, Braconidae), was introduced from Pakistan, and later from India (Overholt, 1998). The same parasitoid had been released by CIBC in the earlier biological control programme. Host-range testing revealed that *C. flavipes* would attack *C. partellus*, *C. orichalcociliellus*, *Sesamia calamistis* Hampson (Noctuidae) (Plate 21), *B. fusca* and *E. saccharina*, but could only complete development in the first three species (Overholt *et al.*, 1994a; Ngi-Song *et al.*, 1995). Thus, it was hypothesized that the parasitoid would establish in areas where the suitable borers occurred, but not in areas where unsuitable borers were abundant. The first releases of *C. flavipes* were made in the coastal area of Kenya in 1993, and surveys during the same season showed that the parasitoid readily colonized the fields where it was released (Overholt *et al.*, 1994b); however, only a few individuals were recovered in 1994 and 1995. The number of recoveries increased from 1996, and by 1997, *C. flavipes* was the dominant parasitoid of stem borers in the southern coastal area of Kenya (Overholt, 1998). The impact of *C. flavipes* on borer populations was examined in 1999, and it was found that the total borer population (*C. partellus*, *C. orichalcociliellus* and *S. calamistis*) had declined by approximately 30%, whereas the *C. partellus* population had declined more than 50% (Zhou *et al.*, 2001a).

C. flavipes was later released in the Eastern Province of Kenya in 1997 and levels of parasitism of around 10% were found 1 year later (Songa *et al.*, 2001). In western Kenya, *C. flavipes* was never purposely released, but was found to be present (Omwega *et al.*, 1997). However, parasitism in western Kenya has remained low compared with the Coast and Eastern Provinces, which is likely due to the presence of two unsuitable hosts, *B. fusca* and *E. saccharina*, both of which are thought to serve as a sink for population growth (Zhou and Overholt, 2001). From 1996, the programme expanded regionally, and releases have now been made in several countries. Establishment has been confirmed in Mozambique, Malawi, Uganda, Zanzibar, Tanzania and Ethiopia (Kfir *et al.*, 2002).

ICIPE imported a second parasitoid, *Xanthopimpla stemmator* (Thunberg) (Hymenoptera, Ichneumonidae) (Plate 24) in 2001. *X. stemmator* is an idiobiont pupal parasitoid, which was successfully established in Mauritius and Reunion for biological control of *C. sacchariphagus* in sugarcane, and is currently being released in Mozambique against the same host (Conlong and Goebel, 2002). Host-range studies at ICIPE revealed that *X. stemmator* attacks and develops in all major borer species in eastern and southern Africa (C.W. Gitau, Nairobi, 2002, unpublished results). Additionally, *X. stemmator* uses a 'drill and sting' foraging strategy (Smith *et al.*, 1993), a behaviour that is not shared by any common native pupal parasitoids, and thus *X. stemmator* may fill a largely vacant niche in Africa.

Habitat management

Recent work in Kenya has shown that stem borer populations in maize can be decreased by planting certain native African grasses that are highly attractive to ovipositing female borers as a border around maize fields (Khan *et al.*, 2001). Studies have shown that both *C. partellus* and *B. fusca* prefer to lay their eggs on *Sorghum sudanense* Stapf or *Pennisetum purpureum* Schumach. (both *Poaceae*) as compared with maize. Moreover, Khan *et al.* (1997) demonstrated that parasitism of stem borers increases when these grasses are present around maize. Work in West Africa has shown similar results (Ndemah *et al.*, 2003). Additionally, Khan and co-workers have found three plants, *Melinis munitiflora* Beauv. (*Poaceae*), *Desmodium uncinatum* Jacq., and *Desmodium intortum* (Mill.) Urb. (*Fabaceae*) that are repellent to stem borers and can be intercropped between maize rows to drive stem borers away from the crop. Together, these two approaches have been referred to as the 'push–pull' strategy (Khan *et al.*, 2001). Efforts are now under way to integrate this strategy with classical biological control.

West Africa

Maize and other cereals such as rice, sorghum and pearl millet are important food crops for millions of people in western Africa. Lepidopteran stem and cob borers are considered to be the major biotic constraints limiting both quality and quantity of yield (Bosque-Pérez and Schulthess, 1998). The most commonly reported stem borers attacking maize include the noctuids *B. fusca*, *S. calamistis* and *Sesamia botanephaga* Tams & Bowden, and the pyralid *E. saccharina*. The cob borer *Mussidia nigrivenella* Ragonot (Lepidoptera, Pyralidae) is also an important pest and occurs in various parts of mainland Africa, Réunion, Madagascar and Asia (Janse, 1941; LePelley, 1959), but is only considered a pest in West Africa, where it attacks maize cobs (Bosque-Pérez and Schulthess, 1998) and cotton bolls (Silvie, 1990). It is highly polyphagous and has been recorded from 20 plant species belonging to 11 families (Sétamou *et al.*, 2000).

On sorghum, *B. fusca* is considered to be most important (Bosque-Pérez and Schulthess, 1998), while in pearl millet, *Coniesta ignefusalis* (Hampson) (Lepidoptera, Pyralidae) is the only important species (Harris and Youm, 1998). *Sesamia* spp. (mainly *S. penisetii* Tams & Bowden and *S. botanephaga*) and *E. saccharina* are reported to be damaging to sugarcane (Scheibelreiter, 1980), and *Chilo zacconius* Bleszynski (Crambidae), *Maliarpha separatella* Ragonot (Pyralidae) and *S. calamistis* are typically the most important borers on rice (Heinrichs, 1998).

Introduction biological control of cereal stem borers

Probably because all the stem and cob borers in West Africa are indigenous, biological control has received little attention. Between 1972 and 1981, five new association larval and pupal parasitoids from Asia were released in Senegal and Côte d'Ivoire against the crambid rice borer, *C. zacconius* (Bordat *et al.*, 1977; Bordat, 1983). There is no information on the numbers of parasitoids released, nor the results of releases. Moreover, the parasitoids were only tested for compatibility with the target pest, although several other borers occur in the same cropping system (Heinrichs, 1998). There appears to be no published information on post-release studies, and it is unlikely that any species established.

In the early 1990s, three *Cotesia* spp. were introduced from ICIPE into the laboratories of the IITA in Bénin: the exotic *C. chilonis* (Matsumura) and *C. flavipes*, both previously introduced into Ghana, Senegal and Côte d'Ivoire, and the indigenous *C. sesamiae* (Cameron) (Plate 25). Suitability studies predicted that the two exotic species would not establish because they readily parasitized unsuitable host species (Hailemichael *et al.*, 1997), which was verified in subsequent releases. The *C. sesamiae* imported by the IITA originated from a coastal Kenyan population, which was reared on *S. calamistis*, but did not successfully reproduce on a Kenyan population of *B. fusca* (Ngi-Song *et al.*, 1995). In eastern and southern Africa, *C. sesamiae* is one of the most common parasitoids of *B. fusca* and *S. calamistis*, and is thought to keep the latter under control in most areas. By contrast, this species is exceedingly rare in West Africa (Conlong, 2000; Ndemah *et al.*, 2001). The Kenyan *C. sesamiae* strain became established in southern Bénin (Schulthess *et al.*, 1997), but releases in Nigeria were not successful and the work was stopped. This emphasizes the need for proper pre- and post-release studies in order to understand the reasons for lack of establishment.

Introduction biological control of the cob borer M. nigrivenella

Country-wide surveys in West Africa found large differences in parasitism of *M. nigrivenella* on various host plants. Significant parasitism was only obtained from four host plants, all tree species, whereas on annual plants, parasitoids were largely absent (Sétamou *et al.*, 2002). By contrast, several egg and pupal parasitoids were recovered from *M. nigrivenella* on maize in Cameroon (Ndemah *et al.*, 2001). Since *M. nigrivenella* is reported from different parts of Africa, but only mentioned as a pest of importance in western Africa, it is possible that *M. nigrivenella* populations in eastern and southern Africa do not recognize maize and cotton as host plants and/or *M. nigrivenella* in eastern and southern Africa is under biological control in its wild habitats, thereby minimizing its movement into crop fields. If this is the case, it would open opportunities for redistributing natural enemies to areas where they do not exist, or for new association biological control against *M. nigrivenella* in West Africa.

Habitat management

Wild grasses are believed to be a reservoir for stem borers and responsible for pest outbreaks on crops (Bowden, 1976). Surveys in several West African countries, however, showed that the higher the wild grass abundance around maize fields, the lower the pest incidence on maize (Schulthess *et al.*, 1997). Ovipositional preference and life-table studies revealed that wild grass species such as *Pennisetum polystachion* (L.), *Panicum maximum* Jacq and *Sorghum arundinaceum* (Desv.) Stapf (all *Poaceae*), were highly attractive to ovipositing female moths, although mortality of immature stages was close to 100%, vs. 70%–80% on maize, and fecundity of adult offspring was considerably lower on wild grasses than on maize (Shanower *et al.*, 1993; S. Sekloka, 1996, unpublished results; A.K. Sémeglo, 1997, unpublished results; Schulthess *et al.*, 1997). Thus, some of the grass species form a reproductive sink for the borers. In addition, Schulthess *et al.* (2001) found high parasitism of *S. calamistis* eggs by *Telenomus* spp. (Hymenoptera, Scelionidae) during the dry season on wild hosts in the inland valleys. Similarly, Ndemah *et al.* (2001), working in the forest zone of Cameroon, found a higher parasitoid species diversity on *P. purpureum* than on maize. Multi-seasonal studies in Bénin, which involved several grass species planted as border rows around maize, gave significantly higher yields, had considerably lower pest densities and increased parasitism compared with fields where no grass borders were present (Ndemah *et al.*, 2003). It was suggested that wild grass habitats play an important role in maintaining stable parasitoid populations during the off-season, and thereby lower pest incidence in crop fields during the growing season (Khan *et al.*, 1997; Ndemah *et al.*, 2001). In the past, most work on the impact of biological control on crops has been aimed at the crop itself, and the role of alternative wild host plants has been widely neglected. Future attempts at solving borer problems in crop fields have to take into consideration the role grass habitats play in maintaining stable populations of natural enemies.

South Africa

Maize is the principal agricultural crop sown in South Africa. It is produced on roughly 4 million ha with an annual production of 12 million t. Sorghum is another important crop, especially in the marginal grain areas, because it is drought resistant (van Rensburg and van Hamburg, 1975). Altogether 300,000 ha of grain sorghum with a yield of 0.5 million t are planted. The principal pests of cereals in South Africa are *B. fusca* and *C. partellus* (Skoroszewski and van Hamburg, 1987).

The profit margin for maize and sorghum in South Africa is relatively low due to high production costs; pest control alone can amount to 56% of the production costs (van Hamburg, 1987). The timing of chemical control is crucial, as chemical measures are only effective against exposed young borer larvae. Older larvae penetrate the stalks and are difficult to control. Estimated yield losses from borer damage range from 10% to total crop loss in South Africa (van Rensburg and Bate, 1987). The economic threshold for applying chemical control is normally when as much as 40% of plants show signs of damage (Bate and van Rensburg, 1992).

Native natural enemies

Early investigations on *B. fusca* in South Africa by Mally (1920) and by Du Plessis and Lea (1943) identified the gregarious, larval, braconid parasitoid, *C. sesamiae,* as an important mortality factor. In more recent studies, the relative abundance and seasonality of the parasitoids of *B. fusca* (Kfir, 1990; Kfir and Bell, 1993) and *C. partellus* (Kfir, 1992b, 1995), as well as certain predators (Watmough and Kfir, 1995) and microbial pathogens (Hoekstra and Kfir, 1997), were clarified. The most abundant larval parasitoids of *B. fusca* in South Africa were *C. sesamiae* (Cameron) (Braconidae) and *Bracon sesamiae* Cameron (Braconidae). The most abundant pupal parasitoid was *Procerochasmias nigromaculatus* (Cameron) (Hymenoptera, Ichneumonidae) (Kfir, 2000). On *C. partellus* in South Africa, the most abundant larval parasitoid was again *C. sesamiae* and the most abundant pupal parasitoids were *Dentichasmias busseolae* Heinrich (Ichneumonidae) and *Pediobius furvus* (Gahan) (Eulophidae) (Kfir, 1992b).

Most parasitoids from *B. fusca* and *C. partellus* are indigenous. Their association with *B. fusca* is a long-standing one, whereas the exotic *C. partellus* is a more recent host (Kfir, 1992b). These parasitoids are more habitat-specific with regard to host selection than they are taxonomically specific (Kfir, 1997). This may explain why several native parasitoids have been recorded from both stem borers in South Africa (Kfir, 1990, 1995, 1997). Using insecticide-exclusion trials, Kfir (2002) demonstrated the importance of these indigenous parasitoids in controlling stem borers in sorghum: infestations of *B. fusca* and *C. partellus* were significantly higher in treated plots, while levels of parasitism were significantly higher in untreated plots.

Introduction biological control

Despite high mortalities by native parasitoids, natural enemies were unable to prevent economic damage (Kfir and Bell, 1993). This factor, together with ineffective chemical control, stimulated a biological control programme involving the introduction of exotic parasitoids into South Africa. Altogether, four braconids, one eulophid, three ichneumonids, two trichogrammatids and four tachinids were introduced from 12 countries between 1977 and 2000. Special attention was given to *C. flavipes* introduced from Pakistan, Mauritius and Indonesia (Skoroszewski and van Hamburg, 1987; Kfir, 1991, 1992a, 1994, 1997, 2001). *X. stemmator* was first released in South Africa in the early 1990s against *Eldana* in sugarcane (by Conlong) and against *C. partellus* in maize (by Kfir). In both cases the parasitoid did not establish. Similarly, *C. flavipes* and *Trichogramma chilonis* Ishii (Hymenoptera, Trichogrammatidae) were recovered for short periods following releases, but have not become permanently established (Kfir, 1997).

Most of these introductions concerned parasitoids originating from the tropics, which may not have been able to adapt successfully to the seasonally cool climate. The main summer grain-producing areas of South Africa lie at high elevations, which experience dry winters with freezing temperatures, necessitating long diapause periods of borer populations that are not conducive to survival of non-adapted exotic tropical parasitoids. Additional releases of *C. flavipes* are planned, but this time releases will be concentrated in the warmer lowland areas of Mpumalanga and Kwazulu-Natal Provinces, where the probability of establishment should be much higher than in the cooler high-elevation areas.

It is unlikely that additional releases of parasitoids from the tropics would result in establishment in the high-elevation areas of South Africa. A logical alternative is to introduce stem borer parasitoids that have obligatory diapause from temperate areas of the world (Kfir, 2000). *Macrocentrus grandi* Goidanich (Hymenoptera, Braconidae), a parasitoid of the European corn borer, *Ostrinia nubilalis* (Hubner) (Lepidoptera, Pyralidae), was introduced from Minnesota and Illinois, USA, and reared in quarantine on *C. partellus* (Kfir, 2001). Unfortunately the culture was lost, but other introductions are being considered.

Another option that has been considered is to introduce stem borer parasitoids that occur in high-elevation areas in other African countries, but are absent in South Africa. For example, *Sturmiopsis parasitica* Curran (Diptera, Tachinidae) is the most common parasitoid of *B. fusca* in the Harare area of Zimbabwe (Smithers, 1959). The tachinid was recently introduced into South Africa (Conlong, 2000, and see section on sugarcane) and releases in maize are planned. This species, collected from Bénin, has already been released into South African sugarcane since 1999 (Conlong, 2000).

Sugarcane Stem Borers

Modern sugarcane (*Saccharum officinarum* L. (Poaceae)) is derived from two wild ancestors, *Saccharum robustum* L. from Melanesia, and *Saccharum spontaneum* L. from East Africa and the southern part of Asia including the islands of South-East Asia and Melanesia (Stevenson, 1965). It is grown throughout the tropical and subtropical areas of the world, including several countries in Africa, the Mascarene Islands (Mauritius and Réunion) and Madagascar.

The insect pests of sugarcane are characteristically of more limited geographical distribution than sugarcane, generally being native insects that adopted sugarcane following its cultivation (Pemberton and Williams, 1969). However, with sugarcane industry expansion, the distribution of local insects was increased. The pest fauna of any region is thus a mixture of indigenous and alien species. On oceanic islands, the latter predominate because limited insular faunas of islands have provided few agricultural pests. On continents, the former are more common. On the islands of Mauritius and Hawaii, 84% of insects feeding on sugarcane are exotic (Pemberton and Williams, 1969), whereas in South Africa, only one of 35 insect pests recorded on sugarcane is exotic. Stem borers found in sugarcane in Africa and the Mascarene Islands reflect this tendency.

Because of the various origins of sugarcane pests, classical biological control has been most practised on islands. A combination of classical biological control, as well as augmentation of local populations of natural enemies, relocation of indigenous natural enemies and habitat management has been used on continents.

Until 1992, 14 species of lepidopteran stem borers had been recorded as pests of sugarcane in Africa. The majority of these are indigenous. Leslie (1994) ranked four as pests: *E. saccharina*, *Chilo agamemnon* Bleszynski, *S. cretica* and *S. calamistis*. More recently, *B. fusca* was recorded from West Africa. In 1999, a 15th species, *C. sacchariphagus* (Bojer), was confirmed to be attacking sugarcane in Mozambique (Way and Turner, 1999). This is the first record of an exotic lepidopteran becoming an economic pest on African sugarcane.

Introduction biological control

Conlong (1998) lists exotic and indigenous parasitoids already tested against the most common sugarcane stem borers in Africa and the Mascarene Islands. It is clear from published literature that most work in mainland Africa has involved *E. saccharina* (Conlong, 2000), although there were a series of introductions into Ghana against a complex of stem borers from 1974 to 1977, in which several new association parasitoids were released. No pre-release suitability studies were conducted, and no parasitoids established (Scheibelreiter, 1980). On the Mascarene Islands, where *C. sacchariphagus* is also an introduced species, it has been the target of much research (Williams, 1983; Ganeshan and Rajabalee, 1997; Goebel, 1999). More recently, research has commenced against *C. sacchariphagus* in Mozambique (Conlong and Goebel, 2002).

From 1979 to 1995, eight egg, 12 larval and two pupal new association parasitoids have been imported into South Africa and tested against *E. saccharina*. Conlong (1997) lists these and provides reasons for their unsuccessful establishment. During the same period, in South Africa, seven larval parasitoids were found attacking *E. saccharina* in *Cyperus papyrus* L. (Cyperaceae), and three egg parasitoids were imported from West Africa where they were collected from *E. saccharina* eggs. Some of these redistributed biological control agents showed early promise, but establishment in sugarcane in South Africa could not be shown (Conlong, 1997).

From 1992 to 1999, more emphasis was placed on redistribution of indigenous parasitoids of *E. saccharina* from other African areas, and 23 species were collected from eight countries (Conlong, 2001). It was evident that numerous indigenous parasitoids of *E. saccharina* were available in Africa. Unfortunately, biosystematic, life cycle and phenological knowledge of these species is poor, as illustrated by the fact that only 30% of species collected could be identified to species level. It is not surprising therefore, that laboratory colonization of these has not been successful. At present, only *S. parasitica* is being mass reared and released.

In South Africa, the association of *E. saccharina* with sugarcane is quite recent (first reported from South African sugarcane in 1939, whereas it was described from sugarcane in West Africa in 1865). Thus an effective natural enemy complex, adapted to exploit this pest in a sugarcane habitat, may not yet have evolved. In situations where there has been a longer association of a pest with sugarcane, as is the case with *S. calamistis,* effective natural enemies are already present, such as *Cotesia sesamiae* and *Stenobracon* sp. Additionally, *C. sesamiae* from Kenya has been successfully introduced into Mauritius for classical biological control of *S. calamistis*.

It is further hypothesized that when sugarcane is attacked by *E. saccharina,* parasitoids do not recognize sugarcane as a potential host habitat, possibly because this plant does not emit volatiles that parasitoids respond to when searching for a host (Conlong and Kasl, 2000). This may account for the lack of successful establishment of the many parasitoids introduced into the sugarcane ecosystem. The lack of success may be exacerbated by the occurrence of different strains of *E. saccharina*, further limiting the acceptability of the host in different regions of Africa (Conlong, 2001; King *et al.*, 2002).

In the Mascarene Islands, attempts to control the exotic *C. sacchariphagus* through parasitoid introductions started in the 1940s, with variable results. In Mauritius and Réunion, introduction and large-scale releases of parasitoids, mainly from India, did not control *C. sacchariphagus*, despite the successful establishment of *T. chilonis* and *C. flavipes*. In Réunion, several attempts to establish tachinid flies in the 1970s also failed (Goebel, 1999). At present, no new biological control agents are being released in the Mascarenes, but previously introduced species are being monitored (Ganeshan and Rajabalee, 1997).

Augmentation biological control

In Réunion, a new biological control programme using *Trichogramma* spp. is currently being implemented (Goebel, 1999). Morphological and molecular characterization of numerous strains from different sites around the country confirmed the presence of *T. chilonis.* All biotypes identified were evaluated for parasitism and the most suitable were mass-reared for inundative releases (Goebel, 1999). Similarly, in Mozambique, *Trichogramma bournieri* Pintureau and Babault has recently been identified from *C. sacchariphagus* eggs collected from sugarcane, and thus may provide a candidate for augmentation biological control in that country (Conlong and Goebel, 2002).

References

Bate, R. and van Rensburg, J.B.J. (1992) Predictive estimation of maize yield loss caused by *Chilo partellus* (Swinhoe) (Lepidoptera: Pyralidae) in maize. *South African Journal of Plant and Soil* 9, 150–154.

Bordat, D. (1983) Mise au point de l'élevage de masse d'*Apanteles chilonis* Matsumura et d'*Apanteles flavipes* (Cameron) (Hymenoptera : Braconidae) sur trois lépidoptères Pyralidae foreurs des graminées (*Chilo zacconius* Bleszynski ; *Chilo partellus* (Swinhoe) et *Diatrea saccharalis* (Fabricius) dans un objectif de lutte biologique. Thèse, Université des Sciences et Techniques du Languedoc, Académie de Montpellier.

Bordat, D., Brenière, J. and Coquard, J. (1977) Foreurs de graminées Africaines: parasitisme et techniques d'elevage. *Agronomie Tropicale* 32, 391–399.

Bosque-Pérez, N.A. and Schulthess, F. (1998) Maize: West and Central Africa. In: Polaszek, A. (ed.) *African Cereal Stem Borers: Economic Importance, Taxonomy, Natural Enemies and Control*. CAB International, Wallingford, UK, pp. 11–24.

Bowden, J. (1976) Stem borer ecology and strategy for control. *Annals of Applied Biology* 84, 107–134.

CIBC (1968–1972) Annual Reports of the Commonwealth Institute of Biological Control, Farnham Royal, UK.

Conlong, D.E. (1997) Biological control of *Eldana saccharina* Walker in South African sugarcane: constraints identified from 15 years of research. *Insect Science and its Application* 17, 69–78.

Conlong, D.E. (1998) Biological control in sugarcane. In: Saini, R.K. (ed.) *Tropical Entomology. Proceedings of the 3rd International Conference on Tropical Entomology*, Nairobi, Kenya. ICIPE Science Press, Nairobi, Kenya, pp. 73–92.

Conlong, D.E. (2000) Indigenous African parasitoids of *Eldana saccharina* (Lepidoptera: Pyralidae). *Proceedings of the South African Sugar Technologists Association* 74, 201–211.

Conlong, D.E. (2001) Biological control of indigenous African stemborers: what do we know? *Insect Science and its Application* 21, 267–274.

Conlong, D.E. and Kasl, B. (2000) Stimulo-deterrent diversion to decrease infestation in sugarcane by *Eldana saccharina* (Lepidoptera: Pyralidae). *Proceedings of the South African Sugar Technologists Association* 74, 212–214.

Conlong, D.E. and Goebel, F.R. (2002) Biological control of *Chilo sacchariphagus* (Lepidoptera: Crambidae) in Moçambique: the first steps. *Proceedings of the South African Sugar Technologists Association* 76, 310–320.

Du Plessis, C. and Lea, H.A.F. (1943) The maize stalk borer, *Calamistis fusca* (Hmps). *Bulletin, Department of Agriculture and Forestry, Union of South Africa*, 238.

Ganashan, S. and Rajabalee, A. (1997) Parasitoids of the sugarcane spotted borer, *Chilo sacchariphagus* (Lepidoptera: Pyralidae), in Mauritius. *Proceedings of the South African Sugar Technologists Association* 71, 87–90.

Girling, D.J. (1980) *Eldana saccharina* as a crop pest in Ghana. *Tropical Pest Management* 26, 152–156.

Goebel, F.R. (1999) Caractéristiques biotiques du foreur de la canne à sucre *Chilo sacchariphagus* (Bojer, 1856) (Lepidoptera: Pyralidae) à l'île de la Réunion. Facteurs de régulation de ses populations et conséquences pour la lutte contre ce ravageur. PhD thesis, University Paul Sabatier of Toulouse, France.

Hailemichael, Y., Schulthess, F., Smith, J.W., Jr and Overholt, W.A. (1997) Suitability of West African gramineous stemborers for the development of *Cotesia* species (Hymenoptera: Braconidae). *Insect Science and its Application* 17, 89–95.

Harris, K.M. and Youm, O. (1998) Millet: West Africa. In: Polaszek, A. (ed.) *African Cereal Stemborers: Economic Importance, Taxonomy, Natural Enemies and Control*. CAB International, Wallingford, UK, pp. 29–37.

Heinrichs, E.A. (1998) Rice: West Africa. In: Polaszek, A. (ed.) *African Cereal Stemborers: Economic Importance, Taxonomy, Natural Enemies and Control*. CAB International, Wallingford, UK, pp. 29–37.

Hoekstra, N. and Kfir, R. (1997) Microbial pathogens of the cereal stem borers *Busseola fusca* (Fuller) (Lepidoptera: Noctuidae) and *Chilo partellus* (swinhoe) (Lepidoptera: Pyralidae) in South Africa. *African Entomology* 5, 161–163.

Hokkanen, H.M.T. and Pimental, D. (1989) New associations in biological control: theory and practice. *Canadian Entomologist* 121, 829–840.

Janse, A.J.T. (1941) Contribution to the study of the Phycitinae (Pyralidae: Lepidoptera) Part 1. *Journal of the Entomological Society of South Africa* 4, 134–166.

Kfir, R. (1990) Parasites of the spotted stalk borer, *Chilo partellus* (Lepidoptera: Pyralidae) in South Africa. *Entomophaga* 35, 403–410.

Kfir, R. (1991) Selecting parasites for biological control of lepidopterous stalk borers in summer grain crops in South Africa. *Redia* 74, 231–236.

Kfir, R. (1992a) Alternative, non-chemical control methods for the stalk borers, *Chilo partellus* (Swinhoe) and *Busseola fusca* (Fuller), in summer grain crops in South Africa. *Technical Communication, Department of Agricultural Development, Republic of South Africa* 232, 99–103.

Kfir, R. (1992b) Seasonal abundance of the stem borer *Chilo partellus* (Lepidoptera: Pyralidae) and its parasites on summer grain crops. *Journal of Economic Entomology* 85, 518–529.

Kfir, R. (1994) Attempts at biological control of the stem borer, *Chilo partellus* (Swinhoe) (Lepidoptera: Pyralidae), in South Africa. *African Entomology* 2, 67–68.

Kfir, R. (1995) Parasitoids of the African stem borer, *Busseola fusca* (Lepidoptera: Noctuidae), in South Africa. *Bulletin of Entomological Research* 85, 369–377.

Kfir, R. (1997) Natural control of the cereal stemborers *Busseola fusca* and *Chilo partellus* in South Africa. *Insect Science and its Application* 17, 61–67.

Kfir, R. (2000) Seasonal occurrence, parasitoids and pathogens of the African stem borer, *Busseola fusca* (Fuller) (Lepidoptera: Noctuidae), on cereal crops in South Africa. *African Entomology* 8, 1–14.

Kfir, R. (2001) Prospects of biological control of *Chilo partellus* in grain crops in South Africa. *Insect Science and its Application* 21, 275–280.

Kfir, R. (2002) Increase in cereal stem borer populations through partial elimination of natural enemies. *Entomologia Experimentalis et Applicata* 104, 299–306.

Kfir, R. and Bell, R. (1993) Intraseasonal changes in populations of the African maize stem borer, *Busseola fusca* (Fuller) (Lepidoptera: Noctuidae), in Natal, South Africa. *Journal of African Zoology* 107, 543–553.

Kfir, R., Overholt, W.A., Khan, Z.R. and Polaszek, A. (2002) Biology and management of economically important lepidopteran cereal stem borers in Africa. *Annual Review of Entomology* 47, 701–31.

Khan, Z.R., Ampong-Nyarko, K., Chiliswa, P., Hassanali, A., Kimani, S., Lwande, W., Overholt, W.A., Pickett, J.A., Smart, L., Wadhams, L.J. and Woodstock, C.M. (1997) Intercropping increases parasitism of pests. *Nature* 388, 631–632.

Khan, Z.R., Pickett, J.A., Wadhams, L. and Muyekho, F. (2001) Habitat management strategies for the control of cereal stemborers and striga in maize in Kenya. *Insect Science and its Application* 21, 375–380.

King, H., Conlong, D.E. and Mitchell, A. (2002) Genetic differentiation in *Eldana saccharina* (Lepidoptera: Pyralidae): evidence from the mitochondrial cytochrome oxidase I and II genes. *Proceedings of the South African Sugar Technologists Association* 76, 321–328.

LePelley, R.H. (1959) *Agricultural Insects of East Africa.* East Africa High Commission, Nairobi, Kenya.

Leslie, G.W. (1994) Pest status, biology and effective control measures of sugar cane stalk borers in Africa and surrounding islands. In: Carnegie, A.J.M. and Conlong, D.E. (eds) *Biology, Pest Status and Control Measure Relationships of Sugar Cane Insect Pests.* Proceedings of Second Sugarcane Entomology Workshop of the International Society of Sugar Cane Technologists. 30 May–3 June 1994, S.A. Sugar Association Experiment Station, Mount Edgecombe, KwaZulu-Natal, South Africa, pp. 61–70.

Maes, K. (1998) Pyraloidea: Crambidae, Pyralidae. In: Polaszek, A. (ed.) *African Cereal Stemborers: Economic Importance, Taxonomy, Natural Enemies and Control.* CAB International, Wallingford, UK, pp. 87–98.

Mally, C.W. (1920) The maize stalk borer, *Busseola fusca*, Fuller. *Bulletin, Department of Agriculture, Union of South Africa* 3.

Ndemah, R., Schulthess, F., Poehling, H.-M. and Borgemeister, C. (2001) Natural enemies of lepidopterous borers on maize and elephant grass in the forest zone of Cameroon with special reference to *Busseola fusca* (Fuller) (Lepidoptera: Noctuidae). *Bulletin of Entomological Research* 91, 205–212.

Ndemah, R., Schulthess, F., Poehling, H.-M., Borgemeister, C. and Cardwell, K.F., (2003) Factors affecting infestations of the maize stalk borer *Busseola fusca* (Lepidoptera, Noctuidae) on maize in the forest zone of Cameroon with special reference to Scelionid eggs parasitoids. *Environmental Entomology* 32, 51–60.

Ngi-Song, A.J., Overholt, W.A. and Ayertey, J.N. (1995) Suitability of African gramineous stemborers for development of *Cotesia flavipes* and *Cotesia sesamiae* (Hymenoptera: Braconidae). *Environmental Entomology* 24, 978–984.

Nye, I.W.R. (1960) The insect pests of graminaceous crops in East Africa. *Colonial Research Studies* 31, 1–48.

Omwega, C.O., Overholt, W.A., Mbapila, J.C. and Kimani–Njogu, S.W. (1997) Establishment and dispersal of *Cotesia flavipes* Cameron (Hyemenoptera: Braconidae), an exotic endoparasitoid of *Chilo partellus* Swinhoe (Lepidoptera: Pyralidae) in northern Tanzania. *African Entomology* 5, 71–75.

Overholt, W.A. (1998) Biological control. In: Polaszek, A. (ed.) *African Cereal Stemborers: Economic Importance, Taxonomy, Natural Enemies and Control.* CAB International, Wallingford, UK, pp. 349–362.

Overholt, W.A., Ngi-Song, A.J., Kimani, S.K., Mbapila, J., Lammers, P. and Kioko, E. (1994a) Ecological considerations of the introduction of *Cotesia flavipes* Cameron (Hymenoptera: Bracondiae) for biological control of *Chilo partellus* (Lepidoptera: Pyralidae), in Africa. *Biocontrol News and Information* 15, 19N-24N.

Overholt, W.A., Ochieng, J.O., Lammers, P.M. and Ogedah, K. (1994b) Rearing and field release methods for *Cotesia flavipes* Cameron (Hymenoptera: Braconidae), a parasitoid of tropical gramineous stemborers. *Insect Science and its Application* 15, 253–259.

Pemberton, C.E. and Williams, J.R. (1969) Distribution, origins and spread of sugarcane insect pests. In: Williams, J.R., Metcalfe, J.R., Mungomery, W. and Mathes, R. (eds) *Pests of Sugarcane.* Elsevier, Amsterdam, The Netherlands, pp. 1–9.

Scheibelreiter, G.K. (1980) Sugar cane borers (Lep.: Noctuidae and Pyralidae) in Ghana. *Zeitschrift für Angewandte Entomologie* 89, 87–99.

Schulthess, F., Bosque-Perez, N.A., Chabi-Olaye, A., Gounou, S., Ndemah, R. and Goergen, G. (1997) Exchange of natural enemies of lepidopteran cereal stemborers between African regions. *Insect Science and its Application* 17, 97–108.

Schulthess, F., Chabi-Olaye, A. and Goergen, G. (2001) Seasonal fluctuations of noctuid stemborer egg parasitism in southern Bénin with special reference to *Sesamia calamistis* Hampson (Lepidoptera: Noctuidae) and *Telenomus* spp. (Hymenoptera: Scelionidae) on maize. *Biocontrol Science and Technology* 11, 765–777.

Seshu Reddy, K.V. (1998) Maize and sorghum: East Africa. In: Polaszek, A. (ed.) *African Cereal Stemborers: Economic Importance, Taxonomy, Natural Enemies and Control*. CAB International, Wallingford, UK, pp. 25–27.

Sétamou, M., Schulthess, F., Poehling, H.-M. and Borgemeister, C. (2000) Host plants and population dynamics of the cob borer *Mussidia nigrivenella* Ragonot (Lepidoptera: Pyralidae) in Bénin. *Environmental Entomology* 29, 516–524.

Sétamou, M., Schulthess, F., Goergen, G., Poehling, H.-M. and Borgemeister, C. (2002) Natural enemies of the maize cob borer, *Mussidia nigrivenella* Ragonot (Lepidoptera: Pyralidae) in Bénin, West Africa. *Bulletin of Entomological Research* 92, 343–349.

Shanower, T.G., Schulthess, F. and Bosque-Pérez, N.A. (1993) The effect of larval diet on the growth and development of *Sesamia calamistis* Hampson (Lepidoptera: Noctuidae) and *Eldana saccharina* Walker (Lepidoptera: Pyralidae). *Insect Science and its Application* 14, 681–685.

Silvie, P. (1990) *Mussidia nigrivenella* Ragonot (Pyralidae, Phycitinae): un ravageur mal connu du cotonnier. *Coton et Fibres Tropicales* 45, 323–333.

Skoroszewski, R.W. and van Hamburg, H. (1987) The release of *Apanteles flavipes* (Cameron) (Hymenoptera: Braconidae) against stalk-borers of maize and grain-sorghum in South Africa. *Journal of the Entomological Society of Southern Africa* 50, 249–255.

Smith, J.W., Wiedenmann, R.N. and Overholt, W.A. (1993) *Parasites of Lepidopteran Stemborers of Tropical Gramineous Plants*. ICIPE Science Press, Nairobi, Kenya.

Smithers, C.N. (1959) Some recent observations on *Busseola fusca* (Fuller) (Lep., Noctuidae) in Southern Rhodesia. *Bulletin of Entomological Research* 50, 809–819.

Songa, J.M., Overholt, W.A., Mueke, J.M. and Okello, R.O. (2001) Colonization of *Cotesia flavipes* (Hymenoptera: Braconidae) in stemborers in the semi-arid Eastern Province of Kenya. *Insect Science and its Application* 21, 289–295.

Stevenson, G.C. (1965) *Genetics and the Breeding of Sugarcane*. Longmans, London.

Tams, W.H.T. and Bowden, J. (1953) A revision of the African species of *Sesamia* Guenée and related genera (Lepidoptera : Agrotidae). *Bulletin of Entomological Research* 43, 645–678.

van Hamburg, H. (1987) A biological control approach to pest management on grain crops with special reference to the control of stalkborers. *Technical Communication, Department of Agriculture and Water Supply, Republic of South Africa* 212, 52–55.

van Rensburg, G.D.J. and Bate, R. (1987) Preliminary studies on the relative abundance and distribution of the stalk borers *Busseola fusca* and *Chilo partellus*. *Technical Communication, Department of Agriculture and Water Supply, Republic of South Africa* 212, 49–52.

van Rensburg, N.J. and van Hamburg, H. (1975) Grain sorghum pests: an integrated control approach. *Proceedings 1st Congress, Entomological Society of Southern Africa*, Entomological Society of South Africa, Pretoria, South Africa, pp. 151–162.

Watmough, R.H. and Kfir, R. (1995) Predation on pupae of *Helicoverpa armigera* Hbn. (Lep., Noctuidae) and its relation to stem borer numbers in summer grain crops. *Journal of Applied Entomology* 119, 679–688.

Way, M.J. and Turner, P.E.T. (1999) The spotted sugarcane borer, *Chilo sacchariphagus* (Lepidoptera: Pyralidae: Crambinae), in Mozambique. *Proceedings of the South African Sugar Technologists Association* 73, 112–113.

Williams, J.R. (1983) The sugarcane stem borer (*Chilo sacchariphagus*) in Mauritius. *Revue Agricole et Sucrière de l'Ile Maurice* 62, 5–23.

Zhou, G. and Overholt, W.A. (2001) Spatial-temporal population dynamics of *Cotesia flavipes* (Hymenoptera: Braconidae) in Kenya. *Environmental Entomology* 30, 869–876.

Zhou, G., Baumgärtner, J. and Overholt, W.A. (2001a) Impact of an exotic parasitoid on stemborer (Lepidoptera) population dynamics in Kenya. *Ecological Applications* 11, 1554–1562.

Zhou, G., Overholt, W.A. and Mochiah, B.M. (2001b) Changes in the distribution of lepidopteran maize stemborers in Kenya from the 1950s to the 1990s. *Insect Science and its Application* 21, 395–402.

10

Biological Control and Management of the Alien Invasive Shrub *Chromolaena odorata* in Africa

James A. Timbilla,[1] Costas Zachariades[2] and Haruna Braimah[1]

[1] *Biological Control Unit, Crops Research Institute, Kumasi, Ghana;*
[2] *Agricultural Research Council – Plant Protection Research Institute, Hilton, South Africa*

Introduction

Chromolaena odorata (L.) King and Robinson (*Eupatorium odoratum* L.) (*Asteraceae*), also known as chromolaena or Siam weed, is native to the Americas, from southern Florida to northern Argentina, including the Caribbean islands (McFadyen, 1988a). *C. odorata*, a pioneer species where it is indigenous (McFadyen 1988a), is one of the worst alien invasive plant species in the humid tropics and subtropics of the Old World (Holm *et al.*, 1977), where it is a menace to agriculture, human health and biodiversity. The weed has invaded western, central and southern Africa; Asia, from India in the west to China, Papua New Guinea and the Philippines in the east; and parts of Oceania (McFadyen, 1988b; Muniappan and Marutani, 1988). In its native and most of its introduced ranges, it is limited to latitudes between 30°N and 30°S, and altitudes of up to 1000 m (McFadyen, 1988a; Muniappan and Marutani, 1988). It is generally also limited to areas receiving over 1000 mm rainfall annum^{-1} (Hoevers and M'Boob, 1996), but in southern Africa this limit declines to 500 mm annum^{-1} (Goodall and Erasmus, 1996).

Introduction and Spread

Two centres of introduction of *C. odorata* exist in Africa. It is thought that the invasion in West and Central Africa is secondary, via Asia, whereas that in southern Africa is primary, originating from the Caribbean. Two distinct biotypes of the species are involved in these two centres of invasion, which also differ in their climatic characteristics, the West and Central African centre being tropical and the southern African, subtropical.

(eds P. Neuenschwander, C. Borgemeister and J. Langewald)

West and Central Africa

In West Africa, *C. odorata* was first introduced into Nigeria around 1937, probably with saplings of a plantation crop from Sri Lanka (McFadyen, 1988b). The biotype is both morphologically and genetically identical with that in Asia (I. von Senger, Grahamstown, South Africa, 2002, unpublished results). By the late 1960s or early 1970s, the weed had spread to the western forest zone and southern Ghana (Hall *et al.*, 1972), southern Côte d'Ivoire (Zebeyou, 1991), southern and central Nigeria, western and central southern Cameroon, southern Togo and Bénin, northern Democratic Republic of Congo (DRC), and southwestern Central African Republic (CAR) (McFadyen, 1988b; Hoevers and M'Boob, 1996). By 1980, it had become common along railways, young tree crop plantations, crop and pasturelands and clearings around villages in most of these countries. The first record of *C. odorata* in Sierra Leone was in 1987. The weed occurs now from eastern Guinea to the central regions of the DRC and the CAR, southwards to northern Angola. Its spread further north is limited by low precipitation and unimodal rainfall pattern. *C. odorata* spreads by wind, man (inter- and intra-regional trade, and transport of agricultural produce) and animals, as well as through runoff water (Timbilla and Braimah, 1996).

Southern Africa

The circumstances of the arrival of *C. odorata* into southern Africa remain unclear, although it was probably introduced either deliberately as an ornamental, or accidentally as seed through Durban harbour (Henderson and Anderson, 1966; Liggitt, 1983). The first herbarium records are from around Durban in the 1940s. The *C. odorata* invading southern Africa is distinct in morphology, chemistry and ecology from that in west and central Africa, Asia and Oceania, and is certainly a different biotype introduced directly from the Americas, probably from one of the northern Caribbean islands (Von Senger *et al.*, 2002).

From the Durban area, *C. odorata* spread both northwards and southwards along the subtropical coastal belt. By the 1970s, it was present throughout the subtropical areas of KwaZulu-Natal province (KZN), and in the 1980s it spread further south into the Eastern Cape province and appeared in Limpopo province, hundreds of kilometres north of previous infestations. It has subsequently spread to Mpumalanga province, situated between Limpopo and KZN. Much of the long-range spread has been anthropogenic, with wind dispersal accounting only for short-range spread (Blackmore, 1998; Wilson and Widayanto, 2002).

The lowveld region of the neighbouring country of Swaziland has become densely infested with *C. odorata*, and unconfirmed reports from Mozambique indicate that *C. odorata* may be present near Cahora Bassa (J. Findlay, personal communication). Other reports of *C. odorata* come from Zimbabwe (Gautier, 1992) and Mauritius (L. Strathie, personal communication). It is not known whether the weed entered Zimbabwe from Central or southern Africa, but Mauritius possesses the West African/Asian biotype, which probably arrived directly from Asia, as explained by the strong cultural links between these countries. It is certainly only a matter of time before the two invasion fronts in Africa cross, as the intervening region has been shown to be climatically highly suitable (McFadyen and Skarratt, 1996). The consequences of such an interaction remain unknown.

Problems caused by *Chromolaena odorata*

C. odorata has an extremely high growth rate and reproductive output, in the form of thousands of light, wind-dispersed seeds. It has a smothering habit and allelopathic properties. It is plastic in the habitat types it invades and takes advantage of both natural and anthropogenic disturbances. It increases the intensity, range and frequency of fires. In all areas, *C. odorata* impacts on cropping and pastoral agriculture, on biodiversity and on human welfare. However, the relative magnitude of the problems that it causes, or is seen to cause, vary depending on the economy of the region or country concerned. Thus in countries where subsistence/small-scale farmers constitute a significant sector of rural society, the effects of chromolaena on these farmers becomes important (Wilson and McFadyen, 2000). In contrast, in a country such as South Africa, where the majority of farms are commercial and large tracts of land have been set aside as conservation areas, the effects of chromolaena in these situations becomes significant (Wilson and McFadyen, 2000). Clearly, the individual crops affected by *C. odorata* also vary from region to region.

West and Central Africa

Throughout the region, *C. odorata* invades cropping fields, grazing areas and young or neglected plantations, particularly of timber, cocoa, citrus, rubber and oil palm (M'Boob, 1991; Prasad *et al.*, 1996; Timbilla, 1996). The weed thus decreases agricultural productivity and increases management costs at both subsistence and commercial scales (Prasad *et al.*, 1996). Outbreaks of *Zonocerus variegatus* (L.) (Orthoptera, Pyrgomorphidae) are made more severe by a non-nutritional association with *C. odorata* that the grasshopper employs for the protection of its diapausing eggs (Boppré, 1991). *C. odorata* causes acute diarrhoea in cattle when browsed. Recent extensive bush fires in the forest regions of Ghana are attributed to the presence of *C. odorata*, as is the case in other parts of the world (Muniappan and Marutani, 1988; Napompeth *et al.*, 1988; Goodall and Erasmus, 1996). These fires cause damage to agricultural land, endanger human life, and further damage the ecology of a biodiverse region already under great environmental pressure (Myers *et al.*, 2000). Some people are allergic to the leaves of the weed and have reported skin rashes and irritation. *C. odorata* has been shown to prevent the regeneration of indigenous secondary forest (Honu and Dang, 2000), and to decrease the carrying capacity and species diversity in grassland and forests (Liggitt, 1983; Erasmus, 1985).

As is, however, often the case with the appearance of a new dominant plant species, positive attributes have also accrued to *C. odorata,* particularly in rural areas in West and Central Africa. The plant is claimed to have medicinal properties (treatment of flesh wounds, ulcers, boils, malaria, jaundice and clotting of blood) and uses in embalming. In agriculture, it enhances nutrient recycling, erosion control, and in some cases is used as fuel wood (Timbilla and Braimah, 1996). However, the most important benefit ascribed to chromolaena in this region, and the one that has caused a blockage of the biological control programme in most of the region (McFadyen, 1996), is its importance as a fallow crop in shifting agriculture systems (De Foresta and Schwartz, 1991; Herren-Gemmill, 1991; Prasad *et al.*, 1996). This conflict-of-interest controversy reached its peak around the time of the Third International *Chromolaena* Workshop held in Abidjan in 1993, resulting in the promulgation of a series of research recommendations to resolve the issue (Prasad *et al.*, 1996).

Southern Africa

In the subtropical regions of South Africa, *C. odorata* is considered the worst alien invasive plant. It is perceived mainly as a threat to natural areas, affecting biodiversity, conservation activities and ecotourism. For example, its invasion and shading of nesting sites of the Nile crocodile, *Crocodylus niloticus* (Laurenti) (Reptilia, Crocodylidae), has been shown to affect their sex ratio by lowering the environmental temperature (Leslie and Spotila, 2001) and it replaces browse of the endangered black rhinoceros, *Rhinocerotis bicornis* (L.) (Mammalia, Rhinocerotidae). It is a serious invasive species in three centres of floral endemism in southern Africa (Cowling and Hilton-Taylor, 1994). In uncultivated areas, *C. odorata* also impacts negatively on pastoral agriculture. In cropping agriculture, *C. odorata* is a problem mainly to timber cultivation, particularly for young pine and eucalypt trees. Other intensive cropping activities that coincide geographically with *C. odorata* infestations, such as sugarcane and fruit farming, are mostly run as commercial enterprises. They are thus not significantly affected by *C. odorata*, as these farmers can afford the costs of conventional, i.e. chemical and mechanical, clearing. The 20th century history of South Africa, with the earlier, black inhabitants being deprived of their land and confined to 'homelands' until recently, has created a different land-use pattern from the other parts of Africa in which *C. odorata* is a problem, and the impact of *C. odorata* on these South African communities has not yet been well recorded. However, it is known to dramatically affect subsistence grazing in parts of KZN (L. Anthony, personal communication) and is likely to affect subsistence cropping agriculture. Land previously belonging to the State or to white, commercial farmers is being given back to black communities in a degraded, infested state. These communities often lack the knowledge or resources to clear these lands. There are currently also a number of community–industry initiatives being spearheaded by the sugar and timber industries, in which communities are paid for what they produce. This small-scale commercial production will be more affected by *C. odorata* than large commercial enterprises because of narrower profit margins.

South Africa is a water-stressed country, and alien plants have been estimated to reduce runoff by up to 7% per annum (Versveld *et al.*, 1998). Since an analysis indicated that large-scale clearance of alien plants from catchments would be more cost-effective than building more dams the South African Department of Water Affairs and Forestry launched a major water-conservation and employment initiative, the Working-for-Water (WfW) Programme (WfW URL) in 1995. Although *C. odorata* is not considered to be one of the top water users, it has been targeted for clearance by the WfW programme as it acts as a secondary invader species in cleared areas, and as part of WfW's mandate to conserve biodiversity (Zachariades and Goodall, 2002).

In Swaziland, conservation groups and commercial agriculture have expressed concern about chromolaena. However, there can be no doubt that chromolaena impacts on various other land-use practices and social groupings. The situation in Mozambique is unknown, and Mauritius does not consider *C. odorata* among its top 10 weeds (P.L. Campbell, personal communication).

Non-biological Control

Extensive work has been conducted on 'conventional' management methods, namely mechanical and chemical control of *C. odorata*. In West and Central Africa, this is mainly in the context of cropping agriculture (e.g. Utulu, 1996; Ikuenobe and Ayeni, 1998), while in South Africa, *C. odorata* is also controlled as an environmental weed (Goodall and Erasmus, 1996; Goodall *et al.*, 1996). Fire has been shown to be an effective control method in situations where it can be applied safely (Gautier, 1996; Goodall and Zacharias, 2002).

Despite the development of these control methods, *C. odorata* remains a serious problem due to the cost-ineffectiveness of its control over larger areas. It has long been recognized that biological control has a critical role to play in any successful general integrated control strategy for *C. odorata* by reducing its growth and reproductive output (Bennett and Rao, 1968; Goodall and Erasmus, 1996; Hoevers and M'Boob, 1996).

Biological Control

Although research on biological control of *C. odorata* was initiated over three decades ago, the relative lack of progress can be attributed to the fact that the weed is a problem mainly for tropical, developing countries that often lack the resources to carry out the expensive initial stages of biological control (McFadyen, 1996), rather than any intrinsic difficulty in suppressing the weed biologically (Kluge, 1990). In Africa, the most successful, sustained programmes have been in Ghana and South Africa, both starting in the late 1980s. Therefore these are the ones we discuss in detail in this chapter.

West and Central Africa

The first biological control programme for *C. odorata* worldwide originated in West Africa when, from 1966 to 1972, an investigation into possible agents in the West Indies was financed by the Nigerian Institute for Oil Palm Research (Bennett and Rao, 1968; Cruttwell, 1974). The leaf-feeding moth *Pareuchaetes pseudoinsulata* Rego Barros (Lepidoptera, Arctiidae) (Plate 26a), initially misidentified as *Ammalo insulata* (Walker), and the seed-feeding weevil *Apion brunneonigrum* Béguin-Billecocq (Coleoptera, Apionidae) were released in Ghana and Nigeria in the early 1970s, but neither species established (Cock and Holloway, 1982; Julien and Griffiths, 1998). *P. pseudoinsulata* did, however, establish and effect some degree of control of *C. odorata* in parts of Asia and the Pacific, with the most spectacular control in Guam in the 1980s (Seibert, 1989). Similar control attempts of *C. odorata* were made in Côte d'Ivoire, with the release of *P. pseudoinsulata* in 1991 (Julien and Griffiths, 1998). The failure of the insect to establish was attributed to predation by natural enemies in the field (M. Zebeyou, personal communication).

A biological control programme for *C. odorata* was re-initiated in Ghana in 1989. *P. pseudoinsulata* was imported from Guam, its biology determined, and its host specificity confirmed (Timbilla and Braimah, 1991). Releases of *P. pseudoinsulata* on *C. odorata* infestations were done at the Fumesua (Kumasi) station of the Crops Research Institute in September 1991, and continued until October 1993. A total of 119,256 larvae and 6256 adults, released between 1991 and 1993, led to an initial establishment of the insect confirmed in 1994. A further 192,325 larvae and 7591 adults were released between 1995 and 1997 at Abesewa and Assin Dadieso. Ghana is thus the first country in Africa to achieve establishment of *P. pseudoinsulata* in the field (Timbilla and Braimah, 1996). Initially, *P. pseudoinsulata* disappeared over the dry season, but it can now be found in *C. odorata* infestations throughout the year. Defoliation and the destruction of buds of *C. odorata* by *P. pseudoinsulata* covered an estimated 2025 km^2 in 1994 and caused the death of a high proportion of chromolaena plants. By December 1998, *P. pseudoinsulata* was established in 37.9% of the total area infested by *C. odorata*, giving considerable control of *C. odorata* on arable-crop farms, plantations, fallow lands, roadsides, forest margins and open spaces within forests. Between 1995 and 1998, *C. odorata* was reduced from an average cover of 85.0% in infested fields to 32.9–40.2% in fields where the insect was present and suppressed the weed. The feeding activities of the insect reduced the competitiveness of *C. odorata*, giving rise to the growth of *Pennisetum purpureum* Schumacher (*Poaceae*) and *Aspilia africana* (Pers.) CD Adams (*Asteraceae*), which are fodder plants for game and domestic animals. Over the same period, grass cover increased from an estimated 2.0% in *C. odorata*-infested fields to between 24.2% and 28.3%. Broadleafed plants increased from an initial estimate of 13.0% in *C. odorata*-infested fields to 38.0% in 1995, 36.9% in 1996, 31.5% in 1997 and 39.4% in 1998 (Timbilla, 1998; Timbilla and Braimah, 2000). Some secondary forests have since developed in areas with successful control of *C. odorata* by *P. pseudoinsulata* (Timbilla, 1996).

The revived biological control programme on *C. odorata* in the 1990s, using *P. pseudoinsulata*, has been extremely successful in terms of effectiveness and cost. As local natural enemies may eventually adapt to *P. pseudoinsulata* as a host, the strategy of using more than one natural enemy (Harris, 1986) is further pursued. Other agents, such as stem and root attackers and disease-causing organisms, would increase the effectiveness of the effort. As West and Central Africa share a *C. odorata* biotype with Asia, the stem-galling fly *Cecidochares connexa* Macquart (Diptera, Tephritidae), which has proved very successful in Indonesia (McFadyen, 2002), is a priority in this regard.

South Africa

In southern Africa, only South Africa is involved in biological control of *C. odorata*, although Swaziland has expressed an interest in obtaining agents. Apart from a project run in Indonesia by the Australian Centre for International Agricultural Research (ACIAR), the South African biological control programme is currently the only one worldwide to be importing, rearing and testing novel agents from the Americas. This programme was initiated by the Plant Protection Research Institute of the Agricultural Research Council (ARC – PPRI) in 1988, when *P. pseudoinsulata* was tested for host-specificity and released in small numbers around the greater Durban area. Neither it nor the congeneric *Pareuchaetes aurata aurata* Butler (Plate 26b), released in the early 1990s, established in the field (Zachariades *et al.*, 1999). *Pareuchaetes insulata* (Walker) (Plate 27) was also tested in the early 1990s and found to be safe, but no application for release was made at the time because of the failure of the first two species. A number of exploratory trips to the Americas resulted in several other agents being imported between 1988 and 1991, but none were released, usually either because they were incompatible with the southern African *C. odorata* biotype or because facilities were inadequate to rear them successfully (Zachariades *et al.*, 1999). From 1992 to 1995, a lack of personnel and funding resulted in little progress being made, after which the project was revived (reviews by Zachariades *et al.*, 1999; Strathie and Zachariades, 2002).

Recent and current work on new candidate agents

A strategic plan for the project was drawn up in 1996–1997, in which a suite of biological control candidates attacking leaves, stems and roots were prioritized for importation and host-specificity testing. Determining the origin of the southern African *C. odorata* biotype was also considered important.

The butterfly *Actinote thalia pyrrha* Fabricius (Lepidoptera, Nymphalidae) (Plates 28, 29) was imported from eastern Brazil. The larvae, which defoliate *C. odorata*, performed equally well on two indigenous asteraceous vines (*Mikania natalensis* DC. and *Mikania capensis* DC.). Governmental permission to release this agent was thus not sought, although a culture was forwarded to the *C. odorata* biological control programme in Indonesia. *C. connexa* was imported but proved incompatible with the southern African chromolaena, and a culture was not established. The stem-boring weevil *Lixus aemulus* Petri (Coleoptera, Curculionidae) (Plate 30), imported from western Brazil, was tested comprehensively for host specificity and found to be safe. The weevil appears quite damaging, and permission to release this insect will soon be sought from the South African government. The leaf-mining fly *Calycomyza eupatorivora* Spencer (Diptera, Agromyzidae) was also tested and found to be safe, and permission to release it is currently being awaited. The stem tip-galling weevil *Conotrachelus reticulatus* Champion (Coleoptera, Curculionidae) and the root-boring flea beetle *Longitarsus horni* Jacoby (Coleoptera, Chrysomelidae), both collected in Venezuela, are now being prioritized for host-specificity testing. Based on their biologies and provenance, they are expected to be tolerant to the dry winter season in southern Africa as well as to fire. In addition, killing *C. odorata* at a seedling stage through the use of *L. horni* should result in greatly decreased recruitment of new plants. Both species have been cultured successfully in the quarantine laboratory, and tests undertaken so far on *C. reticulatus* indicate that this weevil is highly specific to *C. odorata*. The stem-galling fly *Polymorphomyia basilica* Snow (Diptera, Tephritidae), which fills the same niche on the Greater Antilles as *C. connexa* on mainland America, was imported as a more compatible substitute for *C. connexa*, but a laboratory culture has not yet been established. Finally, several leaf-attacking pathogen species, collected mainly in Jamaica, have also been successfully cultured in quarantine. Two of these, *Pseudocercospora eupatorii-formosani* (Sawada) J.M. Yen and *Mycovellosiella perfoliata* (Ellis and Everh.) Munt.-Cvetk. (both *Deuteromycotina*, *Hyphomycetes*), are currently being tested for host specificity and appear promising (den Breeÿen, 2002).

Over the past 5 years, diverse, strong lines of evidence have accumulated to indicate that the southern African *C. odorata* biotype originates from the Greater Antilles (Jamaica, Puerto Rico, Cuba) or the Bahamas. The resolution of this problem has improved the prospects for this biological control programme.

Agents released

Small numbers of adult *P. pseudoinsulata* obtained from Guam were released at several sites in 1989, but did not establish a field population. Egg predation was shown to be high (Kluge, 1994), but was not necessarily the reason for non-establishment. By comparison, Trinidad, from where the culture originated, has a similarly high ant biodiversity as mainland South America (Wilson, 1988). Following the success of this species in establishing and suppressing *C. odorata* in Ghana and Sumatra, a cooperative venture was set up between ARC-PPRI and WfW in Limpopo province to mass-rear and release large numbers of insects, using similar techniques to those employed in the above-mentioned countries (Zachariades *et al.*, 1999). Between December 1998 and April 1999 (summer), approximately 350,000 larvae, reared from a starter culture obtained from Indonesia, were released at two sites in the Letsitele valley, resulting in a substantial reduction of the height and density of the chromolaena stands in the immediate vicinity. The insect persisted in at least one site in large numbers for one to two generations, but from mid-winter (July 1999) onwards could not be found at either site or in surrounding areas, despite regular surveys until April 2001. The reasons for this apparent failure are not known, and may have included climatic or biotype incompatibility, or inadequately long releases to overcome dispersal or predation. After the summer of 1999, the culture in the mass-rearing laboratory became increasingly infected with a microsporidian disease, and eventually died out. Inadequately hygienic rearing conditions and the absence of an on-site entomologist were probably responsible for this.

P. aurata aurata from Tucuman, Argentina, was selected because its biology was thought to make it less susceptible to predation than *P. pseudoinsulata* and the area from which it was collected was more climatically similar to KZN. Approximately 150,000 insects were released at 17 sites around this province in 1993 and 1994, after being mass-reared under hygienic conditions at the South African Sugar Experiment Station (SASEX), KZN (Conlong and Way, 2002). The insects persisted for a few months at several sites, causing significant damage to chromolaena, but disappeared during winter or after burning. Again, the reasons for their disappearance were unclear (Conlong, 2002).

P. insulata from Florida, USA, was selected for testing and release because of the seasonally dry, subtropical climate of its region of origin. Permission for its release was obtained in 1997, but a culture was only re-imported in 2000, once the use of facilities for mass-rearing under hygienic conditions (again at SASEX) and releases (by a WfW biocontrol implementation officer) had been negotiated. Over the first 10 months of 2001, over 200,000 larvae and a few thousand adults were released at eight sites around KZN (20,000–30,000 per site, released over a 1–2-month period). A post-release survey in October 2001 showed that insects had persisted through the exceptionally dry winter only at one site. At several of the sites, *C. odorata* had completely lost its leaves. A new release strategy was thus adopted, using only three, widely spaced sites at which chromolaena remains in good condition year-round, and releasing large numbers of *P. insulata* into these sites over at least an 18-month period. Simultaneously, studies are being conducted to determine climatic tolerances, predation and parasitism levels, biotype preference and adult dispersal patterns. A new culture, better suited to the South African climate and biotype, may be imported from Cuba or Jamaica.

Prospects

Long-term prospects for the biological control of *C. odorata* in South Africa are still good (Zachariades *et al.*, 1999), despite the failure of several agents to establish. For the southern African *C. odorata* biotype, efforts will be concentrated in the northern Caribbean (Jamaica, Cuba), on candidate agents that are either endemic to the region (e.g. *P. basilica*) or biotypes of more widespread species (Cruttwell, 1974; C. Zachariades, personal observation). However, if key taxa such as *Longitarsus* and *Conotrachelus* are absent from this region, it may be necessary to use species from other geographical areas that have a sufficiently broad host range so as not to be affected by intraspecific variation of *C. odorata*. Use of this latter category of species would be beneficial to biological control programmes on the West African/Asian *C. odorata* biotype. In addition, Cuba and Jamaica, which at 17–23°N are less tropical than most other parts of the range of *C. odorata*, may be climatically more closely matched to southern Africa, where the main *C. odorata* infestation currently lies between 22 and 33°S. Drier parts of Cuba may yield agents that are well adapted to southern African conditions.

Integration of Biological Control with Other Methods

In Ghana, extension work on *C. odorata* biological control is currently being undertaken with farming communities through the Integrated Pest Management/Farmer Field School (IPM/FFS) concept, recently introduced into the country (Entsie, 2002). About 106 extensionists and 2470 farmers have been trained nationwide and will facilitate the biological control programme in the future. In South Africa, biological control of *C. odorata* is still at too early a stage to consider its full integration into other control methods. However, once effective agents are established, they can be used in a number of circumstances, e.g. where WfW clearing operations do not reach, after follow-up clearance is complete in an area, in areas that cannot be burned regularly and once the WfW programme has been scaled down in about 2020 (Olckers *et al.*, 1998; Olckers, 2000).

Regional Cooperation

The conflict of interest in West and Central Africa concerning the usefulness of *C. odorata* versus its weediness appears to be easing. This is partly due to recent research from the region (e.g. Honu and Dang, 2000; Timbilla *et al.*, 2002). It is also generally understood that biological control does not seek to eradicate chromolaena, but to reduce its competitiveness and further spread into new areas, thus its positive attributes would not be lost. This raises hope for a regional programme (Bani, 2002; Timbilla and Braimah, 2002).

Within the southern African subregion, it is important to ascertain the extent of the chromolaena infestation, to determine where each of the two biotypes is present, and to increase awareness of the threat of *C. odorata* at an early stage of invasion. As the southern African chromolaena biotype is different from that in West and Central Africa, some of the agents imported by South Africa from the northern Caribbean to overcome incompatibility problems with the former biotype may not be fully compatible with the latter. However, others may have a sufficiently broad host range or may be collected from other biotypes of *C. odorata*, and in these cases no incompatibility problem would arise.

Conclusions

C. odorata, originating in the neotropics, is one of the most serious plant invaders in Africa, where it has two centres of invasion: West and southern Africa. *C. odorata* was first introduced into West Africa from an invasive population in South-East Asia in the late 1930s. It spread rapidly throughout most of West and Central Africa. The southern African invasion consists of plants of a different biotype, apparently originating from the northern Caribbean islands. It now covers much of southern Africa's subtropical regions and is likely to become contiguous with the West and Central African infestation. Chromolaena is considered one of the worst alien invasive plants in the humid tropics and subtropics of the Old World, where it causes problems to agriculture, biodiversity and human health. Although there are effective, well-established methods to control it, these are not generally cost-effective.

Biological control was attempted in Nigeria and Ghana in the early 1970s using the defoliating arctiid moth *P. pseudoinsulata* and other insects, but failed. A renewed effort in Ghana in the 1990s involving continual release of large numbers led, however, to successful establishment of *P. pseudoinsulata* in the area of Kumasi. Today, *P. pseudoinsulata* has effectively controlled *C. odorata* in many parts of the Ashanti region and has spread to the Central, Eastern and Brong Ahafo regions. The feeding activities of the insect have reduced the populations of *C. odorata* from an average of 85% to 33%, while populations of other herbaceous species have increased. This success notwithstanding, the further reduction of *C. odorata* will require other host-specific natural enemies like the stem-galling fly *C. connexa*, which has proved a successful biological control agent in South-East Asia.

The South African biological control programme on *C. odorata*, initiated in 1988, imported, tested and released three *Pareuchaetes* species. Neither *P. pseudoinsulata* nor *P. aurata aurata* established field populations. At present, *P. insulata* is being released in large numbers throughout KZN province. Explorations in the native range of *C. odorata* in the Americas tentatively located the parent population of the southern African *C. odorata* biotype in the northern Caribbean islands and focused the search for novel candidate agents (insects and pathogens) for import into South Africa. The aim is to release a number of species of agents attacking different plant parts. Several additional insect candidates and pathogens have recently been found to be safe or are undergoing host-specificity testing in quarantine.

References

Bani, G. (2002) Status and management of *Chromolaena odorata* in Congo. In: Zachariades, C., Muniappan, R. and Strathie, L.W. (eds) *Proceedings of the Fifth International Workshop on Biological Control and Management of* Chromolaena odorata, October 2000, Durban. ARC – PPRI, Pretoria, South Africa, pp. 71–73.

Bennett, F.D. and Rao, V.P. (1968) Distribution of an introduced weed *Eupatorium odoratum* L. in Asia and Africa and possibilities for its biological control. *PANS Section C* 14, 277–281.

Blackmore, A.C. (1998) Seed dispersal of *Chromolaena odorata* reconsidered. In: Ferrar, P., Muniappan, R. and Jayanth, K.P. (eds) *Proceedings of the Fourth International Workshop on the Biological Control and Management of* Chromolaena odorata, October 1996, University of Guam, Mangilao, Bangalore, India, pp. 16–21.

Boppré, M. (1991) A non-nutritional relationship of *Zonocerus* (Orthoptera) to *Chromolaena odorata* (Asteraceae) and implications for weed control. In: Muniappan, R. and Ferrar, P. (eds) *Proceedings of the Second International Workshop on Biological Control of* Chromolaena odorata, February 1991, Bogor, Indonesia. BIOTROP Special Publication 44, pp. 154–155.

Cock, M.J. and Holloway, J.D. (1982) The history of and prospects for the control of *Chromolaena odorata* (Compositae) by *Pareuchaetes pseudoinsulata* Rego Barros and allies (Lepidoptera: Arctiidae). *Bulletin of Entomological Research* 72, 193–205.

Conlong, D.E. (2002) Release and impact of *Pareuchaetes aurata* (Butler) (Lepidoptera: Arctiidae) on populations of *Chromolaena odorata* in KwaZulu-Natal, South Africa. In: Zachariades, C., Muniappan, R. and Strathie, L. (eds) *Proceedings of the Fifth International Workshop on Biological Control and Management of* Chromolaena odorata, October 2000, Durban. ARC – PPRI, Pretoria, South Africa, p. 177.

Conlong, D.E. and Way, M. (2002) Mass rearing of *Pareuchaetes aurata* (Butler) (Lepidoptera: Arctiidae). In: Zachariades, C., Muniappan, R. and Strathie, L.W. (eds) *Proceedings of the Fifth International Workshop on Biological Control and Management of* Chromolaena odorata, October 2000, Durban. ARC – PPRI, Pretoria, South Africa, p. 153.

Cowling, R.M. and Hilton-Taylor, C. (1994) Patterns of plant diversity and endemism in southern Africa: an overview. In: Huntley, B.J. (ed.) *Botanical Diversity in Southern Africa*. National Botanical Institute, Pretoria, South Africa, pp. 31–52.

Cruttwell, R.E. (1974) Insects and mites attacking *Eupatorium odoratum* in the neotropics. 4. An annotated list of the insects and mites recorded from *Eupatorium odoratum* L., with a key to the types of damage found in Trinidad. *Technical Bulletin of the Commonwealth Institute for Biological Control* 17, 87–125.

De Foresta, H. and Schwartz, D. (1991) *Chromolaena odorata* and disturbance of natural succession after shifting cultivation: an example from Mayombe, Congo, Central Africa. In: Muniappan, R. and Ferrar, P. (eds) *Proceedings of the Second International Workshop on Biological Control of* Chromolaena odorata, February 1991, Bogor, Indonesia. BIOTROP Special Publication 44, pp. 23–41.

Den Breeÿen, A. (2002) *Chromolaena odorata*: biological control using plant pathogens – a South African perspective. In: Zachariades, C., Muniappan, R. and Strathie, L.W. (eds) *Proceedings of the Fifth International Workshop on Biological Control and Management of* Chromolaena odorata, October 2000, Durban. ARC – PPRI, Pretoria, South Africa, pp. 167–169.

Entsie, P.-K. (2002) Ecologically sustainable *Chromolaena odorata* management in Ghana: past, present and the future role of Farmers' Field Schools. In: Zachariades, C., Muniappan, R. and Strathie, L.W. (eds) *Proceedings of the Fifth International Workshop on Biological Control and Management of* Chromolaena odorata, October 2000, Durban. ARC – PPRI, Pretoria, South Africa, pp. 128–132.

Erasmus, D.J. (1985) Achene biology and the chemical control of *Chromolaena odorata*. PhD thesis, Department of Botany, University of Natal.

Gautier, L. (1992) Taxonomy and distribution of a tropical weed: *Chromolaena odorata* (L.) R. King and H. Robinson. *Candollea* 47, 645–662.

Gautier, L. (1996) Establishment of *Chromolaena odorata* in a savannah protected from fire: an example from Lamto, central Côte d'Ivoire. In: Prasad, U.K., Muniappan, R., Ferrar, P., Aeschliman, J.P. and de Foresta, H. (eds) *Proceedings of the Third International Workshop on Biological Control and Management of* Chromolaena odorata, November 1993, Abidjan, Côte d'Ivoire. Agricultural Experiment Station, University of Guam, Publication 202, pp. 54–67.

Goodall, J.M. and Erasmus, D.J. (1996) Review of the status and integrated control of the invasive alien weed, *Chromolaena odorata*, in South Africa. *Agriculture, Ecosystems and Environment* 56, 151–164.

Goodall, J.M. and Zacharias, P.J.K. (2002) Managing *Chromolaena odorata* (chromolaena) in subtropical grasslands in KwaZulu-Natal, South Africa. In: Zachariades, C., Muniappan, R. and Strathie, L.W. (eds) *Proceedings of the Fifth International Workshop on Biological Control and Management of* Chromolaena odorata, October 2000, Durban. ARC – PPRI, Pretoria, South Africa, pp. 120–127.

Goodall, J.M., Kluge, R.L. and Zimmermann, H.G. (1996) Developing integrated control strategies for *Chromolaena odorata* in southern Africa. In: Brown, H., Cussans, G.W., Devine, M.D., Duke, S.O., Fernandez-Quintanilla, C., Helweg, A., Labrada, R.E., Landes, M., Kudsk, P. and Streibig, J.C. (eds) *Proceedings of the Second International Weed Control Congress*, June 1996, Copenhagen, Denmark. Department of Weed Control and Pesticide Ecology, Slagelse, pp. 729–734.

Hall, J.B., Kumar, R. and Enti, A.A. (1972) The obnoxious weed *Eupatorium odoratum* (Compositae) in Ghana. *Ghana Journal of Agricultural Sciences* 5, 75–78.

Harris, P. (1986) Biological control of weeds. In: Franz, J.M. (ed.) *Biological Plant and Health Protection*. Fischer Verlag, Stuttgart, Germany, pp. 123–138.

Henderson, M. and Anderson, J.G. (1966) Common weeds in South Africa. *Memoirs of the Botanical Survey of South Africa,* No 37. Botanical Research Institute, Pretoria, South Africa.

Herren-Gemmill, B. (1991) The ecological role of the exotic asteraceous *Chromolaena odorata* in the bush fallow farming system of West Africa. In: Muniappan, R. and Ferrar, P. (eds) *Proceedings of the Second International Workshop on Biological Control of* Chromolaena odorata, February 1991, Bogor, Indonesia. BIOTROP Special Publication 44, pp. 11–22.

Hoevers, R. and M'Boob, S.S. (1996) The status of *Chromolaena odorata* (L.) R.M. King and H. Robinson in West and Central Africa. In: Prasad, U.K., Muniappan, R., Ferrar, P., Aeschliman, J.P. and de Foresta, H. (eds) *Proceedings of the Third International Workshop on Biological Control and Management of* Chromolaena odorata, November 1993, Abidjan, Côte d'Ivoire. Agricultural Experiment Station, University of Guam, Publication 202, pp. 1–5.

Holm, L.G., Plucknett, D.L. and Pancho, J.V. (1977) *The World's Worst Weeds: Distribution and Biology*. University Press of Hawaii, Honolulu, pp. 212–216.

Honu, Y.A.K. and Dang, O.L. (2000) Response of tree seedlings to the removal of *Chromolaena odorata* Linn. in a degraded forest in Ghana. *Forest Ecology and Management* 137, 75–82.

Ikuenobe, C.E. and Ayeni, A.O. (1998) Herbicidal control of *Chromolaena odorata* in oil palm. *Weed Research* 38, 397–404.

Julien, M.H. and Griffiths, M.W. (1998) *Biological Control of Weeds: a World Catalogue of Agents and Their Target Weeds*, 4th edition. CAB International, Wallingford, UK.

Kluge, R.L. (1990) Prospects for the biological control of triffid weed, *Chromolaena odorata*, in southern Africa. *South African Journal of Science* 86, 229–230.

Kluge, R.L. (1994) Ant predation and the establishment of *Pareuchaetes pseudoinsulata* Rego Barros (Lepidoptera: Arctiidae) for biological control of triffid weed, *Chromolaena odorata* (L.) King and Robinson, in South Africa. *African Entomology* 2, 71–72.

Leslie, A.J. and Spotila, J.R. (2001) Alien plant threatens Nile crocodile (*Crocodylus niloticus*) breeding in Lake St. Lucia, South Africa. *Biological Conservation* 98, 347–355.

Liggitt, B. (1983) *The Invasive Alien Plant* Chromolaena odorata, *with Regard to its Status and Control in Natal.* Monograph 2. Institute of Natural Resources, University of Natal, Pietermaritzburg, South Africa.

M'Boob, S.S. (1991) Preliminary results of a survey and assessment of *Chromolaena odorata* (Siam weed) in Africa. In: Muniappan, R. and Ferrar, P. (eds) *Proceedings of the Second International Workshop on Biological Control of* Chromolaena odorata, February 1991, Bogor, Indonesia. BIOTROP Special Publication 44, pp. 51–56.

McFadyen, R.E.C. (1988a) Ecology of *Chromolaena odorata* in the neotropics. In: Muniappan, R. (ed.) *Proceedings of the First International Workshop on Biological Control of* Chromolaena odorata, February–March 1988, Bangkok, Thailand. Agricultural Experiment Station, Guam, pp. 13–20.

McFadyen, R.E.C. (1988b) History and distribution of *Chromolaena odorata* (L) R.M. King, H. Robinson. In: Muniappan, R. (ed.) *Proceedings of the First International Workshop on Biological Control of* Chromolaena odorata, February–March 1988, Bangkok, Thailand. Agricultural Experiment Station, Guam, pp. 7–12.

McFadyen, R.C. (1996) Biocontrol of *Chromolaena odorata*: divided we fail. In: Moran, V.C. and Hoffmann, J.H. (eds) *Proceedings of the Ninth International Symposium on Biological Control of Weeds*. Stellenbosch, South Africa, University of Cape Town Press, Cape Town, South Africa, pp. 455–459.

McFadyen, R.E.C. (2002) *Chromolaena* in Asia and the Pacific: spread continues but control prospects improve. In: Zachariades, C., Muniappan, R. and Strathie, L.W. (eds) *Proceedings of the Fifth International Workshop on Biological Control and Management of* Chromolaena odorata, October 2000, Durban. ARC – PPRI, Pretoria, South Africa, pp. 13–18.

McFadyen, R. and Skarratt, B. (1996) Potential distribution of *Chromolaena odorata* (Siam Weed) in Australia, Africa, and Oceania. *Agriculture, Ecosystems and Environment* 59, 89–96.

Muniappan, R. and Marutani, M. (1988) Ecology and distribution of *C. odorata* in Asia and the Pacific. In: Muniappan, R. (ed.) *Proceedings of the First International Workshop on Biological Control of* Chromolaena odorata, February–March 1988, Bangkok, Thailand. Agricultural Experiment Station, Guam, pp. 21–24.

Myers, N., Mittermeier, R.A., Mittermeier, C.G., da Fonseca, G.A.B. and Kent, J. (2000) Biodiversity hotspots for conservation priorities. *Nature* 403, 853–858.

Napompeth, B.N., Thi Hai and Winotai, M. (1988) Attempts on biological control of Siam weed, *Chromolaena odorata* in Thailand. In: Muniappan, R. (ed.) *Proceedings of the First International Workshop on Biological Control of* Chromolaena odorata, February–March 1988, Bangkok, Thailand. Agricultural Experiment Station, Guam, pp. 57–62.

Olckers, T., Zimmermann, H.G. and Hoffmann, J.H. (1998) Integrating biological control into the management of alien invasive weeds in South Africa. *Pesticide Outlook* December, 9–16.

Olckers, T. (2000) Integrating biological control technology into the management of alien invasive weeds: South African experiences and challenges. In: Copping, L.G. (ed.) *Predicting Field Performance in Crop Protection*, September 2000,

Canterbury. British Crop Protection Council Symposium Proceedings 74, pp. 111–122.

Prasad, U.K., Muniappan, R., Ferrar, P., Aeschliman, J.P. and de Foresta, H. (eds) (1996) *Proceedings of the Third International Workshop on Biological Control and Management of* Chromolaena odorata, November 1993, Abidjan, Côte d'Ivoire. Agricultural Experiment Station, University of Guam, No. 202.

Seibert, T.F. (1989) Biological control of the weed, *Chromolaena odorata* (Asteraceae), by *Pareuchaetes pseudoinsulata* (Lepidoptera: Arctiidae) on Guam and the Northern Mariana Islands. *Entomophaga* 35, 531–539.

Strathie, L.W. and Zachariades, C. (2002) Biological control of *Chromolaena odorata* in South Africa: developments in research and implementation In: Zachariades, C., Muniappan, R. and Strathie, L.W. (eds) *Proceedings of the Fifth International Workshop on Biological Control and Management of* Chromolaena odorata, October 2000, Durban. ARC – PPRI, Pretoria, South Africa, pp. 74–79.

Timbilla, J.A. (1996) Status of *Chromolaena odorata* biological control using *Pareuchaetes pseudoinsulata* in Ghana. In: Moran, V.C. and Hoffmann, J.H. (eds) *Proceedings of the Ninth International Symposium on Biological Control of Weeds.* University of Cape Town Press, Cape Town, South Africa, pp. 327–331.

Timbilla, J.A. (1998) Effect of biological control of *Chromolaena odorata* on biodiversity: a case study in the Ashanti region of Ghana. In: Ferrar, P., Muniappan, R. and Jayanth, K.P. (eds) *Proceedings of the Fourth International Workshop on the Biological Control and Management of* Chromolaena odorata, October 1996, University of Guam, Mangilao, Bangalore, pp. 97–101.

Timbilla, J.A. and Braimah, H. (1991) Highlights of work done on *Chromolaena odorata* in Ghana. In: Muniappan, R. and Ferrar, P. (eds) *Proceedings of the Second International Workshop on Biological Control of* Chromolaena odorata, February 1991, Bogor, Indonesia. BIOTROP Special Publication 44, pp. 105–112.

Timbilla, J.A. and Braimah, H. (1996) A survey of the introduction, distribution and spread of *Chromolaena odorata* in Ghana. In: Prasad, U.K., Muniappan, R., Ferrar, P., Aeschliman, J.P. and de Foresta, H. (eds) *Proceedings of the Third International Workshop on Biological Control and Management of* Chromolaena odorata, November 1993, Abidjan, Côte d'Ivoire. Agricultural Experiment Station, University of Guam, Publication 202, pp. 6–18.

Timbilla, J.A. and Braimah, H. (2000) Establishment, spread and impact of *Pareuchaetes pseudoinsulata* (Lepidoptera: Arctiidae) an exotic predator on the Siam Weed, *Chromolaena odorata* (Asteraceae: Eupatorieae) in Ghana. In: Spencer, N.R. (ed.) *Proceedings of the Tenth International Symposium on Biological Control of Weeds.* July 1999, Bozeman, Montana, USA, pp. 105–111.

Timbilla, J.A. and Braimah, H. (2002) Successful biological control of *Chromolaena odorata* (L.) King and Robinson in Ghana: a potential for a regional programme in Africa. In: Zachariades, C., Muniappan, R. and Strathie, L.W. (eds) *Proceedings of the Fifth International Workshop on Biological Control and Management of* Chromolaena odorata, October 2000, Durban. ARC – PPRI, Pretoria, South Africa, pp. 66–70.

Timbilla, J.A., Weise, S.F., Addai, A. and Neuenschwander, P. (2002) Changing fallow vegetation following the introduction of *Pareuchaetes pseudoinsulata* to control *Chromolaena odorata* in southern Ghana. In: Zachariades, C., Muniappan, R. and Strathie, L.W. (eds) *Proceedings of the Fifth International Workshop on Biological Control and Management of* Chromolaena odorata, October 2000, Durban. ARC – PPRI, Pretoria, South Africa, p. 178.

Utulu, S.N. (1996) Controlling regrowths of *Chromolaena odorata* (L.) R.M. King and H.

Robinson, using herbicide mixtures in young oil palm plantations in Nigeria. In: Prasad, U.K., Muniappan, R., Ferrar, P., Aeschliman, J.P. and de Foresta, H. (eds) *Proceedings of the Third International Workshop on Biological Control and Management of* Chromolaena odorata, November 1993, Abidjan, Côte d'Ivoire. Agricultural Experiment Station, University of Guam, Publication 202, pp. 148–156.

Versveld, D.B., le Maitre, D.C. and Chapman, R.A. (1998) Alien invading plants and water resources in South Africa: a preliminary assessment. *Water Research Commission Report* TT 99/98, C.S.I.R. No. ENV/S-C 97154.

Von Senger, I., Barker, N. and Zachariades, C. (2002) Preliminary phylogeography of *Chromolaena odorata*: finding the origin of South Africa's pest. In: Zachariades, C., Muniappan, R. and Strathie, L.W. (eds) *Proceedings of the Fifth International Workshop on Biological Control and Management of* Chromolaena odorata, October 2000, Durban. ARC – PPRI, Pretoria, South Africa, pp. 90–99.

Wilson, E.O. (1988) The biogeography of West Indian ants (Hymenoptera: Formicidae). In: Liebherr, J.K. (ed.) *Zoogeography of Caribbean Insects*. Cornell University Press, Ithaca, pp. 214–230.

Wilson, C.G. and McFadyen, R.E.C. (2000) Biological control in the developing world: safety and legal issues. In: Spencer, N.R. (ed.) *Proceedings of the Tenth International Symposium on Biological Control of Weeds*. July 1999, Bozeman, Montana, USA, pp. 505–511.

Wilson, C. and Widayanto, E. (2002) The biological control programme against *Chromolaena odorata* in eastern Indonesia In: Zachariades, C., Muniappan, R. and Strathie, L.W. (eds) *Proceedings of the Fifth International Workshop on Biological Control and Management of* Chromolaena odorata, October 2000, Durban. ARC – PPRI, Pretoria, South Africa, pp. 53–57.

Working-for-Water URL: http://www-dwaf.pwv.gov.za/wfw/

Zachariades, C. and Goodall, J.M. (2002) Spread, impacts and management of *Chromolaena odorata* in southern Africa. In: Zachariades, C., Muniappan, R. and Strathie, L.W. (eds) *Proceedings of the Fifth International Workshop on Biological Control and Management of* Chromolaena odorata, October 2000, Durban. ARC – PPRI, Pretoria, South Africa, pp. 34–39.

Zachariades, C., Strathie-Korrûbel, L.W. and Kluge, R.L. (1999) The South African programme on the biological control of *Chromolaena odorata* (L.) King and Robinson (Asteraceae) using insects. *African Entomology Memoir* 1, 89–102.

Zebeyou, M.G. (1991) Distribution and importance of *Chromolaena odorata* in Côte d'Ivoire. In: Muniappan, R. and Ferrar, P. (eds) *Proceedings of the Second International Workshop on Biological Control of* Chromolaena odorata, February 1991, Bogor, Indonesia. BIOTROP Special Publication 44, pp. 67–70.

11

Aquatic Weeds in Africa and their Control

Carina J. Cilliers,[1] Martin P. Hill,[2] James A. Ogwang[3] and Obinna Ajuonu[4]

[1]*Agricultural Research Council – Plant Protection Research Institute, Biological Control of Weeds, Pretoria, South Africa;* [2]*Department of Zoology and Entomology, Rhodes University, Grahamstown, South Africa;* [3]*Biological Control Unit, Namulonge Agricultural Research Institute, Kampala, Uganda;* [4]*International Institute of Tropical Agriculture, Cotonou, Bénin*

Introduction

The invasion of rivers, dams and lakes throughout Africa by introduced aquatic vegetation represents one of the largest threats to the socioeconomic development of the continent. Currently there are five aquatic weeds that are especially problematic in Africa: water hyacinth, *Eichhornia crassipes* (Mart.) Solms-Laubach (*Pontederiaceae*); red water fern, *Azolla filiculoides* Lam. (Azollaceae); parrot's feather, *Myriophyllum aquaticum* (Vell.) Verdc. (*Haloragaceae*); water lettuce, *Pistia stratiotes* L. (*Araceae*); and salvinia, *Salvinia molesta* Mitchell (*Salviniaceae*). All are native to South America and, except for the rooted *M. aquaticum*, are free-floating macrophytes. The exact date and mode of introduction of these plants in some countries is obscure, but water hyacinth has been present in Africa since the late 1800s, while water lettuce was used as a medicinal plant in ancient Egypt (Holm *et al.*, 1977). These plants were sought after ornamentals, which would have aided their dispersal to new areas. In the absence of natural enemies and, invariably, the presence of nutrient-enriched waters, these aquatic weeds have proliferated and become problematic. Water hyacinth is the most damaging of these weeds and has increased in importance since the late 1980s.

The status of the distribution and impact of aquatic weeds in Africa and their control has been well reviewed in the proceedings of two workshops, one held in Zimbabwe in 1991 (Greathead and de Groot, 1993) and the other held in Nairobi in 1997 (Navarro and Phiri, 2000). Here, we review the status of aquatic weeds in Africa, focusing on recent research into the biological control of these plants.

Water Hyacinth

Water hyacinth is the most widespread and damaging aquatic weed throughout the world and it impacts many of the aquatic ecosystems in Africa. Dense mats of this weed degrade rivers, lakes, wetlands and dams, and limit their utilization.

Mechanical and chemical control

Manual removal of the weed is invariably the first control option practised in most countries where water hyacinth occurs. Barriers in the form of booms or cables have been effectively used in Côte d'Ivoire, Zambia and South Africa to protect hydropower water intake pumps and water extraction pumps. Although often tried, the use of harvesters has not been highly successful in Africa, as the extent and the reproductive rate of the weed and the shallow waters, in which it often occurs, make this technique impractical. Herbicides have been used for the control of water hyacinth in Ghana, Nigeria, South Africa, Zambia and Zimbabwe. The herbicides used include those with glyphosate, diquat and terbutryn and 2,4-D amine, as active ingredients. Creating barriers where water hyacinth can accumulate makes costly herbicidal control more effective as a whole block can thus be sprayed, and saves time and energy following small or individual plants targeted for spraying. Herbicidal control can relieve congestion of areas where access is obstructed (Cilliers *et al.*, 1996a).

Biological control

Biological control is considered the only sustainable control option for water hyacinth and several natural enemy species have been released in Africa (Table 11.1). The first successful biological control programme on the continent was in the Sudan (Beshir and Bennett, 1985) and the most recent and dramatic has been on Lake Victoria where expansive mats of the plant have collapsed (Cock *et al.*, 2000).

Regional summaries

East Africa

Water hyacinth in Tanzania dates back to 1955, when it was recorded on the Sigi and Pangani rivers, which empty into the Indian Ocean. The weed was reported from the Ugandan inland Lake Kyoga in 1987. However, water hyacinth gained importance as an aquatic weed in 1989 when it was first noticed on Lake Victoria. Initial infestations in Kenya were limited to a few bays but they later increased to cover most of the fishing beaches (Mailu *et al.*, 1999), stretching into Uganda. By 1990, the weed had covered most of the

Table 11.1. The natural enemy species established on water hyacinth in Africa.

Country	Natural enemy species established					
	Neochetina eichhorniae Warner Coleoptera: Curculionidae	*Neochetina bruchi* Hustache Coleoptera: Curculionidae	*Eccritotarsus catarinensis* (Carvalho) Heteroptera: Miridae (Plates 35, 36)	*Niphograpta albigutallis* Warren Lepidoptera: Pyralidae (Plates 33a, 33b)	*Othogalumna terebrantis* Wallwork Acarina: Galumnidae (Plate 34)	*Cercospora piaropi* Tharp. *Hyphomycetes*
Bénin	X	X				
Burkina Faso	X	X				
Congo	X	X				
Côte d'Ivoire	X	X				
Egypt	X	X				X
Ghana	X	X				X
Kenya	X	X				X
Malawi	X	X	X	X	X	
Mali	X	X				
Mozambique	X	X			X	
Niger Rep.	X					
Nigeria	X	X				X
Rwanda	X	X				
South Africa	X	X	X	X	X	X
Sudan	X	X		X		
Tanzania	X	X				
Togo	X	X				
Uganda	X	X				
Zambia	X	X			X	
Zimbabwe	X	X			X	

southern, Tanzanian, shoreline of Lake Victoria, with dense infestations occurring on the sheltered bays of Mwanza, Bauman, Emin Pasha, Mara and Rubafu (Mallya, 1999). Upstream, in Rwanda, water hyacinth thrived in a number of small lakes that feed into the Kagera River, which flows into the lake.

At the peak of infestation, water hyacinth covered an estimated 15,000 ha along Lake Victoria's shoreline in Kenya, Tanzania and Uganda, thus creating socio-economic and health problems for the lakeside residents (Mailu, 2001). Problems associated with this massive infestation, among others, included obstruction of both urban, rural and industrial water intake points, reduced fishing and hence fish exports, increase in human diseases and a reduced industrial output due to frequent interference with hydropower generation.

Biological control was considered the only viable method for reducing the impact of water hyacinth on the lake, and the two weevils, *Neochetina eichhorniae* Warner and *Neochetina bruchi* Hustache (Coleoptera, Curculionidae) (Plates 31a, 31b, 32) were released there in 1995. Numerous weevil-rearing stations were erected around the lake and, with the aid of the local fishing communities, several million weevils were released.

In a combined effort of both mechanical and mainly biological control, the water hyacinth biomass was reduced by an estimated 80% within a period of 4 years on the Ugandan part of the lake (Cock *et al.*, 2000). On the Kenyan shores of the lake, some 4 years after introduction, adult weevil numbers varied from 0 to 32 per plant with an average of six adults per plant and the weed coverage had been reduced by up to 80% (Ochiel *et al.*, 2001). In Tanzania, an integrated control approach, which included manual removal of the weed from fishing beaches and the introduction of the two weevils, resulted in a 70% reduction of water hyacinth within 3 years (Mallya *et al.*, 2001). The source of the water hyacinth coming into the lake has also been targeted with the construction of three weevil-rearing stations along the Kagera River in Rwanda (Moorehouse *et al.*, 2001), ensuring that biological control agents have been released at the source of the water hyacinth infestation.

Since the second half of 2000, there have been reports of a resurgence of water hyacinth in the Ugandan waters of Lake Victoria (Ogwang, 2001). This has been attributed to germination of seeds in the sediment stimulated by light penetration following the decline of the water hyacinth mat. The weevils have dispersed on to these new mats and it is envisaged that these will also be brought under control.

Elsewhere in this region, the weevils reduced the water hyacinth coverage by 80% on Lake Kyoga in Uganda (Ogwang and Molo, 1999) and the release of the weevils on the Sigi and Pangani rivers in Tanzania in 1995 reduced the amount of manual removal required to keep the river channels open (Mallya, 1999).

West Africa

Water hyacinth was observed in Bénin in 1977 (van Thielen *et al.*, 1994) and in Ghana and Nigeria in 1984 (Akinyemiju, 1987; De Graft Johnson, 1988). In Bénin, of the four species released, only the two weevil species have established, which is similar to most of the other countries within the region (Table 11.1). *N. eichhorniae* is the dominant species, having dispersed widely, while *N. bruchi* is confined to the release localities. At Tévèdji, Lihu and Kafedji on the Ouémé river, water hyacinth cover was reduced from 100% to 5% within 8 years, while the same level of control was achieved in just 5 years on Lake Azili, where the weevils had dispersed from the nearest release site 15 km away (Ajuonu *et al.*, 2003). When waterways at Tévèdji and Lake Azili became free, the communities celebrated the return of fishing activities, during which there was lots of dancing, wining and eating that was witnessed by government officials. Overall, water hyacinth in Bénin has been reduced from a serious to a moderate pest and the economic return of this project was 149:1 for southern Bénin alone (De Groote *et al.*, 2003).

However, at some sites like Savalou and Kpokissa, the weevils have not been as effective because the host plants were stranded in shallow mud banks when water receded, thus affecting the root-dwelling weevil pupae (Ajuonu *et al.*, 2003).

Nigeria receives an annual influx of water hyacinth from the Niger River, which threatens hydroelectric power generation. The weevil *N. eichhorniae* was released in Nigeria in 1994, while *N. bruchi* was released in 1995. In addition, a series of booms were used to collect the weed for manual removal and by 2001, water hyacinth infestation was visibly reduced compared with 1995 (Femi Daddy, 2002, New Bussa, personal communication).

In Niger, *N. eichhorniae* spread from the releases on the Niger River in Nigeria about 700 km upstream to the capital Niamey, where the weevils have had a visible impact on water hyacinth infestations. Reports from Ghana indicate that there has been a decline in water hyacinth populations following the release of the weevils in 1994 (de Graft Johnson, 2001, Accra, personal communication). A similar decline has been reported from Côte d'Ivoire (G. Mesmer Zebeyou, 2001, Abidjan, personal communication). Biological control projects on water hyacinth have also been initiated in Burkina Faso, Togo and Mali.

As part of the International Mycoherbicide Programme for *E. crassipes* Control in Africa (IMPECCA) field surveys and laboratory experiments have identified those fungal pathogens across the continent that are specific to water hyacinth and cause the greatest levels of disease (A. den Breeÿen, 2002, Stellenbosch, personal communication).

Southern Africa

Since the late 1970s, the Plant Protection Research Institute of South Africa has released five arthropods and one pathogen as natural enemies against the weed in that country (Table 11.1) (Cilliers, 1991a; Hill and Cilliers, 1999). In some areas of South Africa, good control has been achieved, while in other areas, low temperates in high-altitude climatic areas and interference from other control options have retarded biological control. It has been shown that under a rigorous, successful herbicide programme, biological control agents are suppressed. Once continuous spraying stops, agents can build up and control water hyacinth.

In South Africa, active research on pathogens on water hyacinth is ongoing. The rusts *Cercospora piaropi* Tharp. (Hyphomycetes), *Alternaria eichhorniae* Nag Raj & Ponnappa (*Ascomycotina*) and *Acremonium zonatum* (Sawada) W. Gams (*Ascomycotina*) are present in the field and *Cercospora rodmanii* Conway (*Hyphomycetes*) (= *C. piaropi*) has been released (Table 11.1). Elsewhere in the region, Malawi has had a successful water hyacinth biological control programme on the Shire River (Phiri *et al.*, 2001), the weed is under good control on Lake Kariba (Chikwenhere *et al.*, 1999), but there are still sporadic outbreaks on Lake Chivero, in Zimbabwe, which are attributed to high pollution levels (Chikwenhere and Phiri, 1999). The weevils, although present in some areas, have recently been further released in Mozambique (T. Chiconela, 2000, Maputo, personal communication; G. Phiri, 1997, Nairobi, personal communication).

Unfortunately, due to political unrest, little is known of the situation in Angola. In northern Egypt, *N. eichhorniae* and *N. bruchi* were released in August 2000 on two lakes. By July 2002, water hyacinth on Lake Edko was reduced by 90%. On Lake Mariout, reduction was slower due to water pollution (Y. Fayad, 2002. Giza, personal communication). The pathogen *Alternaria alternata* (Fr.) Keisser (Ascomycotina) has been utilized (Fayad, 1999).

Red Water Fern

The genus *Azolla* incorporates heterosporous aquatic fern species, which have a symbiotic association with the heterocystous cyanobacterium (blue–green alga) *Anabaena azollae* Strasburger, which grows within the dorsal leaf lobe cavities (Peters and Calvert, 1987). The alga can fix atmospheric nitrogen and is able to fulfil the nitrogen requirements of the plants, making them successful in nitrogen-deficient waters (Ashton, 1992). Various *Azolla* spp. have therefore been distributed as biological fertilizers in many developing countries, which has increased the distribution of this genus to many parts of the world (Lumpkin and Plucknett, 1982).

Origin and distribution

A. filiculoides is native to South America (Lumpkin and Plucknett, 1982) and has become a weed of small dams and slow-moving rivers in a number of countries around the world. It was first recorded in South Africa in 1948 (Oosthuizen and Walters, 1961). The fern was confined to small streams and farm dams in the centre of the country for many years. However, phosphate-enriched waters, the lack of natural enemies, and dispersal between water bodies by man and waterfowl facilitated an increase in its distribution and abundance. It is also know to be invasive in Zimbabwe, Zambia, Malawi, Mozambique and Uganda (Hill, 1999).

Impact

Red water fern is able to undergo rapid vegetative reproduction throughout the year by elongation and fragmentation of small fronds. Under ideal conditions, the daily rate of increase can exceed 15%. This translates to a doubling time for the fern of 5–7 days (Lumpkin and Plucknett, 1982). The fern can also reproduce sexually via the production of spores, especially when the plant is stressed. Spores can overwinter and are resistant to desiccation, allowing re-establishment of the fern after drought.

The increasing abundance of *A. filiculoides* in conservation, agricultural, recreational and suburban areas over the last 20 years has aroused concern. Among the major consequences of the dense mats (5–30 cm thick) of the weed on still and slow-moving water bodies in South Africa are: reduced quality of drinking water caused by bad odour, colour and turbidity; increased water-borne, water-based and water-related diseases; increased siltation of rivers and dams; reduced water surface area for recreation (fishing, swimming and water-skiing) and water transport; deterioration of aquatic biological diversity (Gratwicke and Marshall, 2001); clogging of irrigation pumps; drowning of livestock; and reduced water flow in irrigation canals.

Mechanical and herbicidal control

Mechanical and herbicidal control options have been suggested for red water fern. However, mechanical control is labour intensive. Small infestations of the weed in accessible areas can be removed with rakes and fine meshed nets, and the plants used as cattle and pig fodder, or compost. The disadvantages of this method are that the rate of increase of the plant is such that a concerted effort is required to keep up with the daily production of even a small infestation, and if eradication was achieved, re-establishment of the weed from spores resistant in substrate of the water body is inevitable. Glyphosate, paraquat and diquat-based herbicides have been used in the control of red water fern (Steÿn *et al.*, 1979; Axelsen and Julien, 1988; Ashton, 1992). The disadvantages of herbicidal control for *A. filiculoides* are that it is expensive, especially in view of the extensive follow-up programme required to eradicate the plants that continually germinate from spores. Also, there is a danger of spray drift on to non-target vegetation and a need for well-trained personnel. As mechanical control was impractical and herbicide use undesirable in the aquatic environment, the need for a biological control programme against *A. filiculoides* was elevated.

Biological control

The pre-introductory survey of the insect fauna associated with *A. filiculoides* (Hill, 1998a), and host records from elsewhere in the world, indicated that the genus *Azolla* is mostly attacked by generalist herbivorous insects and that very few specialist insect species have evolved on this genus (Lumpkin and Plucknett, 1982). Despite this, the weevil *Stenopelmus rufinasus* Gyllenhal (Curculionidae) and the flea beetles *Pseudolampsis guttata* (LeConte) and *Pseudolampsis darwini* (Scherer) (Coleoptera, Chrysomelidae, Alticinae) appear to have specialized on the genus *Azolla* (Habeck, 1979; Buckingham and Buckingham, 1981; Casari and Duckett, 1997) and were thus prioritized as potential biological control agents for red water fern. All three species were introduced into quarantine in South Africa between 1995 and 1998; *S. rufinasus* and *P. guttata* were subjected to host-specificity testing, and the culture of *P. darwini* was terminated as there was concern that its populations might become mixed with those of *P. guttata*, a species similar in morphology. The laboratory host range of *P. guttata* proved to be too wide to consider it for release in Africa (Hill and Oberholzer, 2002), while the host range of the frond-feeding weevil, *S. rufinasus* was far more restricted and it was cleared for release in late 1997 (Hill, 1998b). The second flea beetle, *P. darwini*, has not been re-imported as it was decided first to fully evaluate the impact of the weevil and determine if an additional agent was required to bring the weed under control.

Impact of Stenopelmus rufinasus

The first release was made on a 1-ha dam at a bird sanctuary in Pretoria in December 1997. Nine hundred weevils were released on the dam, which was 100% covered by a 5-cm-thick mat of the weed. Within 2 months, the mat had collapsed and from a 2-m^2 sample of decaying material in excess of 30,000 weevils were reared (Hill, 1999).

By late 2001 the weevils had been released (usually in batches of 100 adults) at some 110 sites throughout South Africa. The information available on these sites was that the weevil had been responsible for completely clearing 72 of them in less than 1 year (Plates 37, 38). For the remaining 38 sites, either the weed had been washed away during flooding, they had not been revisited or it was too early to tell. In addition to this, the weevil had migrated to other sites, up to 300 km away from the release sites. At 7% of the sites, the weed returned up to 2 years after the initial clearing, but the weevil located 90% of these and brought the weed under control again (A.J. McConnachie, 2001, Johannesburg, unpublished data). Five years after the first release of *S. rufinasus* in South Africa, the weed no longer poses a threat to aquatic ecosystems. An economic analysis on the biological control of red water fern in South Africa revealed a benefit to cost ratio of around 70:1 (A.J. McConnachie, 2001, Johannesburg, unpublished data). Unfortunately there was no attempt to quantify monetarily the benefit to biodiversity, but this has no doubt been substantial.

The weevil has been exported to Zimbabwe, where it has been responsible for clearing a number of farm dams (Table 11.2) (G. Chikwenhere, 1998, Harare, personal communication). It has also dispersed to Mozambique from South Africa. This project has been extremely successful in a short period of time and there is no need for an additional agent and no reason why *A. filiculoides* should pose a threat to any aquatic ecosystem in Africa any more.

Table 11.2. Natural enemy species released and established on water lettuce, red water fern, parrot's feather and salvinia in countries of Africa.[a]

	Natural enemy species released						
Country	*Neohydronomus affinis*[b]	*Stenopelmus rufinasus*[c]	*Lysathia* sp.[d]	*Cyrtobagous salviniae*[e]	*Cyrtobagous singularis*[e]	*Paulinia acuminata*[e]	*Samea multiplicalis*[e]
Bénin	X						
Botswana	X				X	X	X
Burkina Faso							
Congo	X			X			
Cotê d'Ivoire	X			X			
Egypt							
Ghana	X			X			
Kenya				X		X	
Malawi							
Mali							
Mozambique		X					
Namibia				X			
Niger Rep.							
Nigeria	X						
Rwanda							
South Africa	X	X	X	X			
Senegal	X						
Sudan							
Tanzania							
Togo	X						
Uganda							
Zambia	X			X	X	X	X
Zimbabwe	X	X		X		X	

[a]Data taken from Julien and Griffiths (1998).
[b]Agent for water lettuce.
[c]Agent for red water fern.
[d]Agent for parrot's feather.
[e]Agent for salvinia.

Parrot's Feather

There are two species of *Myriophyllum*, *M. aquaticum* (parrot's feather) and *M. spicatum* L. (spiked water-milfoil), that are weeds in southern Africa, but only *M. aquaticum* is troublesome. Parrot's feather roots in sediments up to a depth of 1.5 m below the water surface. In South Africa, propagation is entirely vegetative because there are no male plants in the country and no seeds are produced (Henderson, 2001). The plants grow throughout the year in subtropical regions, but at high altitudes the extremities of the aerial shoots are killed by winter frost. The dead parts provide shelter for new shoots, which sprout from the nodes during the following spring. Problems caused by dense mats of *M. aquaticum* are similar to those caused by other aquatic weeds. In South Africa, the threat posed by *M. aquaticum* was already noted in the 1960s, when the weed was considered to be of greater importance than water hyacinth.

Origin and distribution

The plant originates in South America and has been recorded in a number of countries, including Australia (Sainty and Jacobs, 1994). A small yet troublesome infestation was recorded from Lesotho, and it is known as a problem in Zimbabwe (Chikwenhere and Phiri, 1999). This plant has not become a major weed in other African countries.

Mechanical and chemical control

No herbicides have been registered for use against *M. aquaticum* in South Africa. Two glyphosate-based herbicides were used experimentally and brought temporary control, but these herbicides did not affect the root zone of the plants. Mechanical control is not practical, because any plant fragment not removed is capable of growing.

Biological control

A biological control programme was initiated against the weed in South Africa in 1991, which resulted in the release of the flea beetle *Lysathia* n. sp. (Coleoptera, Chrysomelidae) (Table 11.2). The release sites were located in a range of climatic zones, including Mediterranean, subtropical, temperate and cool temperate regions. Releases were made by introducing allotments of 50 adults on to infestations of the weed or by placing out plants with unknown numbers of larvae and pupae among *M. aquaticum* plants in the field. In most cases, establishment was confirmed following single introductions of the beetles in an area, but in a few cases, especially in the Mediterranean climatic region, the initial introductions failed and additional releases were needed before establishment was achieved (Cilliers, 1999a).

Lysathia n. sp. completes some seven to nine generations during summer and has survived through winter throughout South Africa, even in areas where there is frequent frost (e.g. in the Vaal River catchment). Emergent stems of *M. aquaticum* are killed by frost, but the beetles shelter on the undamaged leaves and shoots that are surrounded and protected by the outer layers of dead plant material. The beetle populations decline during winter, but survive in sufficient numbers to rebound rapidly in spring and summer. This induces a cyclical sequence of defoliation followed by partial recovery of the plants. Feeding by adults and larvae of *Lysathia* n. sp. has a cumulative effect on the growth of *M. aquaticum* and as beetle populations and feeding frequencies increase, the time taken for the plants to recover increases. Evaluations at one site indicated that the surface area covered by *M. aquaticum* declined slowly from 50% to 20% over 3 years. Continual resurgence of *M. aquaticum* infestations may necessitate the release of additional biological control agents. The stem-boring weevil, *Listronotus marginicollis* (Hustache), is a promising candidate to supplement the damage caused by *Lysathia* n. sp. The fact that *M. aquaticum* does not produce viable seeds in South Africa should restrict its ability to recover from prolonged periods of damage caused by introduced agents and the prospects for bringing *M. aquaticum* under biological control are very favourable (Cilliers, 1999b).

Water Lettuce

Water lettuce is a free-floating macrophyte. The perennial plants consist of rosettes of leaves with a tuft of long fibrous roots and resemble floating heads of lettuce. The flowers are pale and inconspicuous, and mainly present in mid summer, but can be found throughout the year. Seeds are viable and are borne underneath the older leaves and released into the water. They survive in the dry mud and germinate in the next wet season (Cilliers, 1991b).

Origin and distribution

Water lettuce originates in South America, and has been known in Africa since the time of Pliny (Holm *et al.*, 1977). It is recorded as a weed in a number of countries around the world, including Australia and the USA, and several countries throughout Africa (Julien and Griffiths, 1998).

Mechanical and chemical control

Mechanical control is impractical due to the plant's rapid growth rate. The only known registered herbicide is terbutryn in South Africa, although glyphosate has been used successfully. In small impoundments which are regularly subjected to alternate wet and dry regimes, herbicide control was used to deplete the seed bank as soon as seedlings reappeared after a dry spell (Cilliers *et al.*, 1996b). This proved impractical, as water lettuce infestations often escaped detection until after the plants had set seed.

Biological control

Biological control of water lettuce using the weevil *Neohydronomus affinis* (Hustache) (Coleoptera, Curculionidae) was first attempted in South Africa with positive results within 9 months on a seasonal pool, but control took longer (up to 3 years) on a fast-flowing river (Cilliers, 1991b; Cilliers *et al.*, 1996b). In northern Zimbabwe, control was achieved within 8 months (Chikwenhere, 1994) (Table 11.2).

Where a large seed reserve occurred at a man-made impoundment in the Kruger National Park in South Africa, *N. affinis* successfully controlled resurgence from dormant seeds each year. Subsequently, biological control has succeeded in Ghana, Côte d'Ivoire, Senegal and Kenya (A. Pieterse, 1998, Amsterdam; O. Diop, 2000, Dakar; A. Mailu, 1999, Nairobi, personal communications). In Bénin, the impact of *N. affinis* was monitored for 7 years and the total collapse of the water lettuce biomass was documented at two ponds (Fig. 11.1). The weevil spread on its own across most of the country (Ajuonu and Neuenschwander, 2002).

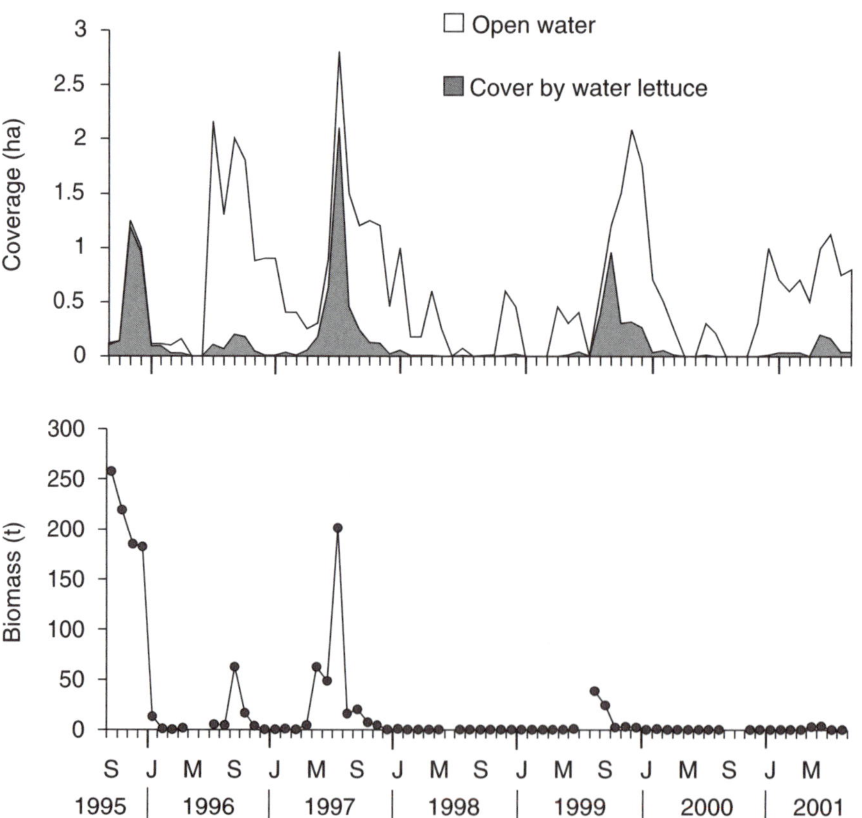

Fig. 11.1. Impact of *Neohydronomus affinis* on coverage (ha) and biomass (t ha^{-1}) of water lettuce, *Pistia stratiotes*, at Sé, in Bénin.

Salvinia

Salvinia is a free-floating aquatic fern species. It is sterile, producing sporocarps without viable spores, and thus only reproduces vegetatively.

Origin and distribution

This weed is of South American origin and has spread to many countries in Africa. In South Africa it was only surpassed by water hyacinth as an aquatic weed and was particularly troublesome in regions with a subtropical climate. It has also been recorded as a weed in a number of African countries. Salvinia was a major problem on Lake Kariba. As the dam was filling up in the early 1960s, the water was enriched due to organic-rich runoff and decay of plant material. By 1963, salvinia covered some 22% of the lake surface and threatened the infrastructure of the hydropower generation plant at the Kariba dam (Mitchell, 1972).

Mechanical and chemical control

Herbicide applications (diquat, terbutryn and glyphosate) were used to control this weed. Manual removal has been used, but the weed invariably outgrows all efforts.

Biological control

A number of species have been introduced for the control of salvinia in Africa (Table 11.2). Initially the plant was misidentified as *Salvinia auriculata* Aublet and the weevil *Cyrtobagous singularis* Hustache was introduced to Botswana and Zambia, but had little impacts on the weed populations (Julien and Griffiths, 1998). Following success in Australia, the weevil *Cyrtobagous salviniae* Calder and Sands (Coleoptera, Curculionidae) was introduced to a number of African countries (Table 11.2) where it has controlled this weed (Cilliers, 1991c). In addition, the grasshopper *Paulinia acuminata* (Degeer) (Orthoptera, Paulinidae) was introduced to Botswana, Kenya, Zimbabwe and Zambia. It did not establish in Kenya and had little impact on the weed in the other three countries (Julien and Griffiths, 1998). The pyralid moth *Samea multiplicalis* (Guenée) was also introduced to Botswana and Zambia, but failed to establish.

In subtropical areas, control of salvinia using *C. salviniae* was obtained within 18 months, regardless of the extent of the infestation, provided the tissue nitrogen content of salvinia was above 1% (dry weight). If the plants had a nitrogen concentration of lower than 1%, the weevils failed to establish (Room and Thomas, 1985). In the cooler areas, control took longer, but usually no more than 3 years. When infestations occurred on rivers, it usually took longer to achieve control. On closed systems, populations of both the weed and the biological control agent often collapsed to the point that neither the plant nor

C. salviniae could be found for years. One to 2 years after flooding, these impoundments could again be infested from nearby rivers or other infestations and the resultant unhindered growth of the weed necessitated reintroduction of the biological control agent. On rivers, a balance was achieved with some host plants remaining to sustain a population of the biological control agent. Equilibrium was reached and thus permanent control was attained on rivers, but not on smaller man-made impoundments.

Discussion and Conclusion

The biological control of aquatic weeds has been highly successful in Africa. Most of these weeds no longer pose as much of a problem as they did 10 years ago, before the introduction of biological control agents. Recently, new agents on water hyacinth and parrot's feather were found in South America and finally released in South Africa, and have proved useful additions to the already existing control agents on these weeds.

Several factors have contributed to the success of these programmes, including the tropical climate of some regions, which allowed rapid increases in biological control agent populations. In more temperate and cooler high altitude regions it has taken longer to attain biological control. There are simple techniques available for the mass-rearing of these agents for waterweeds, which allows for the release of large populations, facilitating their establishment. Biological control has been most successful on large water bodies where wind and wave action have aided in the break up of the waterweed mats already stressed by biological control agents, as was demonstrated on Lake Victoria. Also, the lack of herbicide application, which has been shown to interfere with biological control efforts in the USA and South Africa (Center *et al.*, 1999; Hill and Olckers, 2000) but which has not been used in most of Africa, has also contributed to these successes. However, eutrophication of freshwater ecosystems does contribute to the problem by promoting the luxurious growth of these weeds and this may extend the time it takes to control the weed.

Most of these weed biological control programmes have relied on technology that has been developed in other countries, notably Australia and the USA. However, as institutes throughout the continent become more experienced in this line of research, the reliance on developed countries will diminish and local technology will be able to solve local problems.

References

Ajuonu, O. and Neuenschwander, P. (2003) Release, establishment, spread and impact of the weevil *Neohydronomus affinis* (Coleoptera: Curculionidae) on water lettuce (*Pistia stratiotes*) in Bénin, West Africa. *African Entomology* (in press).

Ajuonu, O., Schade, V., Veltman, B., Sedjro, K. and Neuenschwander, P. (2003) Impact of the exotic weevils *Neochetina* spp: (Coleoptera: Curculionidae) on water hyacinth, *Eichhornia crassipes* (Lil: Pontederiaceae) in Bénin, West Africa. *African Entomology* (in press).

Akinyemiju, O.A. (1987) Invasion of Nigerian waters by water hyacinth. *Journal of Aquatic Plant Management* 25, 24–26.

Ashton, P.J. (1992) *Azolla* infestations in South Africa: history of the introduction, scope of the problem and prospects for management. *Water Quality Information Sheet* Department of Water Affairs and Forestry, Pretoria, South Africa.

Axelsen, S. and Julien, M.H. (1988) Weed control in small dams. Part II, control of salvinia, azolla and water hyacinth. *Queensland Agricultural Journal* 1988, 291–298.

Beshir, M.O. and Bennett, F.D. (1985) Biological control of water hyacinth on the White Nile, Sudan. In: Delfosse, E. (ed.) *Proceedings of the Sixth International Symposium on the Biological Control of Weeds*, 1980, Brisbane. Canadian Government Publishing Centre, Ottawa, Canada, pp. 491–496.

Buckingham, G.R. and Buckingham, M. (1981) A laboratory biology of *Pseudolampsis guttata* (LeConte) (Coleoptera: Chrysomelidae) on waterfern, *Azolla caroliniana* Willd. (Pteridophyta: Azollaceae). *The Coleopterists Bulletin* 35, 181–188.

Casari, S.A. and Duckett, C.N. (1997) Description of immature stages of two species of *Pseudolampsis* (Coleoptera: Chrysomelidae) and the establishment of a new combination in the genus. *Journal of the New York Entomological Society* 105, 50–64.

Center, T.D., Dray, F.A., Jubinsky, G.P. and Grodowitz, M.J. (1999) Biological control of water hyacinth under conditions of maintenance control: can herbicides and insects be integrated. *Environmental Management* 23, 241–156.

Chikwenhere, G.P. (1994) Biological control of water lettuce in various impoundments of Zimbabwe. *Journal of Aquatic Plant Management* 32, 27–29.

Chikwenhere, G.P. and Phiri, G. (1999) History of water hyacinth and its control efforts on Lake Chivero in Zimbabwe. In: Hill, M., Julien, M. and Center, T. (eds) *Proceedings of the First IOBC Global Working Group Meeting for the Biological and Integrated Control of Water Hyacinth*, 16–19 November 1999, Harare, Zimbabwe. ARC – PPRI, Pretoria, South Africa, pp. 91–100.

Chikwenhere, G.P., Keswani, C.L. and Liddel, C. (1999) Control of water hyacinth and its environmental and economic impacts at Gache Gache in the eastern reaches of Lake Kariba, Zimbabwe. In: Hill, M., Julien, M. and Center, T. (eds) *Proceedings of the First IOBC Global Working Group Meeting for the Biological and Integrated Control of Water Hyacinth*, 16–19 November 1999, Harare, Zimbabwe. ARC – PPRI, Pretoria, South Africa, pp. 30–38.

Cilliers, C.J. (1991a) Biological control of water hyacinth, *Eichhornia crassipes* (Pontederiaceae), in South Africa. *Agriculture, Ecosystems and Environment* 37, 207–217.

Cilliers, C.J. (1991b) Biological control of water lettuce, *Pistia stratiotes* (Araceae), in South Africa. *Agriculture, Ecosystems and Environment* 37, 225–229.

Cilliers, C.J. (1991c) Biological control of water fern, *Salvinia molesta* (Salviniaceae), in South Africa. *Agriculture, Ecosystems and Environment* 37, 219–224.

Cilliers, C.J. (1999a) Biological control of parrot's feather, *Myriophyllum aquaticum* (Vell.) Verdc. (Haloragaceae), in South Africa. In: Olckers, T. and Hill, M.P. (eds) *Biological Control of Weeds in South Africa (1990–1998). African Entomology Memoir* 1, 113–118.

Cilliers, C.J. (1999b) *Lysathia* n. sp. (Coleoptera: Chrysomelidae), a host-specific beetle for the control of the aquatic weed *Myriophyllum aquaticum* (Haloragaceae) in South Africa. *Hydrobiologia* 415, 271–226.

Cilliers, C.J., Campbell, P.L., Naude, D. and Neser, S. (1996a) An integrated water hyacinth control programme on the Vaal River, in a cool, high altitude area in South Africa. In: Charudattan, R., Labrada R., Center, T.D. and Kelly-Begazo, C.

(eds) *Strategies for Water Hyacinth Control*. Report of a Panel of Experts Meeting, 11–14 September 1995, Fort Lauderdale, Florida, USA. FAO, Rome, pp. 87–103.

Cilliers, C.J., Zeller, D. and Strydom, G. (1996b) Short- and long-term control of water lettuce (*Pistia stratiotes*) on seasonal water bodies and on a river system in the Kruger National Park, South Africa. *Hydrobiologia* 340, 173–179.

Cock, M.W., Day, R., Herren, H., Hill, M.P., Julien, M.H., Neuenschwander, P. and Ogwang, J. (2000) Harvesters get that sinking feeling. *Biocontrol News and Information* 21, N1-N8.

De Graft Johnson, A.K.K. (1988) Aquatic plant invasion and water resource – the Ghanaian experience. In: Oke, O.L., Imevbore, A.M.A. and Fari, T.A. (eds) *Water Hyacinth, Menace and Resource, Proceedings of an International Workshop*, 7–12 August 1988, Lagos, Nigeria, ARC – PPRI, Pretoria, South Africa, pp. 32–38.

De Groote, H., Ajuonu, O., Attignon, S., Djessou, R. and Neuenschwander, P. (2003) Economic impact of biological control of water hyacinth in Southern Bénin. *Ecological Economics* 45, 105–117.

Fayad, Y.H. (1999) Water hyacinth infestations and control in Egypt. *Proceedings of the First IOBC Global Working Group Meeting for the Biological and Integrated Control of Water Hyacinth,* 16–19 November 1999, Harare, Zimbabwe. ARC – PPRI, Pretoria, South Africa, pp. 106–110.

Gratwicke, B. and Marshall, B.E. (2001) The impact of *Azolla filiculoides* Lam. on animal biodiversity in streams in Zimbabwe. *African Journal of Ecology* 38, 1–4.

Greathead, A. and De Groot, P. (1993) Control of Africa's floating water weeds. *Proceedings of a Workshop held in Zimbabwe*, June 1991. Commonwealth Science Council, CAB International, Wallingford, UK, pp. 13–20.

Habeck, D.H. (1979) Host plants of *Pseudolampsis guttata* (LeConte) (Coleoptera: Chrysomelidae). *The Coleopterists Bulletin* 33, 150.

Henderson, L. (2001) Alien weeds and invasive plants. A complete guide to declared weeds and invaders in South Africa. *Plant Protection Research Institute Handbook* No. 12. Pretoria, South Africa.

Hill, M.P. (1998a) Herbivorous insect fauna associated with *Azolla* species in southern Africa. *African Entomology* 6, 370–372.

Hill, M.P. (1998b) Life history and laboratory host range of *Stenopelmus rufinasus* Gyllenhal (Coleoptera: Curculionidae), a natural enemy for *Azolla filiculoides* Lamarck (Azollaceae) in South Africa. *BioControl* 43, 215–224.

Hill, M.P. (1999) Biological control of red water fern, *Azolla filiculoides* Lamarc (Pteridophyta: Azollaceae), in South Africa. In: Olckers, T. and Hill, M.P. (eds) *Biological Control of Weeds in South Africa (1990–1998). African Entomology Memoir* 1, 119–124.

Hill, M.P. and Cilliers, C.J. (1999) A review of the arthropod natural enemies, and factors that influence their efficacy, in the biological control of water hyacinth, *Eichhornia crassipes* (Mart.) Solms-Laub. (Pontederiaceae), in South Africa. In: Olckers, T. and Hill, M.P. (eds) *Biological Control of Weeds in South Africa (1990–1998). African Entomology Memoir* 1, 103–112.

Hill, M.P. and Oberholzer, I.G. (2002) Laboratory host range testing of the flea beetle, *Pseudolampsis guttata* (LeConte) (Coleoptera: Chrysomeliadae), a potential natural enemy for red water fern, *Azolla filiculoides* Lamarck (Pteridophyta: Azollaceae) in South Africa. *The Coleopterists Bulletin* 56, 79–83.

Hill, M.P. and Olckers, T. (2001) Biological control initiatives against water hyacinth in South Africa: constraining factors, success and new courses of action. In: Julien, M.H., Hill, M.P., Center, T.D. and Jianqing, D. (eds) *Biological and Integrated Control of Water Hyacinth*, Eichhornia crassipes. *Proceedings of the Second Global*

Working Group Meeting for the Biological and Integrated Control of Water Hyacinth, 9–12 October 2000, Beijing, China. ACIAR, Canberra, Australia, Proceedings 102, 33–38.

Holm, L.G., Plucknett, D.L., Pancho, J.V. and Herurger, J.P. (1977) *The World's Worst Weeds, Distribution and Biology.* The University Press of Hawaii, p. 609.

Julien, M.H. and Griffiths, M.W. (1998) *Biological Control of Weeds. A World Catalogue of Agents and Their Target Weeds*, 4th edn. CAB International, Wallingford, UK.

Lumpkin, T.A. and Plucknett, D.L. (1982) *Azolla* as a green manure: use and management in crop production. *Westview Tropical Agriculture Series* 5. Westview Press, Boulder, Colorado, USA.

Mailu, A.M., Ochiel, G.R.S., Gitonga, W. and Njoka, S.W. (1999) Water hyacinth: an environmental disaster in the Winam Gulf of Lake Victoria and its control. In: Hill, M., Julien, M. and Center, T. (eds) *Proceedings of the First IOBC Global Working Group Meeting for the Biological and Integrated Control of Water Hyacinth,* 16–19 November 1999, Harare, Zimbabwe. ARC – PPRI, Pretoria, South Africa, pp. 101–106.

Mailu, A.M. (2001) Preliminary assessment of the social, economic and environmental impacts of water hyacinth in Lake Victoria Basin and status of control. In: Julien, M.H., Hill, M.P. and Jianqing, D. (eds) *Proceedings of the Second Meeting of Global Working Group for the Biological and Integrated Control of Water Hyacinth,* 9–12 October 2000, Beijing, China. ACIAR, Canberra, Australia, Proceedings 102, 130–139.

Mallya, G.A. (1999) Water hyacinth in Tanzania. In: Hill, M., Julien, M. and Center, T. (eds) *Proceedings of the First IOBC Global Working Group Meeting for the Biological and Integrated Control of Water Hyacinth*, 16–19 November 1999, Harare, Zimbabwe. ARC – PPRI, Pretoria, South Africa, pp. 25–29.

Mallya, G.A., Mjema, P. and Ndunguru, J. (2001) Water hyacinth control through integrated pest management strategies in Tanzania. In: Julien, M.H., Hill, M.P. and Jianqing, D. (eds) *Proceedings of the Second Meeting of Global Working Group for the Biological and Integrated Control of Water Hyacinth*, 9–12 October 2000, Beijing, China. ACIAR, Canberra, Australia, Proceedings 102, 120–122.

Mitchell, D.S. (1972) The Kariba weed: *Salvinia molesta. British Fern Gazette* 10, 251–252.

Moorehouse, T.M., Agaba, P. and McNabb, T.J. (2001) Recent efforts in biological control of water hyacinth in the Kagera River headwaters of Rwanda. In: Julien, M.H., Hill, M.P. and Jianqing, D. (eds) *Proceedings of the Second Meeting of Global Working Group for the Biological and Integrated Control of Water Hyacinth,* 9–12 October 2000, Beijing, China. ARC – PPRI, Pretoria, South Africa, pp. 39–42.

Navarro, L.A. and Phiri, G. (2000) *Water Hyacinth in Africa and the Middle East. A Survey of Problems and Solutions.* International Development Research Centre, Ottawa. (ISBN 0–88936–933-X). hhtp:www.idrc.ca.books. 24 September 2002.

Ochiel, G.S., Njoka, S.W., Mailu, A.M. and Gitonga, W. (2001) Establishment, spread and impact of *Neochetina* spp. weevils (Coleoptera: Curculionidae) on water hyacinth in Lake Victoria, Kenya. In: Julien, M.H., Hill, M.P. and Jianqing, D. (eds) *Proceedings of the Second Meeting of Global Working Group for the Biological and Integrated Control of Water Hyacinth*, 9–12 October 2000, Beijing, China. ACIAR, Canberra, Australia, Proceedings 102, 89–95.

Ogwang, J.A. (2001) Is there a resurgence on Lake Victoria? *Water Hyacinth News* 4, 1–2.

Ogwang, J.A. and Molo, R. (1999) Impact studies on *Neochetina bruchi* and *Neochetina eichhorniae* in Lake Kyoga, Uganda. In: Hill, M., Julien, M. and Center, T. (eds) *Proceedings of the First IOBC Global Working Group Meeting for the Biological and Integrated Control of Water Hyacinth*, 16–19 November 1999, Harare, Zimbabwe. ARC – PPRI, Pretoria, South Africa, pp. 10–13.

Phiri, P.M., Day, R.K., Chimatiro, S., Hill M.P., Cock, M.J.W., Hill, M.G. and Nyando, E. (2001) Progress with biological control of water hyacinth in Malawi. In: Julien, M.H., Hill, M.P. and Jianqing, D. (eds) *Proceedings of the Second Meeting of Global Working Group for the Biological and Integrated Control of Water Hyacinth*, 9–12 October 2000, Beijing, China. ACIAR, Canberra, Australia, Proceedings 102, 47–52.

Oosthuizen, G.J. and Walters, M.M. (1961) Control of water fern with diesoline. *Farming in South Africa* 37, 35–37.

Peters, G.A. and Calvert, H.E. (1987) The *Azolla–Anabaena azollae* symbiosis. In: Subba Rao, N.S. (ed.) *Advances in Agricultural Microbiology*. Oxford and IBH Publishing Company, New Delhi, India, pp. 191–218.

Room, P.M. and Thomas, P.A. (1985) nitrogen and establishment of a beetle for biological control of the floating weed *Salvinia molesta* in Papua New Guinea. *Journal of Applied Ecology* 22, 139–156.

Sainty, G.R. and Jacobs, S.W.L. (1994) *Waterplants in Australia.* Sainty and Associates, Darlinghurst, Brisbane, Australia, p. 327.

Steÿn, D.J., Scott, W.E., Ashton, P.J. and Vivier, F.S. (1979) Guide to the use of herbicides on aquatic plants. *Technical Report TR 95*. Department of Water Affairs, South Africa, Pretoria, South Africa.

van Thielen, R., Ajuonu, O., Schade, V., Neuenschwander, P., Adité, A. and Lomer, C.J. (1994) Importation, releases, and establishment of *Neochetina* spp. (Col.: Curculionidae) for the biological control of water hyacinth, *Eichhornia crassipes* (Lil.: Pontederiaceae), in Bénin, West Africa. *Entomophaga* 39, 179–188.

12

Endophytic Microbial Biodiversity and Plant Nematode Management in African Agriculture

Richard A. Sikora,[1] Björn Niere[2] and John Kimenju[3]

[1]*Soil Ecosystem Phytopathology and Nematology, University of Bonn, Bonn, Germany;* [2]*International Institute of Tropical Agriculture – Eastern and Southern Africa Regional Center (ESARC), Namulonge, Kampala, Uganda;* [3]*University of Nairobi, Nairobi, Kenya*

Introduction

Plant parasitic nematodes are ubiquitous soil-borne pests that cause significant damage to most, if not all, crops growing in the tropics and subtropics. In cases where they interact simultaneously with root-rot or wilt pathogens, they can cause enormous yield loss. Although actual losses on a per country basis are not known, estimates of annual losses exceeding US$100 billion worldwide have been made (Luc *et al.*, 1990). Crop damage caused by nematodes for a number of important crops has been estimated and the figures can be used to extrapolate yield loss experienced by African growers (Table 12.1). In particular, horticultural crops are heavily damaged by plant parasitic nematodes. In many cases, these losses are circumvented for short periods of time by annual applications of expensive and highly toxic soil fumigants or non-fumigant nematicides. These pesticides reduce early root penetration, but do not eradicate the nematode from the soil. For practical purposes, nematode management can be divided into a number of categories based on the major control technology used.

Rotation-based control is the classical method of nematode management in field crops. It is used by many African farmers, whether or not they have a nematode problem. Rotation can be effective when used in a planned systems approach with non-host crops and resistant cultivars, when available. It should be noted, however, that the use of nematode resistant cultivars is limited to only a few crops. If a haphazard non-systems approach is taken to rotation then nematode control is for the most part ineffective. However, even with proper rotation, some yield loss occurs due to the fact that most rotations are seldom broad enough to reduce nematode densities below damage threshold levels. In addition, the mixed-cropping often used by African farmers includes

(eds P. Neuenschwander, C. Borgemeister and J. Langewald)

Table 12.1. Summary of estimated annual yield losses worldwide due to damage by plant parasitic nematodes (Sasser and Freckman, 1987).

Crop	Loss (%)	Crop	Loss (%)
Banana	19.7	Citrus	14.2
Potato	12.2	Coffee	15.0
Cassava	8.4	Cotton	10.7
Groundnut	12.0	Aubergine	16.9
Corn	10.2	Forages	8.2
Field bean	10.9	Melons	13.8
Rice	10.0	Okra	20.4
Sorghum	6.9	Ornamentals	11.2
Soybean	10.6	Papaya	15.1
Sugarcane	15.3	Pepper	12.2
Sweet potato	10.2	Pineapple	14.9
Wheat	7.0	Tobacco	14.7
Millet	11.8	Tomato	20.6
Chickpea	13.7	Yam	17.7

host crops that maintain populations at near threshold levels. Other cultural control methods that can be incorporated into rotation systems, relevant to Africa, have been outlined by Kerry (1992) and Bridge (1996).

Nematicide-based approaches are mainly important on high-value, peri-urban and/or export crops such as tobacco, pineapple, banana, vegetables and ornamental crops. Their use is usually limited to large-scale farmers and commercial companies that have adequate financial resources. The mainstay of this approach has been the fumigant methyl bromide and to a lesser extent non-fumigant nematicides. The worldwide phasing out of methyl bromide to be completed by the year 2003 due to its effects on atmospheric ozone has removed the charm many growers relied upon for over 40 years to control nematodes (Sikora, 2002). Furthermore, the targeted integration of the remaining non-fumigant nematicides into IPM systems is still a 'figment of one's imagination'. In most cases, if nematicides are used, other control technologies are ignored, because the reduction attained in nematode early root-penetration ensures good yield. Banana production for export, for example, is highly dependent upon repeated nematicide treatment. For these reasons there has been little integration of nematicide-based approaches into true IPM systems.

Biological-based control of nematodes in Africa is for the most part limited to stimulation of naturally occurring antagonists with organic amendments (Stirling, 1991). This approach works even more effectively when properly integrated into IPM programmes. A great deal of ongoing applied research in African universities and government organizations is now designed to find, optimize and release biological control agents for field use to control nematodes on a broad spectrum of crops.

IPM is the most effective nematode management strategy used by farmers who have knowledge of effective rotations, resistant cultivars and nematicides, or who have the equipment needed to incorporate other cultural control methodologies into their crop management systems. In many cases, however, an 'ignore them' approach is used by farmers who do not have the resources to purchase modern inputs or who do not realize they have a nematode problem. This is the situation with the vast majority of subsistence growers and resource-poor farmers. Therefore, ignoring nematodes is the norm and not the exception in subsistence agriculture.

All forms of crop production have one thing in common – the need for sound control technologies that reduce the impact of pests and diseases and increase yield. Biological control is a promising alternative that can be used to improve crop yield. However, the following questions must be raised with regards to what type of biological control is appropriate for African growers: (i) what biological agent is going to be most effective under local conditions; (ii) which crops lend themselves to biological control technologies; (iii) which nematode species should be targeted; and (iv) how can biological control be strategically positioned in IPM systems for optimum levels of control?

Biodiversity, Biological Control and IPM

The importance of specific microorganisms that are part of the microbial biodiversity of a soil-ecosystem and how to use them for development of innovative biological control technologies has been only superficially researched in Africa. Conversely, the level of microbial biodiversity in African soils is enormously large. The soil is home to still unknown forms of antagonistic potential that can be used for nematode and disease control. The total biomass of micro- and macro-organisms in 1 ha of soil can reach 12 t (Gisi, 1990). The actual number of bacteria in soil has been estimated to range from 10^6 to 10^9 colony forming units per gram and there is an estimated 300 m of fungal mycelium in 1 l of soil. Many of these organisms are responsible for the antagonistic activity that is reflected in the suppressive nature of the soil that leads to reduced nematode and disease incidence. Of even greater significance is the fact that approximately 10% of all the bacteria and fungi isolated from the root have been shown to have significant levels of biological control activity (Sikora, 1997). There are three schools of thought as to how microbial biodiversity in agroecosystems can be used successfully for biological control in practical agriculture.

The inundative or direct soil treatment approach is probably the best-known and studied technique for nematode control. It requires application of highly specific pathogens or parasites to the soil before planting. Control is usually based on fungal antagonists of nematode eggs, juveniles or females that kill the nematode and thereby reduce population densities. Because of poor establishment in the soil, repeated treatments with high levels of inoculum prior to planting are required for effective control.

Biological enhancement of seed and/or transplanting material with mutualistic microorganisms, which are plant health promoting and live either in the rhizosphere or mutualistically in the root tissue, is a new approach that attempts to target application to increase efficacy and reduce treatment costs. In this case, planting material is treated with antagonists prior to planting. The mechanism of action of antagonism is usually related to disruption in juvenile penetration, induced resistance and/or to competition for nutrients that ultimately reduces nematode fecundity (Sikora and Hoffmann-Hergarten, 1993).

Management of the antagonistic potential is a non-invasive approach that leads to nematode or disease suppression. It requires a thorough knowledge of the target pest and the beneficial organisms occurring naturally in the soil, in order to manage and increase activity of the antagonists. In some cases, an initial application of an antagonist is sufficient to establish the antagonist in the soil for management purposes (Sikora, 1997).

The different types of nematode antagonists being studied worldwide that we consider important for biological control, and an example of each, are as follows: bacterial parasites attacking female nematodes (*Pasteuria penetrans*), egg-pathogens (*Verticillium chlamydosporium* Goddard), antagonistic rhizobacteria (*Bacillus subtilis*), endophytic bacteria (*Rhizobium etli*), predatory fungi (*Arthrobotrys* spp.), mutualistic endophytic fungi (*Fusarium oxysporum* Schlecht.: Fries) and arbuscular mycorrhizal fungi (*Glomus* spp.). It should be noted that there are many other organisms in these groups that have been shown to be effective in nematode control (Stirling, 1991). Many of these organisms still need to be tested under African conditions.

Although all of the above organisms are important to biological control in African agriculture, endophytic fungi probably have the greatest potential for effective nematode control. These fungi are found in the endorhiza of all major crops. By targeting the endorhiza as the site for application of biological control agents, efficacy is promoted by increasing the level of intimate contact between the biological control agent and the pest. We are convinced that proper use of this group of biological control agents will have a major plant health-promoting impact on a broad spectrum of African crops.

The level of root health at any one time is directly related to the degree of colonization and to the structure of beneficial microbial communities present in the root zone, before or during the infection process. The fact that roots of all plants are only partially infected by fungal pathogens and nematodes, but are not destroyed by them, indicates the existence of a natural buffering system in the root tissue that works independently of genetically bound host defence mechanisms. The fact that nematodes and diseases can still infect or attack plant roots is an indication that these beneficial organisms are not omnipresent throughout the root system, but are heterogeneously distributed in the root tissue. This heterogeneous colonization is the reason why the use of these biological control agents must be seen as a part of IPM systems.

Strategy

The two main questions raised in this section are: what is the potential of using beneficial microorganisms living in the root system for improving root health and overall plant growth and how can they be integrated into sustainable and high-input production systems in Africa?

Fungal antagonists associated with the rhizosphere can be placed in two main groups: rhizosphere competent (found on the surface of the root) or mutualistic endophytes (found inside the root system). These two groups have been shown to be successful in reducing fungal, bacterial and nematode disease incidence on a wide range of crops. In most cases, these organisms are targeted for use on specific crops and toward specific deleterious organisms. Plants of high market value or cash crops are often targeted, because of the cost of production and formulation of microbial preparations. These two groups of antagonists are effective, because they are targeted at specific nematodes or diseases that cause severe injury in the early stages of plant growth or directly after germination or transplanting. The treatment of planting material is economically advantageous in that it leads to a reduction in costs due to lower amounts of inoculum needed for effective control.

Although a broad spectrum of microbial agents for nematode control is known, we shall concentrate on the importance of endophytic fungi to African agriculture. We believe this group of biological control agents has a broad range of activity to many pests and diseases, a long duration of activity and a higher level of control stability than other microbial antagonists because they grow endophytically in the root. Two groups of mutualistic endophytes are considered to be important in root health promotion: mutualistic endophytic fungi (MEF) that colonize living plant tissue and can also grow saprophytically on organic matter in the soil, and arbuscular mycorrhizal fungi (AMF) that grow inside plant tissue and in the soil but that are obligate symbionts and therefore require a plant host for growth.

Strategies for incorporation of these two groups of fungi into IPM systems for specific crop groups and toward specific nematodes important in Africa will be outlined below. The success of this biological control approach to improving root health is directly related to a sound strategy that includes proper selection of: (i) highly sensitive nematode pests; (ii) suitable host crops; (iii) isolation of effective antagonists; (iv) commercial production; and (v) integration into IPM systems.

Endophytic Fungi for the Biological Control of Plant Parasitic Nematodes

MEF can be defined as fungi that live at some time in their life cycle in the plant without producing symptoms of disease, but simultaneously demonstrate antagonistic activity toward one or more pests or diseases affecting the root system. Endophytes are natural inhabitants of the root system in all field crops. A plant always harbours more than one MEF. Endophytes can be isolated from healthy plant tissue after surface sterilization. To date, the grass endophytes remain the best-studied group, having become well known because of their mutualistic symbiosis with their host and for their effects on insects. However, MEF have now been isolated also from roots of many important crop plants, such as maize, rice, wheat, barley, tomato and banana. More recently, the level of biodiversity and frequency of occurrence of these fungi in crop plants important to African agriculture for the biological control of nematodes, insects and fungal pathogens have been investigated (Hallmann and Sikora, 1994; Kimenju, 1998; Niere *et al.*, 1999; Sikora and Schuster, 1999; Griesbach, 2000; Sikora *et al.*, 2000; Waceke *et al.*, 2001a). Research on the importance of endophytes for nematode biological control is concentrated on maize, ryegrass, tall fescue, tomato and banana (Table 12.2).

A large proportion of studies conducted with MEF have dealt with endophytes of banana. Bananas play an important role in African agriculture, especially in the East African highlands, where bananas are the main food staple and an important source of income. Banana production is threatened worldwide by a number of pests and diseases that are very difficult to control, the most important soil-borne problem being migratory endoparasitic nematodes and their interactions with soil diseases (Luc *et al.*, 1990). In a study of endophytic fungi of healthy banana roots, a large number of fungi were isolated from asymptomatic root and rhizome tissue (Table 12.3). Fungi of the genus *Fusarium* were frequently isolated and seem to form an essential component of the endophytic mycobiota of banana. Similar fungal diversity was detected in studies conducted in Central America (Pocasangre, 2000; Pocasangre *et al.*, 2000). Some endophytic *Fusarium* spp. were also shown to produce metabolites *in vitro* that led to the inactivation of *Radopholus similis* (Cobb) Thorne (Tylenchida, Pratylenchidae) or *Cosmopolites sordidus* Germar (Coleoptera, Curculionidae) (Schuster *et al.*, 1995; Griesbach, 2000).

In greenhouse studies in unsterilized field soil, tissue-cultured banana plantlets were inoculated with endophytic fungi to detect biological control activity toward nematodes, the banana weevil or *Fusarium* wilt (Amin, 1994; Niere *et al.*, 1999; Griesbach, 2000; Pocasangre, 2000). Inoculation of endophytic strains of *F. oxysporum* on to tissue-cultured banana plants after deflasking and before potting, and subsequent challenge with *R. similis*, led to lowered numbers of nematodes in the roots of banana (Table 12.4). Nematode numbers in endophyte-treated plants in these experiments were reduced by more than 30% over the controls. Similar reductions in nematode numbers in banana roots were found for other banana clones, but not all clones responded

Table 12.2. Endophytic fungi in different plants and their influence on populations of plant parasitic nematodes in different host plants.

Endophyte	Nematode species	Host plant	Nematode population
Neotyphodium lolii	*Meloidogyne naasi*	Ryegrass	Unaffected
N. lolii	*M. naasi*	Ryegrass	Reduced
Neotyphodium coenophialum	*Helicotylenchus dihystera*	Tall fescue	Reduced
N. coenophialum	*Helicotylenchus pseudorobustus*	Tall fescue	Unaffected
N. coenophialum	*Meloidogyne marylandi*	Tall fescue	Reduced
N. coenophialum	*Paratrichodorus minor*	Tall fescue	Reduced
N. coenophialum	*Pratylenchus scribneri*	Tall fescue	Reduced
N. coenophialum	*Tylenchorhynchus acutus*	Tall fescue	Unaffected
Fusarium moniliforme	*Pratylenchus zeae*	Maize	Reduced
Fusarium oxysporum	*Radopholus similis* *Helicotylenchus multicinctus*	Banana	Reduced

Table 12.3. Fungal endophyte biodiversity in banana root and rhizome tissue of Pisang Awak (*Musa* ABB) from Thailand.

Fungal species	Roots (%)[a]	Central cylinder (%)	Rhizome cortex (%)	Overall (%)
Acremonium spp.	15.4	6.7	7.5	9.9
Acremonium stromaticum	11.5	6.7	5.0	7.7
Aspergillus spp.	5.4	5.3	2.5	4.4
Colletotrichum musae	2.3	9.3	11.2	7.6
Cylindrocarpon spp.	3.1	1.3	5.0	3.1
Fusarium Sect. Arthrosporiella	5.4	10.7	11.3	9.1
Fusarium Sect. Liseola	2.3	4.0	3.7	3.3
Fusarium oxysporum	7.7	6.7	15.0	9.8
Fusarium solani	0.8	6.7	6.3	4.6
Fusarium spp.	3.1	4.0	1.2	2.8
Gongronella spp.	7.7	6.7	3.8	6.1
Penicillium spp.	21.5	10.7	5.0	12.4
Zygomycetes	4.6	4.0	1.2	3.3
Other species[b]	9.2	17.2	21.3	15.9

[a] Percentage of total number of fungal isolates. A total number of 285 isolates was identified. Number of fungi isolated from roots, central cylinder and rhizome cortex were 130, 75 and 80, respectively.
[b]Mainly fungi belonging to the genera *Cladosporium*, *Cylindrocladium*, *Drechslera*, *Lasiodiplodia*, *Plectosporium*, *Thielaviopsis* and *Trichoderma*.

Table 12.4. Multiplication of *Radopholus similis* in roots of East African Highland bananas, *Musa* AAA-EA, cv. Enyeru, inoculated with endophytic isolates of *Fusarium oxysporum* compared with control plants.

	Radopholus similis per g root			Root necrosis		
Experiment[a]	Control (%)	Endophyte[b] (%)	Change (%)	Control (%)	Endophyte[b] (%)	Change (%)
1	62	41	−34	49	35	−28
2	39	20	−49	29	19	−33
3	33	23	−31	40	23	−42

[a]Plants in experiments 1, 2 and 3 were 22, 25 and 27 weeks old, respectively, and grown in pots. Nematode challenge inoculum consisted of 1000, 900 and 700 nematodes per plant and challenge periods were 14, 17 and 12 weeks for experiments 1, 2 and 3, respectively.
[b]Endophytic isolates of *F. oxysporum* were V2w2, V1w7 and Eny 7.11 o in experiments 1, 2 and 3, respectively. Adapted from Niere (2001).

equally to the same isolate (Niere, 2001). Nematode damage, expressed as a percentage of necrotic root tissue, was similarly reduced (Table 12.4).

The mechanisms leading to nematode control have not been clearly identified. Production of secondary metabolites, competition for nutrients or induced resistance may play a role. Fungal infection of juvenile or adult nematodes or their eggs has not been observed, but depending on MEF isolate is not unthinkable. Since the frequency of re-isolation of MEF from the roots of banana plants decreases over time (Table 12.5), it is concluded that long duration control into the following crop-cycle, as has been shown in the case of some of the MEF of perennial grasses, may not be strong. Initial results suggest that endophytes can provide protection in the first growth cycle. This may be sufficient to give a perennial crop such as banana a head start. Until MEF isolates are found that effectively colonize the suckers of the ratoon crop and until the extent to which the surrounding soil influences the composition and distribution of root endophytes is known, durable biological control in banana will not be possible.

The presence of the fungus *Fusarium moniliforme* Sheld. in maize roots led to significant levels of biological control of the lesion nematode, *Pratylenchus zeae* Graham (Tylenchida, Pratylenchidae), a major parasite of maize in all of Africa (Table 12.6). Variability in colonization of maize cultivars by an endophytic strain of *F. moniliforme* was demonstrated and shows that utilization of the endophytes in plant pest management will require cultivars that are compatible with them (Kimenju, 1998). Similar observations were made by Niere (2001) on banana.

Relatively little is known about the biodiversity of endophytes in the roots of horticultural crops in Africa. Endophytic fungi, however, have been isolated from tomato roots in Africa (Hallmann and Sikora, 1994, 1996). Of the large number of fungi obtained, *F. oxysporum* proved to be the most common species in the root tissue with biological control activity toward nematodes. An assessment of the effect of non-pathogenic *F. oxysporum* strains toward root-knot nematodes revealed that the fungus has high potential as a biological control agent. Galling was reduced by up to 75% in endophyte-harbouring tomato plants (Table 12.7) (Plate 39). These results are similar to those obtained with banana in Africa where *F. oxysporum* was also shown to be the dominant species. Species of *Chaetomium* and *Colletotrichum* growing endophytically in tomato in Kenya were also effective in controlling root-knot nematodes. Astonishingly, approximately 10% of the fungi isolated from the root tissue in Kenya had biological control properties toward the root-knot nematode on tomato.

Table 12.5. Colonization of root and corm tissue of tissue cultured *Musa* AAA plantlets, cv. Gros Michel, by *Fusarium oxysporum* 1 and 5 months after weaning and inoculation with endophytic isolates of *F. oxysporum.*

Endophytic isolate of *F. oxysporum*	Isolation of *F. oxysporum*[a]			
	After 1 month		After 5 months	
	Root	Corm	Root	Corm
Control	0	30	16	33
V4w5	66	83	0	0
III3w3	100	100	66	66
III4w1	100	100	50	33
V5w2	100	100	16	33

[a]Percentage of surface sterilized and plated tissue from which *F. oxysporum* could be isolated.

Table 12.6. Effect of *Fusarium moniliforme* (K75) on plant growth, fungal colonization and *Pratylenchus zeae* density in roots of three maize cultivars.

Cultivar[1]	Treatment	Root weight (g)	Shoot weight (g)	Nematodes per gram root	Fungal colonization (%)
DLCI	Control	7.9 a	3.5 a		
	Fungus + nematode	7.9 a	3.4 a	89 a	42.2
	Fungus	6.4 b	0.8 a	96 a	
H511	Control	23.8 a	1.7 a		
	Fungus + nematode	23.9 a	1.1 a	78 b	36.7
	Fungus	21.4 a	6.9 a	86 a	
Katumani	Control	12.5 a	0.3 a		
	Fungus + nematode	12.2 ab	9.6 a	100 a	23.3
	Fungus	9.6 b	4.9 a	102 a	

[1]Dryland Composite 1 (DLC1), Hybrid 511 and Katumani Composite. Treatments followed by the same letter along a column, within a cultivar, are not significantly ($P < 0.05$) different according to Duncan's Multiple Range Test and Student's t-test for the nematode numbers, $n = 5$.

Table 12.7. Effect of *Fusarium oxysporum* inoculation on plant growth and *Meloidogyne incognita* population development on tomato.

Treatment	Shoot weight (g)	Root weight (g)	Root length (m)	Galls	Egg masses	Galls per m root	Egg masses per m root
Control	11.8 bc	5.66	38.3	165 a	114 a	4.3 a	2.99 a
Fo146	13.2 a	6.01	40.2	120 b	71 bc	3.03 b	1.78 b
Fo153	11.8 ab	5.71	42.4	97 b	55 bc	2.18 b	1.24 bc
Fo162	11.9 bc	5.54	41.7	85 b	56 bc	2.02 b	1.21 c
Fo163	10.4 d	4.75	34.5	94 b	45 c	1.77 b	1.28 bc
LSD	1.1			36.5	24.9	0.93	0.73

Means with the same letter are not significantly different according to Duncan's Multiple Range Test, $P < 0.05$ (Hallmann and Sikora, 1994).

Arbuscular Mycorrhizal Fungi

AMF are obligate fungal symbionts that are beneficial to most plants. They have been found on almost all higher plants. These fungi grow inter- and intracellularly and form specific fungal structures called arbuscules in cortical cells. They are responsible for the exchange of phosphate and other nutrients between the plant and fungal cell membranes. AMF build an extensive hyphal net into the soil that extends the overall reach and efficacy of the root system. The hyphal net along with the arbuscules are responsible for AMF growth in the plant and for uptake and transport of nutrients from the soil to the plant.

The arbuscular mycorrhizas constitute a group of over 130 species in seven genera. Little is known about the diversity of these fungi in African soils, due to difficulty in identification. These symbionts have been found in most tropical and subtropical plants and colonize most agronomic crops. AMF are important for phosphate and nitrogen uptake, transplant establishment, tolerance to salt and heavy metals, reduced drought stress and in increasing plant resistance to both nematode and root diseases (Sieverding, 1991; Sikora, 1995; Waceke *et al.*, 2001a,b).

AMF are produced on living plants where the spores needed for application are recovered from the growth substrate, dried, powdered and then formulated. This makes the approach expensive and limits AMF use to high-value crops. The dried substrate containing the spores and root segments is then applied to the seedling trays before seeding the crop. Adding the AMF to the seedling production units gives the endophyte time to establish in the root system before transplanting into the field where nematode control takes place.

The level of control is determined by the level of AMF colonization. It is estimated that at least 40% of the root length needs to be colonized for significant levels of control, with higher levels of colonization leading to improved control. AMF are effective in reducing sedentary nematodes in that they compete for root nutrients with these highly specialized parasites. Direct parasitism has not been observed. The immobility of the parasite in the root leads to competition with the AMF and ultimately reduced nematode growth and reproductive capacity. If and how they affect migratory parasites, for example *R. similis* on banana or lesion nematodes on field crops, is still poorly studied. Because of slow growth in the roots, acceptable levels of control can only be attained when AMF are given a head start that promotes strong root colonization and ultimately leads to competition and antagonism towards the pest.

To promote AMF efficacy in the field two methods are used: (i) an inundative approach that requires inoculation of seedlings before transplanting into infested fields; and (ii) the management of the antagonistic potential or field stimulation of naturally occurring AMF populations (Sikora, 1992). Management practices, e.g. rotation with AMF-sensitive intercrops that promote endophyte development, must be looked at as an alternative to mass-production, especially for low-value crops.

Integration into IPM

Because both MEF and AMF are slow-growing fungi, nematode control is usually effective only if the fungus becomes well established in the root system before nematode penetration. Therefore, integration of biological enhancement of planting material with these two groups of fungi into sound IPM systems will promote successful nematode control in the field. This is especially true where nematode pre-plant population densities greatly exceed economic threshold levels.

A number of control methodologies can be used to reduce high nematode densities and thereby support the efficacy of endophytic fungi under field conditions (Sikora, 2002). The following approaches can be used to reduce nematode populations before endophyte colonized seedlings are transplanted into the field:

1. Rotation with non-host or resistant crops.
2. Planting of antagonistic plants that trap or are toxic to nematodes.
3. Application of organic amendments to the planting hole to stimulate the antagonistic potential in the rhizosphere.
4. Solarization with plastic mulch to kill nematodes through exposure to high temperature.
5. Biofumigation with a combination of organic amendments and plastic mulch to control through toxin production and stimulation of microbial antagonists.
6. Suppression of nematode densities by long-term flooding of fields (paddy rice) in the season before susceptible crop production.
7. Tilling soil to induce nematode desiccation.
8. Grafting of resistant or tolerant root stocks on to susceptible plants.
9. Application of systemic nematicides that reduce nematode early root penetration and give the endophyte time to establish for long-term biological control.

The health of all crops attacked by nematodes can be enhanced with MEF. Crops produced by tissue culture (banana, coffee and pineapple) lend themselves well to this concept of biological control. These crops have in some cases lost their natural endophytes due to the sterile nature of tissue culture. By re-introducing these beneficial organisms into the plant, root health can be upgraded and reliance on pesticides decreased.

We consider the use of endophytic fungi extremely important for control of plant parasitic nematodes in vegetable crops and other high-value horticultural crops. This is especially true for seedlings that will be transplanted into the field. These types of crops are economically attractive for this form of biological control, in that the amount of inoculum needed is greatly reduced due to targeted application directly on to the plant. Many MEF can be produced efficiently in both liquid and solid state fermentation and AMF on living plants. They then can be applied in granular form to the seedling trays or even pelleted on to seed. Fungal establishment in such systems is promoted due to reduced levels of active competitors. This leads to strong root colonization, which ensures good biological control when the transplants reach the field.

When it comes to field crops, the cost of production of these biological control agents is a limiting factor. Therefore, the selection of cultivars that favour endophyte colonization may lead to higher levels of natural colonization under field conditions and thereby nematode control. This has been observed with AMF, where cereal cultivars have been examined for improved colonization. Breeding programmes that select cultivars that support high levels of endophyte colonization should be considered in regions where nematodes cause severe damage and alternative control measures are non-existent or too expensive.

The effectiveness of endophytes on nematode-tolerant cultivars should also be examined. Control would probably be greater than on non-tolerant and highly susceptible cultivars. This would make the use of tolerants more acceptable to the growers and to the breeders. Because nematode-resistant cultivars are often extremely sensitive to initial nematode penetration, endophytes could be used in place of nematicides to protect this important germplasm during establishment. Another advantage of endophytes over the use of systemic nematicides is that they actually control nematodes, whereas nematicides only inhibit nematode activity for a few weeks.

Research currently focuses on the inoculation of single strains on to endophyte-free planting material such as seed, seedling transplants or tissue-cultured plantlets. However, as our knowledge of microbial biodiversity *in situ* increases, it may lead to the use of combinations of different fungal strains or species that increase efficacy and extend control to a number of pests simultaneously.

Conclusions

All higher plants host endophytic fungi with biological control activity. However, the importance of this segment of microbial biodiversity for the biological control of soil-borne pests and diseases has not been adequately examined. We consider these organisms to have significant potential in the biological control of plant parasitic nematodes and root diseases. These plant protection attributes have been shown on a number of crops and toward a broad spectrum of nematodes and diseases important to African agriculture. Endophytes are well adapted to life inside the plant and often share the same ecological niche with endoparasitic nematodes, fungal pathogens and in some cases insect borers. Because of this intimate relationship, they can be effective at the exact site of pest or disease attack. Endophytes are usually only applied once and persist within the host plant to ensure protection during the infection process and, ideally, for the entire cropping season.

Control mechanisms are not well understood, but competition, direct parasitism as well as induced resistance are known modes-of-action. Long-term control will require persistence of the fungal endophyte in the plant or the activation of effective and durable plant defence responses. Endophytes have a major advantage over other biological control agents applied directly to the soil in that the latter, due to the high levels of inoculum needed to treat the soil, are more costly, have to be applied more frequently, and their efficacy is often strongly influenced by environmental factors. Another advantage of the use of endophytic fungi living in plant tissue is the reduced risk of side-effects on non-target organisms, crops and humans. The application of indigenous strains further limits environmental risks. From both an environmental and economic point of view, endophytic fungi have many advantages over conventional control systems. Application of endophytes for nematode control must be seen as just one component of IPM. In addition, optimizing management strategies to promote endophyte biological control activity is required to obtain effective control to a specific nematode parasite.

References

Amin, N. (1994) Untersuchungen über die Bedeutung endophytischer Pilze für die biologische Bekämpfung des wandernden Endoparasiten *Radopholus similis* (Cobb) Thorne an Bananen. PhD thesis, University of Bonn, Germany.

Bridge, J. (1996) Nematode management in sustainable and subsistence agriculture. *Annual Review of Phytopathology* 34, 201–225.

Gisi, U. (1990) *Bodenoekologie*. Thieme Publishing, New York, USA.

Griesbach, M. (2000) Occurrence of mutualistic fungal endophytes in bananas (*Musa* spp.) and their potential as biocontrol agents of the banana weevil *Cosmopolites sordidus* (Germar) (Coleoptera: Curculionidae) in Uganda. PhD thesis, University of Bonn, Germany.

Hallmann, J. and Sikora, R.A. (1994) Occurrence of plant parasitic nematodes and non-pathogenic species of *Fusarium* in tomato plants in Kenya and their role as mutualistic synergists for biological control of root-knot nematodes. *International Journal of Pest Management* 40, 321–325.

Hallmann, J. and Sikora, R.A. (1996) Toxicity of fungal endophyte secondary metabolites to plant parasitic nematodes and soil-borne plant pathogenic fungi. *European Journal of Plant Pathology* 102, 155–162.

Kerry, B. (1992) An assessment of progress toward microbial control of plant parasitic nematodes. *Journal of Nematology – Supplement* 22, 621–631.

Kimenju, J.W. (1998) Identification and biological control of lesion nematodes (*Pratylenchus* spp.) affecting maize production in Kenya. PhD thesis, University of Nairobi, Kenya.

Luc, M., Sikora, R.A. and Bridge, J. (1990) *Plant Parasitic Nematodes in Subtropical and Tropical Agriculture.* CAB International, Wallingford, UK.

Niere, B.I. (2001) Significance of non-pathogenic isolates of *Fusarium oxysporum* Schlecht.: fries for the biological control of the burrowing nematode *Radopholus similis* (Cobb) Thorne on tissue cultured banana. PhD thesis, University of Bonn, Germany.

Niere, B.I., Speijer, P.R., Gold, C.S. and Sikora, R.A. (1999) Fungal endophytes from banana for the biocontrol of *Radopholus similis*. In: Frison, E.A., Gold, C.S., Karamura, E.B. and Sikora, R.A. (eds) *Mobilizing IPM for Sustainable Banana Production in Africa.* INIBAP, Montpellier, France, pp. 313–318.

Pocasangre, L. (2000) Biological enhancement of banana tissue culture plantlets with endophytic fungi for the control of the burrowing nematodes *Radopholus similis* and the Panama disease (*Fusarium oxysporum* f. sp. *cubense*). PhD thesis, University of Bonn, Germany.

Pocasangre, L., Sikora, R.A., Vilich, V. and Schuster, R.-P. (2000) Survey of banana endophytic fungi from Central America and screening for biological control of the burrowing nematodes (*Radopholus similis*). *InfoMusa* 9, 3–5.

Sasser, J.N. and Freckman, D.W. (1987) A world perspective on Nematology: the role of the Society. In: Veech, J.A. and Dickson, D.W. (eds) *Vistas on Nematology.* Society of Nematologists, Hyattsville, Maryland, USA, pp. 7–14.

Schuster, R.-P., Sikora, R.A. and Amin, N. (1995) Potential of endophytic fungi for the biological control of plant parasitic nematodes. *Communications in Applied Biological Sciences* 60, 1047–1052.

Sieverding, E. (1991) *Vesicular-arbuscular Mycorrhiza Management in Tropical Agrosystems*. GTZ, Eschborn, Germany.

Sikora, R.A. (1992) Management of the antagonistic potential in agricultural ecosystems for the biological control of plant parasitic nematodes. *Annual Review of Phytopathology* 30, 245–270.

Sikora, R.A. (1995) Vesicular-arbuscular mycorrhizae: their significance for biological control of plant parasitic nematodes. *Biocontrol* 1, 29–33.

Sikora, R.A. (1997) Biological system management in the rhizosphere an inside-out/outside-in perspective. *Communications in Applied Biological Sciences* 62, 105–112.

Sikora, R.A. (2002) Strategies for biological system management of nematodes in horticultural crops – fumigate, confuse or ignore them. *Communications in Applied Biological Science* 67(2a), 3–15.

Sikora, R.A. and Hoffmann-Hergarten, S. (1993) Biological control of plant parasitic nematodes with plant-health-promoting rhizobacteria. In: Lumsden, R.D. and Vaughn, J.L. (eds) *Pest Management – Biologically Based Technologies*. American Chemical Society, Washington, DC, USA, pp. 166–172.

Sikora, R.A. and Schuster, R.P. (1999) Novel approaches to nematode IPM. In: Frison, E.A., Gold, C.S., Karamura, E.B. and Sikora, R.A. (eds) *Mobilizing IPM for Sustainable Banana Production in Africa.* INIBAP, Montpellier, France, pp. 127–136.

Sikora, R.A., Schuster, R.-P. and Griesbach, M. (2000) Improved plant health through biological enhancement of banana planting material with mutualistic endophytes. *Acta Horticulturae* 540, 409–413.

Stirling, G.R. (1991) *Biological Control of Plant Parasitic Nematodes*. CAB International, Wallingford, UK.

Waceke, J.W., Waudo S.W. and Sikora, R.A. (2001a) Response of *Meloidogyne hapla* to mycorrhiza fungi inoculation of pyrethrum. *African Journal of Science and Technology* 2, 623–70.

Waceke, J.W., Waudo, S.W. and Sikora, R.A. (2001b) Suppression of *Meloidogyne hapla* by arbuscular mycorrhiza fungi (AMF) on pyrethrum in Kenya. *International Journal of Pest Management* 47, 135–140.

13

Species of *Trichoderma* and *Aspergillus* as Biological Control Agents against Plant Diseases in Africa

Ranajit Bandyopadhyay[1] and Kitty F. Cardwell[2]
[1]*International Institute of Tropical Agriculture, Ibadan, Nigeria;*
[2]*Cooperative State Research, Extension, and Education Service, US Department of Agriculture, Washington, DC, USA*

Introduction

Plant diseases cause significant losses in yield and quality of crops in Africa, where a growing population needs more food and income from agriculture-related activities. Sound plant health management is key to sustainable management of diseases. Integrated plant health management has several interactive facets such as appropriate crop germplasm, sound cultural practices and crop management, biologically based inputs and synthetic chemicals. The complex and interactive nature of good crop management requires a robust understanding of the agroecosystem in which individual components of plant health are nested. For example, crop rotation or tillage practices may be less amenable to manipulation in a perennial cropping system than in an annual cropping system where the field site can be changed periodically. Host resistance reinforced by biological control can be a useful component of management in perennial systems (McSpadden *et al.*, 2002). Economics of crop production, economics of losses caused by a disease in a specific situation, and ease and cost of applying disease management methods determine the level of intervention that a farmer is willing to make to realize gains from farming. Usually, the primary foundation for disease control is manipulation of the physical environment and utilizing host resistance. Biological control and synthetic fungicides provide further support to disease management. However, use of fungicides is being discouraged due to economic reasons and growing concern for environment and safety issues. Biological control is potentially a sustainable solution to plant diseases in African agriculture since its effect is long-term with few, if any, undesirable side-effects.

Biological control agents act against plant pathogens through different modes of action. Antagonistic interactions that can lead to biological control include antibiosis, competition and hyperparasitism (Cook and Baker, 1983). Competition occurs when two or more microorganisms require the same resources in excess of their supply. These resources can include space, nutrients and oxygen. In a biological control system, the more efficient competitor, i.e. the biological control agent, out-competes the less efficient one, i.e. the pathogen. Antibiosis occurs when antibiotics or toxic metabolites produced by one microorganism have direct inhibitory effect on another. Hyperparasitism or predation results from biotrophic or necrotrophic interactions that lead to parasitism of the plant pathogen by the biological control agent. Some microorganisms, particularly those in soil, can reduce damage from diseases by promoting plant growth or by inducing host resistance against a myriad of pathogens (Hutchinson, 1998; Cook, 2000; Kerry, 2000). Efficient biological control agents often express more than one mode of action for suppressing the plant pathogens.

Several naturally occurring microorganisms have been identified as biological control agents of plant pathogens. This chapter deals with biological control of fungal diseases of crops with fungal species belonging to the genera *Trichoderma*, *Fusarium* and *Aspergillus* purely in the African context. Discussion on the use of *Fusarium* species for the control of the obnoxious witchweed (*Striga hermonthica* (Del.) Benth.) in maize and sorghum, and burrowing nematode (*Radopholus similis* (Cobb) Thorne) in banana can be found elsewhere in this volume (see Berner *et al.* for *Striga* and Sikora *et al.* for *R. similis*).

Disease Control Using *Trichoderma* Species

The antagonistic ability of *Trichoderma* species was discovered 70 years ago (Weindling, 1932). *Trichoderma* spp. are now the most common fungal biological control agents that have been extensively researched and deployed throughout the world. The primary mechanism of antagonism in *Trichoderma* is mycoparasitism. Lytic activity is the key feature responsible for the expression of mycoparasitism against several fungal pathogens (Chet, 1987). *Trichoderma* spp. are also good competitors in soil, and producers of volatile and non-volatile antibiotics to suppress target pathogens (Chet, 1987). Because of their effectiveness and ease of production for commercial application, at least nine commercial biological control products based on *Trichoderma* species are manufactured and marketed in Belgium, Sweden, Israel, USA, Denmark, India and New Zealand for use on several crops (Navi and Bandyopadhyay, 2002). In Africa too, considerable research has been done on biological control potential of *Trichoderma* spp. against several fungal pathogens that attack seeds, seedlings, roots, stems and leaves of several crops. Some of the diseases that can be potentially controlled by *Trichoderma* species are listed in Table 13.1. Two specific examples are highlighted below to illustrate the potential biological control of seed and seedling blight of cowpea and stalk and ear rot of maize.

Table 13.1. Evidence for successful experimental use of *Trichoderma* spp. as a biological control agent of various crop diseases in Africa.

Host	Disease	Pathogen	Species of *Trichoderma*	Reference
Cowpea	Damping-off	*Macrophomina phaseolina*	*T. harzianum*, *T. koningii*	Adekunle *et al.* (2001)
Cowpea	Web blight	*M. phaseolina*	*T. koningii*	Latunde-Dada (1991)
Cowpea	Leaf smut	*Protomycopsis phaseoli*	*T.* spp.	Adejumo *et al.* (1999)
Maize	Storage seed rot	*Gibberella fujikuroi* and *Aspergillus flavus*	*T.* spp.	Calistru *et al.* (1997)
Soybean	Brown stem rot	*Phialophora gregata*	*T. harzianum*	Yehia *et al.* (1994)
Potato	Stem canker and black scurf	*Rhizoctonia solani*	*T. harzianum*, *T. koningii*	Abada and Abdel-Aziz (2002)
Potato	Leak	*Pythium aphanidermatum*	*T. harzianum*	Triki and Priou (1997)
Tomato	Southern blight	*Sclerotium rolfsii*	*T. koningii*	Latunde-Dada (1993)
Tomato	Basal stem rot	*S. rolfsii*	*T. viridae*	Wokocha (1990)
Lucerne	Damping-off, wilt and root rot	*R. solani* and *Fusarium oxysporum*	*T. harzianum*	Hassanein *et al.* (2000)
Strawberry	Grey mould rot of fruits	*Botrytis cinerea*	*T. harzianum*	El-Zayat *et al.* (1993)
Cucumber	Damping-off	*R. solani*	*T.* spp.	Askew and Laing (1994)
Sugar beet	Damping-off and root rot	Several fungi	*T. harzianum*	Abada (1994)
Table beet	Damping-off	*P. aphanidermatum*	*T. harzianum*	Abdalla and El-Gizawy (2000)
Avocado	Root rot	*Phytophthora cinnamomi*	*T. harzianum* *T. hamatum*	McLeod *et al.* (1995)
Garlic	White rot	*Sclerotium cepivorum*	*T. harzianum*	Rahman *et al.* (1998)
Tobacco	Damping-off and root rot	*R. solani* and *F. solani*	*T. harzianum*	Cole and Zvenyika (1988)

Seed and seedling blight of cowpea

Several diseases affect cowpea (*Vigna unguiculata* Walp., *Papilionaceae*) during its growth and development from the time the seed germinates in soil to the time when seeds are produced and harvested. Some of these diseases are amenable to biological control while others are not. Seed decay and seedling damping-off cause serious losses in cowpea (Emechebe and Shoyinka, 1985). Among the several pathogens associated with these seed and seedling diseases, *Macrophomina phaseolina* (Tassi.) Goid is prevalent in the Sudano-Sahelian areas where cowpea frequently suffers from moisture stress. In addition to seedling diseases, the pathogen also causes ashy stem blight or charcoal rot. *M. phaseolina* is extremely plurivorous and causes diseases in more than 300 plant species. This soil-inhabiting fungus survives for several years as free sclerotia in soil and in infected plant debris. Under favourable infection conditions, the pathogen propagules around the spermosphere and hypocotyl colonize the seed, hypocotyl and epicotyl leading to pre- and post-emergence damping-off of seedlings. In other words, seed decay and damping-off appear early during plant growth in a localized part of the plant. Therefore, disease control methods targeting the seeds have been useful in managing the disease. Seed treatment with systemic fungicides such as benzimidazole compounds is effective in controlling the disease (Kataria and Sunder, 1985). However, these fungicides are not generally available to resource-poor farmers who are the major cowpea producers.

Biological control of seed decay and damping-off of cowpea has been demonstrated using species of *Trichoderma* as antagonists (Adekunle *et al.*, 2001). *T. harzianum* Rifai, *T. koningii* Oudem and an unknown species of *Trichoderma* were tested at different doses to determine the efficacy of the antagonists. The plant stand was significantly improved when seeds were treated with *T. harzianum* and *T. koningii* compared with untreated seeds. Although the protection with the antagonists was lost over time, *T. harzianum* was more effective than *T. koningii* since the protection with the former lasted longer than the latter. Several formulations of the antagonists were also evaluated. The antagonists were grown in liquid culture, harvested, dried in an oven at 30°C for 48 h, and powdered in a blender. The powdered antagonist was suspended in water, and the aqueous suspension used to prepare two formulations: suspension with a sticker (Tween 20) and with cassava starch as an adhesive. The powdered antagonist was also transformed into concentrated slurry with water and uncooked cassava starch powder. Seeds were treated for different duration with each of these formulations. Generally seed treatment with the slurry formulation was not effective in reducing the disease, while soaking the seeds for 10–40 min in the aqueous suspension of the antagonists amended with cassava starch significantly reduced the disease.

Seed application requires only a small quantity of a biological control agent and can be easily combined with fungicidal seed dressing to enhance the efficacy of both for controlling diseases (Harman and Taylor, 1990; Cook, 2000), as has been suggested for cowpea–*M. phaseolina* system (Alagarsamy and Sivaprakasam, 1988). Of course, the fungicide added in the biological control formulation should not be toxic to the biological control agent nor should it be expensive.

Seed dressing is a technology appropriate for African farming systems, and cottage industry production units have been shown to be economically feasible for meeting local or small-scale demands (Cherry *et al.*, 1999). The feasibility of a local biopesticide with *Trichoderma* depends on several factors. The raw materials, adhesive and production substrates need to be plentiful and cheap. The *Trichoderma* isolate would have to be quite robust and grow quickly on local substrate such as rice hulls or coconut shells. The risk of inadvertently increasing potential human pathogen along with the biological control agent must be very low. The dose response cannot be too stringent or safeguards would have to be developed for 'under-dosing' or 'over-dosing'. If these conditions can be met, then development of *Trichoderma*-based seed treatment in Africa will be attractive. *Trichoderma* populations in soil would probably increase with external introduction, particularly in acidic soils in West Africa.

Stalk and ear rot of maize

Species of *Fusarium* belonging to the section *Liseola* can cause seedling diseases, root rots, stalk rots and ear rots of maize in the field, as well as postharvest storage rots. *Fusarium verticillioides* (Saccardo) Nirenberg (*F. moniliforme* J. Sheldon) and other anamorphs belonging to the teleomorph *Giberella fujikuroi* (Sawada) Ito in Ito and K. Kimura, are most frequently isolated from maize plants. Interest in the disease stems from the concern that infection of grain by *G. fujikuroi* can lead to loss of grain quality and potential production of fumonisin and other harmful mycotoxins (Munkvold and Desjardins, 1997). *G. fujikuroi* comprises several mating populations. Among these, those belonging to mating population A are considered as *F. verticillioides*. Other species infecting maize are *Fusarium proliferatum* (Mats.) Nirenberg ex Gerlach and Nirenberg and *Fusarium subglutinans* (Wollenw. and Reinking) Nelson, Toussoun and Marasas.

Members belonging to mating population A are more potent producers of fumonisin and are found more frequently on maize compared with mating population F (e.g. *Fusarium thapsinum* Klittich, Leslie, Nelson and Marasas), which produces little or no fumonisin (Leslie *et al.*, 2001). *F. verticillioides* is closely associated with maize throughout the plant's life living as an endophyte within the plant right from seedling to grain harvest, often without causing any visible symptoms. While many infected plants remain free of symptoms, damage in others can be dramatic. The fungus is transmitted through seed infection that results from vertical spread of the endophytic phase from stalk to the grain. Seed infection cannot be controlled by fungicide sprays since it is transmitted internally through the plant. The fungus also survives in plant debris on the soil surface, and free ambient spores can infect the stalk through the adventitious roots and the ear via the silk channel. Insects play an important role in moving the fungus and opening infection sites in maize stalks and ears (Munkvold and Desjardins, 1997). At the same time, *F. verticillioides* has been shown to attract insects to the plant (Schulthess *et al.*, 2002) resulting in a critical feedback loop of infection and damage. Thus, control of endophytic *F. verticillioides* may exert a collateral effect of reducing attractiveness and susceptibility of the maize plant to insects.

Fumonisin-related restrictions for trade have led to a renewed interest in finding strategies to reduce the levels of contamination of maize with the toxin. Currently, host-plant resistance, insect-pest control and good storage practices are the major strategies for stalk rot and ear rot management. Biological control of stalk rot (Sobowale, 2002) and storage rot (Bacon *et al.*, 2001) by means of *Trichoderma* spp. has also been explored in order to reinforce other management tactics. Sobowale (2002) isolated 52 fungi from different parts of maize plants and tested these against *F. verticillioides* initially in *in vitro* tests. Seven of these fungal isolates, all belonging to *Trichoderma* spp., were further tested against *F. verticillioides* in artificially infested stalks. *T. harzianum* and *Trichoderma pseudokoningii* Rifai were found to occupy the same niche as *F. verticillioides* and were able to competitively displace the pathogen. These two antagonists were able to move within the stalk to internodes further away from the point of introduction to the sites where *F. verticillioides* existed. Significantly, it appeared as if the antagonists sensed and tracked *F. verticillioides* in the stalks. Recovery of the pathogen from stalks co-inoculated with antagonists was significantly lower than from stalks in which the antagonists were absent. However, introduction of the antagonists into stalks was ineffective and did not protect against accumulation of fumonisins in grains.

The potential of *T. harzianum* and *Trichoderma viride* Persoon: Fries to reduce mycotoxin-producing potential of *F. verticillioides* in grain store has been further explored by Bacon *et al.* (2001) and Calistru *et al.* (1997). The latter authors suggested that the aggressive behaviour (towards *G. fujikuroi*) demonstrated by *Trichoderma* spp. could be partly explained by the liberation of extracellular enzymes by these fungi. An isolate of *T. viride* showed amylolytic, pectinolytic, proteolytic and cellulolytic activity. Although management of *Fusarium* stalk rot and grain spoilage in storage are potentially amenable to biological control, more work is required to test the biological activity of various agents, the different potential delivery mechanisms for biological control agents, and the practical feasibility and economy of this approach.

Disease Control Using *Aspergillus* spp.

Species of *Aspergillus* are almost ubiquitously present in soils of tropical areas. Most species of *Aspergillus* are not of much consequence in agriculture, but some species of *Aspergillus* are found in plant products, particularly oil-rich seeds. Contamination of seeds with highly poisonous aflatoxins results from the presence of toxigenic strains of four species of *Aspergillus*: *A. flavus* Link:Fr. (Plate 40), *A. parasiticus* Speare, *A. nomius* Kurtzman, Horn and Hesseltine and *A. bombycis* Peterson, Ito, Horn and Goto (Peterson *et al.*, 2001), each producing a combination of different types of aflatoxins. In Nigeria, Bénin and Togo, for instance, maize is consumed and stored across all agroecological zones. Depending on agroecology, crop management and length of storage, aflatoxin contamination levels averaging >100 p.p.b. have been recorded in up to 50% of grain stores sampled (Hell *et al.*, 2000a; Udoh *et al.*, 2000) and in some samples collected in Bénin, extremely high amounts of up to 2500 p.p.b. of aflatoxin were detected (Sétamou *et al.*, 1997). These levels of aflatoxin are much higher than the maximum permissible limit of 20 p.p.b. in food and feed.

Aflatoxins in Africa have impact on human and animal health and on trade. Thus, alfatoxin has been reported to be associated with exacerbation of the energy malnutrition syndrome Kwashiorkor in children and vitamin A malnutrition in animals, and many other problems (Hall and Wild, 1994). In various animal models, in addition to being hepatotoxic, aflatoxin causes significant growth faltering and is strongly immune-suppressive at weaning. It has been recently shown that 99% of all children weaned from mother's milk to maize-based diets in Bénin and Togo had aflatoxin in their blood, indicating ingestion of aflatoxin-contaminated food (Gong *et al.*, 2002). Aflatoxin exposure in children was associated with stunting (a sign of chronic malnutrition) and being underweight (an indicator of acute malnutrition). In addition to the direct public health impact, there is ample evidence of negative impact on trade in Africa owing to aflatoxin, particularly with respect to trade with the European Community. Therefore, a reduction of aflatoxins in maize and groundnut through appropriate integrated management in field and store is extremely important to reduce losses, increase rural incomes, and improve health and wellbeing of people.

A. flavus invades and infects developing seed of maize and groundnut in the field before harvest, and mature seeds during harvest, and in storage. Preharvest contamination with aflatoxins is aggravated by drought stress and elevated temperature during seed maturation. Damage by insects is another important predisposing factor by providing injury sites through which *A. flavus* can invade seeds in the stores and in the field (Sétamou *et al.*, 1998; Hell *et al.*, 2000a,b). *A. flavus* populations are genetically highly diverse and are composed of large numbers of vegetative compatibility groups (VCGs), sometimes within a restricted geographic area. VCG composition of native strains within a field may, however, be of only minor importance in predicting the efficacy of a non-aflatoxigenic biological control strain (Horn *et al.*, 2000). Populations of the *Aspergillus* group *flavi* also contain mixtures of members that can produce copious amounts of aflatoxins (aflatoxigenic) and those that cannot produce at all or produce insignificant amounts of aflatoxins (non-aflatoxigenic) (Egel *et al.*, 1994; Cotty, 1997). Diversity in the population of *Aspergillus* spp. also exists with respect to their size of sclerotia. S-strain (those producing small sclerotia) isolates in the USA produce high amounts of aflatoxin B, while S-strains from Africa produce both aflatoxin B and G (Cotty and Cardwell, 1999). L-strains (with large sclerotia) on both continents produce a range of aflatoxin B from none to very high. Subsequent work suggested that small sclerotial S-phenotype is not predictive of aflatoxin production (Geiser *et al.*, 2000). Toxigenicity of *A. flavus* is apparently unrelated to a strain's ability to colonize and/or infect living or dead plant tissues (Cotty and Bayman, 1993).

In the USA, biological control has been used to reduce aflatoxin contamination in various crops such as cotton (Cotty, 1994), groundnut (Dorner *et al.*, 1998) and maize (Brown *et al.*, 1991; Dorner *et al.*, 1999). This technique involves the application to soil of a non-aflatoxigenic biological control strain of *A. flavus* or *A. parasiticus,* resulting in a high population density that allows the biological control strain to compete effectively with the native aflatoxigenic strains during invasion under conditions favourable for aflatoxin contamination. Invasion of a seed in soil (e.g. groundnut) solely by the biological control agent would be expected because of its high density relative to the wild-type strain in the soil. This would result in less aflatoxin contamination. For maize and cottonseed, the high population of the non-aflatoxigenic biological control strain in soil produces abundant spores on the soil surface that become airborne to infect grains and seeds (Horn *et al.*, 2001).

The potential to reduce aflatoxin contamination in maize using the biological control tactics mentioned above has been evaluated in Bénin, where 90% of the *Aspergilli* are *A. flavus* (Cardwell and Cotty, 2002). The non-aflatoxigenic strains of *A. flavus* (BN22 and BN 30 from Bénin, and AF36 from the US) were tested against aflatoxigenic strains of *A. flavus* (BN40 from Bénin and AF13 from the US) and *A. parasiticus* (BN48) *in vitro* (Cardwell and Cotty, 2000). For these *in vitro* trials, maize kernels were dipped in 1×10^6 conidia ml^{-1} suspension of one *Aspergillus* spp. or strain and allowed to dry before repeating the dip with either the same or another strain. Kernels were incubated in a saturated environment at 30°C for 5 days, dried at 40°C for 5 days, crushed, extracted in acetone and aflatoxins quantified using thin-layer chromatography (TLC) and a scanning densitometer. All non-aflatoxigenic isolates significantly reduced toxin production by the African *A. parasiticus* isolate BN48. The American non-aflatoxigenic isolate AF36 was effective against the American aflatoxigenic isolate AF13, but not the aflatoxigenic African S-strain, BN40 suggesting that there may be specificity of action of some non-aflatoxigenic strains. The African non-aflatoxigenic L-strain BN30 was the only isolate that reduced toxin production by the aflatoxigenic African S-strain, BN40.

Field tests were also conducted in Bénin in 1998 and 1999 using non-aflatoxigenic African L-strain BN30 and aflatoxigenic African S-strain, BN40 (Cardwell and Cotty, 2000). Silks of 60-day-old maize plants were dipped in 1×10^6 conidia ml^{-1} suspension of the aflatoxigenic S-strain BN40 or water (control) and covered overnight with a paper bag. Seven days later, the silk was dipped again either in the suspension of the same strain or non-aflatoxigenic L-strain BN30. Ears were harvested 2 weeks after maturity and shelled. Grain was milled, extracted with chloroform and aflatoxins quantified using TLC. In these field trials, toxin production was significantly reduced and was not different from the water control (Fig. 13.1A and B). The second isolate inoculant consistently influenced toxin production in the field. In fact it had more impact on toxin production than the first, which was judged to be counter-intuitive. If the mechanism of toxin reduction was competitive displacement, as is the hypothesis in the cotton model, it was expected that the first inoculant would occupy the kernel and the second would be excluded. Nevertheless, this study confirms that non-aflatoxigenic strains are effective in reducing aflatoxin production in maize by competitively displacing the aflatoxigenic strains of *A. flavus* and *A. parasiticus*.

Biological control of *A. flavus* in Bénin is still at a rudimentary stage. Across the country, many isolates of *A. flavus* strains have been identified (Cardwell and Cotty, 2000). There seems to be a gradient in the distribution of these strains from north to south in Bénin (Cardwell and Cotty, 2002), with the north having more of the highly toxigenic S-strains and the south more of the less toxic L-strains (Cardwell and Cotty, 2002). Further work has to determine whether the atoxigenic strains from the south can be used to reduce aflatoxin production in the north. Selected non-aflatoxigenic strains specific for different agroecozones need to be identified and tested in large areas to reduce the impact of aflatoxins in maize and groundnut economies.

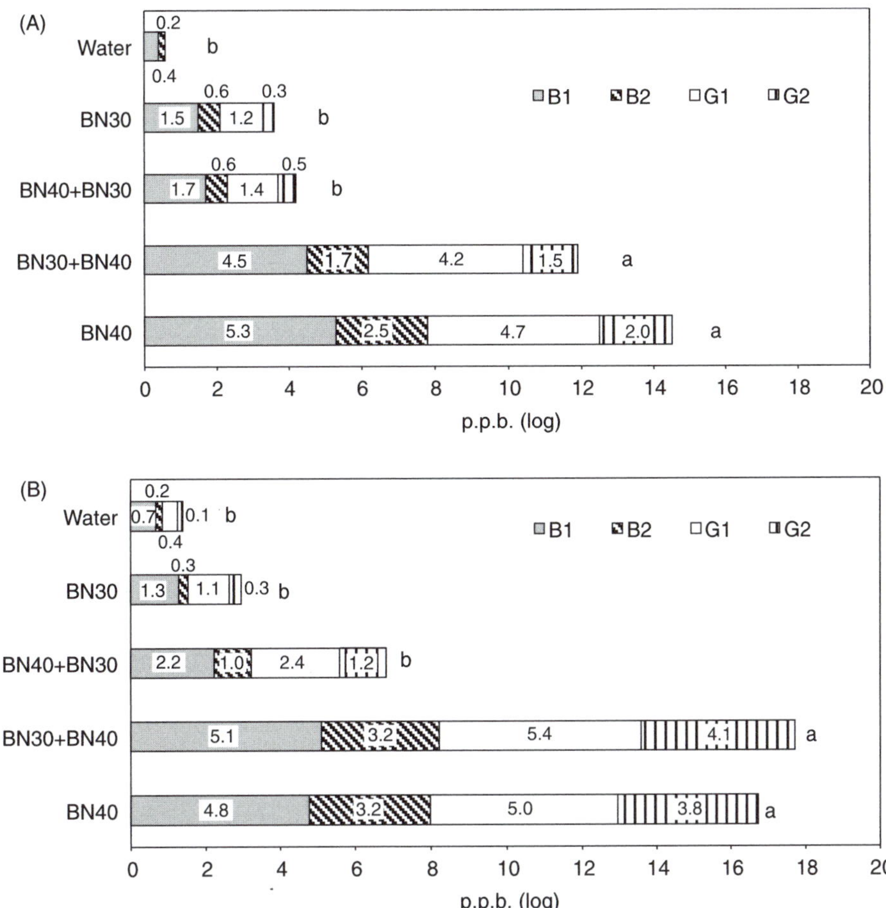

Fig. 13.1. Mean aflatoxin B1, B2, G1 and G2 in maize kernels from cobs inoculated twice at silk stage in the field with either toxigenic (BN40) or atoxigenic (BN30) strains of *Aspergillus flavus*, or one strain inoculated 7 days after another in Bénin. (A) Year 1998; 12 observations per treatment. (B) Year 1999; 24 observations per treatment. Bars not sharing a common letter are significantly different ($P < 0.001$). Log values of aflatoxins found in the samples were plotted. (Source: Cardwell and Cotty, 2000.)

Impact and Prognosis for the Future

The early beliefs that biological control agents offer more variable and less effective protection than fungicides have been refuted (Harman and Taylor, 1990). However, to achieve successful biological control, good knowledge of the host–pathogen–environment interaction is required in specific agroecosystems in which the biological control agent has to act. The interactions between microbial biological control agents, the target species to be controlled, the host and the environment can be complex and require a good research foundation prior to attempting formulation. The development of stable, cost-effective, easy-to-produce and easy-to-apply formulations of biological control agents is another critical research step in order to achieve successful biological control of plant diseases. This is particularly true for resource-constrained situations under which agriculture is practised in Africa. Commercial use of biological control agents for plant disease control is not yet a reality in Africa, unlike the situation with biological control of insect pests that has seen spectacular successes. The final step in the development of microbial biological control agents for disease control in Africa will be to identify and define the economic and policy environment needed for successful increase and deployment of each agent. Depending on the agent, the options could be: (i) cottage industry, i.e. village/regional production as private or public enterprise; (ii) nationally organized production at central laboratories, subsidized by public funding; (iii) internationally organized production either as a one-time, donor-funded programme, or as a business enterprise picked up and exploited by existing private sector companies. For the latter to occur, it is often necessary to obtain patents for the formulation or the isolate, thereby stabilizing the proprietary status and securing the agent as a viable investment against the costs of commercialization. It is important that development of biological control options for plant diseases does not stop at the research laboratory. Commitment to development of this technology for deployment in Africa is needed at policy level, requiring that, as research laboratories enter into biological control agent testing, a conceptual framework for moving the agent to the field be part of the development agenda.

Conclusions

The development of biological control agents as a key component of integrated disease management has tremendous potential for application in the African context for the reduction of losses from plant diseases. Several biological control agents can suppress diseases as effectively as fungicides, an input that is often prohibitively expensive to be of value to resource-poor farmers. In Africa, fungal biological control agents, such as species of *Trichoderma*, *Fusarium* and *Aspergillus*, are efficacious in reducing damage caused by pathogens on maize and cowpea in research station trials. *T. koningii* and *T. harzianum* were effective in controlling damping-off of cowpea caused by *M. phaseolina*, and effective dosage and application methods have been standardized in greenhouse trials to control the disease. *F. verticillioides* is an endophytic fungus that enhances growth of maize, but becomes pathogenic to cause root and stalk rot, damping-off and ear rot when the plants undergo stress. Two strains of *T. harzianum* and *T. pseudokoningii* have been shown to reduce the stalk rot phase caused by the pathogen. These two *Trichoderma* species can penetrate the plant, move systemically within the stalk to occupy the same niche as *F. verticillioides*, and competitively exclude the pathogen. In greenhouse trials, the two species reduced stalk rot either when introduced into the stalk through injured sites or after seed treatment. Aflatoxin contamination of maize and groundnut is a serious problem in West Africa. Aflatoxins are produced by aflatoxigenic strains of *A. flavus*. Non-aflatoxigenic strains of *A. flavus* can colonize the grains, but cannot produce the mycotoxin.

Reduction in soil-borne inoculum of aflatoxigenic strains of *A. flavus* has been demonstrated after soil application with more competitive non-aflatoxigenic strains. As a result, fields receiving aflatoxigenic strain of *A. flavus* had reduced aflatoxin level in grains. These are a few examples that reveal biological control as an effective adjunct in integrated disease management. However, much more work needs to be done to demonstrate field efficacy of biological control agents, their persistence, safety and commercial feasibility, before practical application of biological control agents for plant disease control in Africa becomes a reality.

References

Abada, K.A. (1994) Fungi causing damping-off and root rot on sugar beet and their biological control with *Trichoderma harzianum*. *Agriculture, Ecosystems and Environment* 51, 333–337.

Abada, K.A. and Abdel-Aziz, M.A. (2002) Biological and chemical control of stem-canker and black-scurf on potato in Egypt. *Bulletin of Faculty of Agriculture, Cairo University* 53, 115–128.

Abdalla, M.Y. and El-Gizawy, A.M. (2000) Effect of some biological and chemical seed treatments of table beets on incidence of *Pythium aphanidermatum* damping-off in the greenhouse. *Alexandria Journal of Agricultural Research* 45, 167–177.

Adejumo, T.O., Ikotun, T. and Florini, D.A. (1999) Biological control of *Protomycopsis phaseoli*, the causal agent of leaf smut of cowpea. *Journal of Phytopathology* 147, 371–375.

Adekunle, A.T., Cardwell, K.F., Florini, D.A. and Ikotun, T. (2001) Seed treatment with *Trichoderma* species for control of damping-off of cowpea caused by *Macrophomina phaseolina*. *Biological Control Science and Technology* 11, 449–457.

Alagarsamy, G. and Sivaprakasam, K. (1988) Effect of antagonists in combination with carbendazim against *Macrophomina phaseolina* infection in cowpea. *Journal of Biological Control* 2, 123–125.

Askew, D.J. and Laing, M.D. (1994) Evaluating *Trichoderma* bio-control of *Rhizoctonia solani* in cucumbers using different application methods. *Journal of the Southern African Society for Horticultural Sciences* 4, 35–38.

Bacon, C.W., Yates, I.E., Hinton, D.M. and Meredith, F. (2001) Biological control of *Fusarium moniliforme* in maize. *Environmental Health Perspectives* 109, 325–332.

Brown, R.L., Cotty, P.J. and Cleveland, T.E. (1991) Reduction in aflatoxin content of maize by atoxigenic strains of *Aspergillus flavus*. *Journal of Food Protection* 54, 623–626.

Calistru, C., McLean, M. and Berjak, P. (1997) Some aspects of the biological control of seed storage fungi. In: Ellis, R.H., Black, M., Murdoch, A.J. and Hong, T.D. (eds) *Basic and Applied Aspects of Seed Biology. Proceedings of the Fifth International Workshop on Seeds*, 1995, Reading. Kluwer Academic Publishers, Dordrecht, The Netherlands, pp. 755–762.

Cardwell, K.F. and Cotty, P.J. (2000) Interactions among U.S. and African *Aspergillus* spp. strains: influence on aflatoxin production. *Phytopathology* 90, 11.

Cardwell, K.F. and Cotty, P.J. (2002) Distribution of *Aspergillus* section *Flavi* among field soils from the four agroecological zones of the Republic of Bénin, West Africa. *Plant Disease* 86, 434–439.

Cherry, A.J., Jenkins, N.E., Héviefo, G., Bateman, R. and Lomer, C.J. (1999) Operational and economic analysis of a West African pilot-scale production plant for aerial conidia of *Metarhizium* spp. for use as a mycoinsecticide against locusts and grasshoppers. *Biological Control Science and Technology* 9, 35–51.

Chet, I. (1987) *Trichoderma* – application, mode of action, and potential as a biological control agent of soilborne plant pathogenic fungi. In: Chet, I. (ed.) *Innovative Approaches to Plant Disease Control*. John Wiley & Sons, New York, USA, pp. 137–160.

Cole, J.S. and Zvenyika, Z. (1988) Integrated control of *Rhizoctonia solani* and *Fusarium solani* in tobacco transplants with *Trichoderma harzianum* and triadimenol. *Plant Pathology* 37, 271–277.

Cook, R.J. (2000) Advances in plant health management in the 20th century. *Annual Review of Phytopathology* 38, 95–116.

Cook, R.J. and Baker K.F. (1983) *The Nature and Practice of Biological Control of Plant Pathogens*. American Phytopathological Society, St Paul, Minnesota, USA.

Cotty, P.J. (1994) Influence of field application of an atoxigenic strain of *Aspergillus flavus* group fungi on the populations of *A. flavus* infecting cotton bolls and on the aflatoxin content of cottonseed. *Phytopathology* 83, 1283–1287.

Cotty, P.J. (1997) Aflatoxin-producing potential of communities of *Aspergillus* section *Flavi* from cotton producing areas in the United States. *Mycological Research* 101, 698–704.

Cotty, P.J. and Bayman, P. (1993) Competitive exclusion of a toxigenic strain of *Aspergillus flavus* by an atoxigenic strain. *Phytopathology* 83, 1283–1287.

Cotty, P.J. and Cardwell, K.F. (1999) Divergence of West African and North American communities of *Aspergillus* section *Flavi*. *Applied and Environmental Microbiology* 65, 2264–2266.

Dorner, J.W., Cole, R.J. and Blankenship, P.D. (1998) Effect of inoculum rate of biological control agents on preharvest aflatoxin contamination of peanuts. *Biological Control* 12, 171–176.

Dorner, J.W., Cole, R.J. and Wicklow, D.T. (1999) Aflatoxin reduction in corn through field application of competitive fungi. *Journal of Food Protection* 62, 650–656.

Egel, D.S., Cotty, P.J. and Elias, K.S. (1994) Relationships among isolates of *Aspergillus* sect. *flavi* that vary in aflatoxin production. *Phytopathology* 84, 906–912.

El-Zayat, M.M., Okasha, K.H.A., El-Tobshy, Z.M., El-Kohli, M.M.A. and El-Neshway, S.M. (1993) Efficiency of *Trichoderma harzianum* as a biological control agent against gray mold rot *Botrytis cinerea* of strawberry fruits. *Annals of Agricultural Science Cairo* 38, 283–290.

Emechebe, A.M. and Shoyinka, S.A. (1985) Fungal and bacterial diseases of cowpea in Africa. In: Singh, S.R. and Rachie, K.O. (eds) *Cowpea Research, Production, and Utilization*. John Wiley & Sons, Chichester, UK, pp. 173–193.

Geiser, D.M., Dorner, J.W., Horn, B.W. and Taylor, J.W. (2000) The phyllogenetics of mycotoxin and sclerotium production in *Aspergillus flavus* and *Aspergillus oryzae*. *Fungal Genetics and Biology* 31, 169–179.

Gong, Y.Y., Cardwell, K., Hounsa, A., Egal, S., Turner, P.C., Hall, A.J. and Wild, C.P. (2002) Dietary aflatoxin exposure and impaired growth in young children from Bénin and Togo: cross sectional study. *British Medical Journal* 325, 20–21.

Hall, A.J. and Wild, C.P. (1994) Epidemiology of aflatoxin-related disease. In: Eaton, D.A. and Groopman, J.D. (eds) *The Toxicology of Aflatoxins: Human Health, Veterinary, and Agricultural Significance*. Academic Press, New York, USA, pp. 233–258.

Harman, G.E. and Taylor, A.G. (1990) Development of an effective biological seed treatment system. In: Hornby, D. (ed.) *Biological Control of Soil-Borne Plant Pathogens.* CAB International, Wallingford, UK, pp. 415–426.

Hassanein, A.M., El-Barougy, E., Elgarhy, A.M., Parikka, P. and El-Sharkawy, T.A. (2000) Biological control of damping-off, root-rot/wilt diseases of alfalfa in Egypt. *Egyptian Journal of Agricultural Research* 78, 63–71.

Hell, K., Cardwell, K.F., Sétamou, M. and Poehling, H.M. (2000a) The influence of storage practices on aflatoxin contamination in maize in four agroecological zones of Bénin, West Africa. *Journal of Stored Product Research* 36, 365–382.

Hell, K., Cardwell, K.F., Sétamou, M. and Schulthess, F. (2000b) Influence of insect infestation on aflatoxin contamination of stored maize in four agroecological regions in Bénin. *African Entomology* 8, 169–177.

Horn, B.W., Greene, R.L. and Dorner, J.W. (2000) Inhibition of aflatoxin B_1 production by *Aspergillus parasiticus* using non-aflatoxigenic strains: role of vegetative compatibility. *Biological Control* 17, 147–154.

Horn, B.W., Greene, R.L., Sorensen, R.B., Blankenship, P.D. and Dorner, J.W. (2001) Conidial movement of nontoxigenic *Aspergillus flavus* and *A. parasiticus* in groundnut fields following application to soil. *Mycopathologia* 151, 81–92.

Hutchinson, S.W. (1998) Current concepts of active defence in plants. *Annual Review of Phytopathology* 36, 59–90.

Kataria, H.R. and Sunder, S. (1985) Effect of micronutrients on the efficacy of fungicides against *Rhizoctonia solani* on cowpea seedling. *Pesticide Science* 16, 453–456.

Kerry, B.R. (2000) Rhizosphere interactions and the exploitation of microbial agents for the biological control of plant-pathogenic fungi. *Annual Review of Phytopathology* 38, 423–441.

Latunde-Dada, A.O. (1991) The use of *Trichoderma koningii* in the control of web blight disease caused by *Rhizoctonia solani* in the foliage of cowpea (*Vigna unguiculata*). *Journal of Phytopathology* 133, 247–254.

Latunde-Dada, A.O. (1993) Biological control of southern blight disease of tomato caused by *Sclerotium rolfsii* with simplified mycelial formulations of *Trichoderma koningii*. *Plant Pathology* 42, 522–529.

Leslie, J.F., Zeller, K.A. and Summerell, B.A. (2001) Icebergs and species in populations of *Fusarium*. *Physiological and Molecular Plant Pathology* 59, 107–117.

McLeod, A., Labuschagne, N. and Kotze, J.M. (1995) Evaluation of *Trichoderma* for biological control of avocado root rot in bark medium artificially infested with *Phytophthora cinnamomi*. *Yearbook of the South African Avocado Growers' Association* 18, 32–37.

McSpadden Gardener, B.B. and Fravel, D.R. (2002) Biological control of plant pathogens: research, commercialization, and application in the USA. Online. *Plant Health Progress doi*:10.1094/PHP-2002–0510–01-RV. http://apsnet.org/online/feature/biological control/

Munkvold, G.P. and Desjardins, A.E. (1997) Fumonisins in maize: can we reduce their occurrence? *Plant Disease* 81, 556–565.

Navi, S.S. and Bandyopadhyay, R. (2002) Biological control of fungal plant pathogens. In: Waller, J.M., Lenné, J.M. and Waller, S.J. (eds) *Plant Pathologists' Pocketbook*. CAB International, Wallingford, UK, pp. 354–365.

Peterson, S.W., Ito, Y., Horn, B.W. and Goto, T. (2001) *Aspergillus bombycis*, a new aflatoxigenic species and genetic variation in its sibling species, *A. nominus*. *Mycologia* 93, 689–703.

Rahman, T.M.A., Zayed, G.A. and Asfour, H.E. (1998) Interrelationship between garlic cultivars and the biological control of white rot disease. *Assiut Journal of Agricultural Sciences* 29, 1–13.

Schulthess, F., Cardwell, K.F. and Gounou, S. (2002) The effect of endophytic *Fusarium verticillioides* on infestation of two maize varieties by lepidopterous stemborers and coleopteran grain feeders. *Phytopathology* 92, 120–128.

Sétamou, M., Cardwell, K.F., Schulthess, F. and Hell, K. (1997) *Aspergillus flavus* infection and aflatoxin contamination of pre-harvest maize in Bénin. *Plant Disease* 81, 1323–1327.

Sétamou, M., Cardwell, K.F., Schulthess, F. and Hell, K. (1998) Effect of insect damage to maize ears, with special reference to *Mussidia nigrivenella* (Lepidoptera; Pyralidae), on *Aspergillus flavus* (Deuteromycetes; Monoliales) infection and alfatoxin production in maize before harvest in the Republic of Bénin. *Journal of Economic Entomology* 91, 433–438.

Sobowale, A.A. (2002) Biological control of *Fusarium moniliforme* Sheldon on maize stems by some fungal isolates from maize phyllosphere and rhizosphere. PhD thesis, University of Ibadan, Nigeria.

Triki, M.A. and Priou, S. (1997) Using chemical and biological treatments to reduce the potential of potato leak caused by *Pythium aphanidermatum* in Tunisia. *Potato Research* 40, 391–398.

Udoh, J.M., Cardwell, K.F. and Ikotun, T. (2000) Storage structures and aflatoxin content of maize in five agroecological zones of Nigeria. *Journal of Stored Products Research* 36,182–201.

Weindling, R. (1932) *Trichoderma lignorum* as a parasite of other soil fungi. *Phytopathology* 22, 837–845.

Wokocha, R.C. (1990) Integrated control of *Sclerotium rolfsii* infection of tomato in the Nigerian Savanna: effect of *Trichoderma viride* and some fungicides. *Crop Protection* 9, 231–234.

Yehia, A.H., Abd-El-Kader, D.A., Salem, D.E. and Sayed-Ahmed, A.A. (1994) Biological soil treatment with *Trichoderma harzianum* to control brown stem rot of soybean in Egypt. *Egyptian Journal of Phytopathology* 22, 143–157.

14

Towards the Registration of Microbial Insecticides in Africa: Non-target Arthropod Testing on Green Muscle™, a Grasshopper and Locust Control Product Based on the Fungus *Metarhizium anisopliae* var. *acridum*

Jürgen Langewald,[1] Ine Stolz,[2] James Everts[3] and Ralf Peveling[2]

[1] *International Institute of Tropical Agriculture, Cotonou, Bénin;* [2] *Institut für Natur, Landschafts und Umweltschutz (NLU), Biogeographie, Basel, Switzerland;* [3] *CERES-LOCUSTOX Foundation, Dakar, Senegal*

Introduction

For many years, microbial control has been incorporated in pest management strategies in developing countries, enjoying particular success in Asia and South America, although successful approaches for microbial control have evolved worldwide (Fuxa, 1987). Many of these control strategies were developed in response to environmental constraints related to the use of synthetic pesticides. In Africa too, there is a long history of projects that have sought microbial solutions to pest problems, but rarely were these approaches developed into commercial products as in Europe and North America, where development of microbial control agents typically followed the chemical pesticide model, taking a commercial route (Langewald and Cherry, 2000).

In the West African Sahel, since the desert locust outbreak in the late 1980s, there was increasing evidence of negative environmental impact associated with large-scale applications of synthetic insecticides (Everts, 1990; Peveling *et al.*, 1999; Peveling, 2001). Furthermore, there were indications that insecticide applications might even have aggravated grasshopper infestations, due to adverse effects on natural enemies of acridids (van der Valk and Niassy, 1997; van der Valk *et al.*, 1999).

In response to these environmental constraints, the LUBILOSA project (Lutte Biologique contre les Locustes et les Sauteriaux), executed by the IITA, Bénin, CABI *Bioscience*, UK, and Centre Régional de Formation et d'Application en Agrométéologie et Hydrologie Opérationelle pour les Pays du Sahel (AGRHYMET), Niger, developed Green Muscle™, an industrially produced fungal insecticide for grasshopper and locust control (Lomer *et al.*, 2001). Green Muscle™ is the first microbial control product specifically developed for the African market. The pathogen on which Green Muscle™ is based is a strain of *Metarhizium anisopliae* var. *acridum* (IMI 330189) Driver and Milner (*Deuteromycotina*, *Hyphomycetes*). Green Muscle™ (Plates 41–44) has been successfully tested against many locust and grasshopper species, particularly *Zonocerus variegatus* (L.) (variegated grasshopper), *Hieroglyphus daganensis* Krauss (rice grasshopper), *Oedaleus senegalensis* (Krauss) (Senegalese grasshopper), *Anacridium melanorhodon melanorhodon* (Walker) (tree locust) and *Schistocerca gregaria* (Forskål) (desert locust) (Lomer *et al.*, 2001).

Registration

The development of any commercial product for plant protection has strong regulatory implications. As a result of growing public concern with respect to side-effects of pesticides and persistent industrial pollutants in the 1960s (Carson, 1962), registration schemes were established for the protection of human health and the environment. The Organisation of Economic Co-operation and Development (OECD) was the first to lay down common Guidelines for Testing of Chemicals (GTC) that aim at harmonizing testing requirements and procedures among member countries and facilitate mutual recognition of results (OECD, 1982 onward). During the late 1970s, international agencies and governments began to examine the need for guidelines and regulations for the evaluation, development and registration of microbial control agents (Greathead and Prior, 1990), and only very recently governments started to develop particular registration regulations for microbial pesticides. In 1983, the US Environmental Protection Agency (EPA) developed the first comprehensive regulatory scheme for such products (EPA Subsection M guidelines), and ecotoxicological studies are now required for microbial control agents, despite them being considered *a priori* environmentally safe (Goettel *et al.*, 1990). It is important to realize that microorganisms used for biological control of pests and diseases have a broad spectrum of properties. At one extreme are the highly specific and virulent pathogens such as the Entomophthoralean fungi (e.g. *Neozygites* sp. on cassava green mite, *Entomophthora muscae* (Chon) Fresenius and *Entomophthora maimaiga* Humber, Shimazu, Soper and Hajek) and some of the nucleopolyhedroviruses (NPV). At the other extreme are microbial control agents such as *Bacillus thuringiensis* (Berliner) that may have quite wide host ranges, but are not expected to multiply, spread or persist in the environment, and in these respects are similar to chemical pesticides. For these agents, which are used only as inundative biopesticides, the issues of mammalian toxicology and residue analysis, etc., are not very different from those for chemical pesticides. Intermediate between these two extremes are species such as *M. anisopliae* var. *acridum* or *Beauveria brongniartii* (Saccardo) Petch or many of the NPVs. They are often applied in industrial quantities characteristic of 'inundative biopesticides', but we expect to see moderate persistence at least within the season during which they are applied, if not for several consecutive seasons. Here both toxicological and environmental issues need to be considered.

The registration of microbial pesticides encompasses a similar range of tests to the registration of chemical pesticides. To date, the most detailed and comprehensive documents available are the Microbial Pesticide Test Guidelines (MPTG) of the EPA (EPA, 1996). Their use has also been recommended by the European Union as long as 'acceptance of specific guidelines at international level is pending' (EU, 2001). In West Africa, the Comité Sahélien des Pesticides (CSP), a subcommittee of CILSS, also requests data equivalent to EPA or EU standards. On the basis of a dossier provided for the registration of Green Muscle™, this organization has now developed their own framework for the registration of microbial pesticides in the Sahel, and also South Africa and Madagascar have developed such legislations.

The MPTG follow a tier-testing approach. This idea was first developed by the US National Academy of Sciences (NAS, 1981). The methodology is characterized by a sequence of tests within a dichotomous decision-making scheme in which evidence of low risk on a lower level of biological and ecological complexity precludes further testing on a higher level. Applied to ecotoxicological studies, this means at low tier-level simple laboratory set-ups, and at high tier-level testing under complex natural conditions in the field. The rationale of tier testing is 'to ensure that only the minimum data sufficient to make scientifically sound regulatory decisions will be required' (EPA, 1996). In the case of non-target testing, the determination of the host range is the main objective of first tier testing. Special attention is given to organisms used for biological control and playing an important role in IPM (EU, 2001). It can also be important to include other pests in host-range screening studies, thereby widening the potential target spectrum and improving commercial prospects. Higher tier laboratory or field tests under environmentally more realistic exposure conditions are only envisaged if first tier tests show adverse effects. In case adverse effects are substantiated, further ecotoxicological tests are carried out under operational conditions in natural environments. The main advantage of tier testing to commercial companies is that it reduces the cost for registration. In this context, the option to waive data requirements on a case-by-case basis is important and makes the process less expensive and fast. While the registration of synthetic pesticides in the USA takes approximately 3 years, the review of microbial pesticide registration dossiers takes on average only 12 months. Requests for waivers have to be based on scientific grounds. A simplified procedure has been developed for the registration of products with limited uses (niche products). In this case, data requirements are less than for extensively used products (EPA-Interregional Research Program no. 4).

In this chapter, we focus on non-target arthropod toxicity and pathogenicity testing. Apart from proven beneficial arthropods, we are also reporting on arthropods that are not necessarily beneficial, but still not a target for locust and grasshopper control.

Testing Green Muscle™

Contrary to commercial producers, LUBILOSA did not limit itself to minimum data requirements. The reason is that LUBILOSA was planned and executed as a development assistance project. Development and registration of a mycopesticide against locusts and grasshoppers, i.e. a marketable product, was the principal, but not the only, aim of the LUBILOSA programme. Other goals were to establish collaboration and equitable partnership with African research and plant protection institutions (non-government organizations, as well as government organizations), to build and foster regional biological control capacities, and to conduct participatory on-farm research. Within the scope of this collaborative research, there was a much wider window of opportunity to explore benefits and risks of fungal control than within the scope of commercial pesticide development. Large-scale field applications for efficacy testing of Green Muscle™ were opportunities to carry out large-scale ecotoxicological studies simultaneously – which provided large amounts of additional information on the environmental impact of Green Muscle™ under real field conditions and in comparison with synthetic insecticides.

Most environmental data were generated by the LUBILOSA programme itself, but other partners contributed important data as well. Thus, studies were conducted on various tiers at a time, depending on opportunities in terms of access to research facilities, interest of partners and available financial, technical as well as human resources.

With respect to registration, it is important to note that the genus *Metarhizium* has been reviewed several times. What used to be a distinct group of acridid isolates of *Metarhizium flavoviride* Gams and Rozsypal (Bridge *et al.*, 1993) is now classified as *M. anisopliae* var. *acridum* (Driver *et al.*, 2000). IMI 330189 – the Green Muscle™ isolate – obtained from *Ornithacris cavroisi* Finot (Orthoptera, Acrididae) in 1989, near Niamey, Niger, became the type material for this group. Naturally occurring infections of grasshoppers and locusts with *Metarhizium* are uncommon. A few cases of epizootics, however, have been reported (Lomer *et al.*, 2001). Moreover, observations of LUBILOSA collaborators suggest that a low-level background infection may be common in locusts and grasshoppers in the Sahel and other semiarid biomes.

Green Muscle™ Arthropod Host Range – Laboratory Studies

The arthropod host range and pathogenicity of *M. anisopliae* (reviewed in Goettel *et al.*, 1990; Prior, 1997) varies greatly among genotypes. It is therefore difficult to predict effects of new isolates on non-target arthropods on the basis of susceptibilities to previously tested strains. Previous virulence screening tests using IMI 330189 revealed that 94% of 17 different orthopteran species were susceptible, and 59% – all within Acrididae – highly susceptible (Prior, 1997). In contrast, the corresponding percentages for non-orthopteran insects ($n = 16$) were only 44% and 0%, respectively. The author concluded that high virulence of this isolate seems confined to acridoids. More host-range studies have been conducted since this first appraisal, partly motivated by findings that some hymenopteran parasitoids were highly susceptible to IMI 330189 and other *M. anisopliae* isolates. An updated and extended list of non-orthopteran arthropods challenged with IMI 330189 (Table 14.1) indicates that while some non-target insects were susceptible under lab conditions, these results were largely due to unrealistic exposure and test conditions. For example, two beneficial

Table 14.1. Susceptibility of non-target arthropods to IMI 330189 (Green Muscle™) (Orthoptera not included; laboratory results).

Taxon	Susceptibility	References	Remarks
Branchiopoda			
Cladocera			
Daphnia magna Straus	–	a	
Anostraca			
Streptocephalus sudanicus Daday	–	b	High acute toxicity but no mycosis
Acari			
Phytoseiidae			
Neoseiulus idaeus Denmark and Muma	–	c	
Isoptera			
Termitidae			
Coptotermes spp.	+	d	
Nasutitermes spp.	+	d	
Psammotermes hybostoma Desneux	+	e	Controlled field test
Blattaria			
Blattidae			
Blatta sp.	–	d	
Neuroptera			
Myrmeleonidae			
Indet.	–	d	
Myrmeleon sp.	+	f	
Heteroptera			
Notonectidae			
Anisops sardeus Herrich-Schäffer	+	b,e	Floating layer of formulation can impede breathing
Lygaeidae			
Cosmopleurus sp.	–	f	Formulation toxic at high volume[m]
Coreidae			
Clavigralla shabadi Dolling	–	d	
Clavigralla tomentosicollis Stal	–	c,d	
Anthocoridae			
Orius albidipennis Reuter	–	c	
Coleoptera			
Cucurlionidae			
Neochetina eichhorniae Warner	–	d	

Continued

Table 14.1. *Continued.*

Taxon	Susceptibility	References	Remarks
Coccinellidae			
Hyperaspis notata (Mulsant)	–	d	
Chilocorus bipustulatus (L.)	–	c	
Scarabaeidae			
Phyllophaga sp.	–	d	
Coleoptera			
Tenebrionidae			
Pimelia senegalensis (L.)	–	g	
Trachyderma hispida (Forskål)	–	g	
Tenebrio molitor L.	+	d	
Hymenoptera			
Encyrtidae			
Anagyrus lopezi (De Santis)	–, +, ++	e,g,h	High susceptibility at high dose and in stressed conditions, but no effect on beneficial capacity (parasitism) at field rate
Braconidae			
Bracon hebetor Say	+, ++	g	Larvae in host also infected
Cotesia flavipes Cameron	–	i	
Phanerotoma sp.		h	
Apidae			
Apis mellifera L.	+	j	Topical exposure of worker bees
Apis mellifera adansonii Latreille	–	k	Hives exposed in treated crop
Apis mellifera scutellata Lepeletier	–	l	Conidia dusted into hives
Formicidae			
Tapinoma sp.	–	d	
Lepidoptera			
Sphingidae			
Hyles livornica (Esper)	–	f	
Noctuidae			
Sesamia calamistis Hampson	+	i	

–, not infected; +, low or moderate susceptibility; ++ high susceptibility.
[a]Confidential registration results.
[b]Lahr *et al.* (2001).
[c]cf. Table 14.3.
[d]Prior (1997).
[e]Everts *et al.* (2001), unpublished results.
[f]J. Tabel (1994) Saarbrücken, unpublished data.
[g]Danfa and van der Valk (1999).
[h]Stolz *et al.* (2002), unpublished results.
[i]V.S. Ayitchehou (1996), unpublished data.
[j]Ball *et al.* (1994).
[k]P. Byrne (1995), unpublished results.
[l]R.E. Price (1997), unpublished results.
[m]8 l ha^{-1}.

insects, *Bracon hebetor* Say (Hymenoptera, Braconidae) and *Anagyrus lopezi* (de Santis) (Hymenoptera, Encyrtidae), were highly susceptible in some assays (Danfa and van der Valk, 1999; J. Everts, 2001, Senegal, unpublished results). It appears that the susceptibility varies depending on the actual exposure regime. For example, laboratory spray-tower treatments (Danfa and van der Valk, 1999) caused higher infectivity and mortality in *A. lopezi* than treatments with spinning-disk sprayers (Stolz *et al.*, 2002). By contrast, assays simulating field-exposure of *A. lopezi* and *Cotesia flavipes* Cameron (Hymenoptera, Braconidae) did not reveal adverse effects on the beneficial capacity (reproduction) of these parasitoids (V.S. Ayitchehou, Bénin, 1996, unpublished results; Stolz *et al.*, 2002). This suggests that the risk under field conditions may be low, despite evidence of high virulence in the laboratory. The first results from the laboratory assays with termites suggest that this group may be more susceptible to IMI 330189 than other non-target insects (J. Langewald, 2000, unpublished data). However, field experiments conducted by the CERES-LOCUSTOX Foundation, near Dakar, Senegal on *Macrotermes subhyalinus* Rambur at recommended field dose for grasshopper control did not reveal any reduction in termite activity (Everts *et al.*, 2001, unpublished data). CERES-LOCUSTOX was particularly interested in studying aquatic non-target organisms. An oil-based formulation of Green Muscle™ had a negative effect on *Anisops sardeus* Herrich-Schäffer (Heteroptera, Notonectidae), as had the blank formulation without *M. anisopliae* spores. *A. sardeus* is a water surface breather and the oil layer on the surface resulting from the application impeded breathing. These effects did, however, not occur under field conditions, due to the breaking-up and dispersal of the oil layer by the wind and the motion of the water.

Until now, *M. anisopliae* var. *acridum* has proved harmful only to non-target organisms in first tier tests, i.e. under maximum exposure conditions. In contrast, higher tier studies under operational conditions (at recommended dose rates) provided no evidence of significant adverse effects of *M. anisopliae* var. *acridum* on non-target organisms (Table 14.2). Significant reductions were either below hazard levels (<25%; Hassan, 1998) or related to overdosing (the tenebrionid *Gonocephalum setulosum*). The host specificity of different strains of entomopathogenic hyphomycete fungi seems to be dose dependent, with the host range increasing with dose. Higher tier test protocols should mainly focus on realistic dose applications, which could in many cases still be done under inexpensive laboratory conditions.

Laboratory case studies

A wide range of insect taxa has been screened for their susceptibility to *M. anisopliae* var. *acridum* strain IMI 330189. Some laboratory case studies are described in the following as examples of the LUBILOSA first tier testing approach. The non-target organisms included in these tests were one potential pest and three beneficial arthropods. All of these species are important elements in natural or newly established food webs of West African agroecosystems, including locust and/or grasshopper habitats. Thus, they are likely to be directly affected by control operations.

Table 14.2. Susceptibility of non-target arthropods to IMI 330189 (Green Muscle™) under field conditions (Orthoptera not included).

Taxon	Max. rate (conidia ha^{-1})	Sign. impact	Remarks	References
Arachnida (except Acari)	5×10^{12}	–	Niger	a
Acari	5×10^{12}	–	Niger	a
Collembola	5×10^{12}	–	Niger	a
Homoptera	2.5×10^{12}	–	Niger	a
Cicadellidae	2×10^{13}	+	Mauritania (22% reduction)[e]	b
Cicadellidae	5×10^{12}	+	Mauritania (18% reduction)	b
Coleoptera				
Carabidae	5×10^{12}	–	Niger	a,c
Tenebrionidae	5×10^{12}	–	Niger	a,c
Pimelia angulata angulosa F.	2×10^{13}	–	Mauritania[e]	b
Zophosis posticalis Deyrolles	2×10^{13}	–	Mauritania[e]	
Gonocephalum setulosum Faldermann	2×10^{13}	+	Mauritania (86% reduction)[e]	b
Scarabaeidae	2.5×10^{12}	–	Niger	a
Anthicidae				
Notoxus sp.	2×10^{13}	–	Mauritania[e]	b
Chrysomelidae				
Euryope rubra F.	2.5×10^{12}	–	Niger	a
Curculionidae				
Dereodus marginellus Boheman	2.5×10^{12}	–	Niger	a
Lepidoptera				
Microlepidoptera	2×10^{13}	–	Mauritania[e]	b
Hymenoptera				
Chalicidoidea	5×10^{12}	–	Niger	d
Encyrtidae	2×10^{13}	–	Mauritania[e]	b
Scelionidae	5×10^{12}	–	Niger	d
Mymaridae	5×10^{12}	–	Niger	d
Trichogrammatidae	5×10^{12}	–	Niger	d
Ichneumonoidea	5×10^{12}	–	Niger	d
Chrysidoidea	5×10^{12}	–	Niger	d
Dryinidae	5×10^{12}	–	Niger	d
Apoidea	5×10^{12}	–	Niger	d
Apoidea	2×10^{13}	–	Mauritania[e]	b
Sphecidae				
Podalonia spp.	2×10^{13}	–	Mauritania[e]	b
Vespoidea	5×10^{12}	–	Niger	d
Formicidae	5×10^{12}	–	Niger	a,c,d
Monomorium areniphilum Santschi	2×10^{13}	–	Mauritania[e]	b
Diptera				
Tachinidae				
Indet.	2×10^{13}	–	Mauritania[e]	b
Ephydridae				
Actocetor margaritatus (Wiedemann)	5×10^{12}	–	Niger	a,c

[a]J. Tabel (1994), Saarbrücken, unpublished results.
[b]This chapter, see Table 14.1.
[c]Peveling *et al.* (1999).
[d]Stolz (1999).
[e]High-volume application rate of 8 l ha^{-1}.

The neotropical mite *Neoseiulus idaeus* (Denmark and Muma) (Acari, Phytoseiidae) has been introduced to West Africa to control the cassava green mite (Yaninek *et al.*, 1993; Yaninek and Hanna, this volume). The pod bug *Clavigralla tomentosicollis* Stal (Heteroptera, Coreidae) is a major pest of cowpea in West Africa (Singh and Allen, 1979). Both *Beauveria bassiana* (Bals.) Vuill (Deuteromycotina, Hyphomycetes) and *M. anisopliae* showed promise as fungal control agents in virulence screening assays (Ekesi, 1999). Flower or minute pirate bugs are among the most abundant beneficials in African cropping systems (Hernández and Stonedahl, 1999). The species *Orius albidipennis* Reuter (Heteroptera, Anthocoridae) is particularly important in the natural control of lepidopteran eggs (e.g. *Helicoverpa armigera* (Hübner)), mites and thrips (*Frankliniella* spp.). The coccinellid *Chilocorus bipustulatus* L. var. *iranensis* (Coleoptera, Coccinellidae), a predator of scale insects, is widely distributed in the South Palearctic. In Mauritania, *C. bipustulatus* var. *iranensis* was released in the late 1960s to control *Parlatoria blanchardi* Targioni-Tozzetti, a scale of date palm, *Phoenix dactylifera*.

Details on the experimental procedures and design are summarized in Table 14.3. In all tests, desert locust nymphs (Bénin) or immature adults (Mauritania) were used as positive controls, and subjected to the same treatments as the non-target organisms (Table 14.4).

In all tests compared with the controls, *M. anisopliae* did not increase mortality in any of the non-target organisms; it also had no effect on reproduction

Table 14.3. Effects (mortality and reproduction) of *Metarhizium anisopliae* var. *acridum* (IMI 330189) on non-target arthropods.

	Test 1 *Neoseiulus idaeus* (Phytoseiidae)[a]	Test 2 *Clavigralla tomentosicollis* (Coreidae)[b]	Test 3 *Orius albidipennis* (Anthocoridae)[c]	Test 4 *Chilocorus bipustulatus* (Coccinellidae)[d]
Test organisms				
Stage (age)	Nymph	Adult (variable)	Adult (2 days)	Adult (7 days)
No. of tests (series)	2	1	2	1
Treatments	2	3	3	2
Replicates per test	4	5	4	3
No. per replicate	25	10	10	30
Total no. tested	400	150	240	180
Test conditions				
Application	Conidia mixed with 25 host eggs[e]	Micro-Ulva field treatment[f]	Micro-Ulva field treatment[f]	Burkard microapplicator
Exposure	Contact	Residual[g]	Residual[g]	Topical
Dose	6×10^8 conidia g^{-1} (fresh weight)[h]	5×10^{12} conidia ha^{-1}	5×10^{12} conidia ha^{-1}	2.5×10^3 conidia per beetle[i]
Duration (days)	8	14	14	14

Continued

Table 14.3. *Continued.*

	Test 1 *Neoseiulus idaeus* (Phytoseiidae)[a]	Test 2 *Clavigralla tomentosicollis* (Coreidae)[b]	Test 3 *Orius albidipennis* (Anthocoridae)[c]	Test 4 *Chilocorus bipustulatus* (Coccinellidae)[d]
Mortality (% ± SE)				
(1) Untreated control	3.0 (1.0)[1]	28.0 (8.6)[1]	50.0 (7.6)[1]	–
(2) Carrier control	–	18.0 (5.8)[1]	81.2 (5.8)[2]	5.6 (3.0)[1]
(3) *M. anisopliae*	3.0 (1.5)[1]	36.0 (7.5)[1]	77.5 (4.5)[2]	7.8 (4.0)[1]
Statistics (ANOVA)	$F_{1,14} = 0.0$[j] $P = 1.0$	$F_{2,12} = 1.49$[k] $P = 0.265$	$F_{2,21} = 7.84$[j] $P = 0.003$	$F_{1,4} = 0.18$[k] $P = 0.690$
Efficacy (%)[l]				
(1) Carrier control	–	−13.9	62.5	–
(2) *M. anisopliae*	0.0	11.1	55.0	2.3
Reproduction (±SE)[m]				
(1) Untreated control	68.6 (8.4)[a]	–	–	–
(2) *M. anisopliae*	75.5 (11.3)[a]	–	–	–
Statistics (ANOVA)	$F_{1,14} = 0.24$ $P = 0.633$	–	–	–
Efficacy (%)[l]	−22.0	–	–	–

Means within columns in successive lines with different superscript numbers are significantly different at $P < 0.05$.
[a]Phytoseiids reared according to protocols of Friese *et al.* (1987) and Mégevand *et al.* (1993).
[b]*C. tomentosicollis* was reared using the protocol of Ekesi (1999). The general outline of the bioassay was similar to the one with *O. albidipennis*.
[c]*O. albidipennis* were collected in the field and reared according to Fritsche and Tamò (2000).
[d]Field-collected *C. bipustulatus* var. *iranensis* were reared under conditions similar to those described for the coccinellid *Pharoscymnus anchorago* F. (Peveling and Demba, 1997).
[e]Alternative host eggs (*Tetranychus urticae* Koch).
[f]Recommended field dose formulation and application according to protocols by Bateman *et al.* (1992).
[g]Permanently exposed to spray residues as described by Stolz *et al.* (2002).
[h]Maximum challenge dose.
[i]Approximate maximum field exposure (equal volumes of Shellsol T® and Ondina®).
[j]Differences among test series were not significant. Therefore, replicates from different series were pooled and analysed with one-way ANOVA.
[k]One-way analysis of variance (ANOVA).
[l]Abbott's formula.
[m]Combined number of viable eggs and protonymphs.

of *N. idaeus*. Furthermore, no sporulation occurred on cadavers. Hence, *M. anisopliae* was not pathogenic to any of the species tested. The carrier oils caused some increased mortality in *O. albidipennis*, indicating toxic and/or physical (e.g. clogging of tracheae) effects of the oil formulation (Table 14.3), but the average longevity of females (controls) is only about 14 days anyway (Fritsche and Tamò, 2000) (controls, Table 14.3). Nevertheless, a medium risk is assumed according the IOBC risk classification scheme

Table 14.4. Mortality of desert locust, *Schistocerca gregaria*, and percentage mycosis tested as positive control under the same conditions as the non-target test organisms listed in Table 14.3.

	Test 1[a]	Tests 2 and 3[a]	Test 4[a]
Test locusts			
Stage (age of imago)	5th instar	5th instar	Adult (variable)
Group size	20	4–10	12
Replicates	4	1	3
Treatments	2	3	2
No. of tests (series)	2	2	1
Total number in test	320	42	72
Test conditions			
Application	Conidia mixed with wheat bran	See Table 14.3	See Table 14.3
Exposure	See Table 14.3	See Table 14.3	See Table 14.3
Dose	5×10^6 conidia g^{-1} (dry weight)	See Table 14.3	See Table 14.3
Duration (days)	8	7–14	14
Mortality (% ± SE)			
(1) Untreated control	0.6 (0.6)[1]	14.3[1]	–
(2) Carrier control	–	14.3[1]	33.3 (0.0)[1]
(3) *M. anisopliae*	97.5 (1.6)[2]	71.4[2]	100.0 (0.0)[2]
Statistics (ANOVA or χ^2 test)	$F_{1,14} = 3057.7$[c] ($P < 0.001$)	$\chi^2 = 13.7$ ($P = 0.001$)	No variance
Efficacy (%)[b]			
(1) Carrier control	–	0.0	–
(2) *M. anisopliae*	97.5	66.6	100.0
Sporulation (mycosis) (%)			
Untreated control	0	0	–
Carrier control	–	0	11.1
M. anisopliae	95.6	71.4[c]	77.8

Means within columns in successive lines with different superscript numbers are significantly different at $P < 0.05$.

[a]Locusts were kept as previously described (Peveling and Demba, 1997; Stolz *et al.*, 2002). All tests were conducted under ambient laboratory conditions at the IITA (Bénin) or Akjoujt field station (Mauritania).

[b]Abbott's formula.

[c]Test series pooled.

[d]Pairwise comparisons with Fisher's exact test of pooled series.

(Hassan, 1998). In the target organism (desert locust), *M. anisopliae* treatments caused high mortality, and sporulation of the fungus on cadavers confirmed its pathogenicity for treated locust in all tests (Table 14.4). In conclusion, the isolate IMI 330189 proved not pathogenic to the non-target organisms, but showed high virulence to one of the principal target acridids, desert locust.

Green Muscle™ Arthropod Host Range – Field Tests

Studies on the impact of Green Muscle™ on non-target arthropods were usually conducted within the scope of efficacy testing (e.g. Langewald *et al.*, 1999; Fig. 14.1). The spatial scale varied from ≈ 0.1 ha in Mauritania (Table, 1994, unpublished data) to 800 ha in Niger (Stolz, 1999), and the temporal scale from 5 weeks (Table, 1994, unpublished data) to 1 year post-treatment (Peveling *et al.*, 1999). Most studies used the standard dose rate of 5×10^{12} (100 g) conidia ha^{-1}, but one also tested four times this dose (Table, 1994, unpublished data) and another one only half of it (Stolz, 1999; field case study, see below). In medium- to large-scale field trials (Langewald *et al.*, 1999), synthetic insecticides were included as toxic standards (fenitrothion and fipronil) (Fig. 14.1). This not only allowed comparison of product efficacy but also provided a means to validate monitoring methods. Fenitrothion was applied according to label rates. However, rates of fipronil changed several times since its registration for locust and grasshopper control. In Dogo (Niger) 1998, LUBILOSA used a rate of 2 g a.i. ha^{-1}, which was lower than the former label rate but higher than specifications for grasshopper control today. In 1996, Peveling *et al.* (1999) studied non-target effects of Green Muscle™ and the organophosphate fenitrothion at Maine Soroa in Niger. While both products were highly efficient in reducing grasshopper populations (Langewald *et al.*, 1999) (Fig. 14.1), only fenitrothion caused reductions in non-target ground-dwelling species. Overall, 75% of the observed taxa, including Carabidae, Tenebrionidae, Formicidae and Ephydridae, were significantly reduced by fenitrothion. Most of the non-target fauna recovered

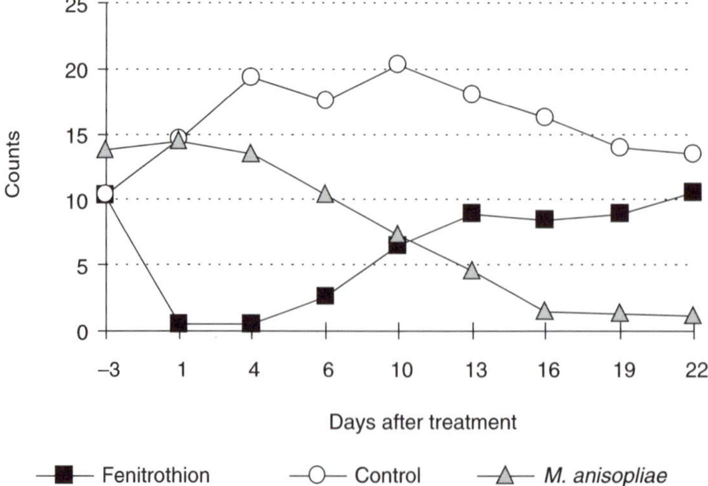

Fig. 14.1. Mean counts of grasshoppers per square metre over a period of 22 days in untreated plots, plots treated with an oil based formulation of *Metarhizium anisopliae* var. *acridum* (Green Muscle™) and with fenitrothion during the 1997 rainy season in east Niger (Langewald *et al.*, 1999).

after 51 days, except one ant species, which had still not recovered 1 year after application (Peveling *et al.*, 1999). During the 1997 and 1998 seasons in Niger, Stolz (1999) studied the impact of Green Muscle™ and fenitrothion compared with untreated sites on parasitoid Hymenoptera in 50 and 800 ha plots. Hymenoptera were monitored using yellow mini sticky traps (160 per study) and malaise traps (32 per study). Treatment effects were compared with the pre-treatment situation and the untreated control plots. She did not find any negative impact of any of the products on this non-target group. Non-target effects on arthropods of IMI 330189 under field conditions are summarized in Table 14.2. The existing evidence clearly confirms the impression from laboratory bioassays that the host range of this genotype of *Metarhizium* among non-orthopteran arthropods is very narrow. However, non-target orthopterans are *a priori* highly at risk, and use restrictions may be necessary in areas where particular grasshopper species or overall grasshopper diversity warrant protection.

Field case study

As an example for ecotoxicological work in the field, the following study is presented in more detail. The study was conducted during the rainy season in the most important millet producing zone in Niger. After conducting field experiments in the dryer parts of the millet belt, where millet is grown under marginal conditions (see above), this site was selected because of its importance for the millet production in Niger. Grasshoppers, in particular *O. senegalensis* Krauss, are a major problem in most years in this area and may cause severe damage to both crops and pastures (Cheke, 1990; Nwanze and Harris, 1992; Jago, 1993). In 1998, the grasshopper fauna consisted of *Acrotylus* spp. and *O. senegalensis*, but densities remained exceptionally low, despite good breeding conditions. Thus, experimental conditions were not ideal for efficacy testing. The effect of *M. anisopliae* var. *acridum* IMI 330189 (Green Muscle™) on non-target arthropods was compared with fipronil (Adonis® 4 UL) and an unsprayed control (for details see Table 14.5). Different taxa of arthropods were studied in cropland (millet) and pastures. In millet, the emphasis was on beneficial organisms, in particular hymenopteran parasitoids. A detailed account of this study was given by Stolz (1999). In pastures, the focus was on ground-dwelling arthropods – including grasshoppers – and two arboreal, herbivorous beetles, the chrysomelid *Euryope rubra* Fabricius, and the curculionid *Dereodus marginellus* Boheman. Here we present results from the study of non-target arthropods in pastures.

Epigeal arthropods were sampled with pitfall traps (Table 14.5). The two herbivorous beetles live on the succulent shrub *Leptadenia pyrotechnica* (*Asclepiadaceae*), which was by far the most abundant woody species. *D. marginellus* feeds on the flowers and *E. rubra* on young shoots. The beetles are easily visible due to their large size and conspicuous colour, and due to the peculiar morphology of the shrub. *L. pyrotechnica* has only tiny leaves and is

Table 14.5. Application and monitoring details for a field study on non-target effects of the application of *Metarhizium anisopliae* var. *acridum* IMI 330189 (Green Muscle™) and fipronil on ground dwelling arthropods and arboreal beetles.

Location (year)	Dogo, 45 km south of Zinder, Niger (1998)	
Crop	Millet (heads developed, milky grain)	
Product	*M. anisopliae* var. *acridum* (Strain IMI 330189)	Fipronil (Adonis® 4 UL)
Plot size × (replication)	50 ha × (3)	50 ha × (3)
Specification	2.5×10^{12} cfu ha^{-1}	2 g a.i. l^{-1}
Volume application rates	0.5 l ha^{-1}	0.5 l ha^{-1}
Swath width	30 m	30 m
Sprayer	Micron ULVA mast mark II	Micron ULVA mast mark II
Wind speed	3–5 m s^{-1}	3–5 m s^{-1}
Non-targets	Ground-dwelling arthropods including grasshoppers; arboreal beetles (*E. rubra* and *D. marginellus*)	
Monitoring[a,b]	Pitfall traps[c] (ground-dwellers); random counts on 50 shrubs of *L. pyrotechnica* (arboreal beetles)	

[a]Arthropods were monitored for about 8 weeks, starting 1 week before treatment.
[b]Only the main effects were significant.
[c]Sampling scheme described by Peveling *et al.* (1999).

generally seen in leafless condition, with narrow, green and juicy branches for assimilation. Another consequence of its morphology is that spray drift is unhampered by stems or foliage and leads to homogenous deposits all over. For example, 10 days post-treatment, tracer-marked droplets were found in >50% of all shoots collected in plots treated with Green Muscle™. Thus, the insecticide–milkweed–herbivore system appeared to be an ideal model to assess effects in a maximum exposure scenario (Table 14.5).

The grasshopper abundance decreased in all treatment groups as the season progressed, including the control (Fig. 14.2). Nevertheless, the decline was highest in plots treated with Green Muscle™. The efficacy of IMI 330189 against Sahelian grasshoppers at standard dose rates was found to be lower than in previous studies (Kooyman *et al.*, 1997; Langewald *et al.*, 1999; Fig. 14.1). The low efficacy in the present study resulted from low grasshopper densities rather than insufficient dose rates. This can be inferred from an equally 'poor performance' of fipronil, an agent that has shown nearly 100% efficacy even at lower dose rates than those tested here (Balança and de Visscher, 1997).

More than 40 different epigeal non-target arthropods, in addition to the two beetles mentioned before, were monitored on species level. For conciseness, we focus on higher taxonomic levels (Arachnida (except Acari), Acari, Collembola, Homoptera, Carabidae, Scarabaeidae, Tenebrionidae, Formicidae, Ephydridae). The study revealed no adverse effects of Green Muscle™ on any of these non-targets. With the exception of Formicidae

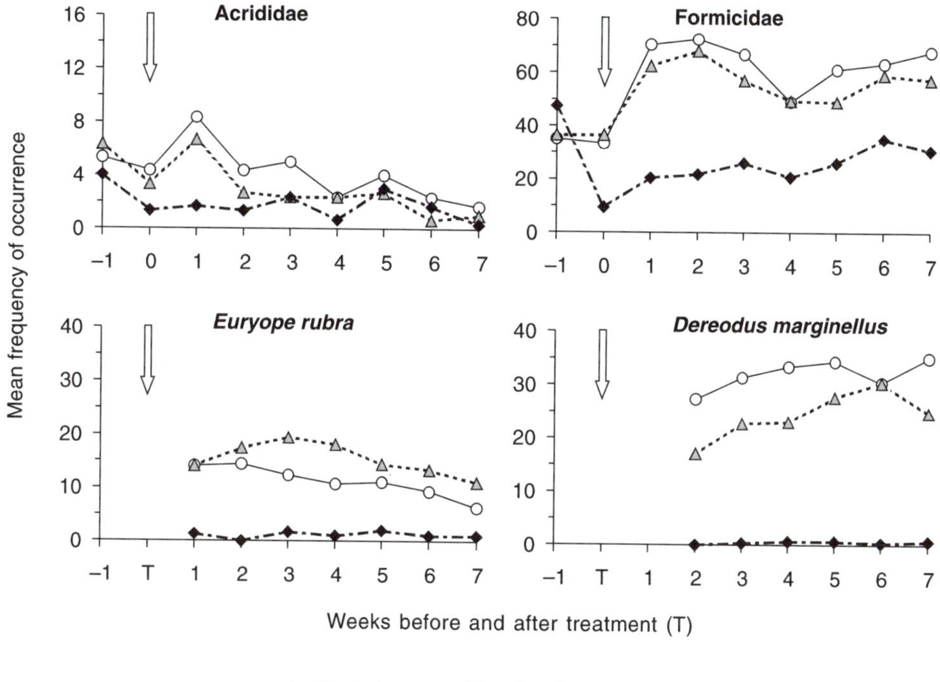

Fig. 14.2. Effects of Green Muscle™ (*Metarhizium anisopliae* var. *acridum*) and Adonis® (fipronil) on target grasshoppers and non-target epigeal (Formicidae) and arboreal (*Euryope rubra* and *Dereodus marginellus*) insects. The frequency of occurrence (presence/absence sampling) was monitored at different points in time as a measure of relative abundance. Repeated measures analyses revealed significant effects of the main factor treatment in all cases. Sidak multiple comparison of means shows that both target and non-target taxa were significantly reduced by fipronil, whereas *M. anisopliae* only affected target grasshoppers. The arrow indicates the time of treatment; error bars are not shown to avoid the cluttering of graphs.

(Fig. 14.2), fipronil had no detectable effects either. Similar results had been obtained by Balança and de Visscher (1997). Further detailed studies on the environmental impact of fipronil for locust control have been carried out by Tingle and McWilliam (2001) and Peveling *et al.* (2001) in Madagascar.

Green Muscle™ had no significant effect on *E. rubra* and *D. marginellus* (Fig. 14.2). In contrast, fipronil had a highly significant adverse effect, with 90 and 99% population reduction, respectively. Neither of the species re-colonized the treatment plots during the observation period. This appeared to be related to their limited dispersal capacity and the high persistence of fipronil. Results confirm the hypothesis that insecticide exposure of insects feeding on the milkweed was particularly high. Sokolov (2000) also noted a strong effect of fipronil on Chrysomelidae and other, in particular herbivorous, insects. The results confirm that Green Muscle™ poses a lower risk than the widely used chemical control agents.

Conclusions

While a vast majority of pest control strategies using microbials such as *Metarhizium* and similar products in developing countries is conducted with non-registered products, showing large differences in quality and efficacy, there is a worldwide tendency towards improving product quality and towards reducing use of non-registered products. In terms of ecological impact under Sahelian conditions, Green Muscle™ is probably the most thoroughly studied insecticide. For the development of registration frameworks for microbial pesticides in Africa, Green Muscle™ played a major role, too. From a purely academic point of view, the amount of knowledge achieved is remarkable, and hopefully, these studies will help to support the development of similar products. On the other hand, this immense scientific work could only be carried out through donor support. A small private producer of microbial pesticides serving a small niche market will hardly be able to finance ecotoxicological and ecopathological studies to such an extend. Therefore, Green Muscle™ is probably a rather untypical example with respect to the generation of environmental impact data. Registration authorities cannot expect small companies producing microbial pesticides to develop registration dossiers with a similar amount of data.

Apart from environmental safety, mammalian toxicity is a major concern of pesticide regulations. Deuteromycete fungi, at least *in vitro*, can metabolize a wide range of compounds. As many fungi, *Metarhizium* produces mycotoxins, too. Mycotoxins enable the fungus to combat the immune reaction of its host. It is important to test mammalian safety for each single *Metarhizium* strain to be registered as a product. However, destruxins, the common mycotoxins in *M. anisopliae*, appear to be mainly produced by isolates of the *M. anisopliae* var. *anisopliae* variety. Apart from ARSEF 324, Kershaw *et al.* (1999) found no destruxins produced by *M. anisopliae* var. *acridum* isolates, including the Green Muscle™ isolate IMI 330189. These findings are supported by tests with IMI 330189 on ring-necked pheasant, which revealed no pathological responses (Smits *et al.*, 1999). Likewise, *Metarhizium* spp. have only rarely been associated with fungal pathologies in mammals (C.J. Lomer, J.E. Eilenberg, J. Langewald, C. Nielsen, S. Vestergaard and H.H. Strasser, Swansea, UK, April 2001, unpublished results). For the future of microbial control, it is essential that registration costs remain substantially lower than the ones for synthetic pesticides, a request supported by EPA regulations. Decision-making schemes developed for microbial control products might be complex, but the key to fast and less expensive registration consists in waiving required tests, if scientifically justifiable. In Africa, registration authorities need to develop a knowledge base to make case-by-case decisions, and here the Green Muscle™ data can be very helpful. To facilitate the process, institutions like the CERES-LOCUSTOX Foundation, CABI *Bioscience*, IITA or NLU, which have a longstanding experience in microbial control and ecotoxicology, can provide scientific support.

References

Balança, G. and de Visscher, M.N. (1997) Effects of very low doses of fipronil on grasshoppers and non-target insects following field trials for grasshopper control. *Crop Protection* 16, 553–564.

Ball, B.V., Pye, B.J., Carreck, N.L., Moore, D. and Bateman, R.P. (1994) Laboratory testing of a mycopesticide on non-target organisms: the effects of an oil formulation of *Metarhizium flavoviride* applied to *Apis mellifera*. *Biocontrol Science and Technology* 4, 289–296.

Bateman, R.P., Godonou, I., Kpindou, D., Lomer, C.J. and Paraiso, A. (1992) Development of a novel field bioassay technique for assessing mycoinsecticide ULV formulations. In: Lomer, C.J. and Prior, C. (eds) *Biological Control of Locusts and Grasshoppers*. CAB International, Wallingford, UK, pp. 255–263.

Bridge, P.D., Williams, M.A.J., Prior, C. and Paterson, R.R.M. (1993) Morphological, biochemical and molecular characteristics of *Metarhizium anisopliae* and *M. flavoviride*. *Journal of General Microbiology* 139, 1163–1169.

Carson, R. (1962) *Silent Spring*. Houghton Mifflin Company, Boston, Massachusetts, USA.

Cheke, R.A. (1990) A migrant pest in the Sahel: the Senegalese grasshopper *Oedaleus senegalensis*. *Philosophical Transactions of the Royal Society London B* 328, 539–553.

Danfa, A. and van der Valk, H. (1999) Laboratory testing of *Metarhizium* spp. and *Beauveria bassiana* on Sahelian non-target arthropods. *Biocontrol Science and Technology* 9, 187–198.

Driver, F., Milner, R.J. and Trueman, J.H.W. (2000) A taxonomic revision of *Metarhizium* based on a phylogenetic analysis of ribosomal DNA sequence data. *Mycological Research* 104, 135–151.

Ekesi, S. (1999) Selection of virulent isolates of entomopathogenic Hyphomycetes against *Clavigralla tomentosicollis* Stal. and evaluation in cage experiment using three cowpea varieties. *Mycopathologia* 148, 131–139.

EPA (1996) *Microbial Pesticide Test Guidelines* – Series 885. United States Environmental Protection Agency, Office of Prevention, Pesticides and Toxic Substances (OPPTS), Washington, DC, USA (continuously updated).

EU (2001) Commission Directive 2001/36/EC. *European Union, Official Journal* L 164, 20/06/2001, 1–38.

Everts, J. (1990) *Environmental Effects of Chemical Locust and Grasshopper Control – A Pilot Study*. Project ECLO/SEN/003/NET, Food and Agriculture Organization of the United Nations, Rome, Italy.

Friese, D.D., Mégevand, B. and Yaninek, J.S. (1987) Culture maintenance and mass production of exotic phytoseiids. *Insect Science and its Application* 8, 875–878.

Fritsche, M.E. and Tamò, M. (2000) Influence of thrips prey species on the life-history and behaviour of *Orius albidipennis*. *Entomologia Experimentalis et Applicata* 96, 111–118.

Fuxa, J.R. (1987) Ecological considerations for the use of entomopathogens in IPM. *Annual Review of Entomology* 32, 225–251.

Goettel, M.S., Poprawski, T.J., Vandenberg, J.D., Li, Z. and Roberts, D.W. (1990) Safety to nontarget invertebrates of fungal biocontrol agents. In: Laird, M., Lacy, L.A. and Davidson, E.W. (eds) *Safety of Microbial Insecticides*. CRC Press, Boca Raton, Florida, USA, pp. 209–231.

Greathead, D.J. and Prior, C. (1990) The regulation of pathogens for biological control with special reference to locust control. In: Somme, L. and Bie, S. (eds) *Proceedings of the Workshop on Health and Environmental Impact of Alternative*

Control Agents for Desert Locust Control. NORAGRIC Occasional Papers Series C: Development and Environment 5, 65–80.

Hassan, S.A. (1998) Pesticides and beneficial organisms. International Organization for Biological and Integrated Control of Noxious Animals and Plants – West Palearctic Regional Section (IOBC/WPRS). *IOBC/WPRS Bulletin* 21.

Hernández, L.M. and Stonedahl, G.M. (1999) A review of the economically important species of the genus *Orius* (Heteroptera: Anthocoridae) in East Africa. *Journal of Natural History* 33, 543–568.

Jago, N.D. (1993) *Millet Pests of the Sahel: Biology, Monitoring and Control.* Natural Resources Institute, Overseas Development Administration, Chatham, UK.

Kershaw, M.J., Moorhouse, E.R., Bateman, R., Reynolds, S.E. and Charnley, A.K. (1999) The role of destruxins in the pathogenicity of *Metarhizium anisopliae* for three species of insects. *Journal of Invertebrate Pathology* 74, 218–223.

Kooyman, C., Bateman, R.P., Langewald, J., Lomer, C.J., Ouambama, Z. and Thomas, M. (1997) Operational-scale application of entomopathogenic fungi for control of Sahelian grasshoppers. *Proceedings of the Royal Society of London* B 264, 541–546.

Lahr, J., Badji, A., Marquenie, S., Schuiling, E., Ndour, K.B., Diallo, A.O. and Everts, J.W. (2001) Acute toxicity of locust insecticides to two indigenous invertebrates from Sahelian temporary ponds. *Ecotoxicology and Environmental Safety* 48, 66–75.

Langewald, J. and Cherry, A. (2000) Prospects for microbial control in West Africa. *Biocontrol News and Information* 21, 51N–56N.

Langewald, J., Ouambama, Z., Mamadou, A., Peveling, R., Stolz, I., Bateman, R., Attignon, S., Blanford, S., Arthurs, S. and Lomer, C. (1999) Comparison of an organophosphate insecticide with a mycoinsecticide for the control of *Oedaleus senegalensis* Krauss (Orthoptera: Acrididae) and other Sahelian grasshoppers in the field at operational scale. *Biocontrol Science and Technology* 9, 199–214.

Lomer, C.J., Bateman, R.P., Johnson, D.L., Langewald, J. and Thomas, M.B. (2001) Biological control of locusts and grasshoppers. *Annual Review of Entomology* 46, 667–702.

Mégevand, B., Kläy, A., Gnanvossou, D. and Paraiso, G. (1993) Maintenance and mass rearing of phytoseiid predators of the cassava green mite. *Experimental and Applied Acarology* 17, 115–128.

NAS (1981) *Testing for the Effects on Ecosystems.* National Academy Press, Washington, DC, USA.

Nwanze, K.F. and Harris, K.M. (1992) Insect pests of pearl millet in West Africa. *Review of Agricultural Entomology* 80, 1133–1155.

OECD (1982 onward) *Guidelines for Testing of Chemicals.* OECD, Paris, France.

Peveling, R. (2001) Environmental conservation and locust control – possible conflicts and solutions. *Journal of Orthoptera Research* 10, 171–187.

Peveling, R. and Demba, S.A. (1997) Virulence of the entomopathogenic fungus *Metarhizium flavoviride* Gams and Rozsypal and toxicity of diflubenzuron, fenitrothion-esfenvalerate and profenofos-cypermethrin to non-target arthropods in Mauritania. *Archives of Environmental Contamination and Toxicology* 32, 69–79.

Peveling, R., Attignon, S., Langewald, J. and Ouambama, Z. (1999) An assessment of the impact of biological and chemical grasshopper control agents on ground-dwelling arthropods in Niger, based on presence/absence sampling. *Crop Protection* 18, 323–39.

Peveling, R., Rasolomanana, H., Raholijaona, Rakotomianina, L., Ravoninjatovo, A., Randimbison, L., Rakotondravelo, M., Raveloson, A., Rakotoarivony, H., Bezaka,

S., Ranaivoson, N. and Rafanomezantsoa, J.-J. (2001) Effets des traitements aériens de fipronil et de deltaméthrine en couverture totale sur la chaîne alimentaire. In: Zehrer, W. (ed.) *Lutte Antiacridienne à Madagascar – Tome III – Ecotoxicologie*. DPV/GTZ, Ministère de l'Agriculture, Antananarivo, pp. 525–572.

Prior, C. (1997) Susceptibility of target acridoids and non-target organisms to *Metarhizium anisopliae* and *M. flavoviride*. In: Krall, S., Peveling, R. and Ba Diallo, D. (eds) *New Strategies in Locust Control*. Birkhäuser Verlag, Basel, Switzerland, pp. 369–375.

Singh, S.R. and Allen, D.J. (1979) Cowpea pests and diseases. Ibadan. International Institute of Tropical Agriculture (IITA). *IITA Manual Series 2*.

Smits, J.E., Johnson, D.L. and Lomer, C. (1999) Pathological and physiological responses of ring-necked pheasant chicks following dietary exposure to the fungus *Metarhizium flavoviride*, a biocontrol agent for locusts in Africa. *Journal of Wildlife Diseases* 35, 194–203.

Sokolov, I.M. (2000) How does insecticidal control of grasshoppers affect non-target arthropods? In: Lockwood, J.A., Latchininsky, A.V. and Sergeev, M.G. (eds) *Grasshoppers and Grassland Health*. Kluwer Academic Publishers, Dordrecht, The Netherlands, pp. 181–192.

Stolz, I. (1999) The effect of *Metarhizium anisopliae* (Metsch.) Sorokin (=*flavoviride*) Gams and Rozsypal var. *acridum* (Deuteromycotina: Hyphomycetes) on non-target Hymenoptera. PhD thesis, University of Basel.

Stolz, I., Nagel P., Lomer, C. and Peveling, R. (2002) Susceptibility of the hymenopteran parasitoid *Apoanagyrus* (=*Epidinocarsis*) *lopezi* (Encyrtidae) and *Phanerotoma* sp. (Braconidae) to the entomopathogenic fungus *Metarhizium anisopliae* var. *acridum* (Deuteromycotina: Hyphomycetes). *Biocontrol Science and Technology* 12, 349–360.

Tingle, C.C.D. and McWilliam, A.N. (2001) Evaluation de l'effet à court terme sur les organismes non cibles des traitements en barrières à grande échelle dans le cadre de la lutte antiacridienne d'urgence. In: Zehrer, W. (ed.) *Lutte Antiacridienne à Madagascar – Tome III – Ecotoxicologie*. DPV/GTZ, Ministère de l'Agriculture, Antananarivo, pp. 115–182.

van der Valk, H. and Niassy, A. (1997) Side effects of locust control on beneficial arthropods: research approaches used by the LOCUSTOX project in Senegal. In: Krall, S., Peveling, R. and Ba-Diallo, D. (eds) *New Strategies in Locust Control*. Birkhäuser, Basel, Switzerland, pp. 337–344.

van der Valk, H., Niassy, A. and Beye, A.B. (1999) Does grasshopper control create grasshopper problems? Monitoring side effects of fenitrothion application in the western Sahel. *Crop Protection* 18, 139–149.

Yaninek, J.S., Onzo, A. and Ojo, B. (1993) Continent-wide releases of neotropical phytoseiids against the exotic cassava green mite in Africa. *Experimental and Applied Acarology* 17, 145–160.

15

Microbial Control of Termites in Africa

Jürgen Langewald,[1] Janette D. Mitchell,[2] Nguya K. Maniania[3] and Christiaan Kooyman[4]

[1]*International Institute of Tropical Agriculture, Cotonou, Bénin;* [2]*Agricultural Research Council – Plant Protection Research Institute, Pretoria, South Africa;* [3]*International Centre of Insect Physiology and Ecology (ICIPE), Nairobi, Kenya;* [4]*CAB International, Africa Regional Centre, Nairobi, Kenya*

Introduction

In Africa, termites are an important part of the soil fauna playing a major role in the degradation of organic matter and thereby improving soil fertility (Lee and Wood; 1971, Eggelton and Bignell, 1995). Compared with earthworms in the temperate regions, many termite species are active at greater depths, as they tunnel to collect water and building material. They also break down organic matter more effectively, aided by microorganisms in their guts or, in the case of the sub-family Macrotermitinae, through fungi grown on fungus combs made of faecal pellets. Decomposition of organic matter takes place within the nests, and mineral nutrients are therefore concentrated and trapped in these relatively few sites for at least as long as the nests are active (Kooyman and Onck, 1987). According to the same authors, constant activities of burrowing new tunnels in the upper layers of the soil and filling up the old tunnels with soil pellets contribute to the soil's aeration and rootability, making it more friable. The African primary forest environment is very rich in termite species and the highest species diversity ever recorded was from a forest in Cameroon (Eggelton *et al.*, 1995). More than 1000 of the 2600 recognized termite species are found on the African continent.

Termites, with their high nutritional value, play an important role both in natural ecosystems and human societies (Ene, 1963; Logan, 1991; Iroko, 1996). They are an important source of food for many animals including birds, reptiles and mammals. In some African societies, termites are regarded as a delicacy (van Huis, 1996). Termite fungus combs from within the mound as well as the mushroom produced by these fungi are frequently eaten and mound soil is utilized for its mineral content. Soil from termite mounds is used in constructing mud wall, brick making and pottery (Logan, 1991; Iroko, 1996).

Termites as Pests

While termites largely play a beneficial role, a number of species are destructive and are a major threat to crops, trees and houses in developing countries (Plate 46). With the spread of agriculture and the destruction of natural forests, termites have increasingly become a problem (Edwards and Mill, 1986; Wood and Pearce, 1991). Socio-economic studies conducted by the IITA revealed that farmers in moist and dry savannah zones consider termites a priority problem with regard to plant health. Estimates of losses due to termites vary widely. Increased termite attack is frequently associated with various stress factors, particularly drought and prolonged dry spells within the growing season as well as low soil fertility. Studies on crop losses due to termites include yam and cassava (Atu, 1993), sugarcane (Sands, 1977), groundnuts (Johnson *et al.*, 1981), sorghum (Logan *et al.*, 1990) and maize (Wood *et al.*, 1980). In the semiarid savannahh ecosystems of Kenya, termite foraging activities are responsible for the destruction of 800–1500 kg ha^{-1} of pasture per year (Lepage, 1981a,b), while in the countries of southern Africa, harvester termites are a serious pest of rangeland, removing 60% or more of the standing grass biomass during dry years (Coaton, 1958). Apart from damage to crops and rangeland, termites are commonly responsible for mortality of tree seedlings in forestry and cause considerable damage to buildings and other wooden structures like fence posts and utility poles. They also attack grain stores, where apart from the wooden structure, the grain stocks are also destroyed. The most important termite pest genera in Africa include: *Odontotermes*, *Macrotermes*, *Pseudacanthotermes*, *Microtermes*, *Ancistrotermes*, *Allodontermes*, *Amitermes*, *Trinervitermes* and *Hodotermes* (Mailu *et al.*, 1995; Mitchell, 2002). These genera characteristically differ in their biology and mode of attack. *Macrotermes* spp. build large epigeal mounds housing larger centralized colonies from which they move out to forage at distances of up to 50 m. They attack plants at the base of the stem, ring-barking or cutting them completely. *Odontotermes* spp. build both subterranean and epigeal nests. Damage is due to feeding either under soil sheeting on the outer surface of the plants or on the roots. *Microtermes* spp. and *Ancistrotermes* spp. have diffuse subterranean nests and attack plants from below ground by entering the root system and tunnelling up into the stem, hollowing it out and filling it with soil.

Termite Control

Until recently, termites were controlled using persistent organochlorines such as aldrin, lindane, chlordane and dieldrin applied as seed treatments, on seedlings, mature plants and for tree protection (Sands, 1977). With the banning of organochlorines, products like organophosphates, carbamates, synthetic pyrethroids and most recently phenyl-pyrazoles are being used. Due to the cryptic lifestyle, termite control needs sophisticated approaches in order to be effective. Pre-construction and remedial control of infestations by soil treatments with chemical pesticides are mainly used for protection of buildings against subterranean termites. Insecticides in current use, when applied to foundations, can protect houses for between 1 and 19 years (Pearce, 1997). Wood can be protected against termite attack with chemical preservatives like creosote, copper chrome arsenate, pentachlorphenol, boric acid and fluoride salts, although in many parts of Africa people rely on treatment with used engine oil, petrol or kerosene. Fumigation is an important means of eradication of dry-wood termites in buildings, and was also used by plant protection agencies for the direct treatment of termite colonies. Methyl bromide, until it was recently removed from the market owing to its impact on the ozone layer, was an important fumigant (Pearce, 1997). In the middle of the 20th century, baiting was developed as an effective and target-specific method against harvester termites in South Africa. More recently, this method has been employed also against other termite species. The idea is to attract the insects to a food source, which is treated with an insecticide. After feeding on the bait, the insects carry the insecticide back to their colony and, due to the grooming behaviour of their nest mates, contaminate more insects inside the colony, which will eventually die. The effectiveness of this method depends on the foraging behaviour of the target termite species, the attractiveness of the bait and the insecticidal product. Also the spatial distribution of the bait plays an important role (Pearce, 1997). Various traditional methods of control are practised by small-scale farmers and include removal of the queen, crop rotation, intercropping and use of plant extracts.

Crop rotation and intercropping sometimes reduce the buildup of pest populations and are used by small-scale farmers against insects with specific host ranges. Both practices are, however, inappropriate against termites, which feed on a wide variety of plants. For instance intercropping maize and beans, though significantly lowering ground tunnelling by termites, did not reduce termite damage to the plants (Gethi *et al.*, 1995). Mound-building termites can be controlled physically by destroying the mound and removing the termite queen, or by excavating the top parts of the mounds and burning straw to suffocate and kill the colony. This has to be done carefully with termite species that can have several queens in one colony. Destroying mounds is very labour intensive and in areas where the density of termite mounds is very high it does not seem to be practical. For African subsistence farmers, who cannot afford expensive chemical insecticides, a safe alternative solution needs to be developed.

Biological Control

Several natural enemies and pathogens of termites have been under consideration as potential biological control agents, although the full impact of predation on termite populations in nature has never been clearly elucidated. Coaton (1958) suggested a link between escalating harvester termite problems over vast areas of South Africa and overgrazing, but also with the severe decline in natural predators. In an early attempt at using predation in termite control, Hartwig (1965) introduced a number of cockerels on to an area heavily infested with the harvester termite, *Hodotermes mossambicus* Hagen (Isoptera, Kalotermitidae). As a result, all surface termites ceased for a 6-month period.

Predation by ants is a major constraint on termite population growth (Bodot, 1967; Longhurst *et al.*, 1978; Lepage, 1981a; Collins and Wood, 1984; Abe and Darlington, 1985), and Logan *et al.* (1990) and Grace (1997) stressed the importance of manipulating predatory ants as a strategy for biological control of termites. Recently, Sekamatte *et al.* (2001) developed baits for predatory ants based on traditional farmers' practice of using dead animals, meat bones and sugarcane husks to 'poison' *Macrotermes* mounds. The technology was evaluated in maize fields for termite control in Uganda. The protein-based baits attracted significantly larger numbers of ants and more ants established nests near maize plants; this in turn reduced termite damage and increased grain yield.

Entomopathogenic nematodes, while effective in the laboratory, have proved to have little impact under field conditions (Mauldin and Beal, 1989). According to Milner and Staples (1995), two bacteria, *Serratia marcescens* Bizio and *Bacillus thuringiensis* Berliner were observed killing termites in the laboratory, but no reports are available demonstrating their efficacy in the field.

Entomopathogenic Fungi

Among the pathogens (bacteria, viruses, protozoa and fungi), entomopathogenic fungi offer the best prospect as termite control agents. Fungi infect their host through the cuticle and do not need to be ingested; their host range can be broad and the potential for horizontal transmission is high. Two review papers on entomopathogenic fungi have recently been published by Milner and Staples (1996) and Rath (2000), but without addressing the problem of termites in cropping systems. Hänel (1981) was the first to demonstrate the suitability of different strains of the entomopathogenic fungus *Metarhizium anisopliae* Metschnikoff (*Deuteromycotina*: *Hyphomycetes*) for biological control of termites.

The application of entomopathogenic fungi for termite control is being explored by several working groups in Australia, the USA and Africa. Though *M. anisopliae* can infect a wide range of insect hosts, some isolates of this fungus are highly host specific (Lomer *et al.*, 2001). In the USA, one strain of this fungus had been commercialized for termite control under the name Bio-Blast™ (Ecoscience Cooperation, USA), but is no longer being marketed, while the Commonwealth Scientific and Industrial Research Organization (CSIRO) of Australia has developed an improved formulation of *Metarhizium*, commercialized under the name Bio-Green™ (Bio-Care Technology Pty. Limited, Australia) (Milner, 2000).

Like many other insect pathogens, *M. anisopliae* occurs naturally in the soil (Zimmermann, 1986) and has potential as management agent for subterranean insect pests. In the soil, factors negatively affecting fungus survival (high temperature, UV radiation and lack of humidity) are absent or their fluctuations are buffered. Furthermore, the high insect density in termite nests and grooming behaviour of these insects was thought to increase the chance of epizootics caused by such pathogens (Grace and Zoberi, 1992). Grooming serves, however, also to remove any infectious material from the surface of colony members' bodies, including the queen. Many termite species can recognize the presence of fungal spores, which are constantly removed from the colony by ingestion, and subsequently deactivation in the gut (Kramm and West, 1982; Pearce, 1997; Rosengaus *et al.*, 1998). In cases where large areas of the colony are contaminated, these areas are walled off from the rest of the colony and put under quarantine. Some termites like the dampwood termite, *Zooteropsis angusticollis* Hagen, release alarm vibration signals throughout the whole colony if a fungal invasion is detected (Walker, 1999). Most termites keep their colonies extremely clean, and any debris is collected in a special area, where it is usually walled off from the rest of the colony.

Microbial Control of Termites in Africa

A new collaborative effort

Until recently there has been little work on microbial control of termites in Africa. The earliest investigations on the use of entomophagous fungi were undertaken in South Africa in the mid to late 1980s with a view to controlling the harvester termite *H. mossambicus*. A number of local *M. anisopliae* and *Beauveria bassiana* (Balsamo) Vuillemin (*Deuteromycotina*, *Hyphomycetes*) isolates from grass litter samples collected from different localities together with an imported isolate of *M. anisopliae* were assayed against *H. mossambicus* in the laboratory. *B. bassiana* isolates and the imported *M. anisopliae* isolate caused 100% mortality within 6 days. *H. mossambicus* was not susceptible to the majority of local *M. anisopliae* isolates and those to which it was susceptible were not as effective as the imported isolate. Field trials were, however, unsuccessful (G.G. Gouse, unpublished results). Similarly in Kenya, a number of *Metarhizium* and *Beauveria* strains have been collected and tested in bioassays on *M. subhyalinus* (Gitonga *et al.*, 1995). Between 1998 and 2001, a collaborative research project (between IITA Bénin; CAB *International* African Regional Center, Kenya; the ICIPE, Kenya; Service de Protection des Végéteaux, SPV, Bénin; KARI, Kenya) funded by the British government, assessed the potential of *M. anisopliae* for termite control in West and East Africa. The virulence of *Metarhizium* to termites and the behavioural response of the host opened two major options for the use of the fungus. One strategy was aimed at direct or indirect contamination of the termites and their eventual death. A second strategy would make use of the repellent effect of the fungus to be used as seed treatment, seedbed treatment, or as protective paint for wooden structures.

Isolate collection

To obtain large numbers of strains of *M. anisopliae*, soil samples and diseased insect cadavers were collected in different agroecological zones. Various isolation methods including selective media, such as Veen's medium, which had previously been successfully used, were employed (Veen and Ferron, 1966;

Milner and Lutton, 1976), and insects such as *Tribolium castaneum* Herbst, *Acanthocinus aedilis* L. and *Galleria mellonella* L. larvae (Zimmerman, 1986; Vanninen, 1995) were used as baits. Following the surveys, ten *M. anisopliae* isolates were found in Bénin, West Africa, and 33 isolates in Kenya. These isolates were rarely found on termite cadavers, but were collected mostly from termite mounds or samples of humus rich soils.

Comparing virulence of different isolates

Comparing bioassay results concerning subterranean termites ('lower termites') is difficult, as termite survival and response to stress do not only vary according to laboratory conditions, but also between termite groups removed from different sites and colonies (Lenz *et al.*, 1993). This is also the case for the highly organized mound building and fungus-growing 'higher termite' species found in Africa. The normally functioning termite colony is a homeostatically regulated, self-stabilizing system (Emerson, 1956). Its maintenance of optimal conditions for reproduction is based on efficient social integration. The role of grooming (trophallaxis) is not only restricted to the distribution of food, but plays an important role in colony organization. Generally in social insects, the existence of a differentiated cast system, including a large number of sterile workers, leads to greater working efficiency, greater longevity and greater vigour (Grassé, 1946; Krishna and Weesner, 1969).

These factors explain the difficulties in creating laboratory conditions in which the test insects would survive long enough, and which would provide consistent experimental results. Because 'lower termites' (Families: Mastotermitidae, Kalotermitidae, Hodotermitidae and Rhinotermitidae) can be reared comparatively easily they have been used in most experiments. At IITA, attempts were, however, made to maintain workers of the 'higher termite' *Macrotermes subhyalinus* Rambur (Isoptera: Termitidae) long enough to complete the experiments. The survival of *M. subhyalinus* workers was found to be better on crushed humid and loamy termite mound material than on moist filter paper, as recommended by Milner (2000) for the Australian 'lower termites' *Coptotermes* spp. (Tables 15.1 and 15.2). The addition of pieces of fungal comb further helped to extend survival up to a couple of days.

The virulence of fungal isolates was tested using two different methods of bioassay. In the first method, isolates were mixed with the substrate of crushed termite mound material. Isolates collected in Bénin as well as isolates from Australia (CSIRO, FI 23), Kenya (ICIPE 30) and the commercial product, Bio-Blast™ from the USA were compared at four different dose rates. When tested mixed with the substrate, at four different dosages (5×10^7, 1.25×10^7, 1×10^7 and 6.25×10^6 cfu (colony forming units) g^{-1} of substrate, according to a logarithmic scale) the isolates collected in Bénin and the commercial strain (Bioblast™) (LD_{50} at day 2 $<2.5 \times 10^7$ cfu g^{-1} substrate) were not more virulent than the reference strain from Kenya (ICIPE 30). FI 23 from Australia did not cause a dose response in the target insects. Across the four dosages, FI 23 and ICIPE 30 sporulated on average on more than 25% of the cadavers, while

Table 15.1. Comparison of median survival time (days) of 20 *Macrotermes subhyalinus* workers exposed to filter paper and differently treated termite mound materials in Petri dishes, kept under ambient conditions or inside a container. (Experiment with *N* = 3 carried out at IITA.)

	Ambient	Inside container
Non-sterile dried mound material + 3 ml sterile water	3.50 a	3.09 a
Sterile dried mound material + 3 ml sterile water	4.13 a	4.50 b
Non-sterile mound material + 0.5 ml sterile water	6.20 b	5.65 c
Filter paper + 2 ml sterile water	2.88 a	2.73 a
Average	4.34 a	3.73 a
Temperature (°C)	28.1 (SE: 0.2)	29.4 (SE: 0.4)
RH	85% (SE: 3%)	84% (SE: 2%)

Means within a column followed by the same letter do not differ significantly by *P* = 0.05. Means were compared using survival analysis in SPSS (SPSS Inc.,1999, Chicago).

Table 15.2. Comparison of median survival time (days) of 20 *Macrotermes subhyalinus* workers exposed to filter paper and differently treated termite mound materials in Petri dishes, kept under different conditions. (Experiment carried out at IITA.)

N = 3	Ambient	Inside container
Non-sterile dried mound material + 4 ml sterile water	2.33 a	2.07 a
Sterile dried mound material + 4 ml sterile water	2.42 ab	1.94 a
Non-sterile mound material + 0.5 ml sterile water	2.55 b	2.70 b
Filter paper + 2 ml sterile water	2.55 b	2.34 c
Average	2.47 a	2.29 b
Temperature (°C)	28.0 (SE: 0.4)	28.1 (SE: 0.7)
RH	65% (SE: 8%)	70% (SE: 4%)

Means within a column followed by the same letter do not differ significantly by *P* = 0.05. Means were compared using survival analysis in SPSS (SPSS Inc., 1999, Chicago).

all other isolates, including Bioblast™, sporulated on fewer cadavers, which is an important consideration in horizontal transmission in termite colonies. In a second approach, at ICIPE and CAB *International*, the grooming method of Milner *et al.* (1998) was used. At ICIPE, a single *M. subhyalinus* worker was rolled in dry fungal spores and allowed to walk for a few minutes to dislodge loose spores from its body. It was then placed in containers with groups of 20 uncontaminated workers for 24 h. Preliminary observation had revealed that an insect rolled in conidia accumulates on average 1×10^7 conidia on its body. Mortality in the control was 9%. Four of the isolates tested were as effective as ICIPE 30, causing mortalities of between 90% and 100% at 3 days post-treatment. At the CAB *International* Africa Regional Centre laboratories, bioassays were carried out on *Odontotermes amanicus* Sjöstedt and *M. subhyalinus* workers. In addition to the grooming technique, workers were exposed to filter papers soaked in suspensions of *Metarhizium* spores. As in the other experiments, none of the tested isolates performed better than ICIPE 30 and consequently this strain was selected for further studies. At the Plant

Protection Research Institute, South Africa (PPRI), in a similar bioassay, South African isolates of *B. bassiana* proved more effective than *M. anisopliae* against both *Odontotermes latericius* Haviland and *M. natalensis* Haviland. The harvester termite *H. mossambicus* proved less susceptible to the isolates than the other species.

Repellency tests

At ICIPE in Kenya, the repellent action of ten isolates of *M. anisopliae* was evaluated in a choice bioassay system consisting of two 50-ml plastic cages used for the odour source and a connecting tube with a hole for the introduction of insects. Results showed percentage repellency values of between 53% and 78% among the isolates. ICIPE 30 strain was repellent for 53% of the test insects (Table 15.3).

Table 15.3. Mean (± SE) percentage repellency values after exposure of 20 workers of *Macrotermes subhyalinus* workers to various isolates of *Metarhizium anisopliae* for 45 min, in a choice system consisting of two 50-ml volume chambers, connected with a transparent plastic tube (25 cm length), one containing the test substance (experiment carried out at ICIPE).

Treatments[a]	Mean ± SE[b] ($N = 4$)
ICIPE 30	53.4 ± 9.5 cde
NANYUKI 8	77.6 ± 5.7 c
KITUI 4	66.6 ± 11.7 c
KITUI 10	61.9 ± 4.8 cd
KITUI 11	35.3 ± 7.3 ef
KITUI 13	70.1 ± 2.7 c
MERU 50	19.2 ± 10.2
MERU 7	0.0 ± 0.0 g
EMBU 36	23.1 ± 16.2 def
EMBU 12	0.0 ± 0.0 g
EMBU 7	0.0 ± 0.0 g
BLANK	0.0 ± 0.0 g
UNTREATED	0.0 ± 0.0 g

[a]5×10^{10} cfu 29 g crushed maize (treatment), versus 30 g of non-contaminated crushed maize (control);
[b]Mean percentage of the total number of termite workers found in the treated chambers.
Means within a column followed by the same letter do not differ significantly by SNK ($P = 0.05$).
Means were angularly transformed before analysis, but values represent untransformed.

Further investigations were carried out at ICIPE to evaluate the role of water content in fungal conidia on the repellency of fungal spores to termites. The results seem to indicate that high moisture content of active or deactivated spores attracts rather than repels termites (N.K. Maniania, 2000, unpublished results).

Many studies have reported on the repellency of fungal spores to termites (Milner and Staples, 1996; Rosengaus *et al.*, 1999), but its nature has not yet been established (Rath, 2000). Workers of *Reticulitermes flavipes* have been reported to avoid nest mates that have died from fungal pathogens (Kramm *et al.*, 1982; Grace and Zoberi, 1992). Although healthy workers of *R. flavipes* avoided cadavers killed by *B. bassiana*, the 'mycelial mat' cultures were not avoided and resulted in mortality of the workers (Grace and Zoberi, 1992). The results of our study corroborate the findings of Grace and Zoberi (1992), where humidity seems to play an important role in the repellent action. The results of these studies can, however, not be generalized because they concern only a few fungal isolates. In fact, Milner *et al.* (1998) found differences in the repellent action of different isolates of *M. anisopliae* to termites. More studies are therefore required to elucidate the mechanism of repellency of fungal spores to termites.

Control strategies in the field

Both properties of *Metarhizium* on termites, the repellent effect and the direct impact through infection, were tested in the field (Plate 45). In Bénin, field trials were undertaken by IITA and SPV at two neighbouring sites in farmers' maize fields using *M. anisopliae* spores (plain), or formulated either as dusts or granules and the chemical insecticides, fipronil and propoxur. The treatments were applied to planting holes before adding three maize kernels. The treatments with fipronil (recommended field rate, 50 g a.i. ha^{-1}), propoxur (recommended field rate, 40 g a.i. ha^{-1}) and three different dose rates of *M. anisopliae* dry spores (2×10^{12}, 4×10^{12} and 8×10^{12} cfu ha^{-1}, ICIPE 30) had no effect on the germination rate of maize seed. In one site, germination varied between 18% and 30%. In the second site between 61% and 82% of the plants germinated. All the phenological stages of maize were attacked and termite damage increased in all treatments over a 9-week observation period. At the end of the season, only the maize plots treated with 8×10^{12} cfu ha^{-1} resulted in higher yields (142 g m^{-2}) compared with the other treatments (88 g m^{-2} across all other treatments), including the untreated control. In another experiment carried out by IITA and SPV, maize was planted in plastic buckets (21 cm high × 16 cm diameter). Seventeen holes of 2.3 cm diameter were punched in the sides of the buckets before planting to allow free movements of termites. The soil surface inside the buckets was treated with four different formulations of *M. anisopliae*, strain ICIPE 30 post planting. All four formulations were applied at two dosages: Oil-ULV formulation (8×10^7 and 4×10^7 cfu), EC (8×10^7 and 4×10^7 cfu), rice granular (1×10^8 and 2×10^7 cfu) and dry spore powder (1.25×10^8 and 2.5×10^8 cfu). The presence of termites inside the plastic buckets was significantly reduced at higher dose compared with the lower dose treatments across all four formulations. As with the seed-bed treatments, termite damage on the potted maize plants increased with time. Similarly, there was less damage on the maize plants in the high-dose treatments across all formulations. However, when compared at individual treatment level with the untreated controls, none of the treatments protected maize significantly. In Uganda, field experiments were carried out using a granular formulation of *M. anisopliae*, strain ICIPE 30 (A. Russell-Smith, 2002, unpublished results). Two grams of spore dust formulation were applied per maize pocket at planting, which is equivalent to 1×10^{11} cfu. This treatment was compared with non-treated maize and lindane (100 g l^{-1} EC) as maize seed dressing. There was a significant effect of treatments on the yield of maize. A. Russell-Smith (2002, unpublished results) reported 69% greater yield for the *M. anisopliae* treatment and 87% for the lindane treatment compared with the non-treated control. Maniania (2002) also reported that application of the same isolate as granules (4×10^9 cfu, 1 g granulate per hole) at planting significantly reduced maize plant logging and increased grain yield in Kenya.

Common to these field experiments was the idea to use the repellent effect of *M. anisopliae* to protect termite attack to the roots of maize. It could be demonstrated that indeed maize yields were significantly increased in Bénin, Uganda and Kenya, applying pure spores to the planting holes. Common to the experiments was also the rather high amounts of spores needed to achieve a good protection against termites. At the tested dosages, this method will hardly be economically viable. Further research should therefore concentrate on formulations that are less expensive, for instance by making use of the complete growth substrate from the *Metarhizium* mass-production process. Apart from *M. subhyalinus*, maize logging is also caused by *Microtermes* spp. Difference in field test results might be caused by different termite species compositions attacking maize.

Timber protection

Damage by termites to structures and buildings in Africa occurs mainly in rural areas, where timber is a major element for the construction of simple houses. This has resulted in increased deforestation since more timber is cut to rebuild the houses. Assays were carried out in Bénin and Kenya to investigate whether *M. anisopliae* could give protection to wooden structures. In Bénin, wooden pickets were coated with four doses of an oil concentrate of *M. anisopliae* strain ICIPE 30 (5×10^8, 2.5×10^9, 2.7×10^{10} and 2×10^{11} cfu per picket, over a length of 30 cm) and then buried in the soil. A second set of pickets was placed in holes with the surrounding soil mixed with different doses of dry spores of the same strain (9×10^{10}, 1.6×10^{11}, 2.8×10^{11} and 5×10^{11} cfu per hole). Termite damage to the pickets increased significantly over 18 weeks observation time. The dry spores applied to the soil surrounding the pickets protected the timber better than the oil concentrate. Both treatments were effective compared with the untreated controls. It is difficult to compare the two treatment dosages, but it seems that a thin layer of spores is less repellent compared with a thick layer of contaminated soil. At CAB *International* Kenya, a mineral oil formulation (Ondina) of spores protected pickets over 3 months (damage <1%). Dry spores and a kerosene formulation gave less protection, though at most 10% of the wood was affected. Additional tests are needed to evaluate these formulations over periods of several years, because if protection only lasts several months, this approach is not likely to be cost effective. Similar observations were made by Milner *et al.* (1997), who demonstrated that mixing soil with *Metarhizium* spores reduced damage on timber, but did not stop termite damage. These results have to be compared with traditional methods of timber protection. Particularly metal salts can protect impregnated timber up to 5 years (Pearce, 1997).

Direct application of Metarhizium *spores into termite colonies*

A different approach of controlling termites in the field consists in the application of *M. anisopliae* spores directly into termite nests. Hänel and Watson (1983) inoculated field colonies of *Nasutitermes exitiosus* Hill with conidia of *M. anisopliae*. Colonies declined and the fungus persisted for up to 6 weeks. Fernandes (1991) achieved 90% control of the pasture pest termite *Cornitermes cumulans* Koller in Brazil with *M. anisopliae* and *B. bassiana*. Similar results have been achieved by Milner and Staples (1996) in Australia and N.K. Maniania in Kenya. We have also demonstrated that termite colonies can be killed with dry spores of *M. anisopliae*. This requires, however, large amounts of spores (5×10^{11} cfu per colony in the case of *M. subhyalinus*) and takes a long time, as also reported by Milner and Staples (1996). After application of dry spore powder of the *M. anisopliae*, ICIPE strain 30, at dosages between 5×10^{11} and 9×10^{10} cfu per mound, the activity of termite colonies was reduced on average by 60%. Treatments with waste rice from the mass production showed a similar efficacy of about 60% over the controls. A second method to monitor efficacy was to compare the percentage number of termite mounds that remained active. After 1 year, 63% of the colonies treated with different dosages of dry spores of *M. anisopliae* strain ICIPE 30 were killed. When using different dosages of waste rice from the *M. anisopliae* mass production (between 50 and 2.7 g per mound or 1.5×10^9–8.1×10^9 cfu), the reduction in average termite mound activity was similar to the one achieved with the dry spore formulation. By contrast, fewer termite colonies were killed, 1 year after application of spore dust. After the spore extraction, the waste rice still contains about 3×10^7 cfu g^{-1} rice. Additionally the rice grain is covered with fungal hyphae that might continue to produce spores when embedded in soil. This might explain why waste rice can be a rather efficient treatment. Economically, this method would be quite interesting, if the extracted spores could be marketed for a different purpose.

Conclusions

Whether because of global warming or other anthropogenic influences, the fact is that termites are expanding on a worldwide scale, particularly in the tropics. Microbial control using entomopathogenic fungi is an interesting option for termite control in Africa. Common experience shows that there are no specialist strains of *M. anisopliae* or *B. bassiana*, as is the case for grasshoppers and locusts (Lomer *et al.*, 2001). An optimal strain should have two major properties. It should be virulent and kill termites, but this should not occur too rapidly. Contaminated or infected termites should survive long enough to infect a sufficient number of nest mates to trigger a fungal epizootic (Carruthers and Soper, 1987; Ebert, 1994). Similarly, the degree of sporulation of the fungus on the cadaver is important. On *M. subhyalinus*, Bioblast^{TM}, for example, the commercial product, did not sporulate well, and it killed termites too quickly. On the other hand, ICIPE 30 is well suited for termite control, because it satisfies both requirements. A major constraint for microbial termite control is the complex behavioural response of termite colonies to the presence of fungal pathogens. The key to a future success of microbial termite control lies in finding ways of triggering epizootics in termite colonies more effectively and avoiding behavioural defence mechanisms by the termites. It has been demonstrated that the repellent effect of *M. anisopliae* spores could be potentially used to reduce termite damage in the field. This effect is, however, too short-lived to protect timber, compared with the length of protection given by synthetic products. The application of *Metarhizium*-based products to *M. subhyalinus* colonies had a long-term effect, but it took almost a year to see clear differences. It seems to be unlikely that farmers would be willing to wait such a long time. Further research should include synergistic combinations of *Metarhizium* with Imidacloprid (Kaakeh*et al.*, 1997; Ramakrishnan *et al.*, 1999). Low dosages of Imidaclorprid can have an inhibiting effect on termite grooming behaviour, improving the chance for fungal spores to attach on the insects' cuticle successfully.

Another constraint in the use of fungal-based biopesticides for termite control could be the relatively high cost. At the current production costs of approximately US\$200 kg^{-1} of dry spores of *M. anisopliae* (based on the experimental production of Green Muscle^{TM} for locust and grasshopper control in Africa), control strategies using mycoinsecticides are not cost effective even in developed countries, where consumers are able to pay much higher prices than subsistence farmers in Africa. Bioblast^{TM}, which was developed for the protection of houses in the USA, though much more expensive, has not been commercially successful and is no longer available on the US market. The use of waste products from a mycoinsecticide mass production plant could possibly be economically viable, if the extracted spores are also used for a different purpose. It is clear from this review that fungal pathogens have the potential to play a role in the management of termites in Africa. It is likely that they will be only applicable in an integrated approach, combining measures to increase plant resistance through the application of fertilizers; making use of traditional methods such as luring predatory ants into termite colonies and the application of synergistic combinations. Additionally, research is needed to develop packages that are suitable for different termite species, cropping systems and wooden structures.

References

Abe, T. and Darlington, J.P.E.C. (1985) Distribution and abundance of a mound-building termite, *Macrotermes michaelseni*, with special reference to its subterranean colonies and ant predators. *Physiological Ecology Japan* 22, 59–74.

Atu, U.G. (1993) Cultural practices for the control of termite (Isoptera) damage to yams and cassava in south-Eastern Nigeria. *Journal of International Pest Management* 39, 462–446.

Bodot, P. (1967) Étude écologique des termites des savanes de basse Côte d'Ivoire. *Insectes Sociaux* 14, 229–258.

Carruthers, R.I. and Soper, R.S. (1987) Fungal diseases. In: Fuxa, J.R. and Tanada, Y. (eds) *Epizootiology of Insect Diseases*. John Wiley & Sons, New York, USA, pp. 357–416.

Coaton, W.G.H. (1958) *The Hototermitid Harvester Termites of South-Africa*. Union of South Africa, Department of Agriculture, Entomological Series No. 43, Bull. 375.

Collins, N.M. and Wood, T.G. (1984) Termite damage and crop loss studies in Nigeria: assessment of damage to upland sugarcane. *Tropical Pest Management* 30, 26–28.

Ebert, D. (1994) Virulence and local adaptation of a horizontally transmitted parasite. *Science* 265, 1084–1086.

Edward, R. and Mill, A.E. (1986) *Termites in Buildings: Their Biology and Control*. Rentokil Limited, East Grinstead, UK.

Eggelton, P. and Bignell, D.E. (1995) Monitoring the response of tropical insects to changes in the environment: troubles with termites. In: Harrington, R. and Stork, N.E. (eds) *Insects in a Changing Environment*. Academic Press, London, UK, pp. 473–497.

Eggelton, P., Bignell, D.E., Sands, W.A., Waite, B., Wood, T.G. and Lawton, J.H. (1995) The species richness of termites (Isoptera) under differing levels of forest disturbance in the Mbalmayo Forest Reserve, southern Cameroon. *Journal of Tropical Ecology* 11, 85–98.

Emerson, A.E. (1956) Regenerative behavior and social homeostasis of termites. *Ecology* 37, 248–258.

Ene, J.C. (1963) *Insects and Man in West Africa*. Ibadan University Press, Nigeria.

Fernandes, P.M. (1991) Controle microbiano de *Cornitermes cumulans* (Koller, 1832) utilizando *Beauveria bassiana* (Bals.) e *Metarhizium anisopliae* (Metsch.) Sorok. PhD dissertation, ESALQ-Univ. de São Paulo, Pracicaba, Brazil.

Gethi, M., Gitonga, W. and Ochiel, G.R.S. (1995) Termite damage and control options in Eastern Kenya. In: Mailu, A.M., Gitonga, W. and Fugoh, P.O. (eds) *Proceedings of the 2nd Regional Workshop on Termite Research and Control*. KARI, Nairobi, Kenya, pp. 49–51.

Gitonga, W., Maniania, N.K., Eilenberg, J. and Ochiel, G.R.S. (1995) Pathogenicity of indigenous entomopathogenic fungi *Metarhizium anisopliae* and *Beauveria bassiana* to *Macrotermes michaelseni* in Kenya. In: Mailu, A.M., Gitonga, W. and Fugoh, P.O. (eds) *Proceedings of the 2nd Regional Workshop on Termite Research and Control*. KARI, Nairobi, Kenya, pp. 21–28.

Grace, J.K. (1997) Biological control strategies for suppression of termites. *Journal of Agricultural Entomology* 14, 281–289.

Grace, J.K. and Zoberi, M.H. (1992) Experimental evidence for transmission of *Beauveria* by *Reticulitermes* workers (Isoptera: Rhinotermitidae). *Sociobiology* 20, 23–28.

Grassé, P.P. (1946) Sociétés animales et l'effet de groupe. *Experientia* 2, 77–82.

Hänel, H. (1981) A bioassay for measuring the virulence of the insect pathogenic fungus

Metarhizium anisopliae (Metsch.) Sorok. (fungi imperfecti) against the termite *Nasutitermes exitiosus* (Hill) (Isoptera, Termitidae). *Zeitschrift für angewandte Entomologie* 92, 9–18.

Hänel, H. and Watson, J.A. (1983) Preliminary field tests on the use of *Metarhizium anisopliae* for the control of *Nasutitermes exitiosus* (Hill) (Isoptera: Termitidae). *Bulletin of Entomological Research* 73, 305–313.

Hartwig, E.K. (1965) Die nesstesel van die grasdraertermiet *Hodotermes mossambicus* (Hagen) (Isoptera) en aspekte rakende bestryding. *Suid-Afrikaanse Tydskrif vir Landbouw* 8, 643–660.

Iroko, A.F. (1996) *L'homme et les termitières en Afrique*. Editions Karthala, Bénin.

Johnson, R.A. (1981) Termite damage and crop loss studies in Nigeria. The incidence of termites scarified groundnuts pods and resulting kernel contamination of field and market samples. *Tropical Pest Management* 27, 343–350.

Kaakeh, W., Reid, B.L., Bohnert, T.J. and Bennett, G.W. (1997) Toxicity of imidacloprid in the German cockroach (Dictyoptera: Blattellidae), and the synergism between imidacloprid and *Metarhizium anisopliae* (Imperfect Fungi: Hyphomycetes). *Journal of Economic Entomology* 90, 473–482.

Kooyman, C. and Onck, R.F.M. (1987) The interactions between termite activity, agricultural practices and soil characteristics in Kisii District, Kenya. *Agricultural University Wageningen Papers* 87(3).

Kramm, K.R. and West, D.F. (1982) Termite pathogens effects of ingested *Metarhizium*, *Beauveria* and *Gliocladium* conidia on worker termites (*Reticulitermes* spp.). *Journal of Invertebrate Pathology* 40, 7–11.

Kramm, K.R., West, D.F. and Rockenback, P.G. (1982) Termite pathogens: transfer of the entomopathogen *Metarhizium anisopliae* between *Reticulitermes* sp. termites. *Journal of Invertebrate Pathology* 40, 1–6.

Krishna, K. and Weesner, F.M. (1969) *Biology of Termites*, Vol. 1. Academic Press, New York, USA.

Lee, K.E. and Wood, T.G. (1971) *Termites and Soil*. Academic Press, London, UK.

Lepage, M.G. (1981a) The impact of foraging populations of *Macrotermes michaelseni* (Sjostedt) (Isoptera: Macrotermitinae) in a semi-arid ecosystem (Kajiado-Kenya). I – Foraging activity and factors determining it. *Insectes Sociaux* 28, 297–308.

Lepage, M.G. (1981b) The impact of foraging populations of *Macrotermes michaelseni* (Sjostedt) (Isoptera: Macrotermitinae) in a semi-arid ecosystem (Kajiado-Kenya). II – Food gathered, in comparison with large herbivores. *Insectes Sociaux* 28, 309–319.

Lenz, M., Creffield, J.W., Zhong Yun-hong and Miller, L.R. (1993) Establishing standard principles for laboratory bioassays of termiticides with subterranean termites – progress, problems and prospects. Paper prepared for the 24th Annual Meeting, Orlando, Florida, USA, 16–20 May.

Logan, J.W.M. (1991) Damage to sorghum by termites (Isoptera: Macrotermitinae) in the lower Sire Valley, Malawi. *Sociobiology* 19, 2.

Logan, J.W.M., Cowie, R.H. and Wood, T.G. (1990) Termite (Isoptera) control in agriculture and forestry by non-chemical methods: a review. *Bulletin of Entomological Research* 80, 309–330.

Lomer, C.J., Bateman, R.P., Johnson, D.L., Langewald, J. and Thomas, M. (2001) Biological control of locusts and grasshoppers. *Annual Review of Entomology* 46, 667–702.

Longhurst, C., Johnson, R.A. and Wood, T.G. (1978) Predation by *Megaponera foetens* (Fabr.) (Hymenoptera: Formicidae) on termites in the Nigerian southern Guinea savannah. *Oecologia* 32, 101–107.

Mailu, A.M., Gitonga, W. and Fungoh, P.O. (1995) *Second Regional Workshop of Termite Research and Control*. KARI, Nairobi, Kenya, 1995.

Maniania, N.K., Ekesi, S. and Songa, J.M. (2002) Managing termites in maize cropping systems with the entomopathogenic fungus, *Metarhizium anisopliae*. *Insect Science and its Application* 21, 41–46.

Mauldin, J.K. and Beal, R.H. (1989) Entomogenous nematodes for control of subterranean termites, *Reticulotermes* spp. (Isoptera: Rhinotermitidae). *Journal of Economic Entomology* 82, 1638–1642.

Milner, R. and Staples, J.A. (1995) An overview on biological control of termites. In: Mailu, A.M., Gitonga, W. and Fungoh, P.O. (eds) *Second Regional Workshop of Termite Research and Control*. KARI, Nairobi, Kenya, 1995, pp. 21–28.

Milner, R. and Staples, J.A. (1996) Biological control of termites: results and experiences within a CSIRO project in Australia. *Biocontrol Science and Technology* 6, 3–9.

Milner, R., Staples, J.A. and Lutton, G.G. (1997) The effect of humidity on germination and infection of termites by the Hyphomycete, *Metarhizium anisopliae*. *Journal of Invertebrate Pathology* 69, 64–69.

Milner, R., Staples, J.A. and Lutton, G.G. (1998) The selection of an isolate of the Hyphomycete fungus, *Metarhizium anisopliae*, for control of termites in Australia. *Biological Control* 11, 240–247.

Milner, R.J. (2000) Current status of *Metarhizium* as a mycoinsecticide in Australia. *Biocontrol News and Information* 21, 47N–50N.

Milner, R.J. and Lutton, G.G. (1976) *Metarhizium anisopliae*. Survival of conidia in the soil. In: *Proceedings of the 1st International Colloquium on Invertebrate Pathology*. Queen's University Press Kingston, Ontario, Canada, pp. 428–429.

Mitchell, J.D. (2002) Termites as pests of crops, forestry, rangeland and structures in Southern Africa and their control. *Sociobiology* 40, 47–69.

Pearce, M.J. (1997) *Termites, Biology and Pest Management*. CAB International, Wallingford, UK.

Ramakrishnan, R., Suiter, D.R., Nakatsu, C.H., Humber, R.A. and Bennett, G.W. (1999) Imidacloprid-enhanced *Reticulitermes flavipes* (Isoptera: Rhinotermitidae) susceptibility to the entomopathogen *Metarhizium anisopliae*. *Journal of Economic Entomology* 92, 1125–1132.

Rath, A.C. (2000) The use of entomopathogenic fungi for control of termites. *Biocontrol Science and Technology* 10, 563–581.

Rosengaus, R.B., Guldin, M.R. and Traniello, J.F.A. (1998) Inhibitory effect of termite fecal pellets on fungal spore germination. *Journal of Chemical Ecology* 10, 1697–1706.

Rosengaus, R.B., Traniello, J.F.A., Chen, T., Brown, J.J. and Karp, R.D. (1999) Immunity in social insects. *Naturwissenschaften* 86, 588–591.

Sands, W.A. (1977) The role of termites in tropical agriculture. *Outlook on Agriculture* 9, 136–143.

Sekamatte, M.B., Latigo, M. and Russel-Smith, A. (2001) The potential of protein- and sugar-based baits to enhance predatory ant activity and reduce termite damage to maize in Uganda. *Crop Protection* 20, 653–662.

van Huis, A. (1996) The traditional use of arthropods in sub-Saharan Africa. *Proceedings of the Section Experimental and Applied Entomology, the Netherlands Entomological Society* 7, 3–19.

Vanninen, I. (1995) Distribution and occurrence of four entomopathogenic fungi in Finland: effect of geographical location, habitat type and soil type. *Mycological Research* 100, 93–101.

Veen, K.H. and Ferron, P. (1966) A selective medium for the isolation of *Beauveria tenella* and of *Metarhizium anisopliae*. *Journal of Invertebrate Pathology* 8, 268–269.

Walker, M. (1999) How termite head-bangers outwit a fungus. *New Scientist* 13 November, 1999, 17.

Wood, T.G., Johnson, R.A. and Ohiagu, C.E. (1980) Termites damage and crop loss studies in Nigeria: a review of termites (Isoptera) damage to maize and estimation of damage, loss in yield and termite (*Microtermes*) abundance at Mokwa. *Tropical Pest Management* 26, 355–370.

Wood, T.G. and Pearce, M.J. (1991) Termites in Africa: the environmental impact of control measures and damage to crops, trees, rangeland and rural buildings. *Sociobiology* 19, 221–234.

Zimmermann, G. (1986) The 'Galleria bait method' for detection of entomopathogenic fungi in soil. *Journal of Applied Entomology* 102, 213–215.

16

IPM of Banana Weevil in Africa with Emphasis on Microbial Control

Clifford S. Gold,[1] Caroline Nankinga,[1] Björn Niere[1] and Ignace Godonou[2]

[1]*International Institute of Tropical Agriculture – Eastern and Southern Africa Regional Center (ESARC), Namulonge, Kampala, Uganda;* [2]*CAB International, Nairobi, Kenya*

Introduction

The East African highland banana (*Musa* spp., genome group AAA-EA, *Musaceae*) is the principal staple crop of the Great Lakes region of eastern Africa, while plantain (*Musa* spp., genome group AAB) is an important food in West and Central Africa. The banana weevil *Cosmopolites sordidus* (Germar) (Coleoptera, Curculionidae) (Plate 47) is the most important insect pest of highland banana and plantain in Africa. The larvae bore in the corm, reducing nutrient uptake and weakening the stability of the plant.

The banana weevil originated in South-East Asia and was introduced into Africa at least 100 years ago. Salient features of the weevil's biology (reviewed by Gold *et al.*, 2003) include narrow host range, nocturnal activity, long life span (up to 4 years), limited mobility, low fecundity (<2 eggs week^{-1}) and slow population growth. Adults are free-living and most often found in the leaf sheaths, at the base of the banana mat or associated with cut residues. The adults rarely fly and few crawl more than 50 m in a year. The weevils are hydrotrophic and occur in higher numbers in mulched systems. They are attracted to their hosts by plant volatiles, especially those emanating from damaged corms. Males produce an aggregation pheromone that is attractive to both sexes. Eggs are laid in the corm or lower pseudostem. The immature stages are within the host plant, mostly in the corm. Dissemination is through infested planting material. The weevil's biology makes its control difficult.

Banana weevils are monitored by trapping adults with pseudostem or corm residues or by population estimates using mark and recapture methods (Gold *et al.*, 2003). In a survey in Ntungamo district, Uganda, adult populations across farms ranged from 1600 to 149,000 ha^{-1} with a median density of 15 weevils per mat. This wide variation under relatively uniform ecological condi-

(eds P. Neuenschwander, C. Borgemeister and J. Langewald)

tions suggests that management influences weevil density. Although action thresholds based on adult captures in traps are widely reported, pest status is most commonly determined by assessing damage to the corms of recently harvested plants (Gold *et al.*, 2003).

Bananas are herbaceous plants that are vegetatively propagated. A mat consists of an underground corm from which one or more plants (shoots) emerge. The apparent stem or pseudostem is composed of leaf sheaths. The true stem arises from the apical meristem after leaf production has terminated and grows through the centre of the pseudostem. The true stem bears a single terminal inflorescence. After the fruit matures, the stem dies back to the corm. New plants are produced by suckers emerging from lateral buds in the corm. These can be left *in situ* (i.e. ratoon crops) or serve as a source of planting material, in which case they are removed and planted elsewhere. Suckers used to establish new fields are called the plant crop.

Banana weevil attack in newly planted banana stands can lead to poor crop establishment. McIntyre *et al.* (2002) observed 40% mortality of highland banana suckers from weevils remaining from a previous planting. In established fields, weevil damage can result in plant loss due to snapping and toppling, lower bunch weight, mat die-out and shortened plantation life. Damage and yield losses tend to increase with time. For example, Rukazambuga *et al.* (1998) found that yield loss increased from 5% in the plant crop to 47% in the third ratoon (Table 16.1). More than 35% of highland banana mats died out in 5 years in plots infested with weevils, compared with 2% mat loss in controls (C. Gold, 2001, Namulonge, unpublished results). This suggests that the weevil can severely reduce stand life.

Current research results suggest that no single control strategy will provide complete control of banana weevil. Therefore, a well-defined IPM approach might offer the best chance for success. For small-scale African producers of highland banana or plantain, the components of such a programme include cultural, botanical and biological control and host plant resistance.

Table 16.1. Banana weevil damage and associated highland banana (cv. Atwalira) yield loss across four crop cycles in Uganda.

	Crop cycles			
	Plant crop	1	2	3
Damage (%)	4	10	15	29
Plant loss (%)	3	9	19	29
Bunch weight reduction (%)	3	7	7	25
Yield loss (%)	5	15	25	47

Adapted from Rukazambuga *et al.* (1998) and Rukazambuga, unpublished data sets. Damage represents mean cross-section weevil damage by banana mat scored as percentage of banana larvae consumed tissue in the central cylinder of a corm from a freshly (<2 weeks postharvest) harvested banana plant. Damage score method as described by Rukazambuga *et al.* (1998).

Cultural Control and Botanicals

Clean planting material

Infested planting material provides the principal entry point of banana weevils into new plantations. The use of clean planting material can reduce initial weevil infestation levels and, given the insect's low fecundity and slow population growth, can retard pest build-up for several ratoon cycles. As a result, the use of clean planting material has been widely recognized and promoted. The most important methods for ensuring that plants are weevil-free include the use of tissue culture plantlets, selection of weevil-free suckers, paring (i.e. removal of the leaf sheaths and outer surface of the corm) and hot-water treatment. Banana weevils are attracted to cut corms and may quickly re-infest pared or hot-water-treated planting material if these are left in an area exposed to weevils. Therefore, quick planting of treated material is recommended. Re-infestation from nearby fields remains a concern.

Tissue culture material is widely used for commercial banana production to keep new plantings free of weevils and other pests. This technology is being promoted for small farmer production in parts of East Africa but availability, costs and adoption levels are not yet clear. Alternatively, paring of suckers will eliminate most weevil eggs and many first-instar larvae. However, larvae deep within the corm will not be removed by paring. Immersion of pared suckers in hot-water baths (e.g. 52–55°C for 20–27 min) has been recommended to kill these larvae and for nematode control (Gold *et al.*, 2003). In Uganda, however, hot-water baths killed only 32% of the weevil larvae in the corm. In a field trial, plots planted from pared and hot-water-treated suckers had lower weevil damage than controls during the plant crop, but not during the first ratoon. Thus, the use of clean planting material is not likely to provide long-term control of banana weevils.

Trapping

Given the low fecundity of the banana weevil, trapping of adults has been proposed as a control measure. Such trapping is most often carried out with crop residues (Gold *et al.*, 2003). Enhanced trapping, using chemicals, biopesticides or semiochemicals, has also been proposed. In a 1-year on-farm study in Uganda, weevil numbers declined by 61% on farms following monthly trapping (one pseudostem trap per mat) compared with a 38% decline on farms with no trapping; however, results varied greatly among farms and, overall, there was no significant effect of trapping on weevil numbers (Gold *et al.*, 2002). There was only a weak relationship between the number of weevils removed and the change in population density. Trapping success appeared to be affected by management levels and immigration from neighbouring farms. Moreover, few farmers adopted trapping following this study because of labour and material requirements. The use of semiochemicals (i.e. pheromone lures and plant volatiles) in trapping and delivery of entomopathogens is currently under study.

Sanitation

It is widely believed that destruction of crop residues (splitting of harvested pseudostems and/or removal of corms) eliminates adult refuges, food sources and breeding sites, lowers overall weevil populations and reduces damage on standing plants in susceptible clones. Gold *et al.* (1997) found that sanitation had more impact on weevil pest status than any other agronomic practice in a survey in Ntungamo district. Similarly, in Masaka district (W. Tinzaara and C. Gold, 2001, Namulonge, unpublished results) lowest levels of weevils were found on those farms that employed the highest levels of sanitation. The role of crop sanitation on weevil population dynamics and related damage is currently under study in Uganda.

Botanicals

In Kenya, neem (*Azadirachta indica* A. Juss., Meliaceae) seed derivatives reduced banana weevil damage by interfering with each stage of the attack: (i) fewer adults located or remained at treated plants; (ii) oviposition per female was reduced on treated plants; (iii) lower egg-hatching rates were observed; (iv) an antifeedant effect delayed initiation of feeding; and (v) treated plants had higher levels of larval mortality (Musabyimana, 1999). Reduced larval survival suggests that neem derivatives also have a systemic effect. In Cameroon, neem applications were followed by reduced sucker mortality of >70% and total plant mortality by 50% (S. Messiaen, R. Fogain, H. Ysenbrandt and P. Sama Lang, 2000, Njombe, unpublished results). Musabyimana (1999) recommended applications of 60–100 g of neem seed powder and neem cake to the soil at the base of each mat at 4-monthly intervals.

Biological Control with Arthropods

The use of arthropod natural enemies against banana weevil has been reviewed by Gold *et al.* (2003). Searches for natural enemies in South-East Asia during the first half of the 20th century uncovered several generalist predators of which the most important was *Plaesius javanus* Erichson (Coleoptera, Histeridae). Classical biological control attempts with *P. javanus* and seven other predators were unsuccessful. Recently, no parasitoids were found in 19,000 eggs and 1500 larvae that were laboratory-reared following collection from banana stands in Indonesia (A. Abera and C. Gold, 2001, Namulonge, unpublished results).

Koppenhofer *et al.* (1992) listed 12 coleopteran and dermapteran predators of banana weevil in western Kenya. Their potential for controlling banana weevil was limited. In contrast, the myrmicine ants *Tetramorium guineense* (Mayr) and *Pheidole megacephala* (Fabricius) have been reported as reducing weevil populations and damage by >60% in Cuba (Castineiras and Ponce, 1991). The control potential of myrmicine ants in Uganda is currently under study.

Microbial Control with Entomopathogens

The entomopathogens *Beauveria bassiana* (Balsamo) Vuillemin (Plate 48) and *Metarhizium anisopliae* (Metsch.) Sorokin have gained considerable attention as biological control agents for weevils and other agricultural pests. These are especially important for controlling cryptic insects, such as banana weevil, which are not accessible to arthropod natural enemies. The use of entomopathogens against banana weevil has been reviewed by Nankinga (1999) and Gold *et al.* (2003). Strains of *B. bassiana* and *M. anisopliae* were first tested against banana weevil in the 1970s. Since then, numerous strains have been tested using a wide variety of formulations, dosages, application methods and delivery systems under a wide range of conditions. In general, it is more feasible to apply these entomopathogens against adults than to larvae, even though these entomopathogens have been shown to be effective against immature stages in laboratory experiments.

The development of a successful microbial control programme involves a series of steps including: (i) isolation and screening of pathogen strains that have good growth characteristics, sporulation and high level of virulence to the target pest; (ii) development of efficient mass production and formulation systems that can yield spores that remain viable and infective during storage and under field conditions; (iii) development of economical and effective field delivery systems; and (iv) selection of a pathogen strain that is harmless to non-target organisms and the overall ecosystem where it is used. Various studies have been undertaken to address some of the above requirements.

Isolation and pathogenicity

Entomopathogenic fungi tested for microbial control can be isolated from dead banana weevils, from other insects, or from soils in banana fields using soil-baiting techniques. For example, Nankinga (1999) isolated from soil samples 40 strains of *B. bassiana* and 11 of *M. anisopliae* using *Galleria* larvae as bait. Six additional *B. bassiana* and four *M. anisopliae* strains were isolated from dead banana weevils and two *B. bassiana* strains were obtained from the sweet potato weevil *Cylas puncticollis* Boheman.

Numerous strains of entomopathogens have been tested against the banana weevil in Africa, the Americas and Australia. Under laboratory conditions, many strains cause more than 50% mortality within 5–10 days of exposure (Nankinga, 1999; Godonou, 1999), with the best strains resulting in >90% mortality in 3 weeks or less. In general, *B. bassiana* isolates have been found to be more infective to the banana weevil than *M. anisopliae* isolates.

Pathogenicity of fungal isolates has been observed to be influenced by: (i) the source of the isolate; (ii) the method of culturing; (iii) spore dose; (iv) temperature; (v) substrate and formulation; and (vi) method of application (Godonou, 1999). Although *B. bassiana* isolates are often most pathogenic to the original or related hosts, some of the best performing strains against banana weevil have been isolated from non-coleopterans. In Uganda, indigenous isolates consistently gave higher weevil mortalities than most exotic strains. C.M. Nankinga (1994, Kampala, unpublished results) found weevil mortality levels to be influenced by spore dose. For example, a dose of 3.35×10^7 spores ml^{-1} for one isolate resulted in 96% weevil mortality, compared with <20% mortality at 3.35×10^4 spores ml^{-1}.

Some *B. bassiana* isolates are infective to immature stages of the banana weevil. Nankinga (1999) found infected larvae and pupae in banana suckers where maize- and soil-based *B. bassiana* formulations had been applied to the planting hole. Dusting suckers with *B. bassiana* spores prior to planting resulted in 41% egg and 19% larval infection. In Ghana, soaking of corm and pseudostem pieces in water-based spore formulations resulted in 46% egg and 27% larval mortality (Godonou, 1999). Nevertheless, developing field-delivery systems against weevil immatures, especially in established stands, would be difficult. As a result, targeting of adult weevils remains the principal strategy.

Characterization and strain selection

Candidate fungal strains for banana weevil control should have good characteristics that would allow mass-production, persistence under a range of field conditions and virulence to the banana weevil at reasonable doses, resulting in rapid mortality. The last characteristic is important as *B. bassiana*-inoculated weevils continue to oviposit until infection is advanced (Nankinga, 1999).

Characterization using biochemical and molecular methods can generate genetic profiles that can be related to virulence levels of different strains, leading to more rapid selection of useful isolates. The strain identity can also be useful in monitoring of the pathogen in the field. Preliminary characterization, based on PCR and isoenzyme analysis, showed that *B. bassiana* isolates from Uganda soil baited by the *Galleria* larvae were very similar, and could be distinguished from strains isolated from the banana weevil or other hosts (Nankinga, 1999).

Influence of environmental factors

Entomopathogenic fungi are sensitive to sunlight, high temperatures, soil amendments, organic material and moisture levels. Thus, it is necessary to investigate how such physical and biological factors in banana fields might influence the persistence and efficacy of *B. bassiana*. Nankinga (1999) found that direct exposure of dry maize-based *B. bassiana* formulations to sunlight reduced the germination and viability of the spores. *B. bassiana* conidia grew well between 19 and 30°C but were adversely affected by temperatures above 37°C. This suggests that fungal establishment may be reduced following applications on hot, sunny days or in younger stands with open canopies. However, once the fungus has been established in the field, spores may be protected by cooler microenvironments (e.g. mulches).

B. bassiana applied as a maize-based formulation (grown on maize and applied as conidia on dry substrate) has been observed to quickly degrade in unsterilized soil with high moisture content, suggesting antagonistic action of other soil organisms (Nankinga, 1999). Farmers in Uganda have a belief that application of concoctions made by mixing ash, tobacco, urine and various other types of organic matter can help in controlling the banana weevil and act as fertilizers to improve the health of the banana plant. However, Nankinga (1999) observed that applying *B. bassiana* spores to soils amended with ash and/or urine resulted in reduced spore germination and lower weevil mortality. She postulated that this might be due to an increase in soil pH that could have an adverse effect on conidial survival.

Mass production and formulation

For successful use of entomopathogenic fungi for insect control, it is important to select a strain that displays rapid growth, abundant sporulation and high virulence to the target pest and can be economically mass-produced. There are three well-tried methods of mass-producing entomopathogenic fungi: (i) solid substrate; (ii) liquid fermentation; and (iii) a two-stage system that involves the liquid and solid substrates for mass-production. The IITA, CAB *International* and the Ugandan National Banana Research Programme have adopted a two-phase system utilizing sucrose and yeast-based liquid media and various solid substrates for the production of *B. bassiana* (Godonou, 1999; Nankinga, 1999). Whereas dosage effects are important in producing satisfactory kill rates of banana weevils, systems producing high fungal yields at reasonable costs are critical.

In Ghana, two *B. bassiana* isolates (IMI330194 and 194–907) gave similar yields (i.e. $5 \times 10^8 - 6 \times 10^8$ conidia g^{-1} substrate) when grown on rice, but isolate IMI330194 gave higher conidia viability than 194–907 (Godonou, 1999). Therefore, the former would be a preferred strain for a microbial control programme. These studies also showed that *B. bassiana* can be produced on cassava starch suspension as liquid media and on oil palm kernel cake (OPKC) as a solid substrate yielding up to 1.8×10^9 conidia g^{-1} of the OPKC pre-treated with 100% water. Nankinga (1999) also screened a range of solid substrates such as barley, beans, cracked maize, millet, sawdust and sorghum inoculated with sucrose and yeast-based liquid media containing *B. bassiana* mycelia. Of these, only cracked maize gave acceptable fungal growth and yield (2.3×10^8 conidia g^{-1} of substrate).

Further screening of other locally available substrates has shown that maize bran and ‘machecha’ (millet waste from local breweries) can provide adequate yields (e.g. 2×10^7–10^8 conidia g^{-1} substrate), while cheaper materials such as ‘bagasse’ (residues from extracted sugar cane) and cotton seed waste (following oil extraction) gave unsatisfactorily low yields (C. Nankinga, 2001, Namulonge, unpublished results).

B. bassiana spores can be applied as a dry solid carbohydrate substrate used in culturing or the dry carbohydrate substrate with the fungus can be mixed with dry soil or clay to form dust formulations. Alternatively, the conidia can be harvested from the carbohydrate substrate and mixed with water or vegetable oil to form a liquid formulation that can be applied to the plant and soil around the plant with conventional sprayers.

Application methods can influence the infectivity of *B. bassiana* (Nankinga, 1999; Gold *et al.*, 2003). Nankinga (unpublished) reported that spraying weevils with *B. bassiana* spore suspensions resulted in higher mortality (56–62%) after 4 weeks than exposing weevils to treated pseudostem traps (10–19%) or soil (<10%). Exposing weevils to dry maize- and rice-based *B. bassiana* formulations resulted in nearly 100% mortality within 3 weeks (Nankinga and Ogenga-Latigo, 1996; Nankinga *et al.*, 1998).

In recent bioassays in Uganda, we observed comparable levels of infectivity of *B. bassiana* grown on locally available substrates. Cracked maize- and maize bran-based formulations of *B. bassiana* caused 90% and 55% weevil mortality, respectively, in 14 days. *B. bassiana* grown on maize bran formulated with clay or loam soil caused 50% weevil mortality, while that grown on cracked maize and formulated with clay caused 63% weevil mortality. Further research is being undertaken to find economical substrates for large-scale *B. bassiana* production and delivery.

Delivery systems and field evaluation

The development of economic and effective field delivery systems is among the most important steps in a microbial control programme. Field persistence of the fungus and assuring maximal chance of exposure of the target insect to the fungus are critical requirements in fungal efficacy. The banana weevil occurs at low densities and is a sedentary insect that may spend extended periods buried in the soil, hidden in leaf sheaths or otherwise inactive. Transmission of diseases from infected to non-infected weevils in the field has been demonstrated, but it is not yet clear how important this is (Godonou, 1999; Nankinga, 1999; Godonou *et al.*, 2000).

Nankinga (1999) and Godonou (1999) tested three main methods of delivering *B. bassiana* in pot and on-station field experiments by applying different fungal formulations to: (i) banana planting material; (ii) pseudostem and corm traps; and (iii) the soil around established banana mats. *B. bassiana* in the form of dry mycelium and conidia in dry maize or soil mycelial formulations or as water or oil spore suspensions were found to infect banana weevils when applied to banana suckers (Nankinga, 1999). For example, application of 150 g of dry maize- or soil-based formulations of *B. bassiana* conidia to the soil around banana suckers in pots resulted in 70–80% infection of adults. In contrast, application of water or vegetable spore suspensions resulted in 52–55% adult infection. Although oviposition levels were similar, treated plants supported fewer larvae and had less damage than untreated plants, suggesting a systemic effect of the fungus. Under field conditions, suckers protected by a dry maize-based formulation of *B. bassiana* spores had fewer banana weevil immatures and adults and less damage than untreated plants.

In Ghana, Godonou *et al.* (2000) conducted field studies to evaluate field efficacy of *B. bassiana* conidia on OPKC applied to planting holes and pure conidial powder applied to plantain suckers immediately prior to planting. Marked weevils were then released in each plot. After 28 days, 31% of the marked weevils were recovered from suckers treated with OPKC, of which 24% were infected. In contrast, suckers treated with conidial powder displayed a 23% recovery and 30% infection rate, while controls showed a 50% recovery rate and no infection. Godonou *et al.* (2000) estimated 76% mortality in the two *B. bassiana* treatments, compared with 1% in controls.

In a second experiment, Godonou *et al.* (2000) applied the same treatments to suckers planted among mature plantain plants in an established field. Suckers protected with *B. bassiana* conidia in OPKC substrates had higher rates of infection of adult weevils, lower damage, fewer attacked plants and less sucker mortality than either suckers treated with *B. bassiana* conidia or controls. From these results, Godonou *et al.* (2000) concluded that suckers could be protected at the critical stages of plant establishment by applications of conidial powder on an OPKC substrate. In related studies, Godonou (1999) evaluated the persistence of three formulations of *B. bassiana*, OPKC solid formulation, groundnut oil-based formulation of conidia and conidial powder applied to plantain suckers at planting. The OPKC formulation caused 83% mortality 14 days after application and showed a high level of field persistence. In contrast, the groundnut oil formulation and the conidial powder caused only 16% and 25% weevil mortality, respectively, in 14 days.

B. bassiana and *M. anisopliae* can be added to pseudostem and corm traps as a means of disseminating the fungus. Nankinga (1999) found that application of *B. bassiana* spores on cracked-maize, in vegetable oil or in water-suspension to pseudostem traps caused high infection to the weevils released in these traps. Highest mortality and *B. bassiana* infection (83%) was obtained from weevils collected from traps with maize culture, followed by traps with oil suspensions (60%) and water suspensions (47%). No weevil infection was observed in the control traps.

Under field conditions, application of these *B. bassiana* formulations to traps caused lower infection levels (<15% within 14 days) than those obtained in pot experiments where 20–50% weevils were killed within 7 days. However, further incubation of the field-collected weevils revealed that 70–90% of these weevils were infected, suggesting that the development of mycosis was slower in the field. Moreover, the incidence of *B. bassiana*-infected weevils in the field was higher in the wet season than in the dry season, implying that moisture might influence the performance of *B. bassiana*. The moist conditions under the traps provided a favourable environment for extra sporulation and this probably enhanced fungal efficacy. *B. bassiana* infectivity was observed up to 5 weeks after its application to traps.

In another field trial, Nankinga (1999) applied 500 g of maize-based formulation of *B. bassiana* (10^{15} conidia ha^{-1}), 500 g of a soil-based formulation (10^{14} conidia ha^{-1}) or 30 ml of an oil formulation (10^{15} conidia ha^{-1}) to the topsoil around banana mats in small plots (i.e. eight mats) covered with grass and banana trash mulch. Weekly trapping of weevils was done for 8 months. Mean weevil counts were lowest in plots treated with the maize formulation, followed by plots receiving the soil formulation, oil formulation and controls (Table 16.2). These data demonstrated significant decline of banana weevils in *B. bassiana*-treated banana plots relative to the untreated controls. Maize-based formulations also tended to reduce weevil damage levels in the banana corm central cylinder and cortex (Table 16.2). Although the maize-based *B. bassiana* formulation showed potential for field level control, the application rate employed in this study was not economically affordable (Nankinga and Moore, 2000).

The way forward with entomopathogens

The studies conducted to date suggest that good potential exists for use of *B. bassiana* as a microbial control agent against banana weevil, and that entomopathogens would fit well within the IPM programme being developed for this pest. Further research will be undertaken to establish economically viable mass-production and formulation systems for *B. bassiana* and to explore banana weevil IPM options (e.g. cultural control option) that can be integrated with *B. bassiana* under different ecological conditions. In particular, we are developing economically viable delivery systems to overcome the problems associated with field fungal efficacy, persistence and disease transmission. Studies on behaviour of the weevil, which might influence the chances of the insect contacting pathogen, and the effects of the pathogen on different weevil biotypes will be undertaken. The applicability of integrating *B. bassiana* with pheromone- and kairomone-based traps will be investigated.

Endophytic Fungi for the Biological Control of the Banana Weevil

Endophytes are microorganisms that colonize living plant tissue at some time in their life cycle without causing symptoms of damage. It is assumed that all higher plants host endophytic fungi. Endophytes have been isolated from all plant tissues including roots, corms, stems and leaves. Grass endophytes are the best-studied group of endophytes and have become well known because of

Table 16.2. Banana weevil population counts and damage in the banana field treated with *Beauveria bassiana* at Kawanda Agricultural Research Institute in Uganda.

		Weevil damage to pre-flowered plants	
B. bassiana formulation	Weevils trapped per plot	Cortex (%)	Central cylinder (%)
Maize-based	40 d	1 ± 2.9	0.1 ± 0.6
Soil-based	54 c	2 ± 2.6	0.5 ± 1.02
Oil-based	67 b	2 ± 2.5	0.9 ± 1.18
Untreated (control)	80 a	9 ± 6.6	5.1 ± 4.68

Means followed by same letters are significantly different by separation of log-transformed weevil counts. Source: Nankinga (1999).

their mutualistic symbioses with their hosts (Clay, 1991). Grass endophytes can colonize their host systemically and produce a number of toxic substances that are suspected to act as oviposition repellents, toxins or feeding deterrents.

Endophytes of crop plants and their interaction with other (pest) organisms are generally less well understood than endophytes in grasses. Systemic infection has not been proven in this group of endophytes, but it must not be excluded. Secondary metabolites produced under *in vitro* conditions by endophytic fungi from tomato were shown to have an inactivating effect on nematode pests (Hallmann and Sikora, 1996). Similarly, culture filtrates from endophytes of banana caused high inactivation and mortality of the burrowing nematode *Radopholus similis* Cobb (Schuster *et al.*, 1995). Whether or not the production of toxins is the prevalent mode of action under natural conditions has not been proved. Mutualistic interactions of banana endophytes, their host plant and pests are currently being investigated at IITA in Uganda and at the University of Bonn, Germany.

Griesbach (2000) isolated 200 fungal endophytes from corm tissue of East African highland bananas and Pisang Awak (*Musa* ABB). Most plants supported two to four distinct isolates in the material sampled. More endophytes were isolated from the central cylinder than the corm cortex. In screening experiments involving spore suspensions of these isolates, 12 isolates (eight *Fusarium* spp., three *Acremonium* spp., one *Geotrichum* sp.) caused mortality rates of >80% in weevil eggs. Larvae were less affected by spore suspensions. The most effective isolates tended to originate from the corm cortex and all showed proteolytic activity. This may explain why spore suspensions caused higher mortality rates in eggs than autoclaved culture filtrates. The single most important species appeared to be *Fusarium oxysporum* Schlecht.

F. oxysporum is a cosmopolitan fungus and has been reported as an endophyte of many crop plants including tomato, banana, rice and maize. Although the fungus is notorious as the causal agent of vascular wilts or root rots of many plants, the majority of *F. oxysporum* strains are non-pathogenic soil inhabitants. Non-pathogenic isolates of *F. oxysporum* are some of the best-studied microorganisms used for the biological control of soil-borne plant diseases (e.g. Alabouvette *et al.*, 1998). The fungus is also reported as a pathogen of insect eggs and larvae and endophytic isolates are effective biological control agents of nematodes (see Sikora *et al.*, this volume).

The utilization of endophytic fungi depends on their non-pathogenicity towards the target as well as other susceptible crops grown in the area. Rigorous screening protocols are applied to all candidate strains and aim at excluding pathogenic strains of *F. oxysporum* from further testing. Currently these involve pathogenicity testing and the use of genetic markers (vegetative compatibility groups); molecular markers are being developed to distinguish banana endophytes from *F. oxysporum* f. sp. *cubense*, the causal agent of Panama disease (*Fusarium* wilt) of banana.

Protocols for the successful inoculation of endophytes on to tissue-cultured banana plants have been established. Little inoculum is needed and the inoculation process can easily be integrated in the production of tissue-culture plants. At present colonization rates are difficult to establish precisely due to difficulties in strain differentiation of *F. oxysporum*, but were generally high in 4-week-old plants and lower in 20-month-old plants (Niere, 2001). Inoculation of other planting materials, such as banana suckers, may not lead to successful establishment of the endophyte (Griesbach, 2000). Therefore, the use of fungal endophytes for the control of banana weevil must be combined with the use of tissue-culture plantlets.

The main advantages of an ideal endophyte for weevil control over classical biocontrol agent or biopesticides can be summarized as follows: (i) initial application requires little inoculum and re-applications are not necessary; (ii) endophytes are well adapted to their ecological niche and target the damaging (i.e. larval) stage of the banana weevil; (iii) efficacy of endophytes is not dependent on target pest densities; (iv) endophytes have a broad host range and are antagonistic to different pests and diseases (e.g. to both banana weevils and nematodes).

Side-effects on the environment are greatly reduced because endophytes are delivered together with clean tissue-cultured plants. This approach combines two IPM strategies: exclusion of the pest and delivery of a biological control agent for extended protection.

The mechanisms leading to egg and larval mortality or reduced weevil damage have not been established. It is possible that endophytes have enzymatic activity or produce secondary metabolites. They might also induce resistance in banana plants to pests, but neither of these mechanisms could be demonstrated. Despite promising laboratory experiments, field and pot experiments with endophytes for biological control of the banana weevil have not produced conclusive results so far.

Whether the use of fungal endophytes for banana weevil control in African agriculture will have an impact remains to be seen. The search for new effective strains of endophytic fungi, and improved screening protocols for the selection of candidate strains that give consistent and good weevil control under field conditions, have high priority. The full potential of this technology can only be shown when banana tissue-culture plantlets with endophytes integrated into them are made readily available to African farmers.

Host Plant Resistance

Although most highland banana and plantain cultivars are susceptible to banana weevil, many other *Musa* clones, including Calcutta-4 (AA), Kisubi (AB) Yangambi-Km5 (AAA), Kayinja (ABB), Pisang awak (ABB) and FHIA-03 (AABB), are highly resistant (A. Kiggundu, 2000, Orange Free State University, unpublished results; Gold *et al.*, 2003). Antibiosis appears to be the major pathway by which resistance occurs in these clones. In Kayinja, biochemical compounds within the corm lead to high rates of larval failure (A. Kiggundu and C. Gold, 2001, A. Hassanali, ICIPE, Nairobi, unpublished results). In parts of East Africa, Kayinja, Kisubi and FHIA-03 have gained some acceptance as replacements of highland banana, although the first two are used for beer and not food production. Crop improvement, through conventional and non-conventional means, to deliver acceptable weevil-resistant clones to farmers has been recently initiated. Thanks to new breeding procedures that make crossing of *Musa* possible (Ortiz and Vuylsteke, 1996), this represents a promising long-term strategy.

Conclusions

Highland banana and plantain are important cash and subsistence food crops throughout much of sub-Saharan Africa. Small-scale production, typical of highland banana and plantain systems, is extremely important for many third-world farmers. The banana weevil is the most important insect pest of bananas in Africa. Highland banana and plantain are both highly susceptible to this pest, although many other clones are not.

By virtue of its biology, the banana weevil is a difficult pest to control. The adults are relatively sedentary and are often hidden in the soil or banana leaf sheaths, while the immature stages occur within the host plant. There are no known arthropod natural enemies of the adult stage, while those attacking the egg and larval stages tend to be opportunistic, generalist predators that have limited effect on weevil populations. As of yet, no parasitoids have been reared from banana weevils. Attempts at classical biological control have so far been unsuccessful. Myrmicine ants have been reported to contribute to effective control of banana weevils in Cuba and may offer some possibility in Africa. The potential for manipulating indigenous African species of ants as biological control agents is currently under study.

Cultural control options against banana weevil have been widely promoted and are available to most farmers. The most important of these methods are the use of clean planting material, crop sanitation, treatment with neem extract and trapping. A combination of these methods is likely to provide at least partial control of banana weevil. However, all of these methods have costs and adoption by resource-poor subsistence farmers is often limited.

Microbial control offers promise for the control of banana weevil. Numerous strains of *B. bassiana* and *M. anisopliae* have been demonstrated to kill high percentages (i.e. >70%) of banana weevils in the laboratory and in field trials. To date, research has focused on pathogenicity studies comparing strains, doses, formulations and modes of applications. Current research priorities include the development of economic mass production and delivery systems and evaluation of fungal performance and efficacy under different agroecological conditions. Unless economically feasible mass-production and efficient delivery systems for fungal pathogens (e.g. *B. bassiana*) are developed, the use of entomopathogenic fungi as biopesticides will be beyond the means of most farmers.

Endophytic fungi offer a promising area of research in that: (i) they target the damaging larval stage of the weevil; (ii) they require only single applications with small amounts of inoculum; (iii) they are less affected by environmental factors than other entomopathogenic fungi; and (iv) they are compatible with other crop management methods. Persistence of fungal endophytes in the host plant and efficacy in reducing pests under a range of growing conditions remain important research areas.

To date, there have been no attempts to breed for resistance to banana weevil. Screening trials have demonstrated the availability of many clones, including Calcutta-4, Yangambi-km5 and FHIA-03, that are resistant to be the banana weevil and might be utilized in breeding programmes. Antibiosis appears to be the predominant mechanism conferring resistance to weevils within *Musa* germplasm.

References

Alabouvette, C., Schippers, B., Lemanceau, P. and Bakker, P.A.H.M. (1998) Biological control of *Fusarium* wilts: toward development of commercial products. In: Boland, G.J. and Kuykendall, L.D. (eds) *Plant–Microbe Interactions and Biological Control.* Marcel Dekker, New York, USA, pp. 15–36.

Castineiras, A. and Ponce, E. (1991) Efectividad de la utilizacion de *Pheidole megacephala* (Hymenoptera: Formicidae) en la lucha biologica contra *Cosmopolites sordidus* (Coleoptera: Curculionidae). *Proteccion de Plantas* 1, 15–21.

Clay, K. (1991) *Endophytes as Antagonists of Plant Pests.* In: Andrews, J.H. and Hirano, S.S. (eds) *Microbial Ecology of Leaves.* Springer-Verlag, New York, pp. 331–357.

Godonou, I. (1999) The potential of *Beauveria bassiana* for the management of *Cosmopolites sordidus* (Germar, 1824) on plantain (*Musa*, AAB). PhD thesis University of Ghana, Legon, Ghana.

Godonou, I., Green, K.R., Oduro, K.A., Lomer, C.J. and Afreh-Nuamah, K. (2000) Field evaluation of selected formulations of *Beauveria bassiana* for the management of the banana weevil (*Cosmopolites sordidus*) on plantain (*Musa* spp., AAB group). *Biocontrol Science and Technology* 10, 779–788.

Gold, C.S., Okech, S.H. and Ssendege, R. (1997) Banana weevil population densities and related damage in Ntungamo and Mbarara districts, Uganda. In: Adipala, E., Tenywa, J.S. and Ogenga-Latigo, M.W. (eds) *African Crop Science Conference Proceedings.* 13–17 January 1997, Pretoria. Makerere University, Kampala, Uganda, pp 1207–1219.

Gold, C.S., Okech, S.H. and Nokoe, S. (2002) Evaluation of pseudostem trapping as a control of banana weevil, *Cosmopolites sordidus* (Germar), populations and damage in Ntungamo district, Uganda. *Bulletin of Entomological Research* 92, 35–44.

Gold, C.S., Pena, J.E. and Karamura, E.B. (2003) Biology and integrated pest management for the banana weevil *Cosmopolites sordidus* (Germar) (Coleoptera: Curculionidae). *Integrated Pest Management Reviews* (in press).

Griesbach, M. (2000) Occurrence of mutualistic fungal endophytes in bananas (*Musa* spp.) and their potential as biocontrol agents of the banana weevil *Cosmopolites sordidus* (Germar) (Coleoptera: Curculionidae) in Uganda. PhD thesis, University of Bonn, Germany.

Hallmann, J. and Sikora, R.A. (1996) Toxicity of fungal endophyte secondary metabolites to plant parasitic nematodes and soil-borne plant pathogenic fungi. *European Journal of Plant Pathology* 102, 155–162.

Koppenhofer, A.M., Seshu Reddy, K.V., Madel, G. and Lubega, M.C. (1992) Predators of the banana weevil, *Cosmopolites sordidus* (Germar) (Col., Curculionidae) in Western Kenya. *Journal of Applied Entomology* 114, 530–533.

McIntyre, B.D., Gold, C.S., Kashaija, I.N., Ssali, H., Night, G. and Bwamiki, D.P. (2002) Effects of legume intercrops on soil-borne pests, biomass, nutrients and soil water in banana. *Biology and Fertility of Soils* 34, 342–348.

Musabyimana, T. (1999) Neem seed for the management of the banana weevil, *Cosmopolites sordidus* Germar (Coleoptera: Curculionidae) and banana parasitic nematode complex. PhD thesis, Kenyatta University, Nairobi, Kenya.

Nankinga, C.M. (1999) Characterization of entomopathogenic fungi and evaluation of delivery systems for the biological control of the banana weevil, *Cosmopolites sordidus*. PhD thesis, University of Reading, Reading, UK.

Nankinga, C.M. and Ogenga-Latigo, W.M. (1996) Effect of method of application on the effectiveness of *Beauveria bassiana* against the banana weevil, *Cosmopolites sordidus*. *African Journal of Plant Protection* 6, 12–21.

Nankinga, C.M. and Moore, D. (2000) Reduction of banana weevil populations using different formulations of the entomopathogenic fungus *Beauveria bassiana*. *Biocontrol Science and Technology* 10, 645–657.

Nankinga, C.M., Moore, D., Bridge, P. and Gowen, S. (1998) Recent advances in microbial control of the banana weevil. In: Frisson, E.A., Gold, C.S., Karamura, E.B. and Sikora, R.A. (eds) *Proceedings of a Workshop on Mobilizing IPM for Sustainable Banana Production in Africa*. 23–28 November 1998, Nelspruit, South Africa. INIBAP, Montpellier, France, pp. 73–84.

Niere, B.I. (2001) Significance of *Fusarium oxysporum* Schlecht.:Fries for the biological control of the burrowing nematode *Radopholus similis* (Cobb) Thorne on tissue cultured banana. PhD thesis, University of Bonn, Bonn, Germany.

Ortiz, R. and Vuylsteke, D. (1996) Recent advances in *Musa* genetics, breeding and biotechnology. *Plant Breeding Abstracts* 66, 1355–1363.

Rukazambuga, N.D.T.M., Gold, C.S. and Gowen, S.R. (1998) Yield loss in East African highland banana (*Musa* spp., AAA-EA group) caused by the banana weevil, *Cosmopolites sordidus* Germar. *Crop Protection* 17, 581–589.

Schuster, R.-P., Sikora, R.A. and Amin, N. (1995) Potential of endophytic fungi for the biological control of plant parasitic nematodes. *Communications in Applied Biological Sciences* 60, 1047–1052.

17

The Role of Biological Control in Integrated Management of *Striga* Species in Africa

Dana K. Berner,[1] Joachim Sauerborn,[2] Dale E. Hess[3] and Alphonse M. Emechebe[4]

[1]*US Department of Agriculture, Agricultural Research Service, Foreign Disease, Weed Science Research Unit, Maryland, USA;* [2]*University of Hohenheim, Institute for Plant Production and Agroecology in the Tropics and Subtropics, Stuttgart, Germany;* [3]*Agronomy Department, Purdue University, West Lafayette, Indiana, USA;* [4]*International Institute of Tropical Agriculture, Kano, Nigeria*

Introduction

Parasitic flowering plants of the genus *Striga* are a major constraint to agricultural production in Africa. The parasites are widespread on a vast continent, and the effectiveness of conventional control methods is limited due to numerous factors, including: small farm sizes (frequently less than 3 ha); low farm income; lack of appropriate agricultural extension services; insufficient funds for access to affordable agricultural inputs; and the pernicious nature of the parasites, which reproduce prodigiously, have lengthy seed viability and are difficult to diagnose until irreversible crop damage has occurred. Many years of research into these problems have revealed that an integrated approach is essential in order to manage *Striga* spp. under the constraints of African farming systems. The integrated approach is robust, and encourages local modifications of inputs and management options to suite site-specific requirements. As a result, integrated management is favoured by farmers and has proved effective, not only in controlling the parasites but also in improving crop yields in their absence. Much of this chapter deals with descriptions of integrated control modules and how they interact to provide sustainable *Striga* spp. control in Africa. Biological control is one such module with several components. For the most part, however, the integrated management strategy is composed primarily of procedures that, through direct biological manipulations, affect parasite seed survival, plant infection, host plant resistance and parasite reproduction. Thus, although reliant on some chemical intervention, integrated *Striga* management is biological control in its broadest sense.

(eds P. Neuenschwander, C. Borgemeister and J. Langewald)

The Problem

General description of the parasites

There is a diversity of parasitic seed plants in the savannah zones of Africa that pose substantial problems for both cereal and legume production (Doggett, 1984; Aggarwal and Ouédraogo, 1989; Carson, 1989). The most devastating of these are *Striga* species (Musselman, 1980). *Striga* is a genus of obligate, root-parasitic flowering plants in the family *Scrophulariaceae* which comprises 41 species, most of which are found in Africa (Raynal-Roques, 1994) and with greatest diversification in West and Central Africa. Several species occur in Asia and Madagascar and three are endemic to Australia (Raynal-Roques, 1994). Two species, *Striga asiatica* (L.) Kuntze and *Striga gesnerioides* (Willd.) Vatke, have been unintentionally introduced to the eastern USA (Herbaugh *et al.*, 1980; Sand, 1990).

Life cycle of parasites

Striga spp. seeds are extremely small, each weighing about 7 μg. Germination of seeds requires exposure to an exogenous germination stimulant after an environmental conditioning period in which the seeds imbibe water (Worsham, 1987). Usually this stimulant is a host-root exudate (Worsham, 1987; Hauck *et al.*, 1992) but some non-host-root exudates and synthetic compounds can also stimulate germination (Cook *et al.*, 1972). These stimulants either initiate endogenous ethylene production within the parasite seed (Logan and Stewart, 1991; Babiker *et al.*, 1993) or directly provide ethylene (Eplee, 1975), which is the ultimate compound responsible for stimulating germination. After germination, endosperm nutrients can sustain the seedlings for 3–7 days in the absence of a host (Pieterse and Pesch, 1983; Worsham, 1987). If the seedling does not attach to a host and successfully establish a parasitic link within this period, it dies.

After successful attachment, developing *Striga* plants grow underground for 4–7 weeks prior to emergence. Usually, numerous parasitic attachments develop on the same plant and most of the damage to the host plant occurs during this stage. Symptoms of parasitism are dramatic but nondescript, resembling drought stress, nutrient deficiency or vascular disease, and frequently causing severe stunting of the host plant. Following emergence above the soil surface, *Striga* plants form chlorophyll and begin to photosynthesize (Pieterse and Pesch, 1983). Flowering time is species and environment dependent. *S. gesnerioides* begins to flower as it emerges (Emechebe *et al.*, 1991; Parker and Riches, 1993). *Striga asiatica*, *Striga aspera* (Willd.) Benth. and *Striga hermonthica* (Del.) Benth. (Plate 49) begin to flower about 4 weeks after emergence. Flowering begins basally on the raceme and seeds mature about 4 weeks after flowering (Pieterse and Pesch, 1983). Species of *Striga* produce 50,000 to 500,000 seeds per parasite plant (Pieterse and Pesch, 1983). Sorghum fields heavily infested with *S. hermonthica* have been reported to yield more than 900,000 flowering *S. hermonthica* plants ha^{-1} (Hess and Ejeta, 1992). At this density up to 4.5×10^{10} parasite seeds ha^{-1} could be produced. The seeds remain viable for 7 to 14 years (Bebawi *et al.*, 1984; Sand, 1990), and only a small fraction germinates in any season in the presence of a host.

Hosts

Most *Striga* species are of no agricultural importance, but all of the cultivated food-crop grasses in Africa, e.g. fonio (*Digitaria exilis* (Kippist) Stapf), maize (*Zea mays* L.), pearl millet (*Pennisetum glaucum* (L.) R. Br.), finger millet (*Eleusine coracana* (L.) Gaertn.), rice (*Oryza sativa* L., *Oryza glaberrima* Steud.), sorghum (*Sorghum bicolor* (L.) Moench.) and sugarcane (*Saccharum officinarum* L.), are parasitized by one or more *Striga* species. The species of primary economic importance on cereals are *S. hermonthica* and *S. asiatica.* Of secondary importance are *S. aspera* and *Striga forbesii* Benth. Yield losses on cereals infected by these parasites may reach 100% and levels of infestation are frequently so great that continued cereal production becomes impossible. Farmers often abandon infected fields for less-infested areas (Doggett, 1984; Lagoke *et al.*, 1991).

Unlike the aforementioned species, *S. gesnerioides* parasitizes only dicotyledonous plants including species of *Alysicarpus*, *Euphorbia*, *Indigofera*, *Ipomea*, *Jacquemontia*, *Merremia*, *Tephrosia*, *Vigna* and *Nicotiana tabacum* L. (Parker and Riches, 1993). Of these hosts, cowpea (*Vigna unguiculata* (L.) Walp.) is the most important in Africa. At present, the parasite devastates cowpea mostly in the Guinea and Sudan savannahs and the Sahel region of West and Central Africa. The Sudano-Sahelian zone is generally more affected than the Guinea savannah zone (Singh and Emechebe, 1997), but severe damage to cowpea has also been reported in the coastal savannah of Bénin Republic (Lane *et al.*, 1994) and Togo. With more cowpea monocropping and increasing population pressure, *S. gesnerioides* damage in cowpea has become more acute, particularly in areas with sandy, infertile soils and low rainfall (Singh and Emechebe, 1997).

Strains

Striga diversity is reflected in species differences and also in distinct morphotypes, physiological strains and races within species. Differential virulence of *S. hermonthica* on maize, millet and sorghum has been well documented (King and Zummo, 1977; Bebawi, 1981; Hess, 1994) and is evidenced by the relative inability of *S. hermonthica* isolates collected from any given host in one location to parasitize a different host from the same location or the same host from another location.

Morphotypes of *S. gesnerioides* are distinguishable by stem succulence, internode length, flower colour, and size (Musselman, 1980; Musselman and Parker, 1981; Parker and Riches, 1993). There is also considerable variation in host specificity between isolates of *S. gesnerioides*, and different host species vary in their susceptibility to different parasite isolates (Musselman, 1980; Musselman and Parker, 1981). Mohamed *et al.* (2001) have described eight host-specific strains: *Euphorbia* strain (on *Euphorbia* spp.); *Ipomoea* strain (on *Ipomoea batata* L.); *Indigofera* strain (on *Indigofera* spp.); *Jaquemontia* strain (on *Jaquemontia* spp.); *Merremia* strain (on *Merremia* spp.); *Nicotiana* strain (on tobacco); *Tephrosia* strain (on *Tephrosia* spp.); and *Vigna* strain (on *V. unguiculata*). The *Vigna* strain is by far the most important, causing considerable damage to cowpea, with substantial yield reductions, especially in West and Central Africa (Emechebe *et al.*, 1991).

Races

Lane *et al.* (1997) reported pathogenic variability within and among populations of *S. gesnerioides*. They characterized the virulence of *S. gesnerioides* from 48 sites in seven West African countries using four cowpea varieties as differentials. Five races of *S. gesnerioides* were detected. Race 1, virulent only on Blackeye cultivar, occurs in Burkina Faso, northern Togo, central Nigeria and central Mali. Race 2 is virulent on cultivars Blackeye and 58–57 and appears to be restricted to Mali. Race 3 occurs in Nigeria and Niger and is virulent on all varieties, except B301. Race 4, virulent on Blackeye and B301, was reported to be restricted to the Zakpota area of southern Bénin Republic; but it has also been detected in Kazaure, Gumel and Birnin Kuddu in Jigawa State of Nigeria. Race 5 is virulent on Blackeye and IT81D-994 and was detected in samples from northern Cameroon, central and northeastern Nigeria, southern Burkina Faso and northern Bénin; it has also been detected in Maradi (Niger Republic) by Touré *et al.* (1998). Races of other *Striga* spp. have not been reported.

Crop losses

Striga is a serious problem in all the countries of West and Central Africa. Few attempts have been made to estimate the yield losses caused by *Striga* in farmers' fields. Sauerborn (1991) observed that *Striga* is present on 40–75% of the visited cereal fields that were scattered over the whole of northern Ghana. In the infested fields, 34% of the stands were parasitized. It was further estimated that the yield losses in maize, pearl millet and sorghum in infested areas were 16%, 31% and 29% of the potential yield per hectare, respectively. Based on data available from six West African countries, the loss in total grain production was estimated to be 12%. A conservative estimate by Sauerborn (1991) suggested grain production in Africa to be endangered by *Striga* on 44 million ha of land. Based on Freight on Board prices of 1988, the total annual loss due to *Striga* amounts to approximately US$2.9 billion. These figures may be much higher today, as the rate of spread to *Striga*-free areas is alarming. The growing severity of *Striga* infestation in Africa results from drastic changes in farming systems followed by an intensification of cereal production, principally through monocropping, in an attempt to produce sufficient food for the burgeoning population (Doggett, 1984; Butler, 1995; Berner *et al.*, 1996). In West Africa alone, it is estimated that about 40 million ha in cereal production are severely infested with *Striga* spp., while nearly 70 million ha have moderate levels of infestation (Lagoke *et al.*, 1991). As a result, annual yield losses due to all *Striga* spp. in the savannah regions alone are estimated to impact the lives of over 300 million African people (M'Boob, 1989).

Conventional Control Approaches

Farmer control measures

Controlling root parasites such as *Striga* is particularly difficult because significant damage is caused to the host before the parasite emerges from the soil (Parker and Riches, 1993; Hess and Haussmann, 1999). Approaches that farmers use for control are based on restraining *Striga* development and seed production, and improving soil conditions through crop rotation, intensive fallow, and mineral and organic fertilization. Infestations may be maintained at low levels by removing flowering *Striga* from fields during late weeding (Hess and Haussmann, 1999). Home-grown *Striga* control strategies in northern Ghana consist of hand-pulling, rotation and cow-dung application. From Eritrea, Ethiopia and Sudan, it is reported that catch-cropping, called 'serwala', is employed to combat the parasitic weed (Kroschel *et al.*, 1996). Sorghum is sown at high densities and about 6 weeks after sowing the crop is disc-harrowed or hand-hoed to normal crop levels. However, desired levels of control are seldom achieved by these practices. For example, some farmers in Malawi pulled *Striga* but left the uprooted plants between the crop rows where they could mature and release seeds.

Eradication in the USA

Soon after *S. asiatica* was discovered in maize fields in North and South Carolina in the USA in the 1950s, an extensive eradication programme was initiated with a targeted eradication date in the mid-1990s (Sand and Manley, 1990). During this 45-year period, 17 counties in the two states were quarantined to prevent movement of the parasite to other areas. Initially, control efforts relied mostly on pre- and post-emergence herbicide applications, but these efforts failed to effectively control the infestation. In the mid-1970s, ethylene was found to be a potent stimulant of *S. asiatica* seed germination (Eplee, 1975), and injection of field soils with ethylene gas became the central practice in the eradication programme. The ethylene stimulated premature germination of the *S. asiatica* seeds in the soil prior to planting of a host crop and the young plants died in the absence of suitable hosts. In combination with herbicide applications to prevent the parasite from setting seed, this strategy proved successful in eradicating *S. asiatica* from these areas. Unfortunately, soil injections of ethylene gas and herbicide applications are prohibitively expensive for African farmers, and alternative means of inducing premature germination of *Striga* spp. seeds in the absence of a crop are necessary.

Integrated Management

General

The *Striga* problem in Africa is intimately associated with intensification of land use (Portères, 1952). Traditional African cropping systems included prolonged fallow, rotations and intercropping that kept *Striga* infestations at tolerable levels (Doggett, 1984; Sprich and Sauerborn, 1999). As population pressure increased and markets developed, demand for food production increased and land use intensified. This intensification is reflected in greater use of monocropping with little or no fallow to non-host crops. As a result, the extent and intensity of *Striga* infestations have rapidly increased and become threats to food production. To reverse this trend, a rational system of land management is required. For sustained use, this system must improve crop yield per unit area and be acceptable to farmers even in the absence of *Striga* infestation. What is described in the following sections is an approach to integrated *Striga* management that is designed to meet these objectives.

Striga*-free planting material*

Most movement of *Striga* into previously *Striga*-free areas probably occurs through contaminated crop seed lots (Berner *et al.*, 1994b). These seed lots are usually contaminated with *Striga* seeds at and after harvest. Harvesting frequently entails cutting the plant at ground level and laying the cut plant, including panicles, spikes or cobs, in the field to dry. After a drying period of days, or in some cases weeks, the plants are collected and taken to a common area to thresh the grain from the plants. While collecting the plants from the field, *Striga* plants and seeds that are interspersed with the drying crop are inadvertently collected and taken to the threshing area where allotments of crop seeds become contaminated. During winnowing of the trash, the tiny *Striga* seeds adhere to winnowing baskets and these become an additional source of contamination of the crop seed lots (Berner *et al.*, 1994b). When bagging the final product, farmers inadvertently mix crop seeds from both uncontaminated and *Striga*-contaminated seed lots, and these mixtures are either sold in the market or stored in village homes. In both cases these contaminated seeds frequently change hands and are subsequently used both for food and planting material in the next season. When used as planting material by an unsuspecting purchaser, newly infested sites can quickly result through contamination, sale and redistribution of crop seeds.

The use of *Striga*-free planting material is thus an important first step in preventing new infestations or re-infestations and in making sustainable control feasible (Berner *et al.*, 1994b). Because *Striga* seeds are small and relatively dense (specific gravity = 1.4), they settle to the bottom of stored seed lots. If the uppermost portion of the seed lot, which is relatively free from *Striga* seeds, is saved for planting and the bottom, *Striga*-contaminated, portion reserved for food, a considerable amount of unintentional infestation can be avoided. Purchasing seeds from reputable seed companies, only planting seeds harvested from *Striga*-free fields, drying, threshing and winnowing harvested seeds only in *Striga*-free areas are other methods to help prevent contamination and infestation.

Crop rotation with selected non-host cultivars

To reduce the amount of *Striga* seeds in soil, effective and acceptable alternatives to soil injections of ethylene gas are needed. An alternative means of reducing densities of *Striga* seeds in the soil is through cultivation of non-host crops that stimulate *Striga* seed germination (Berner *et al.*, 1995, 1996; Sauerborn, 1999). As with the use of ethylene gas, the germinated seeds cannot survive without a host, and the seedlings die. Rotating nitrogen-fixing legume non-hosts with cereal hosts can be an effective means of reducing densities of *Striga* seeds in the soil while producing food and improving soil fertility. However, there is considerable variability within legume species among cultivars or accessions in their ability to stimulate *Striga* spp. seed germination (Berner *et al.*, 1996; Berner and Williams, 1998).

The task of characterizing variation in germination stimulant production and legume cultivar efficacy is compounded by the variability within and between *Striga* populations. To ensure that the most effective legume cultivars are placed in rotation systems, cultivars should be screened for efficacy at the local level, e.g. in each ecological zone of a country, with the particular *Striga* population present in the area. Screening for legume cultivar efficacy can be done in the field, but the amount of material that can be screened is limited by land and labour availability. In addition, at least 2 years are required: one for growth of the legume and another for evaluation of the legume effects by host cultivation. An alternative for national programmes is to use a simple laboratory assay (Berner *et al.*, 1996) that is relatively inexpensive and can readily be used for routine testing of promising legume cultivars. Efficacy rankings of soybean (*Glycine max* Merr.) and cowpea cultivars evaluated by the assay correlated significantly with reduced *Striga* parasitism of sorghum and increased sorghum yield in the field after only a single-season cultivation of soybean or cowpea (Berner *et al.*, 1996).

Although this example focuses on soybean and cowpea, cultivars of other grain legumes, such as pigeon pea (*Cajanus cajan* Millsp.), groundnut (*Arachis hypogaea* L.), and bambara groundnut (*Vigna subterranea* (L.) Verdc.), can also be screened and used to stimulate *Striga* seed germination. Participatory research with farmers in northern Ghana revealed that sole cropping with cotton (*Gossypium* sp.), cowpea, groundnut, soybean and sunflower (*Helianthus annuus* L.) planted once or twice, significantly increased yield of the staple foods of the region, maize and sorghum (Sauerborn *et al.*, 2000): lowest yield of maize and sorghum was obtained where these cereals followed maize–sorghum (monoculture).

Resistant host cultivars – cereals

Adapted, productive *Striga*-resistant cereal cultivars are a highly effective management option that requires no additional inputs other than improved seed (Hess and Ejeta, 1992). Efficient and reliable screening methodologies are the main prerequisites for effective transfer of resistance into improved varieties (Hess *et al.*, 1992). Although screening for individual resistance mechanisms may be undertaken under controlled conditions, complex resistance must be assessed in the field at several locations and, or, across several seasons. Integration of molecular marker selection techniques into resistance breeding holds promise of enhanced movement of desirable genes among varieties and will facilitate the transfer of novel genes from related wild species.

The breeding strategies proposed by groups of researchers active in this field have been summarized elsewhere (Kim, 1994; Haussmann *et al.*, 2000). The lack of reliable single-plant screening techniques distinguishing resistant and susceptible plants in segregating populations is a major hindrance to progress. Selection for *Striga* resistance is generally delayed until true breeding progenies are produced, meaning that large numbers of progeny are advanced before the trait of interest can be assessed – a time-consuming and costly procedure. Laboratory assays allowing for non-destructive, rapid and inexpensive evaluation of individual plants would facilitate early generation testing and improve selection efficiency. Presently, the only practical assay screens for the low stimulant character (Hess *et al.*, 1992). The agar-gel assay can be an excellent tool to transfer the low stimulant gene(s) to local adapted cultivars using classical backcross procedures. However, other tests evaluating different resistance mechanisms are needed that are suited to selection programmes with large numbers of entries to be screened.

Selection is traditionally based on plant phenotype, determined not only by the plant's genes, but also by the environment in which it is grown, and therefore an imperfect measure of a plant's genetic potential. Selection for minor genes is particularly difficult. Minor genes of interest can be identified and selected when they are linked to major genes or 'markers' that are easier to score. The International Crops Research Institute for the Semi-Arid Tropics and the University of Hohenheim are collaborating to identify and map genes for qualitative and quantitative resistance to *S. hermonthica* in two recombinant inbred sorghum populations. These findings will facilitate the implementation of marker-assisted selection leading to the eventual cloning and characterization of *S. hermonthica* resistance genes in sorghum.

Resistant host cultivars – cowpea

Cowpea cultivars with different susceptibilities to *S. gesnerioides* infection were first observed in 1981 in Burkina Faso, and two lines (Suvita-2 and 58–57) were found to be completely resistant (Aggarwal, 1985). Further screening of new lines revealed that IT82D-849 (breeding line from IITA) and B301 (a landrace from Botswana) were completely resistant to *S. gesnerioides* populations in Burkina

Faso, Mali, Niger, Nigeria and Cameroon (Aggarwal, 1991; Emechebe *et al.*, 1991). A systematic breeding programme for *S. gesnerioides* resistance in cowpea was started in 1987. The programme used B301 as the resistance source and IT84S-2246–1 as a susceptible but otherwise high-yielding cultivar with several desirable traits. From this breeding programme several lines were obtained that had complete resistance to *S. gesnerioides* in several countries of West and Central Africa (Singh and Emechebe, 1997). Such lines included those that have been released for cultivation in Mali (IT89KD-374–5 and IT89KD-245) and Nigeria, IT90K-76 and IT90K-82–2. They were, however, found to be moderately susceptible to a *S. gesnerioides* strain at Zakpota (in the coastal savannah of Bénin Republic) to which Suvita-2, 58–57 and IT81D-994 were completely resistant. IT81D-994 and 58–57 were used as resistance sources to incorporate resistance to Zakpota strain in B301 derived lines such as IT90K-76 and IT90K-59 to obtain lines such as IT94K-437–1 and IT97K-499–39, with combined resistance to all five races of *S. gesnerioides* found in West and Central Africa. As noted by Lane *et al.* (1997), the overlapping distribution of the five races has necessitated multi-location trials in Nigeria, Bénin and Burkina Faso in order to obtain varieties with resistance against all the five races of *S. gesnerioides*.

Genetic studies on the nature of inheritance in cowpea to *S. gesnerioides* has revealed that resistance in each of B301, IT82D-849 and Suvita-2 is controlled by one dominant gene (designated as Rsg1, Rsg2 and Rsg3, respectively), the genes being independent of each other (Singh and Emechebe, 1990; Atokple *et al.*, 1995).

The possibility of using marker-assisted selection for incorporating *S. gesnerioides* resistance in cowpea was suggested by the recent work of Ouédraogo *et al.* (2001), who confirmed earlier findings by Atokple *et al.* (1995) that resistance to race 1 of *S. gesnerioides* in IT82D-849 is controlled by a single dominant gene (Rsg2–1). They also showed that resistance to race 3 of *S. gesnerioides* in Tvu 14676 is controlled by a single dominant gene (Rsg4–3). They further identified three genetic markers tightly linked to Rsg2–1 as well as six markers linked to Rsg4–3. From the set of genetic markers thus far identified, nine were placed on an improved genetic map of cowpea. These findings will facilitate a marker-assisted selection programme and the eventual cloning and characterization of *S. gesnerioides* resistance genes in cowpea.

Host seed treatments with ALS-inhibitors

A supplement to host plant resistance is host-seed treatment with acetolactate synthase (ALS) inhibitors such as imidazolinones and sulphonylureas. Results of tests with imazaquin for the control of *S. gesnerioides* on cowpea showed that a simple aqueous seed soak solution with low amounts of imazaquin gave excellent protection against the parasite and improved plant yields (Berner *et al.*, 1994a). With the seed treatment, parasite seed germination occurred and protection was achieved by post-attachment mortality of the parasite. An additional control benefit was obtained by reducing parasite seed density in the soil (Berner *et al.*, 1994a).

To use these seed treatments effectively on maize in Africa, tolerance to the ALS inhibitor must first be incorporated into high-yielding adapted African maize cultivars (Berner *et al.*, 1997). Recently it was shown that soil drenches at planting, with low amounts of imazapyr, helped to control *S. hermonthica* on ALS inhibitor-resistant maize (Abayo *et al.*, 1998). These seed treatments and drenches are particularly attractive to African farmers because the treatments use a very small amount of herbicide, which specifically targets only the invading parasite, and are thus inexpensive.

Biological Control

General

Because all economically important *Striga* spp. are endemic to Africa, there is little opportunity for classical biological control through introduction of natural enemies, including pathogens, of *Striga* spp. from areas outside Africa. A workable alternative is the manipulation (through multiplication and periodic release) of indigenous insect and microbial agents for *Striga* control. Because these biological control agents are indigenous to Africa, they also have indigenous antagonists, diseases and competitors that keep their own populations in check. Also, as in normal predator–prey relationships, once populations of the target species drop below a certain threshold, populations of the biological control agent are also likely to decrease, thereby allowing the target species populations to rise again. Thus, it is unlikely that a single release of indigenous biological control agents will provide sustained control. Periodic releases will be needed for sustained control.

Soil suppressiveness

Tests on soil samples taken at random from varying locations in Nigeria showed that natural soil suppressiveness, measured by infesting soil with *S. hermonthica* seeds and growing a susceptible host, occurred throughout the country (Berner *et al.*, 1996). In all locations there was a reduction in the number of attached parasites in the unpasteurized soils (those possessing suppressiveness) compared with parasitism in pasteurized soils (those in which suppressiveness had been eliminated by the pasteurization process). Over all locations there was a significant reduction (47%) in the number of attached parasites in unpasteurized soils compared with pasteurized soils. These results indicate that natural biotic soil suppressiveness probably occurs frequently in Nigeria, including in soils from *S. hermonthica*-infested areas, and will probably be found to be widespread in other *S. hermonthica*-endemic areas throughout Africa.

Because the soil suppressiveness is biotic, a simple means of maximizing suppressiveness is to improve soil conditions to encourage growth of the biotic agents. In the northern region of Ghana, it has been shown that fields close to dwellings benefit from household waste, faeces and dung. Those fields differ significantly in soil organic matter from more distant, so-called bush fields. Although much research is still needed, it seems that increasing the biological activity of the soil enhances the demise of *Striga* seeds.

Although additions of organic matter could improve microbial activity, the logistics of doing so on a large scale are prohibitive. Ahonsi *et al.* (2002a) demonstrated that growing soybean prior to maize, direct application of nitrogen to maize and application of phosphorus to the preceding soybean crop enhanced soil suppressiveness to *S. hermonthica*. As nitrogen is one of the most limiting elements for microbial growth in soils, a manageable means of improving suppressiveness is to rotate cereal hosts with nitrogen-fixing legumes. In addition to providing nitrogen, legume rotations promote nutrient recycling and help to aerate the soil (a necessary condition for good growth by aerobic microbes). Thus, appropriate management of the rotation legume crop becomes a key point in maximizing soil suppressiveness to *Striga* and contributing to integrated management.

Ethylene-producing bacteria

Rotating cereals with legume cultivars selected for the ability to stimulate germination of *Striga* spp. seeds is a low-cost means of reducing *Striga* seeds in soil. Another method is the use of ethylene-producing bacteria. The pathovar *glycinea* of the bacterium species *Pseudomonas syringae* van Hall synthesizes relatively large amounts of ethylene, and some strains stimulate more germination of *Striga* spp. seeds than the synthetic stimulant or ethylene gas itself (Berner *et al.*, 1999). In recent greenhouse tests, *S. hermonthica* parasitism on maize was significantly reduced by incorporating ethylene-producing bacteria into the soil and by growing cowpea and soybean, the seeds of which had been inoculated with the bacteria, prior to planting maize (Ahonsi *et al.*, 2002a). Co-inoculating cowpea and soybean seeds with ethylene-producing bacteria and nitrogen-fixing *Bradyrhizobium japonicum* (Buchanan) Jordan strains augmented this effect. In all cases, there was a concomitant increase, over no-bacteria controls, in maize shoot dry weight when bacteria were used. Reduced parasitism and increased shoot dry weight attributable to the bacteria treatments were additive with the effects attributable to growing either cowpea or soybean prior to planting maize. Thus, combining legume cultivars, selected for ability to stimulate *Striga* spp. seed germination, with inoculations of both types of bacteria would further improve the efficacy of rotations for *Striga* spp. control. Although *P. syringae* pv. *glycinea* (Psg) is a pathogen of soybean, ethylene-producing strains could be used with Psg-resistant soybean or cowpea cultivars or in areas where soybean is not grown. Alternatively, ethylene-producing strains of non-pathogenic bacteria may be found and used (Ahonsi *et al.*, 2002b).

Rhizobacteria

Rhizobacteria capable of suppressing germination of *Striga* seeds or actually destroying the seeds are particularly promising biological control agents since they can be easily and cheaply formulated into seed inoculants (Berner *et al.*, 1995; Miché *et al.*, 2000). Laboratory screening of 460 fluorescent pseudomonad isolates from soil samples collected from four locations in Nigeria yielded 15 *Pseudomonas fluorescens-putida* isolates that significantly inhibited germination of *S. hermonthica* seeds (Ahonsi *et al.*, 2002a). There was a significant reduction in number of *S. hermonthica* plants on maize grown from seeds that were inoculated with any of these 15 bacterial isolates, and maize seed inoculation with six of these isolates resulted in significantly reduced number of *S. hermonthica* plants and in significantly increased maize shoot biomass compared with controls. These bacteria could be immediately used as prophylactic seed treatments on maize and, if compatible, in combination with ALS-inhibitor seed treatments.

Fungi, particularly Fusarium

About 52 fungal species from 16 genera are reported to occur on *Striga* spp. (Kirk, 1993; Abbasher *et al.*, 1995, 1998; Ciotola *et al.*, 1995; Hess *et al.*, 2002), many of which are reported to be pathogenic (Abbasher and Sauerborn, 1992; Ciotola *et al.*, 1995). Results of surveys for fungal pathogens of *Striga* spp. in West Africa indicated that *Fusarium* species were the most prevalent fungal species associated with diseased *Striga* plants. Of these *Fusarium oxysporum* was the predominant (93%) species (Abbasher *et al.*, 1998). In host range tests, indigenous *F. oxysporum* from Burkina Faso, Mali and Niger were found to infect only *Striga* spp. (Abbasher *et al.*, 1998). Pathogenicity tests demonstrated that one *Fusarium* species is able to control more than one *Striga* species: *Fusarium nygamai*, *F. oxysporum* and *Fusarium solani* isolated from infected *S. hermonthica* plants from Sudan and Ghana reduced emergence of *S. asiatica* and *S. gesnerioides* and improved growth of the maize and cowpea host crops (Abbasher *et al.*, 1998).

Research with *Fusarium* spp. (*F. nygamai*, *F. oxysporum*, *Fusarium semitectum* var. *majus*) has demonstrated that, applied pre-sowing, they were able to reduce *Striga* spp. incidence under controlled and field conditions (Abbasher and Sauerborn, 1992; Ciotola *et al.*, 1995, 2000; Kroschel *et al.*, 1996; Sauerborn *et al.*, 1996; Hess *et al.*, 2002). All growth stages, from ungerminated seeds up to flowering stalks, are attacked. Thus, *Fusarium* spp. can be very effective in reducing the parasite seed bank by destruction of the soil seed bank and prevention of seed-set. Since *Striga* seeds can be infected in the field after application of *Fusarium* spp., it might not be necessary to apply the inoculum during crop growth. This could lower the parasite seed bank every season even if there is no host plant for *Striga* in the field and impart additional advantages for the mycoherbicide strategy. Since significant damage to host plants occurs while the parasite is still underground, use of *Fusarium* spp. in biological control can contribute to enhanced crop yield (Sauerborn *et al.*, 1966; Hess *et al.*, 2002).

Microbial delivery systems

Similar application techniques apply to both mycoherbicides and chemical herbicides. Delivery systems must be easy, effective and compatible with available agricultural equipment. With few exceptions, liquid formulations of bioherbicides are best suited for use as post-emergence sprays and are used primarily to incite leaf and stem diseases. However, pathogens that infect at or below the soil surface are probably best delivered in a granular formulation applied to the soil. A simple approach is the use of infested cereal grains, although formulated granules have the advantage of increased shelf life and lower dosage in the field. The development of effective, economical, environmentally friendly and technically feasible bioherbicide formulations is still needed.

A major constraint with the mycoherbicide approach remains with the quarantine regulations of affected countries. Experience so far has shown that countries where the beneficial organism has not been described yet will generally not allow its importation. In consequence, this means that a pathogenic strain of an organism has to be identified in each country and developed independently into a mycoherbicide (but see also Langewald *et al.*, this volume).

Insects

Due to its short lifespan and high seed production, and the great damage caused to the host by un-emerged plants, *Striga* cannot be considered an ideal plant for biological control by insects. However, important insect pests of *Striga* in Africa include *Smicronyx* spp. (Coleoptera, Curculionidae), which are gall-forming weevils that prevent seed production through the development of larvae inside the seed capsules. Another species, *Smicronyx albovariegatus* Faust., from India, forms galls in all plant parts (root, stem, fruit) and attacks *S. asiatica*, *Striga angustifolia* and *Striga densiflora* (Smith *et al.*, 1993; Parker and Riches, 1993). *Smicronyx umbrinus* Hustache, *Smicronyx guineanus* Voss and other, as yet unidentified *Smicronyx* species from Africa, have been described that cause galls mainly on the fruits and stems (Parker and Riches, 1993; Smith *et al.*, 1993).

Conclusions

For implementation, a flow chart of integrated *Striga* management practices is presented in Fig. 17.1. No recommendations are made for specific cultivars, seed-treatment chemicals or isolates of biological control agents. These aspects need to be determined at local levels using national research programmes and farmer input to decide which procedures are best for each particular area. This flexibility allows the integrated management programme to be robust enough to fit local needs. This flexibility virtually guarantees that not all of the outlined practices are used in any given locality.

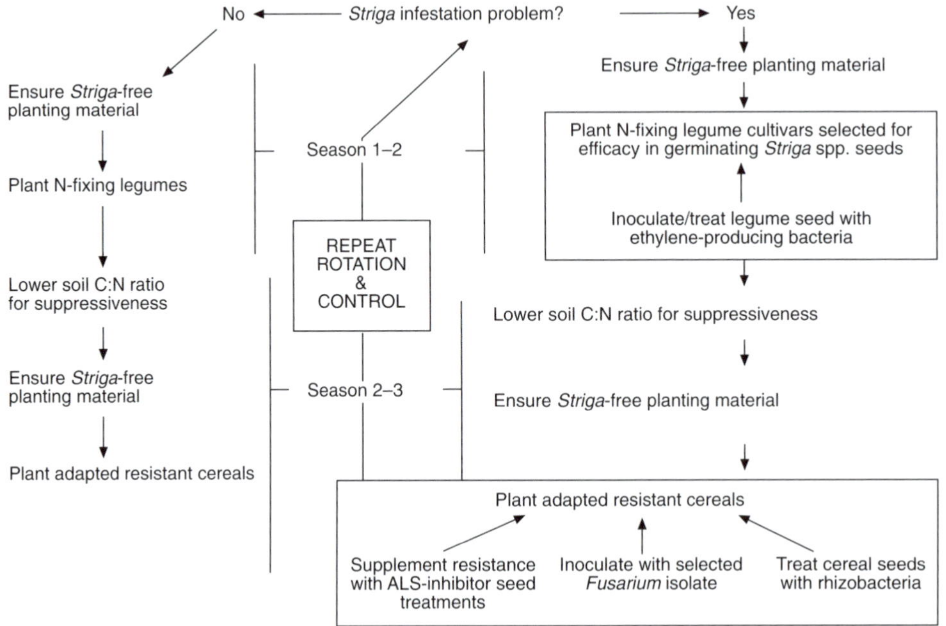

Fig. 17.1. Decision tree and flowchart of elements to be implemented for integrated *Striga* spp. management.

Rotation of cereals with legumes and use of resistant cultivars appear to be easy to implement immediately and are gaining acceptance among farmers where the appropriate planting materials are available (Carsky *et al.*, 2000; Sauerborn *et al.*, 2000). As these farmers become more familiar with the concepts and successes of integrated management, more of the modules, e.g. seed-treatment bacteria, ALS-inhibitors, etc., are expected to be included, depending on support of local research facilities to tailor these modules to local needs.

The entire integrated management programme is designed to improve crop yields and provide a greater economic return to farmers. This return is independent of *Striga* spp. infestation, since the management practices are designed to improve crop yield even in the absence of parasite infestation (Fig. 17.1). By following the suggested practices, farmers will prevent *Striga* spp. infestation while improving the quality of soil microbiota, fertility and water-retention capacity. Introducing adapted, high-yielding resistant cultivars into their cropping system will further improve yields and economic returns. However, *Striga* control methods have to be evaluated from the perspective of the farm household, which is the relevant decision making unit with respect to the control method. Nevertheless, in the case of *Striga* there are external influences since *Striga* occurrence seems to be correlated with low soil fertility. Soil fertility considerations are of a long-term nature, i.e. investments of today will pay off in future while investments neglected today will affect coming generations negatively.

Although integrated *Striga* management is finding success among farmers where the appropriate inputs and technical support are available, these areas are relatively small in comparison with the area of *Striga* infestation in Africa. What is needed, to expand and improve integrated management and sustained control of *Striga* spp., is a dedicated effort among more countries and their agricultural support infrastructure. This will require research to provide locally acceptable inputs, e.g. selected legume cultivars, resistant host cultivars, seed-treatment bacteria, ALS-inhibitors; etc., but one cannot minimize the critical need for active and involved extension services. Mass field days and farmer-managed research trials are needed on a vast scale if *Striga* control is to be successful in Africa. This should not require massive amounts of additional funding for national agricultural programmes, but will require a reorientation of activities toward integrated management, interdisciplinary cooperation and farmer involvement.

References

Abayo, G.O., English, T., Eplee, R.E., Kanampiu, F.K., Ransom, J.K. and Gressel, J. (1998) Control of parasitic witchweeds (*Striga* spp.) on corn (*Zea mays*) resistant to acetolactate synthase inhibitors. *Weed Science* 46, 459–466.

Abbasher, A.A. and Sauerborn, J. (1992) *Fusarium nygamai* a potential bioherbicide for *Striga hermonthica* control in sorghum. *Biological Control* 2, 291–296.

Abbasher, A.A., Kroschel, J. and Sauerborn, J. (1995) Micro-organisms of *Striga hermonthica* in northern Ghana with potential as biocontrol agents. *Biocontrol Science and Technology* 5, 157–161.

Abbasher, A.A., Hess, D.E. and Sauerborn, J. (1998) Fungal pathogens for biological control of *Striga hermonthica* on sorghum and pearl millet in West Africa. *African Crop Science Journal* 6,179–188.

Aggarwal, V.D. (1985) Cowpea *Striga gesnerioides* research. In: Singh, S.R. and Rachie, K.O. (eds) *Cowpea Research, Production and Utilization*. John Wiley & Sons, Chichester, UK, pp. 335–340.

Aggarwal, V.D. (1991) Research on cowpea *Striga* resistance at IITA. In: Kim, S.K. (ed.) *Combating Striga in Africa*. IITA, Ibadan, Nigeria, pp. 90–95.

Aggarwal, V.D. and Ouédraogo, J.T. (1989) Estimation of cowpea yield loss from *Striga* infestation. *Tropical Agriculture* 66, 91–92.

Ahonsi, M.O., Berner, D.K., Emechebe, A.M. and Lagoke, S.T. (2002a) Selection of rhizobacterial strains for suppression of germination of *Striga hermonthica* (Del.) Benth. seeds. *Biological Control* 24, 143–152.

Ahonsi, M.O., Berner, D.K., Emechebe, A.M., Sanginga, N. and Lagoke, S.T.O. (2002b) Selection of non-pathogenic ethylene-producing rhizobacteria for accelerated depletion of *Striga hermonthica*'s seed bank. *African Crop Science Journal* 10, 145–156.

Atokple, I.D.K., Singh, B.B. and Emechebe, A.M. (1995) Genetics of inheritance to *Striga* and *Alectra* in cowpea. *Journal of Heredity* 86, 45–49.

Babiker, A.G.T., Ejeta, G., Butler, L.G. and Woodson, W.R. (1993) Ethylene biosynthesis and strigol-induced germination of *Striga asiatica*. *Physiologia Plantarum* 88, 359–365.

Bebawi, F.F. (1981) Intraspecific physiological variants of *Striga hermonthica*. *Experimental Agriculture* 17, 419–423.

Bebawi, F.F., Eplee, R.E., Harris, C.E. and Norris, R.S. (1984) Longevity of witchweed (*Striga asiatica*) seed. *Weed Science* 32, 494–497.

Berner, D.K. and Williams, O.A. (1998) Germination stimulation of *Striga gesnerioides* seeds by hosts and nonhosts. *Plant Disease* 82, 1242–1247.

Berner, D.K., Awad, A.E. and Aigbokan, E.I. (1994a) Potential of imazaquin seed treatment for the control of *Striga gesnerioides* and *Alectra vogelii* in cowpea (*Vigna unguiculata*). *Plant Disease* 78, 157–164.

Berner, D.K., Cardwell, K.F., Faturoti, B.O., Ikie, F.O. and Williams, O.A. (1994b) Relative roles of wind, crop seeds, and cattle in the dispersal of *Striga* species. *Plant Disease* 78, 402–406.

Berner, D.K., Kling, J.G. and Singh, B.B. (1995) *Striga* research and control – a perspective from Africa. *Plant Disease* 79, 652–660.

Berner, D.K., Carsky, R.J., Dashiell, K.E., Kling, J.G. and Manyong, V.M. (1996) A land management based approach to integrated *Striga hermonthica* control in sub-Saharan Africa. *Outlook on Agriculture* 25, 157–164.

Berner, D.K., Ikie, F.O. and Green, J.M. (1997) ALS-inhibiting herbicide seed treatments control *Striga hermonthica* in ALS-modified corn (*Zea mays*). *Weed Technology* 11, 704–707.

Berner, D.K., Schaad, N.W. and Völksch, B. (1999) Use of ethylene-producing bacteria for stimulation of *Striga* spp. seed germination. *Biological Control* 15, 274–282.

Butler, L.G. (1995) Chemical communication between the parasitic weed, *Striga* and its crop host: a new dimension in allelochemistry. In: Indejit, M.D. and Einhellig, F.A. (eds) *Allelopathy: Organisms, Processes, and Applications*. ACS Symposium Series 582, American Chemical Society, Washington, DC, USA, pp. 158–168.

Carsky, R., Berner, D.K., Dashiell, K., Oyewole, B. and Schulz, S. (2000) Reduction of *Striga hermonthica* parasitism on maize using soybean rotation. *International Journal of Pest Management* 46, 115–120.

Carson, A. (1989) The *Striga* problem in the Sahel. *Sahel Protection des Vegetaux, Information* 10, 11–14.

Ciotola, M., Watson, A.K. and Hallett, S.G. (1995) Discovery of an isolate of *Fusarium oxysporum* with potential to control *Striga hermonthica* in Africa. *Weed Research* 35, 303–309.

Ciotola, M., Di Tommaso, A. and Watson, A.K. (2000) Chlamydospore production, inoculation methods and pathogenicity of *Fusarium oxysporum* M12–4A, a biocontrol for *Striga hermonthica*. *Biocontrol Science and Technology* 10, 129–145.

Cook, C.E., Whichard, L.P., Wall, M.E., Egley, G.H., Coggan, P., Luhan, P.A. and McPhail, A.T. (1972) Germination stimulants. 2. The structure of strigol – a potent seed germination stimulant for witchweed (*Striga lutea* Lour.). *Journal of the American Chemical Society* 94, 6198–6199.

Doggett, H. (1984) *Striga* its biology and control an overview. In: Ayensu, E.S., Doggett, H., Keynes, R.D., Marton-Lefèvre, J., Musselman, L.J., Parker, C. and Pickering, A. (eds) *Striga Biology and Control*. International Council of Science Union, Paris, France and the International Research and Development Center, Ottawa, Canada, pp. 27–36.

Emechebe, A.M., Singh, B.B., Leleji, O.I., Atokple, I.D.K. and Adu, J.K. (1991) Cowpea *Striga* problems and research in Nigeria. In: Kim, S.K. (ed.) *Combating Striga in Africa*. IITA, Ibadan, Nigeria, pp. 18–28.

Eplee, R.E. (1975) Ethylene: a witchweed seed germination stimulant. *Weed Science* 23, 433–436.

Hauck, C., Müller, S. and Schildknecht, H. (1992) A germination stimulant for parasitic flowering plants from *Sorghum bicolor*, a genuine host plant. *Journal of Plant Physiology* 139, 474–478.

Haussmann, B.I.G., Hess, D.E., Welz, H.G. and Geiger, H.H. (2000) Improved methodologies for breeding *Striga*-resistant sorghums. *Field Crops Research* 66, 195–211.

Herbaugh, L., Upton, N.P. and Eplee, R.E. (1980) *Striga gesnerioides* in the United States of America. *Proceedings of the Southern Weed Science Society* 33, 187–190.

Hess, D.E. (1994) Crop-specific strains of *Striga hermonthica* in Niger. *Phytopathology* 84, 1151.

Hess, D.E. and Ejeta, G. (1992) Inheritance of resistance to *Striga* in sorghum genotype SRN39. *Plant Breeding* 109, 233–241.

Hess, D.E. and Haussmann, B.I.G. (1999) Status quo of *Striga* control: Prevention, mechanical and biological control methods and host plant resistance. In: Kroschel, J., Mercer-Quarshie, H. and Sauerborn, J. (eds) *Advances in Parasitic Weed Control at On-farm Level* Vol. 1. *Joint Action to Control Striga in Africa*. Margraf Verlag, Weikersheim, Germany, pp. 75–87.

Hess, D.E., Ejeta, G. and Butler, L.G. (1992) Selecting sorghum genotypes expressing a quantitative biosynthetic trait that confers resistance to *Striga*. *Phytochemistry* 31, 493–497.

Hess, D.E., Kroschel, J., Traoré, D., Elzein, A.E.M., Marley, P.S., Abbasher, A.A. and Diarra, C. (2002) *Striga*: Biological control strategies for a new millennium. In: Leslie, J.F. (ed.) *Sorghum and Millet Diseases 2000*. Iowa State Press, Ames, Iowa, USA, pp. 165–170.

Kim, S.K. (1994) Genetics of maize tolerance of *Striga hermonthica*. *Crop Science* 34, 900–907.

King, S.B. and Zummo, N. (1977) Physiologic specialization in *Striga hermonthica* in West Africa. *Plant Disease Reporter* 61, 770–773.

Kirk, A.A. (1993) A fungal pathogen with potential for control of *Striga hermonthica* (Scrophulariaceae). *Entomophaga* 38, 359–460.

Kroschel, J., Hundt, A., Abbasher, A. and Sauerborn, J. (1996) Pathogenicity of fungi collected in northern Ghana to *Striga hermonthica*. *Weed Research* 36, 515–520.

Lagoke, S.T.O., Parkinson, V. and Agunbiade, R.M. (1991) Parasitic weeds and control methods in Africa. In: Kim, S.K. (ed.) *Combating Striga in Africa*. IITA, Ibadan, Nigeria, pp. 3–14.

Lane, J.A., Moore, T.H.M., Child, D.V., Cardwell, K.F., Singh, B.B. and Bailey, J.A. (1994) Virulence characteristics of a new race of the parasitic angiosperm, *Striga gesnerioides*, from southern Bénin on cowpea (*Vigna unguiculata*). *Euphytica* 72, 183–188.

Lane, J.A., Moore, T.H.M., Child, D.V. and Bailey, J.A. (1997) Variation in virulence of *Striga gesnerioides* on cowpea: new sources of crop resistance. In: Singh, B.B., Mohan Raj, D.R., Dashiell, K.E. and Jackai, L.E.N. (eds) *Advances in Cowpea Research*. IITA/JIRCAS, Ibadan, Nigeria, pp. 225–230.

Logan, D.C. and Stewart, G.R. (1991) Role of ethylene in the germination of the hemiparasite *Striga hermonthica*. *Plant Physiology* 97, 1435–1438.

M'Boob, S.S. (1989) A regional programme for *Striga* control in West and Central Africa. In: Robson, T.O. and Broad, H.R. (eds) *Striga – Improved Management in Africa*. FAO Plant Production and Protection Paper, FAO, Rome, Italy, pp. 190–194.

Miché, L.M., Boillant, L., Rohr, R., Sallé, G. and Bally, R. (2000) Physiological and cytological studies on the inhibition of *Striga* seed germination by the plant growth-promoting bacterium *Azospirillum brasilense*. *European Journal of Plant Pathology* 106, 347–351.

Mohamed, K.I., Musselman, L.J. and Riches, C. (2001) The genus *Striga* (Scrophulariaceae) in Africa. *Annals of the Missouri Botanical Garden* 88, 60–103.

Musselman, L.J. (1980) The biology of *Striga*, *Orobanche*, and other root-parasitic weeds. *Annual Review of Phytopathology* 18, 463–489.

Musselman, L.J. and Parker, C. (1981) Studies on indigo witchweed, the American strain of *Striga gesnerioides* (Scrophulariaceae). *Weed Science* 29, 594–596.

Ouédraogo, J.T., Maheshwari, V., Berner, D.K., St-Pierre, C.A., Belzile, F. and Timko, M.P. (2001) Identification of AFLP markers linked to resistance of cowpea (*Vigna unguiculata* L.) to *Striga gesnerioides*. *Theoretical and Applied Genetics* 102, 1029–1036.

Parker, C. and Riches, C.R. (1993) *Parasitic Weeds of the World: Biology and Control*. CAB International, Wallingford, UK.

Pieterse, A.H. and Pesch, C.J. (1983) The witchweeds (*Striga* spp.) a review. *Abstracts of Tropical Agriculture* 9, 9–35.

Portères, R. (1952) Linear cultural sequences in primitive systems of agriculture in Africa and their significance. *Sols Africains/African Soils* 2, 15–28.

Raynal-Roques, A. (1994) Répartition géographique et spéciation dans le genre *Striga* (Scrophulariaceae parasites). *Memoirs of the Society of Biogeography* 4, 83–94.

Sand, P.F. (1990) Discovery of witchweed in the United States. In: Sand, P.F., Eplee, R.E. and Westerbrooks, R.G. (eds) *Witchweed Research and Control in the United States*. Weed Science Society of America, Champaign, Illinois, USA, pp. 1–6.

Sand, P.F. and Manley, J.D. (1990) The witchweed eradication programme survey, regulatory and control. In: Sand, P.F., Eplee, R.E. and Westerbrooks, R.G. (eds) *Witchweed Research and Control in the United States*. Weed Science Society of America, Champaign, Illinois, USA, pp. 141–151.

Sauerborn, J. (1991) *Parasitic Flowering Plants: Ecology and Management*. Verlag Josef Margraf, Weikersheim, Germany.

Sauerborn, J. (1999) Legumes used for weed control in agroecosystems in the tropics. *Plant Research and Development* 50, 74–82.

Sauerborn, J., Dörr, J., Abbasher, A., Thomas, H. and Kroschel, J. (1996) Electron microscopic analysis of the penetration process of *Fusarium nygamai*, a hyperparasite of *Striga hermonthica*. *Biological Control* 7, 53–59.

Sauerborn, J., Sprich, H. and Mercer-Quarshie, H. (2000) Crop rotation to improve agricultural production in sub-saharan Africa. *Journal of Agronomy and Crop Science* 184, 67–72.

Singh, B.B. and Emechebe, A.M. (1990) Inheritance of *Striga gesnerioides* resistance in cowpea genotype 301. *Crop Science* 30, 879–881.

Singh, B.B. and Emechebe, A.M. (1997) Advances in research on cowpea *Striga* and *Alectra*. In: Singh, B.B., Mohan Raj, D.R., Dashiell, K.E. and Jackai, L.E.N. (eds) *Advances in Cowpea Research*. IITA/JIRCAS, Ibadan, Nigeria, pp. 215–224.

Smith, M.C., Holt, J. and Webb, M. (1993) Population model of the parasitic weed *Striga hermonthica* (Scrophulariaceae) to investigate the potential of *Smicronyx umbrinus* (Coleoptera: Curculionidae) for biological control in Mali. *Crop Protection* 12, 471–476.

Sprich, H. and Sauerborn, J. (1999) Characterization of chiefly cereals-based cropping systems in northern Ghana. *Plant Research and Development* 49, 91–103.

Touré, M., Olivier, A., Ntare, B.R., Lane, J.A. and St-Pierre, C.A. (1998) Reaction of cowpea (*Vigna unguiculata*) cultivars to *Striga gesnerioides* races from Mali and Niger. *Canadian Journal of Plant Science* 78, 477–480.

Worsham, A.D. (1987) Germination of witchweed seeds. In: Musselman, L.J. (ed.) *Parasitic Weeds in Agriculture*, Vol. 1, *Striga*. CRC Press, Boca Raton, Florida, USA, pp. 45–61.

18

Mango-infesting Fruit Flies in Africa: Perspectives and Limitations of Biological Approaches to their Management

Slawomir A. Lux, Sunday Ekesi, Susan Dimbi, Samira Mohamed and Maxwell Billah
International Centre of Insect Physiology and Ecology (ICIPE), Nairobi, Kenya

Introduction

The rapid population growth of sub-Saharan Africa and progressive urbanization (World Bank, 1996; World Resources, 1999) has resulted in increasing rates of malnutrition (IFPRI, 2002) and vitamin deficiency in large sectors of rural and urban populations (WHO, 2002). This, along with developing awareness of the nutritive value of fruit and the increased purchasing power of affluent segments of local populations, translates into increased domestic demand for fresh fruit. In addition, the demand for quality tropical fruit in Europe, America and Japan is also growing. The above factors, combined with increasingly liberal global trade arrangements, have created new and lucrative production and trade opportunities.

Horticulture is the fastest-growing agricultural sector in Africa, providing income and employment. However, profitable fruit production in Africa is greatly hampered by fruit flies. Our assessment reveals that of the 1.9 million t of mangoes produced in Africa annually, about 40% is lost due to fruit flies. Fruit infestation rates vary among countries and seasons, ranging from 5% to 100%. The recent introduction of a uniform and strict quarantine and the Maximum Residue Level regulations in the European Union (EU) are other factors compounding the problem and jeopardizing the lucrative export of fresh mangoes from Africa. According to data presented during the meetings of the FAO Inter-Governmental Sub-Group on Tropical Fruits held in Australia in 1999 and Costa Rica in 2001, mango exports are estimated at 35,000–40,000 t annually and are worth over US$42 million.

Responding to requests from farmers in the region, the ICIPE in 1998 launched the African Fruit Fly Initiative (AFFI) to provide environmentally friendly and affordable technologies and skills for the management of fruit flies.

(eds P. Neuenschwander, C. Borgemeister and J. Langewald)

The AFFI is a collaborative network, linking 12 African countries. Mango, *Mangifera indica* L. (*Anacardiaceae*), was chosen as a target crop out of the many fruits grown in Africa, and mango-infesting fruit flies of the genus *Ceratitis* (Diptera, Tephritidae) became the target pests, hence the focus of this chapter. The authors believe, however, that this example and the derived conclusions are largely representative of other fruit fly problems in Africa.

Mango-infesting Fruit Flies in Africa

Globally, of the 4257 fly species comprising the family of Tephritidae, about 1400 species are known to develop in fruits. Out of these, about 250 species already are, or may become, pests by inflicting severe damage to fruits of economic value (White and Elson-Harris, 1992; Thompson, 1998). Equatorial Africa is the aboriginal home to 915 fruit fly species from 148 genera, out of which 299 species develop in either wild or cultivated fruit. They belong mainly to four genera: *Dacus*, *Ceratitis*, *Trirhithrum* and *Bactrocera* (White and Elson-Harris, 1992; Thompson, 1998). Most of them are highly polyphagous and their host ranges overlap to varying degrees. Though a great number of fruit species had already been reported to harbour fruit flies, numerous new host associations were recently found by Copeland (R. Copeland, 2002, Kenya, unpublished results) in Kenya, revealing that current literature records are far from complete. Accordingly, our understanding of host ranges, environmental reservoirs and patterns of host utilization remains superficial, even for the most common African fruit flies of major economic importance.

Mango is not indigenous to Africa, but originated in South-East Asia, from where it was introduced to all other tropical regions. In each region where it is grown, mango is attacked by fruit flies from different genera: *Bactrocera* (30 species) in Asia, *Anastrepha* (eight species) in America and *Ceratitis* (seven species) in Africa. The genus *Ceratitis* is endemic to the Afrotropical region, and contains about 65 species, the majority of which are highly polyphagous (White and Elson-Harris, 1992). The results of our surveys conducted in Kenya, Tanzania (mainland and Zanzibar), Sudan, Uganda, Côte d'Ivoire, Nigeria, Namibia, South Africa and the Indian Ocean Islands (Réunion) reveal that mango is commonly attacked by varying combinations of four major species across Africa: *Ceratitis cosyra* (Walker) (Plate 50), *Ceratitis fasciventris* (Bezzi) (Plate 52), *Ceratitis rosa* (Karsch) (Plate 53) and *Ceratitis anonae* (Graham) (Plate 51) (Lux *et al.*, 1999), and much less frequently, by *Ceratitis capitata* (Wiedemann). According to the literature, mango can also be infested by *Ceratitis catoirii* (Guérin-Meneville) (in Mauritus, Seychelles and Réunion) or by *Ceratitis flexuosa* (Walker) and *Ceratitis punctata* (Wiedemann) in West and Central Africa, including Uganda (White and Elson-Harris, 1992).

From our surveys, the major pest of mango across Africa is the mango fly also known as the marula fly, *C. cosyra*, formerly known as *Pardalaspis cosyra* (Walker), *Pardalaspis parinarii* (Hering) or *Trypeta cosyra* (Walker). It causes enormous damage, ranging from 10% to 100%. On average, about 20–30% of mango crops are lost due to this pest alone. *C. cosyra* is broadly distributed across eastern, central and western Africa, and also in parts of southern Africa (De Meyer, 2001a). Although it has been recorded from many fruit species, its typical host range is relatively narrow. Apart from mango, it commonly develops in fruits of marula, *Sclerocarya birrea* (A. Rich.) Hochst. (*Anacardiaceae*), the primary endemic host of this fly, in custard apple, *Annona senegalensis* Pers. (*Annonaceae*) and guava, *Psidium guajava* L. (*Myrtaceae*). In extensive surveys conducted in South Africa and Zimbabwe, *C. cosyra* was also recovered from a range of other fruits (White and Elson-Harris, 1992; De Meyer, 2001a). However, some of these records may represent infrequent host associations that had not been recorded elsewhere, even though the same fruits are also widely grown in the other regions where *C. cosyra* is common. Males of the marula fly do not respond well to commercially available parapheromone lures such as Trimedlure (TML), CueLure or methyl eugenol (ME). They do respond, however, to terpinyl acetate, β-caryophylene and several other terpenoids. Females respond to food baits such as Nulure.

Second in terms of economic impact on mango production are the Natal fruit fly, *C. rosa*, and its close relative, *C. fasciventris*. The two were formerly regarded as strains of the same species (Thompson, 1998), but are now considered to be separate species (De Meyer, 2001b). *C. rosa* is broadly distributed throughout eastern and southern Africa, while *C. fasciventris* is scattered throughout central and western Africa (De Meyer, 2001a). Both species occur in Kenya, the former mainly on the coast, the latter mainly in central and western parts of the country. Our surveys in several locations in Kenya revealed that both species co-exist in the same orchard. The two species are largely compatible reproductively, and produce viable fertile crosses. Under laboratory conditions, females of both species sometimes mate with non-conspecific males, even when given access to the conspecific males (S.A. Lux, unpublished data). Both *C. rosa* and *C. fasciventris* are highly polyphagous and capable of inflicting considerable damage. They have been recorded from a broad range of cultivated fruits apart from mango (White and Elson-Harris, 1992; De Meyer, 2001a). The Natal fruit fly *C. rosa* (and, perhaps, to a lesser degree also *C. fasciventris*) is tolerant to a broad range of temperatures and seems capable of establishing in cooler areas than the Mediterranean fruit fly, *C. capitata*. It is a highly competitive species and, in the areas where both species co-occur, it is capable of displacing *C. capitata* from several hosts (Hancock, 1984), hence, the reason for quarantine concerns in Europe and America. Males of both *C. rosa* and *C. fasciventris* respond well to TML, whereas females respond to food baits such as Nulure.

C. anonae is broadly distributed in western and central Africa, and eastwards to western Kenya and the shores of Lake Victoria. In West Africa (Hala N'klo, Côte d'Ivoire, 2000, unpublished results), it attacks mango in succession following the marula fly. Apart from mango, *C. anonae* develops in robusta coffee, *Coffea canephora* Pierre ex. A. Froehner (*Rubiaceae*), tropical almond, *Terminalia catappa* L. (*Combretaceae*), and probably in common guava, *P. guajava*, soursop, *Annona muricata* L. (*Annonaceae*) and strawberry guava, *Psidium littorale* Raddi (White and Elson-Harris, 1992; De Meyer, 2001a). Males respond to TML and females to Nulure.

The Mediterranean fruit fly, *C. capitata*, is of rather minor importance to mango production in Africa, although it is a very important pest of a range of other fruits of economic value. It has perhaps the broadest host range of all fruit flies, being capable of infesting over 300 fruit species (White and Elson-Harris, 1992). Males respond well to TML, and to a certain extent also to terpinyl acetate and other terpenoids. Females respond to Nulure. *C. capitata* is the only species of the *Ceratitis* genus which has invaded other tropical regions and has become permanently established outside Africa, namely in Latin America, Hawaii, the Mediterranean basin and Australia. Other mango-infesting fruit flies from this genus have not yet expanded their range beyond the Afrotropical region. However, as the fruit trade increases, *C. cosyra* is becoming the most frequently intercepted African fruit fly in Europe (I. White, Nairobi, Kenya, 2000, personal communication) and is already listed as a potentially invasive fruit fly of quarantine concern in the USA (Steck, 2000).

Status of Fruit Fly Control in Equatorial Africa

In equatorial Africa, fruit farms vary considerably in size and are seldom grouped into uniform production blocks. They are scattered among areas containing wild and abandoned fruit trees that act as reservoirs for a range of pests, including fruit flies. During our surveys, attempts to control fruit flies were reported by some professional fruit producers, but the majority of the small growers interviewed reported no such attempts. The interviews revealed that small growers tended to harvest fruits early before they matured, a strategy to evade fruit infestation, rather than prevent it by fruit fly management. Indeed, unripe fruits are either not yet infested or contain only the inconspicuous eggs or very young larvae. The ripe fruits are only utilized locally and are usually infested with fruit flies. Limited quantities of un-infested fruit are sold on local urban markets or exported. The rare attempts to control fruit flies are usually based on blanket pesticide sprays or, uncommonly, the use of imported bait sprays. Most of these actions are ineffective due to a lack of basic knowledge about fruit fly biology and management strategies. The current level of development of fruit production, especially its low productivity, fragmentation and general technological simplicity, greatly limits the options for economically sound fruit fly management.

Nevertheless, development of a profitable and modern fruit industry in Africa is possible and the Republic of South Africa (RSA) provides a good example. Remarkably, this has been achieved in an environment where fruit flies are present, but are managed by an integrated approach. The productivity of the fruit industry and the effectiveness of fruit fly management methods in RSA have reached the stage where losses to the industry caused by fruit flies (combined costs of management actions and residual losses caused by fruit flies) are less than 1% of the total value of the produce.

Major Approaches and Methods of Fruit Fly Control and their Applicability in Equatorial Africa

There are two basic approaches to fruit fly management: (i) the IPM approach, i.e. controlling fruit fly populations in order to reduce yield losses (often supplemented by postharvest fruit treatments); and (ii) the eradication approach, i.e. eliminating fruit flies to create certified 'fruit fly-free' zones. Eradication is conducted as an area-wide action that is very costly and only justifiable when a highly productive industry is threatened. Where eradication methods are not available or not justifiable, fruit fly management methods are used to allow quality fruit production in spite of the presence of these pests in the environment. The major components of such an integrated approach are discussed below.

Indigenous Parasitoids of African Fruit Flies

Following the establishment of the Mediterranean fruit fly, *C. capitata*, in other tropical regions (the Americas and Australia), several explorations for its natural enemies were undertaken in Africa in order to seek parasitoids effective in controlling this pest in invaded regions (see review of Wharton in Robinson and Hooper, 1989). There is also considerable current interest in obtaining more effective parasitoids for biological control of fruit flies in the neotropics (see reviews by Messing *et al.*, in McPheron and Steck, 1996). Indeed, out of over 40 parasitoid species collected from Africa, only a few have so far been used to control *C. capitata* (Table 18.1). Several non-African parasitoids, introduced for controlling other tephritid fruit flies, were able to create new host associations and develop in *C. capitata* (Table 18.2). Although a number of natural enemies

Table 18.1. Species of African fruit fly parasitoids that have been introduced and established outside Africa.

Parasitoid species	Origin	Where established/used	Host attacked
Psyttalia concolor (Braconidae)	Tunisia	Mediterranean region	*Bactrocera oleae*, *Ceratitis capitata*
Psyttalia humilis (Braconidae)	South Africa	Hawaii, Bermuda	*C. capitata*
Diachasmimorpha fullawayi (Braconidae)	West Africa	Brazil, Hawaii, Fiji, Spain	*C. capitata*
Tetrastichus giffardianus (Eulophidae)	West and South Africa	Hawaii, Brazil, Fiji, Spain	*C. capitata*, *Bactrocera* spp.
Tetrastichus dacicida (Eulophidae)	Tanzania	Hawaii	*C. capitata*, *Bactrocera dorsalis*
Dirhinus giffardii (Chalcididae)	Nigeria	Hawaii	*B. dorsalis* and *C. capitata*
Dirhinus anthracia (Chalcididae)	East and West Africa	Hawaii, Sri Lanka	*C. capitata*, *Bactrocera* spp.
Coptera silvestrii (Diapriidae)	West and South Africa	Hawaii	*C. capitata*, *Ceratitis anonae*, *Ceratitis giffardi*, *Trirhithrum nigerrimum*

Compiled from: Clausen *et al.* (1965); Wharton (1989) and Waterhouse and Sands (2001).

Table 18.2. Exotic parasitoid species released for biological control of various Tephritidae and also recovered from *Ceratitis capitata*.

Parasitoid species	Natural host(s)	Host stage attacked	Region of origin	Country of introduction
Fopius arisanus (Braconidae)	*Bactrocera* spp.	Egg and early-instar larva	South-East Asia	Hawaii
Fopius vandenboschi (Braconidae)	*Bactrocera dorsalis*	Early-instar larva	South-East Asia	Hawaii
Psyttalia incisi (Braconidae)	*B. dorsalis* complex	Late-instar larva	South-East Asia	Hawaii
Diachasmimorpha longicaudata (Braconidae)	*B. dorsalis* complex	Late-instar larva	Philippines and several other Indo-Pacific localities	Hawaii, Australia, Fiji, Costa Rica, Florida, Trinidad
Diachasmimorpha tryoni (Braconidae)	*Bactrocera* spp.	Late-instar larva	Eastern Australia	Hawaii
Diachasmimorpha kraussii (Braconidae)	*Bactrocera tryoni*, *Bactrocera neohumeralis*, *Bactrocera cacuminata*	Late-instar larva	Eastern Australia	Hawaii
Aceratoneuromyia indica (Eulophidae)	*Bactrocera* spp.	Larval stage	India	Central America

Compiled from: Clausen *et al.* (1965); Wharton (1989) and Waterhouse and Sands (2001).

have been tried in fruit fly biological control programmes outside Africa with some success, in most cases, the impact of the introduced parasitoids on *C. capitata* populations was rather limited (Knipling, 1992; see also review of Montoya and Liedo in Tan, 2000). Recently, the potential of an inundative approach in the control of the Mediterranean fruit fly was successfully demonstrated on coffee in Hawaii (Vargas *et al.*, 2001). It must be mentioned, however, that there are growing concerns about the potential environmental risks associated with 'classical' biological control, especially non-target effects on weed-controlling beneficial fruit flies or on rare endemic fruit flies that merit conservation efforts (see review by Messing in McPheron and Steck, 1996).

During the first phase of AFFI, several parasitoids capable of attacking *C. capitata*, *C. rosa* and *C. cosyra* were collected and identified (Table 18.3). Laboratory colonies of three species, *Psyttalia concolor* (Szépligeti) (Kenyan population), *Psyttalia cosyrae* (Wilkinson) and *Psyttalia phaeostigma* (Wilkinson) (Hymenoptera, Braconidae) were established and their host ranges, host preferences and host suitability studied. Perhaps for the first time, this kind of research was not restricted to the Mediterranean fruit fly, but included other African fruit fly species (*C. rosa* and *C. cosyra*) of local economic importance (S. Mohamed and M. Billah, Nairobi, Kenya, 2002, unpublished). Several of these parasitoids proved capable of attacking the Mediterranean fruit fly and hence were sent for further evaluation to the United States Department of Agriculture (USDA) in Hawaii and Guatemala, and also to RSA (for use on St

Table 18.3. Parasitoids of *Ceratitis* spp. collected by AFFI in equatorial Africa.

Parasitoid species	Country	Possible host fly	Host fruit
Diachasmimorpha fullawayi (Braconidae)	Kenya, Ghana	*Ceratitis capitata*, *Ceratitis rosa*, *Trirhithrum coffeae*	Coffee, mango
Diachasmimorpha sp. (Braconidae)	Kenya	*Ceratitis* spp.	Coffee
Bracon sp. (Braconidae)	Kenya, South Africa	*C. capitata*	Coffee
Fopius ceratitivorus (Braconidae)	Kenya, South Africa	*Ceratitis* spp.	Coffee
Fopius caudatus (Braconidae)	Kenya, Uganda, Cote D'Ivoire	*Ceratitis anonae*, *Trirhithrum* spp.	Coffee, guava, lemon, mango
Fopius silvestrii (Braconidae)	Kenya, Ghana	*Ceratitis* spp.	Coffee, tomato
Psyttalia cosyrae (Braconidae)	Kenya, Tanzania, Uganda, Cote D'Ivoire	*Ceratitis cosyra*	Mango, marula
Psyttalia sp. (Braconidae)	Kenya, South Africa, Uganda	*C. capitata*	Coffee
Psyttalia perproximus (Braconidae)	Ghana	*T. coffeae*	Coffee
Tetrastichus giffardianus (Eulophidae)	Kenya	*Ceratitis* spp., *Dacus* spp.	Coffee, mango, squash

Helena Island). Surveys of collected fruits indicated, however, that the level of natural parasitism of fruit fly larvae is typically quite low, although these observations might not accurately reflect the field situation. Some parasitized late-instar larvae might have left the fruits to pupate in the soil, and thus might have escaped observation. Indeed, in our samples of puparia collected from the soil in coffee fields in Kenya, parasitism by all parasitoid species combined reached over 50% at the peak of the season. Under laboratory conditions, two *Psyttalia* species achieved parasitism rates as high as 65% on both *C. capitata* and *C. cosyra*. However, for most of the year, parasitism rates were much lower, ranging from about 1% to 10%. In general, fruit fly populations in these coffee fields remained extremely high throughout the season.

It appears that natural populations of indigenous parasitoids, although capable of high parasitism rates at certain times during the year, are not able to suppress fruit fly populations significantly. The effectiveness of parasitoids in controlling fruit flies in various cultivated fruits also needs to be critically evaluated. In our surveys in Kenya, the highest parasitism rates were found in coffee, while the parasitism rates of fruit flies infesting other cultivated fruits such as mango was negligible. It appears that the prospects for augmentation of indigenous parasitoids in the aboriginal home of the pest seem limited. The beneficial effects and supportive role of native parasitoids in fruit fly management cannot be questioned, and might justify possible conservation efforts, but are unlikely to result in substantial levels of fruit fly control on their own. The feasibility of using an inundative approach also appears to be limited in Africa because of the prohibitively high investment required for facilities dedicated to mass-rearing parasitoids.

Other Biological Control Agents

Predators from the families Staphylinidae, Carabidae, Chrysopidae, Pentatomidae and several mite species are known to attack tephritids (see Bateman, 1972 for review). The efficiency of two earwig species as predators of *Bactrocera dorsalis* (Hendel) (Diptera, Tephritidae) in Hawaii has been studied by Marucci (1955). Also ants were reported to cause up to 38% mortality of *C. capitata* (Wong *et al.*, 1984). In the laboratory, Poinar and Hislop (1981) reported mortality of Mediterranean fruit fly adults caused by the entomopathogenic nematodes *Neoaplectana* spp. (Rhabditida, Steinernematidae) and *Heterorhabditis* spp. (Rhabditida, Heterorhabditidae), and Beavers and Calkins (1984) showed that *Anastrepha ludens* (Loew) (Diptera, Tephritidae) larvae were also susceptible to steinernematids and heterorhabditids. In field studies in Hawaii exposure of mature larvae of *C. capitata* to 500 infective juveniles cm^{-2} of *Steinernema feltiae* (Filipjev) in the soil resulted in 70% mortality of puparia (Lindegren *et al.*, 1990). In Bermuda, the lizard *Anolis grahami* Gray (Sauria, Iguanidae) was introduced from Jamaica for control of fruit flies, although its role in controlling the pest has not been quantified (Clausen, 1978). Also, birds and rodents have been reported to cause a high level of larval mortality by consuming infested fruits (Drew, 1987).

Similar groups of predators are likely to play a role in restricting fruit fly populations throughout Africa, and their conservation may be of practical importance. However, a thorough assessment of their impact on fruit fly populations in various regions in Africa, and in various production systems, is needed.

Pathogens

One of the first observations of a pathogen attacking fruit flies was made by Fujii and Tamashiro (1972), who reported an infection by the protozoan pathogen, *Nosema tephritidae* (Fujii and Tamashiro) (Microspora, Nosematidae) on *B. dorsalis* and *C. capitata* in Hawaii. Several isolates of *Bacillus thuringiensis* (Berliner) (*Bt*) have also been evaluated in the laboratory and in the field against larvae of *Bactrocera oleae* (Gmelin) and adults of *A. ludens* (Robacker *et al.*, 1996). Field application of a suspension of *Bt* applied at OD_{600} spores and crystals in 4% yeast enzymatic hydrolysate has been reported to significantly reduce damage to olive fruits by *B. oleae* in Greece (Navrozidis *et al.*, 2000). Recently, the use of entomopathogenic fungi has received increasing attention, both in Europe and in the Americas. In Spain, topical application of suspensions of *Metarhizium anisopliae* (Metsch.) Sorok. and *Paecilomyces fumosoroseus* (Wize) Smith and Brown (both *Hyphomycetes*) at 1×10^6 conidia per fly resulted in 100% mortality of adult *C. capitata* in 6 days (Castillo *et al.*, 2001). The same isolates were also reported to reduce fly fecundity and fertility. The virulence of *M. anisopliae* to adult *A. ludens* has also been demonstrated in both laboratory and field tests in Mexico (Lezama-Gutierrez *et al.*, 2000). In field cage experiments, applications of *M. anisopliae* sprayed on the soil at 2.5×10^6 colony forming units (cfu) per ml reduced adult emergence by 33% in loam soil and 49% in sandy loam soil. In Mexico, De La Rosa *et al.* (2002) achieved between 82% and 100% mortality in the laboratory spraying adult *A. ludens* with *Beauveria bassiana* (Bals.) Vuill. (*Hyphomycetes*) at a concentration of 1×10^8 conidia ml^{-1}.

During the first phase of AFFI, two application tactics were evaluated in Kenya: (i) soil inoculation of pathogens to create a hostile environment for pupariating larvae and puparia; and (ii) use of pathogens as an alternative to pesticides commonly used in localized baiting stations as a killing agent for the attracted adult flies. Special attention was given to fungal pathogens as they are able to invade via the external cuticle of the insect and do not require ingestion. Contemporary methods of mass production and conidia formulation that maintain high levels of conidial stability during storage and increase the efficacy of field applications (Inglis *et al.*, 2001) were employed.

A number of strains of *M. anisopliae* and *B. bassiana* were screened and several were found to be effective against adults, larvae and puparia of a wide range of fruit flies (*C. cosyra*, *C. rosa*, *C. fasciventris*, *C. anonae* and *C. capitata*). The evaluated isolates reduced adult emergence from treated soil by 6–68% in the laboratory and in field cages (Ekesi *et al.*, 2002). Mortality of puparia decreased with increasing pupal age. This was attributed to the inability of germ tubes to penetrate the pupal integument, as it hardens due to sclerotization.

Post-emergence mycosis was also observed in adults escaping infection as pupariating larvae, and this may play a significant role in field suppression (Ekesi *et al.*, 2002). From our preliminary evaluations, soil inoculation with a pathogen can effectively reduce the numbers of fruit flies emerging from the treated soil. The pathogen appears to be able to retain its effectiveness even in fairly harsh environmental conditions, such as high temperature and prolonged drought. There are also indications that the isolates selected are quite benign to parasitoids. Soil-borne fungal pathogens are ideal candidates for soil inoculation targeting pupariating larvae and puparia.

Auto-inoculative devices of various designs are currently being tried for control of adult stages of various pests (Vega *et al.*, 2000). The main advantage of the auto-inoculative approach is that the amount of conidia required for field application is greatly reduced relative to large-scale inundative applications. In addition, fungal spores housed in a baiting station are protected from fungicides that may be used in the orchard, and from the harmful effects of UV radiation. In our experiments, a simple baiting station was used as an auto-inoculative device for attracted fruit flies. In earlier experiments, exposure of the target fruit fly species to dry conidia of *M. anisopliae* and *B. bassiana* resulted in mortality ranging from 9% to 100% within 5 days post-inoculation (S. Dimbi and S. Ekesi, 2002, Nairobi, unpublished results). The most pathogenic isolate of *M. anisopliae* was deployed in the baiting stations and evaluated for 2 successive years in small-scale mango orchards at Nguruman, Kenya. Encouragingly, the effectiveness of the baiting stations equipped with the pathogen did not differ from those equipped with an insecticide (malathion). A 70% reduction in fruit fly populations was achieved in these small plots (0.5 ha).

It is unlikely that fungal pathogens can constitute a stand-alone fruit fly management strategy, but they do have the potential to play an important role if integrated with other IPM control tactics such as baiting stations, crop sanitation and conservation of natural enemies (predators and parasitoids). As a component of such programmes, the pathogens would be suitable for control of fruit flies in both subsistence and commercial orchards.

Utilizing Chemical Signals Mediating Fruit Fly Behaviour

Apart from biological control *sensu stricto*, other methods of biological manipulation are commonly used in fruit fly control. These include application of odours that mimic chemical signals used by fruit flies in their reproduction (pheromones and parapheromones), feeding (food attractants) and host finding (kairomones).

Application of pheromones

The reproductive behaviour of tropical fruit flies and its related intra-specific chemical communication is diverse and complex. Males usually aggregate to form leks (groups of displaying males). Pheromones are emitted by males and used to recruit other males during lek formation, and to attract females to the lek. The attracted females interact with individual males within the lek and may choose (or frequently not) one of them for mating. Males approached by females display elaborate courtship behaviour, the quality of which determines their chances of being accepted for mating. Other aspects of fruit fly reproduction such as oviposition are mediated by kairomones that assist females in finding suitable host fruits. There are also host-marking pheromones deposited by females on the fruit surface immediately after an egg is laid. The role of the latter is to reduce intra-specific competition by discouraging other conspecific females from laying eggs in the same fruit. Such oviposition-deterring pheromones are known in many fruit flies from the *Rhagoletis* genus (Diptera, Tephritidae), and also in the Mediterranean fruit fly and a few other species. It is likely that other fruit flies from the genus *Ceratitis* use such pheromones, but no conclusive evidence has yet been collected. Summarizing the status of our knowledge on fruit fly behaviour and chemical ecology goes beyond the scope of this chapter, but the interested reader is referred to chapters in Robinson and Hooper (1989) and McPheron and Steck (1996).

Although an extensive body of information has been accumulated to date, our understanding of fruit fly pheromones still remains superficial. In spite of intense efforts and the huge economic benefits at stake, none of the true sex pheromones of tropical fruit flies have yet been exploited for practical applications. This is in contrast to the control of moths, beetles and other insects, where pheromones have been widely used in IPM for a long time. Nearly nothing is known about the nature of pheromones used by the major mango-infesting fruit flies in Africa and it is only recently that identification of their key components has been undertaken (N. Gikongo and S.A. Lux, Nairobi, Kenya, 2002, unpublished results).

Application of parapheromones

The term 'parapheromone' covers a broad category of highly attractive chemicals that are not components of true pheromones, but have similar activity. They include ME (attracting several *Bactrocera* species), TML (attracting *C. capitata*, *C. rosa*, *C. fasciventris*), Cue-lure (attracting *Bactrocera cucurbitae* Coquillet, and a few other species), VertLure (attracting *Dacus vertebratus* Bezzi.), terpinyl acetate (attracting a range of *Ceratitis* spp.) and other compounds (see chapter by Cunningham in Robinson and Hooper, 1989). These materials attract males, sometimes with great efficacy, as in the case of ME. The biological significance of these substances is not yet fully understood, and the statement of Cunningham (in Robinson and Hooper, 1989) that 'parapheromone is an artificial term to cover our ignorance', remains largely valid. Some of these compounds occur widely in essential plant oils. ME is known from at least 25 plants from several families; other parapheromones do not occur naturally but are derivatives of natural plant products. For example, Cuelure relates to raspberry ketone; TML relates to terpinyl acetate (or more correctly, terpineol acetate; TA) and *α*-copaene. Recent studies suggest that the mysterious attraction of several *Bactrocera* species to ME might have evolved as a consequence of their association with rainforest orchids from the genus of *Bulbophyllum* (*Orchidaceae*). Indeed, males of these species are known to be unique pollinators of these orchids, and by sequestering components of the scent of the pollinated flowers they enhance their attractiveness to females (Shelly and Devire, 1994; see also chapter of Tan in Tan, 2000). Similar enhancement of male sexual attractiveness was demonstrated in Mediterranean fruit flies exposed to TML (Shelly *et al.*, 1996). Interestingly, despite intense screening of thousands of chemicals, no parapheromones have been found for New World fruit flies from the genus *Anastrepha*.

Despite their obscure biological significance, parapheromones constitute exceptionally powerful IPM tools that are widely used in fruit fly detection and monitoring, and some (such as ME) are also used in direct control. The marula fruit fly, *C. cosyra*, and other mango-infesting fruit flies of the same genus (*Ceratitis*) respond to TA and other terpenoids. The response is sufficient for reliable monitoring, but is not strong enough for TA to be used in direct fruit fly control.

Application of food baits

To attain full reproductive potential, females of fruit flies must feed after emergence. Apart from sugars and vitamins, their food must contain the proteins or amino acids necessary for ovigenesis. Natural food sources consist of honeydew, bacteria living on leaf surfaces, bird droppings, etc. Females are guided to food by odours emanating from food sources. Artificial food baits are used to attract and kill food-seeking females before their eggs mature. Food baits are usually made from hydrolysed proteins derived from industrial waste materials such as brewery yeast or corn syrup. Food baits attract mainly females but are neither species-specific nor very powerful. However, since food baits can directly reduce the numbers of pre-reproductive females they constitute a useful tool for fruit fly control. Due to their broad spectrum of attractiveness, food baits are also widely used in fruit fly detection and monitoring, especially where several species are targeted at the same time. For control, food baits can be mixed with pesticides and used as sprays. Typically they are applied only on parts of the field by spraying only every second or third row in the orchard, or in spot applications covering only a small fraction (about 1 m^2) of the tree canopy. More localized application allows considerable reduction in the quantity of pesticide used. Current efforts use traps baited with food attractants, not only for monitoring, but also for mass trapping. The focus is on the development of localized baiting stations composed of bait, a killing agent (usually a pesticide) and a protective housing. Such methods restrict pesticide use even further and limit its contact with fruit and the environment. Standard food baits are produced by several companies in the USA and Europe, and are relatively inexpensive. However, they are usually formulated as a liquid, which not only increases shipment costs, but also decreases convenience of application and durability in the field. A dry bait consisting of putrescine, trimethyamine and ammonium acetate (PTA) was recently developed by USDA in an attempt to overcome these disadvantages. Although PTA is both highly effective and quite durable (lasting several weeks in the field even in harsh conditions), it is relatively expensive. Another type of a semi-dry baiting station was recently developed in RSA, the M3 baiting station. It contains a food attractant mixed with a pesticide, formulated as a paste and placed in a plastic housing. Initial results are promising and it is currently being evaluated in RSA and Europe.

Several food baits were evaluated during the first phase of AFFI. The results confirmed that all mango-infesting fruit flies respond well to Nulure imported from the USA, and an alternative bait developed by AFFI and made from locally available brewery waste. As indicated above in the section on pathogens, our results confirm that baiting stations are effective in controlling fruit flies at the village or even at the farm level. Application of baiting stations, if conducted properly on larger plots or blocks of small farms, can be sufficient to protect fruit grown for domestic markets. These methods should be integrated with recognized postharvest treatments when fruit is destined for export markets. In our opinion, baiting stations are the most promising and appropriate method for fruit fly management in small-scale, fragmented horticulture, especially when integrated with other tools such as pathogens, orchard sanitation and the conservation of natural enemies.

Interfering with Fruit Fly Reproductive Behaviour

Biological manipulation of fruit fly reproduction has been successfully used for eradication and many examples are known from outside Africa. Frequently, a combination of various methods is used for initial reduction of fruit fly populations, although the final eradication is usually achieved by behavioural manipulations that nullify the reproductive potential of wild females. One method is the male annihilation technique (MAT), which is achieved by mass trapping of wild males until they are completely eliminated. Another option is the release of mass-reared sterile males at inundative ratios to the wild population (sterile insect technique; SIT). In the first case, wild females lack males to mate with, while in the latter case, they mate with sterile males. In either case, females lay infertile eggs, ultimately resulting in eradication of the pest population. The efficiency of MAT depends on the power of the attractant and is therefore largely restricted to use against those fruit fly species that respond to ME. The effectiveness of SIT depends on the behavioural quality (mating competitiveness) of the mass-produced males (for current reviews see Cayol *et al.*, Hendrichs *et al.*, Calgano *et al.*, Liedo *et al.*, Lux *et al.*, Robinson *et al.*, in FAO/IAEA Proceedings, 2002). Both approaches can be effective and economically viable when a single fruit fly species causes a problem to a large-scale fruit industry. Area-wide fruit fly control geared towards pest eradication is becoming a common approach in countries where fruit production is highly developed (for review see chapters by Lindquist, De Longo *et al.*, Villasenor *et al.*, Dowell *et al.*, Reyes *et al.*, Hancock *et al.*, in Tan, 2000). Such an approach requires efficient methods for achieving complete eradication and preventing re-infestation.

Our surveys in equatorial Africa have revealed that, in most locations, various combinations of mango-infesting fruit fly species co-exist. These species do not respond to ME, and no other attractants of comparable efficacy are known. SIT has not been developed against African fruit flies other than the Mediterranean fruit fly, and huge investments would be required to develop SIT against these other species that are really the most important in Africa. Even if such methods were developed, the economic viability of using SIT against a complex of several species is yet to be demonstrated. It is recognized that eradication approaches are mainly applicable to single-species scenarios and on the peripheries of their range, rather than within an extensive ancestral range (see overview by Hendrichs in McPheron and Steck, 1996). In our opinion, SIT is not an appropriate tactic for fruit fly control in most locations of equatorial Africa under current circumstances, although it might be locally applicable if the required conditions were met.

Conclusions

The picture emerging from the review presented above is rather grim. As one would expect, the global distribution of expertise follows closely the global distribution of modern horticulture industry and investments into fruit fly research and control. A virtual lack of investment in the African region results in a lack of local expertise and capacity to cope with fruit fly problems. To date, only those African fruit flies that have invaded other regions and threatened highly developed horticulture outside of Africa have been the subjects of substantial research attention. Those fruit flies that are of major economic importance within Africa remain relatively unstudied. The case of the Mediterranean fruit fly versus other fruit fly species from the same genus provides the most vivid example.

Many sophisticated control methods such as SIT developed for modern horticulture are not applicable to most of the fragmented and simplistic production systems prevailing over most of equatorial Africa. Other specific control methods such as those based on powerful parapheromones have never been developed. Generic control methods such as food-based baiting techniques, although potentially applicable in Africa, have not been validated against most of the locally important flies. Africa has been used as a rich source of natural enemies for biological control efforts in other regions, but has never been a beneficiary of such programmes.

The needs of local African populations and producers are changing rapidly, and there is a great need for further adaptation of existing methods for fruit fly control, the development of new methods, and the dissemination of these control techniques within Africa. Several methods of biological control and behavioural manipulation, such as the use of pathogens and baits, orchard sanitation, conservation of natural enemies, and postharvest treatments, could be combined into IPM packages to bring about fruit fly control that is sufficient to permit quality fruit production for domestic and export markets. RSA provides a model example of a highly productive fruit industry that has been developed in an environment containing indigenous fruit fly populations.

References

Bateman, M.A. (1972) The ecology of fruit flies. *Annual Review of Entomology* 17, 493–517.

Beavers, J.B. and Calkins, C.O. (1984) Susceptibility of *Anastrepha suspensa* (Diptera; Tephritidae) to steinernematid and heterorhabditid nematodes in laboratory studies. *Environmental Entomology* 13, 137–139.

Castillo, M.A., Moya, P. and Primo-Yufera, E. (2001) Susceptibility of *Ceratitis capitata* Wiedemann (Dipt: Tephritidae) to entomopathogenic fungi and their extracts. *Biological Control* 19, 274–282.

Clausen, C.P. (1978) Tephritidae (Trypetidae, Trupaeidae). In: Clausen, C.P. (ed.) *Introduced Parasites and Predators of Arthropod Pests and Weeds: a World View.* USDA Handbook 480, 320–325.

Clausen, C.P., Clancy, D.W. and Chock, Q.C. (1965) Biological control of the oriental fruit fly (*Dacus dorsalis* Hendel) and other fruit flies in Hawaii. Agricultural Research Service USA, *Technical Bulletin* 322.

De La Rosa, W., Lopez, F.L. and Liedo, P. (2002) *Beauveria bassiana* as a pathogen of the Mexican fruit fly (Diptera: Tephritidae) under laboratory conditions. *Journal of Economic Entomology* 95, 36–43.

De Meyer, M. (2001a) Distribution patterns and host-relationships within the genus *Ceratitis* MacLeay (Diptea: Tephritidae) in Africa. *Cimbebasia* 17, 219–228.

De Meyer, M. (2001b) On the identity of the Natal fruit fly *Ceratitis rosa* Karsch (Diptera, Tephritidae). *Entomologie* 71, 55–62.

Drew, R.A.I. (1987) Reduction in fruit fly (Tephritidae: Dacinae) populations in their endemic rainforest habitat by frugivorous vertebrates. *Australian Journal of Zoology* 35, 283–288.

Ekesi, S., Maniania, N.K. and Lux, S.A. (2002) Mortality in three African tephritid fruit fly puparia and adults caused by the entomopathogenic fungi, *Metarhizium anisopliae* and *Beauveria bassiana*. *Biocontrol Science and Technology* 12, 7–17.

FAO/IAEA Proceedings (2002) Proceedings of an FAO/IAEA Research Coordination Project on Medfly Mating. *Florida Entomologist* 85, 1.

Fujii, J.K. and Tamashiro, M. (1972) *Nosema tephritidae* sp. a microsporidian pathogen of the oriental fruit fly, *Dacus dorsalis* Hendel. *Proceedings of the Hawaiian Entomological Society* 21, 191–203.

Hancock, D.L. (1984) Ceratitinae (Diptera: Tephritidae) from the Malagasy subregion. *Journal of the Entomological Society of Southern Africa* 47, 277–301.

IFPRI (2002) webpage, Malnutrition and Food Insecurity Projections, 2020. www.ifpri.cgiar.org

Inglis, G.D., Goettel, M.S., Butt, T.M. and Strasser, H. (2001) Use of hyphomycetous fungi for managing insect pests. In: Butt, T.M., Jackson, C. and Magan, N. (eds) *Fungi as Biocontrol Agents: Progress, Problems and Potential.* CAB International, Wallingford, UK, pp. 27–69.

Knipling, E.F. (1992) *Principles of Insect Parasitism Analysed from new Perspectives. Practical Implications for Regulating Insect Populations by Biological Means.* United States Department of Agriculture, Agricultural Research Service, Agricultural Handbook 693, 1–335.

Lezama-Gutierrez, R., Trujillo-De La Luz, A., Molina-Ochoa, J., Rebolledo-Dominguez, O., Pescador, A.R., Lopez-Edwards and Aluja, M. (2000) Virulence of *Metarhizium anisopliae* (Deuteromycotina: Hyphomycetes) on *Anastrepha ludens* (Diptera: Tephritidae): laboratory and field trials. *Journal of Economic Entomology* 93, 1080–1084.

Lindegren, J.E., Wong, T.T. and McInnis, D.O. (1990) Responses of Mediterranean fruit fly (Diptera: Tephritidae) to the entomogenous nematode *Steinernema feltiae* in field tests in Hawaii. *Environmental Entomology* 19, 383–386.

Lux, S.A., Zenz, N. and Kimani, S. (1999) *Economic Role and Distribution of Fruit Flies.* ICIPE Annual Scientific Report 1998/99. ICIPE Science Press, Nairobi, Kenya, pp. 29–30.

Marucci, P.E. (1955) Notes on the predatory habits and life cycle of two Hawaiian earwigs. *Proceedings of Hawaiian Entomological Society* 15, 565–569.

McPheron, B.A. and Steck, G.J. (1996) *Fruit Fly Pests: a World Assessment of Their Biology and Management.* St Lucie Press, Delray Beach, Florida, USA.

Navrozidis, E.I., Vasara, E., Karamanlidou, G., Salpiggidis, G.K. and Koliais, S.I. (2000) Biological control of *Bactrocera oleae* (Diptera: Tephritidae) using a Greek *Bacillus thuringiensis* isolate. *Journal of Economic Entomology* 93, 1657–1661.

Poinar, G.O. Jr and Hislop, R.G. (1981) Mortality of Mediterranean fruit fly adults *Ceratitis capitata* from parasitic nematodes *Neoaplectana* and *Heterorhabditis* spp. *IRCS Journal of Medical Science* 11, 531–641.

Robacker, D.C., Matinez, A.J., Garcia, J.A., Diaz, M. and Romero, C. (1996) Toxicity of *Bacillus thuringiensis* to Mexican fruit fly (Diptera: Tephritidae). *Journal of Economic Entomology* 89, 104–110.

Robinson, A.S. and Hooper, G. (1989) *Fruit Flies: Their Biology, Natural Enemies and Control.* Elsevier, Amsterdam, The Netherlands.

Shelly, T.E. and Devire, A.M. (1994) Chemically mediated mating success in male Oriental fruit flies (Diptera: Tephritidae). *Annals of the Entomological Society of America* 87, 375–382.

Shelly, T.E., Whittier, T.S. and Villalobos, E.M. (1996) Trimedlure affects mating success and mate attraction in male Mediterranean fruit flies. *Entomologia Experimentalis et Applicata* 78, 181–185.

Steck, G. (2000) *Ceratitis cosyra* (Walker) (Diptera: Tephritidae). Florida Department of Agriculture and Consumer Services, Division of Plant Industry. *Entomology Circular* No. 403, November/December 2000.

Tan, K.H. (2000) *Area-Wide Control of Fruit Flies and Other Insect Pests.* Penerbit Universiti Sains Malaysia, Pulau Pinang.

Thompson, F.C. (1998) Fruit fly expert identification system and systematic information database. *Myia* 9, 1–224.

Vargas R.I., Peck, S.L., McQuate, G.T., Jackson, C.G., Stark, J.D. and Armstrong, J.W. (2001) Potential for areawide integrated management of Mediterranean fruit fly (Diptera: Tephritidae) with a braconid parasitoid and a novel bait spray. *Journal of Economic Entomology* 94, 817–825.

Vega, F.E., Dowd, P.F., Lacey, L.E., Pell, J.K., Jackson, D.M. and Klein, M.G. (2000) Dissemination of beneficial microbial agents by insects. In: Lacey, L.E. and Kaya, H.K. (eds) *Field Manual of Techniques in Invertebrate Pathology: Application and Evaluation of Pathogens for Control of Insect and Other Invertebrate Pests.* Kluwer Academic Press, Dordrecht, The Netherlands, pp. 153–177.

Waterhouse, D.F. and Sands, D.P.A. (2001). Classical biological control of arthropods in Australia. *ACIAR Monograph* No. 77.

Wharton, R.A. (1989) Classical biological control of fruit-infesting Tephritidae. In: Robinson, A.S. and Hooper, G. (eds) *Fruit Flies; Their Biology, Natural Enemies and Control.* Elsevier Science Publishers, Amsterdam, The Netherlands, pp. 303–313.

White, I.M. and Elson-Harris, M.M. (1992) *Fruit Flies of Economic Significance: Their Identification and Bionomics.* CAB International, Wallingford, UK.

WHO (2002) webpage, Nutrition. who.int/nut/vad.htm

Wong, T.T.Y., Mochizuki, N. and Nishimoto, J.I. (1984) Seasonal abundance of parasitoids of the Mediterranean and Oriental fruit flies (Diptera: Tephritidae) in the Kula area of Maui, Hawaii. *Environmental Entomology* 13, 140–145.

World Bank (1996) *Towards Environmentally Sustainable Development in Sub-Saharan Africa – a World Bank Agenda*. World Bank, Washington, DC, USA, November, 1996.

World Resources (1999) *U.N. World Resources Institute*. Washington, DC, USA, 1999.

19

Biological Control, a Non-obvious Component of IPM for Cowpea

Manuele Tamò,[1] Sunday Ekesi,[2] Nguya K. Maniania[2] and Andy Cherry[1]

[1]*International Institute of Tropical Agriculture, Cotonou, Bénin;*
[2]*International Centre of Insect Physiology and Ecology (ICIPE), Nairobi, Kenya*

Introduction

Cowpea, *Vigna unguiculata* Walpers (*Papilionaceae*) is the most important grain legume in West Africa, where the most recent FAO statistics give a total production of 2,663,390 t for an estimated area of 9,441,562 ha (FAO, 2001). The resulting overall average yield is 282 kg ha^{-1}, which is just a fraction of the estimated potential yield of over 2 t ha^{-1} (Singh *et al.*, 1997). The reasons for this yield gap are diverse and most of the time a combination of limiting factors, both abiotic (e.g. drought, poor soil fertility) and biotic (e.g. arthropod pests, diseases, birds and rodents). In most of West Africa, insect pests are reported to be the single most important constraint to cowpea production (Singh *et al.*, 1990). Four species are considered key field pests and are dealt with in this chapter: the flower thrips *Megalurothrips sjostedti* Trybom (Thysanoptera, Thripidae) (Plate 54), the pod borer *Maruca vitrata* (Fabricius) (Lepidoptera, Pyralidae) (Plate 58), the cowpea aphid *Aphis craccivora* Koch (Homoptera, Aphididae) and the brown coreid bug *Clavigralla tomentosicollis* Stål (Heteroptera, Coreidae).

From a purely technical standpoint, synthetic pesticides recommended for use in cowpea (e.g. pyrethroids) can effectively control these key pests, thereby increasing the yields (Singh *et al.*, 1990). However, apart from environmental and human health concerns (Jackai and Adalla, 1997), there are also socio-economic implications that make the use of chemical pesticides problematic. Among these are low level of farmers' education, lack of capital, high prices of pesticides, lack of input market and access to recommended pesticides (Coulibaly *et al.*, 2003). All these factors lead to inappropriate and hazardous practices when applying pesticides, such as the non-use of protective equipment and the non-respect of standard spraying dosage and intervals. Moreover, the lack of cash pushes farmers to opt for lower cost solutions (e.g. highly subsi-

(eds P. Neuenschwander, C. Borgemeister and J. Langewald)

dized cotton insecticides), which are then used to spray cowpea and vegetable crops. The most recent, tragic example of what can happen when farmers divert cotton insecticides to food crops has been reported from northern Bénin. In an attempt to break pyrethroid resistance that had recently been detected in *Helicoverpa armigera* (Hübner) (Lepidoptera, Noctuidae) on cotton, Endosulfan was introduced in Bénin during the 1999–2000 cotton season. However, farmers started to spray it also on food crops in the same manner as the much less toxic synthetic pyrethroids they had been used to in the past. The consequences of this sudden change: over 100 cases of acute poisoning, of which 37 were lethal. These were officially reported figures for one of the cotton-producing regions in Bénin only, unofficial figures for the whole country were estimated to be close to 70 fatalities (PAN-UK, 2000). This example is certainly an extreme case of pesticide misuse, but it clearly shows the vulnerability of a pest control approach based solely on synthetic insecticides. As advocated by Jackai and Adalla (1997), pest control practices in cowpea should rely on IPM, where synthetic insecticides are used in a 'firefighting' mode only, e.g. when all other measures fail to keep pests below acceptable levels. Since 1994, IITA, in partnership with farmers, extension and NGO agents, and scientists in nine African countries, has been developing, testing and implementing safer alternatives to chemical pest control through the PRONAF (Projet Niébé pour l'Afrique) project, co-financed by the Swiss Development Cooperation (SDC) and the International Fund for Agricultural Development (IFAD). Among the pest control options proposed by PRONAF, the use of aqueous extracts of various leaves, such as those from neem and papaya trees, has proven to have sufficient activity against the major cowpea insect pests, and is now starting to be adopted by farmers. While this represents an immediate response to pesticide misuse, research at IITA and in other centres like ICIPE is continuing to develop longer-lasting solutions to the cowpea pest problem, of which biological control is one of the pillars.

During the 1980s and early 1990s, research on biological control against cowpea pests had mainly focused on exploiting the naturally occurring interactions between pests and their locally available antagonists (Jackai and Daoust, 1986; Tamò *et al.*, 1993a). In most of the cases, the overall level of pest control exerted by these indigenous antagonists was found to be inadequate for controlling pest populations. There was, in fact, 'no clear evidence of a dominant role played by the parasites and predators of the major insect pests' (Singh *et al.*, 1990). As a consequence, recommendations for biological control were exclusively aimed at preserving the available natural enemies (Ezueh, 1991). However, the first critical review of biological control against two of the key cowpea pests, *M. vitrata* and *M. sjostedti*, indicated some opportunities for biological control interventions (Tamò *et al.*, 1997). While the above review was mainly an *ex ante* feasibility study for biological control, the present chapter is devoted to the assessment of two practical applications of biological control against *M. sjostedti*, which are the first attempts at biological control ever tried on cowpea, and are therefore presented as detailed case studies. Potential biological control interventions against the other three key pests mentioned above are briefly summarized.

Flower Thrips, *Megalurothrips sjostedti*

Management with entomopathogenic fungi

Entomopathogenic fungi can play a significant role in the reduction of insect populations either through their natural occurrence or by inoculative and inundative introductions. Fungal pathogens are the most common pathogens attacking thrips in nature and their mode of action makes them suitable candidates for the control of sucking pests (Butt and Brownbridge, 1997). The development and use of fungi as microbial insecticides have increased considerably and stable commercial products are now available for the control of various insect pests including thrips (Lacey and Goettel, 1995; Butt and Brownbridge, 1997).

The potential of entomopathogenic fungi to control *M. sjostedti* has been studied both at ICIPE and IITA. At ICIPE, two strains of *Beauveria bassiana* (Balsamo) Vuillemin and four strains of *Metarhizium anisopliae* (Metschnikoff) Sorokin were selected as the most pathogenic among 22 fungal strains, causing 100% mortality in the lab in 7 days (Ekesi *et al.*, 1998a). Strains from *Verticillium lecanii* (Zimmerman) Viegas and *Paecilomyces fumosoroseus* (Wize) Brown and Smith were generally less pathogenic. *M. anisopliae* isolate ICIPE 69 is pathogenic across a broad temperature range compared with the other strains and was selected for further studies (Ekesi *et al.*, 1999a). The pathogen's optimum activity at 25–30°C compares favourably with the optimum temperature for thrips development (Alghali, 1991).

The field performance of isolate ICIPE 69 was assessed in field trials conducted at the ICIPE's Mbita Point Field Station, western Kenya. Results showed that three applications of the fungus caused up to 75% reduction in *M. sjostedti* density with an increase in cowpea grain yield (Ekesi *et al.*, 1998b). A high-volume (HV) aqueous formulation performed better than an oil/aqueous ultra low volume (ULV) formulation, possibly because the HV treatment reached thrips feeding sites more efficiently, thus enhancing direct contact of the pathogen with the thrips. The fungus remained active in the field for about 3–4 days and effective management of thrips was achieved by timing one application at flower bud stage and two applications at flowering stage (Ekesi *et al.*, 2000b).

In tropical Africa, the majority of subsistence farmers who practise low-input agriculture are dependent on intercropping to minimize the risk of crop failure and to improve the nutritional quality and yields of their crop (Ofori and Stern, 1987). This form of cultural practice has also been reported to reduce the population of *M. sjostedti* (Kyamanywa and Tukahirwa, 1988). The effect of intercropping cowpea with maize on the performance of *M. anisopliae* ICIPE 69 was evaluated over two field seasons in 1997. Thrips mortality caused by the fungus treatments was significantly higher in intercropped cowpea than in the monocrop leading to a significant decrease in thrips density and damage (Ekesi *et al.*, 1999b). Additionally, grain yields in the intercropped cowpea treated with *M. anisopliae* did not differ from yields in intercropped cowpea treated with Karate (Table 19.1). Although insect reduction was slow in the fungal treated plots (Fig. 19.1), the debilitating effect of infection caused by fungi can reduce an insect's capacity to harm crops several days before death and

Table 19.1. Effect of *Metarhizium anisopliae* and Karate application on mean ± SE cowpea yield, Mbita Point, Kenya, 1997.

	First season		Second season	
Treatment	Yield (g per plant)	Yield (kg ha^{-1})	Yield (g per plant)	Yield (kg ha^{-1})
Monocrop NP	3.7 ± 0.7 d	306.2 ± 34.6	3.8 ± 0.5 d	312.5 ± 23.9 d
Intercrop NP	6.9 ± 0.8 c	283.8 ± 27.5	6.5 ± 0.7 c	542.7 ± 32.1 c
Monocrop PF	9.6 ± 1.2 b	795.8 ± 41.8	9.7 ± 0.8 b	801.5 ± 51.1 b
Intercrop PF	14.0 ± 1.6 a	577.4 ± 38.7	12.3 ± 2.4 a	1017.3 ± 67.5 a
Monocrop PK	12.8 ± 1.5 a	887.5 ± 35.3	14.7 ± 2.1 a	1213.1 ± 56.4 a
Intercrop PK	13.2 ± 2.1 a	542.6 ± 42.3	13.8 ± 1.9 a	1141.8 ± 75.2 a

Means within a column followed by the same letter are not significantly different by Student–Newman–Keuls ($P = 0.05$) tests. NP, non-protected; PF, protected with fungus; PK, protected with Karate.

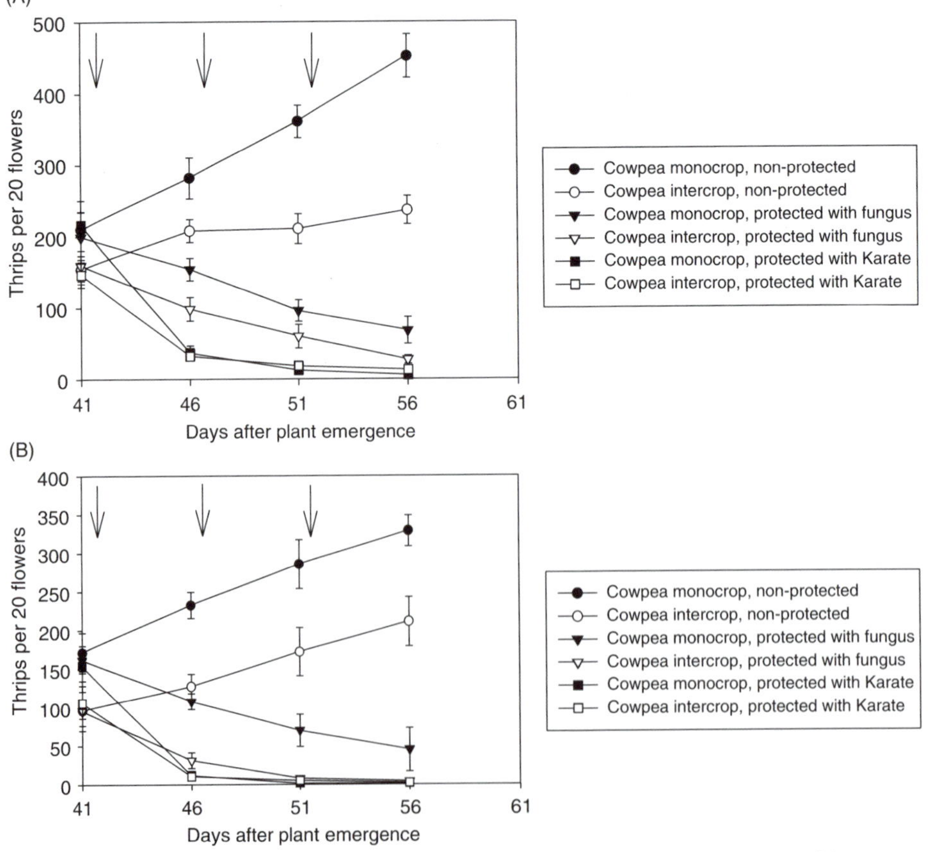

Fig. 19.1. *Megalurothrips sjostedti* population (mean ± SE) following treatment with *Metarhizium anisopliae* ICIPE69 strain and Karate during the first (A) and second (B) seasons. Arrows indicate date of spray (Ekesi and Maniania, 2000).

this has also been demonstrated in the case of *M. anisopliae* infection in *M. sjostedti* (Ekesi and Maniania, 2000). Factors thought to be responsible for good fungal control in intercropped cowpea may include high relative humidity, which is an essential factor for fungal infection, and increased light interception by the maize, which reduces the rate of inactivation of conidia through UV degradation. The impact of the fungus on non-target predators was also tested in this study, since predatory activity is believed to be higher within the intercrop (Matteson, 1982; Ezueh and Taylor, 1984). When compared with the control, *M. anisopliae* was found to have no adverse effect on the population of predatory insects including coccinellids, predatory bugs, ants, spiders, staphylinids and earwigs (Ekesi *et al.*, 1999b).

Following on from the extensive work at ICIPE, the potential for fungal control of *M. sjostedti* with *M. anisopliae* ICIPE 69 was also investigated at IITA in West Africa. Laboratory tests first compared an isolate of *B. bassiana* (IMI 330194) alongside *M. anisopliae* (ICIPE 69) (T. Houndékon, 1998, unpublished results). Despite marginally better laboratory performance of *B. bassiana*, *M. anisopliae* ICIPE 69 was evaluated in a complementary field trial against flower thrips in cowpea at IITA Bénin in 1999 (T. Sissinto, 2000, unpublished results). Though applications of an HV aqueous formulation of *M. anisopliae* (ICIPE 69) led to reductions in the numbers of *M. sjostedti* and an increased yield compared with the control, these differences were not significant, nor were they different to two botanical treatments (neem and papaya leaf extracts). In comparison, the chemical standard deltamethrin (Decis) performed significantly better than all treatments both in terms of population reduction and yield enhancement.

The above studies have demonstrated the potential of fungal pathogens in the management of thrips on cowpea, but the role of mycopathogens in *M. sjostedti* management must be viewed within the context of integrated pest management in the cowpea agroecosystem. Ekesi *et al.* (2000a) have demonstrated the possibility of integrating fungus treatment with host plant resistance, but knowledge of its compatibility with other non-target beneficial species, especially parasitoids, is crucial. An understanding of the complementary activities will not only reduce over-dependence on broad spectrum synthetic insecticides, but would also reduce disruption of the natural enemy complex, while promoting biodiversity and sustaining environmental quality. For mycopesticides to become integrated into IPM systems in Africa, a key challenge is to achieve economically competitive and stable formulations of reliable quality that are widely and consistently available. Without achieving this goal, mycopesticides remain an experimental tool.

Experimental releases and establishment of larval parasitoids

In West Africa, the solitary larval endoparasitoid *Ceranisus menes* Walker (Hymenoptera, Eulophidae), a cosmopolitan species (Loomans and van Lenteren, 1995), has been found attacking first- and second-instar *M. sjostedti* larvae (Tamò *et al.*, 1993b). Subsequent studies, however, have revealed poor field parasitism rates on cowpea, averaging below 1% (Tamò *et al.*, 1997). More recent observations (summarized in Tamò *et al.*, 2002) have also indicated that even on wild host plants, thought to be more conducive to biological control than the cowpea field, parasitism rates were still very low, rarely exceeding 6%, and with an overall mean parasitism of 2.6% (calculated using over 43,000 larvae collected from some 14 different host plants). Two cases of seemingly high parasitism rates were found on *Cochlospermum planchonii* Hook. f. (*Cochlospermaceae*) (19.1%) and *Erythrina senegalensis* DC. (*Fabaceae*) (14.5%), although these were obtained from a rather low number of samples (235 and 558, respectively) and over a short period of time (Zenz, 1999).

Following the hypothesis of a possible foreign origin for *M. sjostedti* (Tamò *et al.*, 1997), explorations for thrips parasitoids available in tropical Asia were carried out between 1994 and 1997, focusing on the regions of Andhra Pradesh (India), Johor, Melaka, Sabah and Sarawak (Malaysia), where the highest diversity of *Megalurothrips* spp. has been described from literature (Palmer, 1987). These surveys revealed several morphologically different strains of *C. menes*, three of which (two from India, one from Malaysia) could be collected in sufficient numbers to be reared in quarantine (at the Wageningen Agricultural University, The Netherlands). They were subsequently introduced into the laboratories of the IITA Bénin research station under standard import permits for evaluating their performance against *M. sjostedti*. These three strains were compared with the local (Bénin) *C. menes* both in terms of behavioural and physiological compatibility in parasitizing *M. sjostedti*. The outcome of these studies, presented in detail by Diop (2000), indicated that physiological incompatibility is the most likely explanation for the low parasitism rates caused by the local strain of *C. menes*. Because of behavioural incompatibility at the encounter and attack phase, none of the Asian strains were considered to be good biological control candidates against *M. sjostedti*. However, it is worth mentioning that one of the Indian strains had a substantially higher physiological compatibility than the local strain.

One of those parasitoids, of which only few individuals could be collected during the above explorations around Hyderabad (India), and subsequently was lost during the quarantine rearing in The Netherlands in early 1996, was a species of *Ceranisus* with completely dark brown metasoma, and whose pupa had a different shape to that of *C. menes*, also lacking the typical orange spot at the centre of the abdomen (Loomans and van Lenteren, 1995). Two years later, however, a survey in southern Cameroon revealed the presence of a morphologically similar species, also with brown metasoma, which had never been reported before from West and Central Africa. Specimens were sent for identification, together with those of the 1996 collected Indian species: both of them revealed to be the same species, identified as *Ceranisus femoratus* Gahan (Triapitsyn, 1998, University of California, Riverside, personal communication) (Plate 55a,b). Adult *C. femoratus* were first encountered in flowers of *Centrosema pubescens* Benth. and *Millettia* sp. (*Fabaceae*) around the IITA Humid Forest Center at Nkolbisson (Yaoundé, Cameroon), but were also obtained in large numbers from parasitized larvae of *M. sjostedti* (M. Tamò and

M. Tindo, 1998, unpublished results). Subsequent studies to assess the presence of this newly discovered parasitoid on cowpea and to measure its possible impact on *M. sjostedti*, revealed that *C. femoratus* was able to parasitize a substantial percentage of *M. sjostedti* larvae collected from cowpea flowers in two locations in Cameroon (H. Ndam, 1998, unpublished results). A comparison of parasitism rates on different host plants (summarized in Fig. 19.2) clearly indicates that *C. femoratus* is a more efficient parasitoid than *C. menes*.

Based on these encouraging results, and following the delivery of standard import permits, *C. femoratus* from Cameroon was introduced into the IITA Bénin laboratories. In collaboration with the national plant protection services, experimental releases were carried out in the coastal savannah in both Bénin (July 1999) and Ghana (December 1999). The coastal savannah was chosen as the agroecological zone for the initial releases because it was our intention to test the capacity of the parasitoid to cope with low and widely dispersed thrips populations, very scarce cowpea crops on problem soils, and low density of alternative host plants. If the parasitoid was able to survive under these difficult conditions, it would also be able to survive anywhere else.

In Ghana, one of the releases sites (near Gomoa Buduatta, 130 *C. femoratus* females released), was hit by a severe bush fire shortly after the release, leaving only few *C. pubescens* plants bearing flowers during the dry season. Unfortunately, these plants were attacked by aphids, and were subsequently colonized by ants (mainly *Crematogaster* spp.), which interfered with the activity of the *C. femoratus*, as it is often observed with other hymenopterous parasitoids (e.g. Cudjoe *et al.*, 1993). In this particular case, parasitism was reduced both as a result of direct predation on thrips, thus reducing the availability of

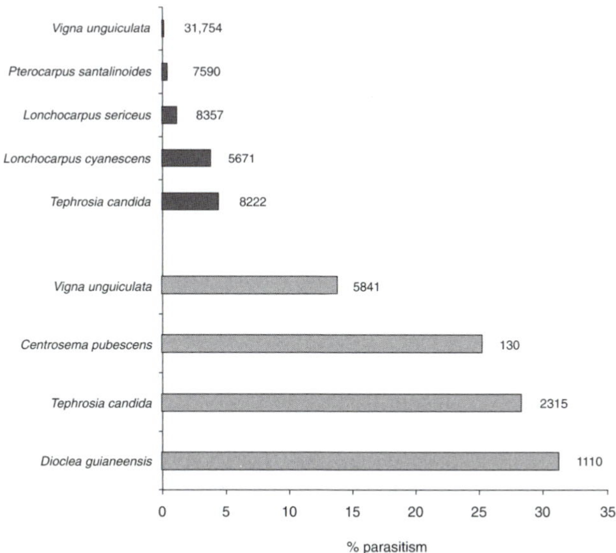

Fig. 19.2. Comparison of field parasitism of *Megalurothrips sjostedti* larvae inflicted by *Ceranisus menes* (■) in Bénin and *C. femoratus* (□) in Cameroon. Numbers after the graph bars indicate the total number of *M. sjostedti* larvae studied (summarized from Tamò *et al.*, 2002).

M. sjostedti larvae, and also because female *Ceranisus* spp. take up to 2–3 min for ovipositing inside a thrips larva (Loomans and van Lenteren, 1995; Diop, 2000), which makes them an easy target for foraging ants. At the other site (near Pokuase, 120 *C. femoratus* females released), the release was followed by initial establishment: *C. femoratus* could be recovered for over 1 year, though at very low levels, reflecting the overall low *M. sjostedti* population on patches of *C. pubescens*, the only host plant present in the area that would sustain *M. sjostedti* development and reproduction. Around February 2001, however, probably because of an unusually severe dry period, virtually all plants of *C. pubescens* around the release site were attacked by aphids, which then attracted ants. As a result, from early 2001 onward, repeated attempts to recover *C. femoratus* from both sites remained unsuccessful.

In Bénin, *C. femoratus* was initially released at two sites around Ouidah (180 *C. femoratus* females released) and at the IITA Bénin research station (150 *C. femoratus* females released). As had been the case for the release sites in Ghana, both sites around Ouidah were also affected by two consecutive severe dry seasons characterized by the same type of problems, i.e. bush fires and interference from ants. Again, the consequence was non-establishment of the released parasitoids on *C. pubescens*.

At the IITA Bénin station, the experimental releases were carried out on planted plots of *Tephrosia candida* (Roxb.) DC. (*Fabaceae*), one of the plants that had been found sustaining high *C. femoratus* numbers at the IITA station of Mbalmayo in southern Cameroon. Within a few months, *C. femoratus* became established and began spreading to adjacent planted plots of another interesting host plant, *Dioclea guianensis* Benth. (*Fabaceae*). During the first 1.5 years following release, the monthly average parasitism by *C. femoratus* remained substantial (Fig. 19.3), though subjected to seasonal variation, largely influenced by the flowering phenology of the available host plant. From March 2001 onwards, following a sharp decline of the thrips larval population, the presence of *C. femoratus* dropped below detectable levels. While the causes for this disappearance are still being investigated, we suspect that both direct factors (e.g. increased presence of ants, particularly in those planted plots, which remained undisturbed for nearly 2 years) and indirect factors (poor nutritional quality of the thrips larvae due to water stress of the host plant, after a very severe dry season) might be responsible. In spite of the re-colonization of the host plants by thrips during the 2001 rainy season, it is only during early 2002 that *C. femoratus* has become re-established (Fig. 19.3).

Two additional experimental releases on *T. candida* were effectuated early September 2001 in southern Bénin (Oueme valley and Mono, with 220 and 160 *C. femoratus* females, respectively), in cowpea-producing areas. Preliminary observations early in 2002 revealed that *C. femoratus* was present on the leguminous trees *Millettia thonningii* (Schum. & Thonn.) Bak. and *Pterocarpus santalinoides* L'Hér. ex DC. (*Fabaceae*) as far as 65 km north of the original release site. All these data from the experimental releases are still being compiled and analysed, while new releases are planned for the near future.

Although it is not possible completely to exclude the risk of bush fires and interference from ants at the release sites, the strategy for future releases is to

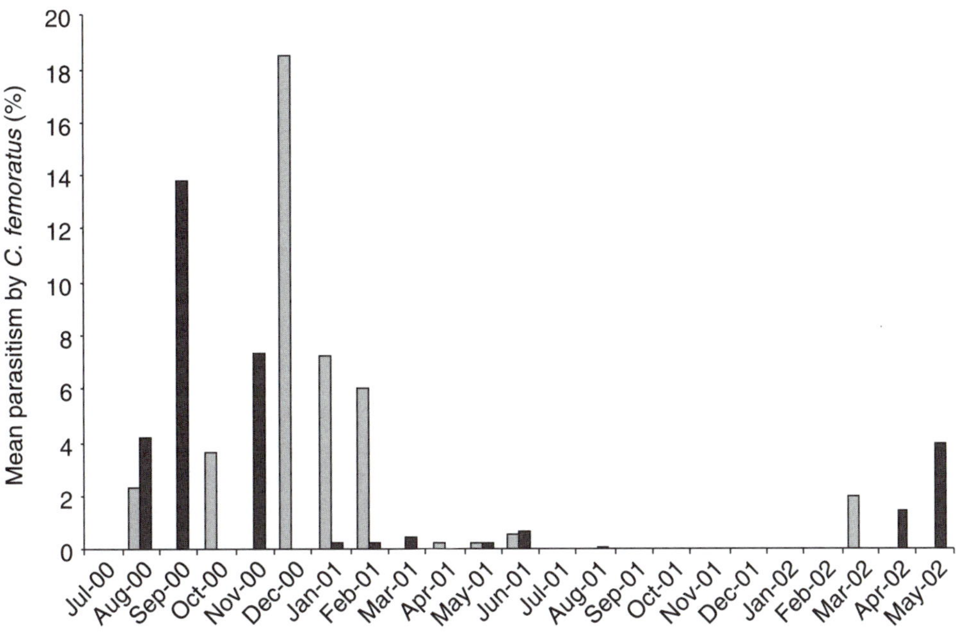

Fig. 19.3. Temporal fluctuations of mean monthly *Ceranisus femoratus* parasitism rates on *Megalurothrips sjostedti* larvae collected at the IITA Bénin station on *Tephrosia candida* (■) and *Dioclea guianensis* (□) (Agboton *et al.*, IITA Bénin, unpublished data).

try to establish initial populations in moister areas and on host plants less susceptible to being invaded by aphids (e.g. trees). The fact that *C. femoratus* is getting established on the natural vegetation is the first sign that we might be on the right track. Citing the example of the introduction of parasitoids against *Thrips tabaci* (Lindeman) (Thysanoptera, Thripidae), Parrella and Lewis (1997) indicated that biological control of thrips attacking field crops has not been very successful during the past 60 years. In fact, some of the released parasitoids eventually became established, but their potential impact might still be masked by the continuous and inappropriate use of pesticides, as is believed also to be the case for the congeneric *Thrips palmi* (Karny) in South-East Asia (Hirose, 1990). In our case, the data collected so far concerning the establishment of *C. femoratus* do not allow any conclusion on its possible impact, although it appears that biological control alone will not be able to control thrips populations in all agroecological zones. However, if combined with improved host plant resistance in a good agronomic background, and punctual applications of either *M. anisopliae* or plant-based insecticides, it will definitely contribute to a sustainable solution for the thrips problem in cowpea.

Pod Borer, *Marnca vitrata*

In areas where *M. vitrata* behaves like a migratory pest (e.g. the Kano region of northern Nigeria), periodically invading the cowpea fields during the rainy season (Bottenberg *et al.*, 1997), biological control interventions need to have a 'knockdown' effect on early stages (eggs or neonate larvae) to reduce feeding damage. This could be achieved by inundative releases of locally available, mass-reared egg parasitoids *Trichogrammatoidea ?eldanae* Viggiani (Hymenoptera, Trichogrammatidae) (Plate 56) (Arodokoun, 1996), if possible in conjunction with pheromone traps used to determine the peak oviposition period (Downham *et al.*, 2002). In the context of an IPM approach to *M. vitrata*, the above releases could be complemented by subsequent sprays of botanical extracts such as neem (Jackai and Adalla, 1997), which has a much less negative impact on hymenopterous parasitoids than synthetic pesticides (Schmutterer, 1997).

In addition to combating *M. vitrata* in cowpea fields, control strategies can be targeted to areas where alternative host plants are abundant and constitute a major factor influencing the dynamics of *M. vitrata* populations. In this case, the objective would be to reduce overall population pressure, which could be obtained by the introduction of presumed more efficient natural enemies such as the parasitoids *Phanerotoma philippinensis* (Ashmead) and *Bassus javanicus* (Bhat & Gupta) (Hymenoptera, Braconidae) observed in South-East Asia (Tamò *et al.*, 1997).

In 1998, a cypovirus (CPV) was found in southern Bénin infecting larvae of *M. vitrata* on *Lonchocarpus sericeus* (Poir.) H. B. & K. and *Sesbania pachycarpa* DC. (both *Papilionaceae*). CPV, entomopathogenic RNA viruses belonging to the family Reoviridae, have received less attention than baculoviruses in biological control programmes because infection of an insect population is in general chronic rather than epidemic (Payne and Mertens, 1984) and less lethal. Nevertheless, disruption of the midgut by CPV infection leads to nutritional deficiencies and a reduction in feeding, and CPV-infected pupae and adults are malformed, thus reducing survival and longevity as well as their mating ability and fecundity. Furthermore, as CPV is usually transmitted vertically to the next generation, the viability of offspring may be compromised (Belloncik, 1996). In laboratory studies at IITA, at least some of these characteristics have already been demonstrated in *M. vitrata* larvae infected with indigenous CPV, and work towards integrating *Mv*CPV into an IPM programme is underway.

Brown Coreid Bug, *Clavigralla tomentosicollis*

This pest has also become the target of investigations into the potential of entomopathogenic fungi as a pest control option. Isolates of both *M. anisopliae* and *B. bassiana* from Nigeria with good activity against eggs, nymphs and adult *C. tomentosicollis* were identified by Ekesi (1999) and Ekesi *et al.* (2002). More recently, *M. anisopliae* isolate ICIPE 69 has been found to be active against adult *C. tomentosicollis* (Plate 57) in Bénin (F. Agonvi, 2002, unpublished results) and in Kenya (N.K Maniania and E. Minja, 2002, unpublished results). This is particularly important in view of the prospects of using this isolate against *M. sjostedti*. Some of the isolates tested above were also active against adult *Ootheca mutabilis* Shalberg (Coleoptera, Chrysomelidae) and eggs of *M. vitrata* (Ekesi, 2001; Ekesi *et al.*, 2002). As yet, however, none of these fungi have been developed into commercially available biopesticides, nor tested in large-scale trials.

Egg parasitoids such as *Gryon fulviventris* Crawford (Hymenoptera, Scelionidae) are very common and have been observed inflicting high mortality to *C. tomentosicollis* egg batches (Asante *et al.*, 2000). However, in most of the cases, they cannot prevent feeding damage by *C. tomentosicollis* in the field, which is caused by migrating adults before oviposition.

Cowpea Aphid, *Aphis craccivora*

Repeated surveys to identify major natural enemies of *A. craccivora* in southern and central Bénin have revealed the total absence of hymenopterous parasitoids attacking this pest (A. Soukossi, 2001, unpublished results). In view of the multitude of *A. craccivora* parasitoids found elsewhere, such as *Trioxys* spp. (Hymenoptera, Braconidae) (e.g. in Pakistan and India), which are reported to exert some control on this pest in a variety of environments and on different crops (Singh and Agarwala, 1992), these findings need to be verified for other agroecological zones in West Africa. If the absence of parasitoids is indeed confirmed, there could be a potential 'off-the-shelf' biological control project, as has been the case for the successful introduction of *Trioxys pallidus* (Haliday) in hazelnut orchards of Oregon (AliNiazee, 1997).

Laboratory testing of Nigerian isolates of *M. anisopliae* and *B. bassiana* indicated good activity against apterous adults of *A. craccivora* (Ekesi *et al.*, 2000c). However, the authors caution the need for further investigation on the effect of these entomopathogens on beneficial arthropods such as syrphids and coccinellids known to attack aphids, before undertaking large-scale field experiments.

Conclusion

During the past 5 years, biological control in cowpea has evolved from a hypothetical management option to an IPM component that has already been tested in experimental field applications against thrips. As already mentioned for the case of *C. femoratus*, however, it is very unlikely that biological control alone will become a cure-all for the cowpea pest problems.

In the first place, as opposed to the use of pesticides, biological control will need to be tailored to each of the major pests. We have seen that in the case of entomopathogens, a single organism might be developed into a product that can be used to control two key pests, thrips and pod bugs, but, in the case of classical biological control or new associations, each pest will require highly specific parasitoids that need to be introduced or mass-released.

All four key pests have alternative hosts, which, in many areas, are far more important for their population dynamics than cowpea, the cultivated host. Again, this factor needs to be taken into consideration when evaluating biological control candidates. While it might not be possible to establish some natural enemies in a short-living habitat such as cowpea fields, their efficacy in regulating the pest population in the natural vegetation, e.g. by lowering the overall population fitness, needs to be carefully examined.

In areas where cowpea pests invade the fields in high numbers like migratory insects, which require an immediate solution, it is particularly important that all other pest management interventions remain compatible with biological control. For instance, plant extracts with insecticidal and/or repellent properties and the application of entomopathogens have negligible activity on most predators and parasitoids, and should consequently be used as a substitute for synthetic insecticides. In the long run, an IPM strategy based on improved host plant resistance (including transgenic plants, see Neuenschwander *et al.*, synopsis, this volume) and biological control as preventive measures, together with applications of entomopathogens or botanical insecticides as curative measures, should enable the African farmer to sustainably manage cowpea pests.

References

Alghali, A.M. (1991) The effect of agrometeorological factors on the fluctuation of the flower thrips, *Megalurothrips sjostedti* Trybom (Thysanoptera: Thripidae), on two cowpea varieties. *Journal of Plant Protection in the Tropics* 8, 63–68.

AliNiazee, M.T. (1997) Biology, impact, and management of *Trioxys pallidus* in hazelnut orchards of Oregon, U.S.A. *Acta Horticolturae (ISHS)* 445, 477–482.

Arodokoun, D. (1996) Importance des plantes-hôtes alternatives et des ennemis naturels indigènes dans le contrôle biologique de *Maruca testulalis* Geyer (Lepidoptera: Pyralidae), ravageur de *Vigna unguiculata* Walp. PhD thesis, University of Laval, Québec, Canada.

Asante, S.K., Jackai, L.E.N. and Tamò, M. (2000) Efficiency of *Gryon fulviventris* (Hymenoptera: Scelionidae) as an egg parasitoid of *Clavigralla tomentosicollis* Stål. (Hemiptera: Coreidae) in Northern Nigeria. *Environmental Entomology* 29, 815–821.

Belloncik, S. (1996) Interactions of cytoplasmic polyhedrosis viruses with insects. *Advances in Insect Physiology* 26, 233–296.

Bottenberg, H., Tamò, M., Arodokoun, D., Jackai, L.E.N., Singh, B.B. and Youm, O. (1997) Population dynamics and migration of cowpea pests in northern Nigeria: implications for integrated pest management. In: Singh, B.B., Mohan Raj, D.R., Dashiell, K.E. and Jackai, L.E.N. (eds) *Advances in Cowpea Research*. Co-publication of International Institute of Agriculture (IITA) and Japan International Center for Agricultural Sciences (JIRCAS). IITA, Ibadan, Nigeria, pp. 271–284.

Butt, T.M. and Brownbridge, M. (1997) Fungal pathogens of thrips. In: Lewis, T. (ed.) *Thrips as Crop Pests*. CAB International, Wallingford, UK, pp. 399–433.

Coulibaly, O., Nkamleu, B. and Tamò, M. (2002) Technical efficiency of cowpea production in western Cameroon. An application of stochastic frontier analysis. *Storage Products Research*.

Cudjoe, A.R., Neuenschwander, P. and Copland, M.J.W. (1993) Interference by ants in biological control of the cassava mealybug *Phenacoccus manihoti* (Hemiptera: Pseudococcidae) in Ghana. *Bulletin of Entomological Research* 83, 15–22.

Diop, K. (2000) The biology of *Ceranisus menes* (Walker) (Hym., Eulophidae), a parasitoid of the bean flower thrips *Megalurothrips sjostedti* (Trybom) (Thys., Thripidae): a comparison between African and Asian populations. PhD thesis, Universtity of Ghana, Legon, Ghana.

Downham, M.C.A., Tamò, M., Hall, D.R., Datinon, B., Dahounto, D. and Adetonah, J. (2002) Development of sex pheromone traps for monitoring the legume podborer *Maruca vitrata* (F.) (Lepidoptera: Pyralidae). In: Fatokun, C.A., Tarawali, S.A., Singh, B.B., Kormawa, P.M. and Tamò, M. (eds) *Challenges and Opportunities for Enhancing Sustainable Cowpea Production. Proceedings of the 3rd World Cowpea Conference*, 5–10 September 2000, Ibadan, Nigeria, IITA, Ibadan, Nigeria, pp. 124–135.

Ekesi, S. (1999) Selection of virulent isolates of entomopathogenic hyphomycetes against *Clavigralla tomentosicollis* Stål. and evaluation in cage experiment using three cowpea varieties. *Mycopathologia* 148, 131–139.

Ekesi, S. (2001) Pathogenicity and antifeedant activity of entomopathogenic hyphomycetes to the cowpea leaf beetle, *Ootheca mutabilis* Shalberg. *Insect Science and its Application* 21, 55–60.

Ekesi, S. and Maniania, N.K. (2000) Susceptibility of *Megalurothrips sjostedti* developmental stages to *Metarhizium anisopliae* and the effects of infection on feeding, adult fecundity, egg fertility and longevity. *Entomologia Experimentalis et Applicata* 94, 229–236.

Ekesi, S., Maniania, N.K., Onu, I. and Löhr, B. (1998a) Pathogenicity of entomopathogenic fungi (Hyphomycetes) to the legume flower thrips, *Megalurothrips sjostedti* (Trybom) (Thysanoptera: Thripidae). *Journal of Applied Entomology* 122, 629–634.

Ekesi, S., Maniania, N.K., Ampong-Nyarko, K. and Onu, I. (1998b) Potential of the entomopathogenic fungus, *Metarhizium anisopliae* (Metsch.) Sorokin for control of the legume flower thrips, *Megalurothrips sjostedti* (Trybom) on cowpea in Kenya. *Crop Protection* 17, 661–668.

Ekesi, S., Maniania, N.K. and Ampong-Nyarko, K. (1999a) Effect of temperature on germination, radial growth and virulence of *Metarhizium anisopliae* and *Beauveria bassiana* on *Megalurothrips sjostedti*. *Biocontrol Science and Technology* 9, 177–185.

Ekesi, S., Maniania, N.K., Ampong-Nyarko, K. and Onu, I. (1999b) Effect of intercropping cowpea with maize on the performance of *Metarhizium anisopliae* against *Megalurothrips sjostedti* (Thysanoptera: Thripidae) and predators. *Environmental Entomology* 28, 1154–1161.

Ekesi, S., Maniania, N.K. and Lwande, W. (2000a) Susceptibility of the legume flower thrips to *Metarhizium anisopliae* on different varieties of cowpea. *Biocontrol* 45, 79–95.

Ekesi, S., Maniania, N.K., Ampong-Nyarko, K. and Akpa, A.D. (2000b) Importance of timing of application of the entomopathogenic fungus, *Metarhizium anisopliae* for the control of legume flower thrips, *Megalurothrips sjostedti* and its persistence on cowpea. *Archives of Phytopathology and Plant Protection* 33, 431–445.

Ekesi, S., Akpa, A.D., Onu, I. and Ogunlana, M.O. (2000c) Entomopathogenicity of *Beauveria bassiana* and *Metarhizium anisopliae* to the cowpea aphid, *Aphis craccivora* Koch (Homoptera: Aphididae). *Archives of Phytopathology and Plant Protection* 33, 171–180.

Ekesi, S., Adamu, R.S. and Maniania, N.K. (2002) Ovicidal activity of entomopathogenic hyphomycetes to the legume pod borer, *Maruca vitrata* and the pod sucking bug, *Clavigralla tomentosicollis*. *Crop Protection* 21(7), pp. 589–595.

Ezueh, M.I. (1991) Prospects for cultural and biological control of cowpea pests. *Insect Science and its Application* 12, 585–592.

Ezueh, M.I. and Taylor, T.A. (1984) Effect of time of intercropping with maize on cowpea susceptibility to three major insect pests. *Tropical Agriculture* 61, 82–86.

FAO (2001) FAOSTAT database. http://apps.fao.org

Hirose, Y. (1990) Prospective use of natural enemies to control *Thrips palmi* (Thysanoptera, Thripidae). In: Bay-Petersen, J., Mochida, O. and Kiritani, K. (eds) *The Use of Natural Enemies to Control Agricultural Pests*. FFTC Book Series, Food and Fertilizer Technology Center for the Asian and Pacific Region, Taipei, Taiwan, pp. 135–141.

Jackai, L.E.N. and Daoust, R.A. (1986) Insect pests of cowpeas. *Annual Review of Entomology*, 31, 9–119.

Jackai, L.E.N. and Adalla, C.B. (1997) Pest management practices in cowpea: a review. In: Singh, B.B., Mohan Raj, D.R., Dashiell, K.E. and Jackai L.E.N. (eds) *Advances in Cowpea Research*. Co-publication of International Institute of Agriculture (IITA) and Japan International Center for Agricultural Sciences (JIRCAS). IITA, Ibadan, Nigeria, pp. 240–258.

Kyamanya, S. and Tukarhirwa, E.M. (1988) Effect of mixed cropping beans, cowpeas and maize on population density of bean flower thrips, *Megalurothrips sjostedti* (Trybom) (Thripidae). *Insect Science and its Application* 9, 255–256.

Lacey, L.A. and Goettel, M. (1995) Current developments in microbial control of insect pests and prospects for the early 21st century. *Entomophaga* 40, 3–27.

Loomans, A.J.M. and van Lenteren, J.C. (1995) Hymenopterous parasitoids of thrips. In: Loomans, A.J.M., van Lenteren, J.C., Tommasini, M.G., Maini, S. and Riudavets, J. (eds) *Biological Control of Thrips Pests*. Wageningen Agricultural University Papers, Wageningen, The Netherlands, pp. 90–201.

Matteson, P.C. (1982) The effect of intercropping and minimal permethrin application on insect pests of cowpea and their natural enemies in Nigeria. *Tropical Pest Management* 28, 372–388.

Ofori, F. and Stern, W.R. (1987) Cereal legume intercropping systems. *Advances in Agronomy* 41, 41–49.

Palmer, J.M. (1987) *Megalurothrips* in the flowers of tropical legumes: a morphometric study. In: Holman, J., Pelikan, J., Dixon, A.G.F. and Weismann, L. (eds) *Population Structure, Genetics and Taxonomy of Aphids and Thysanoptera*. SPB Academic Publishing, The Hague, The Netherlands, pp. 480–495.

PAN-UK (2000) Endosulfan deaths and poisonings in Bénin. *Pesticides News* 47, 12–14.

Parrella, M.P. and Lewis, T. (1997) Integrated pest management (IPM) in field crops. In: Lewis, T. (ed.) *Thrips as Crop Pests*. CAB International, Wallingford, UK, pp. 595–614.

Payne, C.C. and Mertens, P.P.C. (1984) Cytoplasmic polyhedrosis viruses. In: Joklik, W.C. (ed.) *The Reoviridae*. CRC Press, New York, USA, pp. 425–504.

Schmutterer, H. (1997) Side-effects of neem (*Azadirachta indica*) products on insect pathogens and natural enemies of spider mites and insects. *Journal of Applied Entomology* 121, 121–128

Singh, B.B., Chambliss, O.L. and Sharma, B. (1997) Recent advances in cowpea breeding. In: Singh, B.B., Mohan Raj, D.R., Dashiell, K.E. and Jackai, L.E.N. (eds) *Advances in Cowpea Research*. Co-publication of International Institute of

Agriculture (IITA) and Japan International Center for Agricultural Sciences (JIRCAS). IITA, Ibadan, Nigeria, pp. 30–49.

Singh, R. and Agarwala, B.K. (1992) Biology, ecology and control efficiency of the aphid parasitoid *Trioxys indicus*: a review and bibliography. *Biological Agriculture and Horticulture* 8, 271–298.

Singh, S.R., Jackai, L.E.N., Dos Santos, J.H.R. and Adalla, C.B. (1990) Insect pests of cowpea. In: Singh, S.R. (ed.) *Insect Pests of Tropical Food Legumes*. John Wiley & Sons, Chichester, UK, pp. 43–89.

Tamò, M., Baumgärtner, J. and Gutierrez, A.P. (1993a) Analysis of the cowpea (*Vigna unguiculata* Walp.) agroecosystem in West Africa: II. Modelling the interactions between cowpea and the bean flower thrips *Megalurothrips sjostedti* (Trybom). *Ecological Modelling* 70, 89–113.

Tamò, M., Baumgärtner, J., Delucchi, V. and Herren, H.R. (1993b) Assessment of key factors responsible for the pest status of the bean flower thrips *Megalurothrips sjostedti* (Trybom) (Thysanoptera, Thripidae). *Bulletin of Entomological Research* 83, 251–258.

Tamò, M., Bottenberg, H., Arodokoun, D. and Adeoti, R. (1997) The feasibility of classical biological control of two major cowpea insect pests. In: Singh, B.B., Mohan Raj, D.R., Dashiell, K.E. and Jackai, L.E.N. (eds) *Advances in Cowpea Research*. Copublication of International Institute of Agriculture (IITA) and Japan International Center for Agricultural Sciences (JIRCAS). IITA, Ibadan, Nigeria, pp. 259–270.

Tamò, M., Arodokoun, D.Y., Zenz, N., Agboton, C. and Adeoti, R. (2002) The importance of alternative host plants for the biological control of two main cowpea insect pests, the pod borer *Maruca vitrata* (Fabricius) and the flower thrips *Megalurothrips sjostedti* (Trybom). In: Fatokun, C.A., Tarawali, S.A., Singh, B.B., Kormawa, P.M. and Tamò, M. (eds) *Challenges and Opportunities for Enhancing Sustainable Cowpea Production. Proceedings of the 3rd World Cowpea Conference*, 5–10 September 2000, Ibadan, Nigeria, pp. 81–93.

Zenz, N. (1999) Effect of mulch application in combination with NPK fertilizer in cowpea (*Vigna unguiculata* (L.) Walp.; Leguminosae) on two key pests, *Maruca vitrata* F. (Lepidoptera: Pyralidae) and *Megalurothrips sjostedti* Trybom (Thysanoptera: Thripidae), and their respective parasitoids. PhD thesis, University of Hohenheim, Germany.

20

Biological Control and Other Pest Management Options for Larger Grain Borer *Prostephanus truncatus*

Christian Borgemeister,[1] Niels Holst[2] and Rick J. Hodges[3]

[1]*Institute of Plant Diseases and Plant Protection, Hannover, Germany;*
[2]*Department of Crop Protection, Danish Institute of Agricultural Sciences, Research Centre Flakkebjerg, Slagelse, Denmark;*
[3]*Natural Resources Institute, University of Greenwich, Kent, UK*

Introduction

The larger grain borer *Prostephanus truncatus* (Horn) (Coleoptera, Bostrichidae) (Plate 59) is endemic to meso-America and has long been known as a pest of maize grain there (Hodges, 1986; Markham *et al.*, 1991). This situation changed dramatically with the accidental introduction of *P. truncatus* into East Africa in the late 1970s and West Africa in the early 1980s. Spreading out from the two countries of initial introduction, Tanzania and Togo, the pest is now officially recorded from a total of 16 African countries, namely Bénin, Burkina Faso, Burundi, Ghana, Guinea-Conakry, Kenya, Malawi, Namibia, Niger, Nigeria, Rwanda, South Africa, Tanzania, Togo, Uganda and Zambia (Farrell, 2000).

In some, but not all, of these countries, *P. truncatus* has become the major constraint to the storage of maize and, to a lesser extent, dried cassava roots. For example, detailed data on dry weight losses of maize stored in Togo for an average period of 6 months were around 7% prior to the introduction of *P. truncatus*, but rose to more than 30% soon after, thereby causing famine-like conditions in certain parts of southern Togo (Pantenius, 1987). In those African areas where the pest is well established, it is estimated that average losses of farm-stored maize increased from less than 5% to approximately 10% per year (Dick, 1988). Unfortunately, most loss estimates are not cumulative and do not take into account farmer consumption patterns. Although such estimates demonstrate the highly damaging nature of the pest, they do not accurately reflect the threat to the farmer. An exception to this is a study of dried cassava root losses in central Togo by Wright *et al.* (M.A.P. Wright, D. Akou-Edi and A.

 Biological Control in IPM Systems in Africa (eds P. Neuenschwander, C. Borgemeister and J. Langewald)

Stabrawa, Lomé, 1993, unpublished results). They found that 25 farmers, in five villages, sustained average cumulative weight losses of 9.7% after 3 months of storage, rising to 19.5% after 7 months. On a national basis it was estimated that these losses could amount to an average 4% of cassava production (which includes cassava products not infested by the pest) with a value of 0.05% of Togo's GNP.

Biology and Ecology of *Prostephanus truncatus*

P. truncatus takes about 26 days to develop from egg to adult under optimal conditions (32°C, 80% relative humidity (r.h.)) (Bell and Watters, 1982). On maize cobs, its natural rate of increase (*r*) at 27°C and 70% r.h. has been estimated to be 0.73 week^{-1} (Hodges and Meik, 1984) and on shelled maize of three different varieties, under optimum physical conditions of 30°C and 70% r.h., rates of 0.7–0.8 week^{-1} were recorded (Bell and Watters, 1982). These rates are similar to those found for other fast-breeding postharvest pests. After mating, there is a pre-ovipostion period of at least 5–10 days. Females construct tunnels and side chambers in the food source for oviposition and generally seven to eight eggs are laid per chamber (Li, 1988). On average, females lay up to 430 eggs during their lifetime (Li, 1988). The larvae feed internally and complete their development mostly inside the grains. There are three larval instars (Subramanyam *et al.*, 1985). Beetles may live at least 6 months when reproducing (Shires, 1980).

Male *P. truncatus* release a pheromone that is attractive to both females and other males. The pheromone has two components, Trunc-call 1 and Trunc-call 2, that have been identified (Cork *et al.*, 1991) and synthesized for use as a lure in traps. Pheromone-baited traps have proved highly effective for monitoring the pest outside grain stores for both phytosanitary purposes and biological research. The pest does not respond to the pheromone until dispersing from its food source (Pike, 1993). Small populations already feeding on maize in a store can therefore not be detected by pheromone traps placed in them. Only when the population has increased to an extent whereby the infestation is obvious and individuals are starting to disperse will the traps catch beetles. Presently, the only means of assessing infestations in stores is by manual sampling of the produce (Meikle *et al.*, 1998a, 2000).

Monitoring beetle flight activity with such traps revealed the presence of *P. truncatus* in different habitats, including forests, both in the neotropics (Rees *et al.*, 1990) and in East and West Africa (Nang'ayo *et al.*, 1993; Borgemeister *et al.*, 1998a). Laboratory studies have shown that the beetle is able to reproduce on a variety of different wood species (e.g. Nang'ayo *et al.*, 1993). *P. truncatus* has indeed occasionally been recovered in tropical deciduous forests in Kenya (Nang'ayo *et al.*, 1993), Mexico (Ramírez-Martínez *et al.*, 1994) and Bénin (Borgemeister *et al.*, 1998b); but generally it has been very difficult to find breeding *P. truncatus* populations in forest habitats. In Kenya, it was suggested that the beetle was living in twigs resulting from seasonal dieback. In all three countries, *P. truncatus* colonies were found in association with twig-girdling cerambycids. However, Nansen *et al.* (2002) studying *P. truncatus* flight activity in a forest in southern Bénin recorded comparatively low pheromone trap catches in the vicinity of cerambycid-girdled trees. Moreover, the high trap catches of *P. truncatus* observed in forest environments in Africa and Mexico strongly suggest that the beetle is not only breeding in association with twig-girdling cerambycids, but also in other yet to be defined niches. Thus, *P. truncatus* most likely exists in a patchwork of meta-populations possibly reproducing predominately in dead wood, such as seasonal dieback and/or cerambycid-girdled twigs, supplemented by starch-rich maize and cassava in farm stores.

The strong association of *P. truncatus* with woody hosts suggests that the beetle has evolved as a woodborer (Chittenden, 1911; Nang'ayo *et al.*, 1993; Ramirez-Martinez *et al.*, 1994). Many typical storage pests, including the closely related bostrichid *Rhyzopertha dominica* (F.), are strongly attracted from considerable distances to the odours of stored products (Barrer, 1983). This is, however, not the case in *P. truncatus*, which is not attracted to stored food over long ranges, and possibly not even at very short ranges (Hodges, 1994; Scholz *et al.*, 1997a). Host selection behaviour has been investigated in some detail (Hodges, 1994; Scholz *et al.*, 1997a,b; Hodges *et al.*, 1999a) and is diagrammed in Fig. 20.1. Primary selection is a chance event and made, mostly or entirely, by males boring test burrows into anything soft enough, including rubber, soap or plastic. When a male has found a food source, it releases its pheromone. This leads to secondary host selection by the females and other males so attracted. On the arrival of a female at the food source, the male ceases, or greatly reduces, pheromone release. This observation, and other aspects of the beetle's biology, suggests that the male pheromone is released as a sex attractant and that other males are exploiting the signal to gain food resources and mates (Hodges *et al.*, 1999a).

In rural maize stores, *P. truncatus* usually occurs as part of a community of several pest species like *Sitophilus zeamais* Motschulsky (Coleoptera, Curculionidae) and the Angoumois grain moth, *Sitotroga cereallela* (Olivier) (Lepidoptera, Gelechiidae). These are primary pests that can initiate an attack. They are often followed by a large array of secondary pests that are unable to penetrate intact grains themselves. Secondary pests include *Carpophilus* spp. (Coleoptera, Nitidulidae) and *Cathartus quadricollis* Guérin (Coleoptera, Silvanidae) attacking wet and mouldy grains, and *Cryptolestes* spp. (Coleoptera, Cucujidae), *Gnatocerus* spp., *Tribolium* spp. and *Palorus* spp. (all Coleoptera, Tenebrionidae) that feed primarily on the flour produced by the primary pests. Although the humidity and temperature requirements of *P. truncatus* are very similar to those of other important postharvest maize pests, larger grain borer is more tolerant of low grain moisture contents (Hodges *et al.*, 1983). Unlike the maize weevils, *P. truncatus* is more destructive if the maize is stored as cobs than as shelled grain. In several laboratory experiments, competition patterns between *P. truncatus* and *S. zeamais* were investigated. In single grain studies, *S. zeamais* larvae tended to out-compete *P. truncatus* larvae (Giga and Canhao, 1993). However, in laboratory investigations with maize cobs, the tunnelling behaviour of *P. truncatus* females considerably hampered the successful development of *S. zeamais* larvae, leading

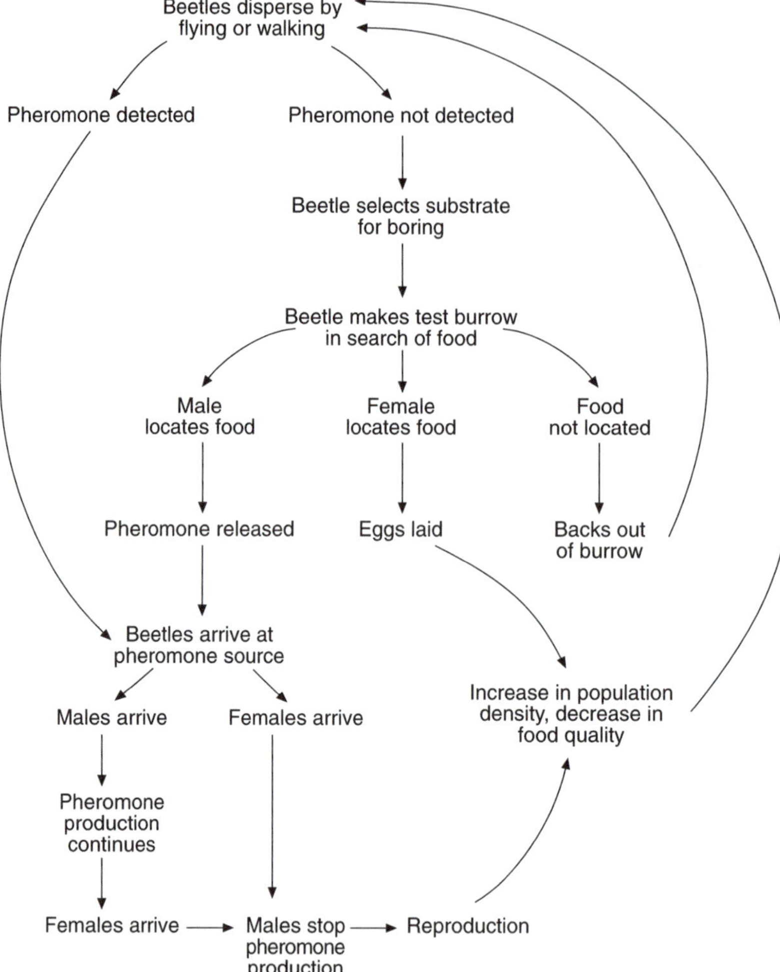

Fig. 20.1. Strategy for the selection of stored products hosts (maize or cassava) by *Prostephanus truncatus* (from Hodges *et al.*, 1999a).

to high infestation levels of *P. truncatus* at the end of the experiments (Vowotor *et al.*, 1998). Likewise, in stores of maize cobs, the population growth rate of *S. zeamais* is negatively affected by *P. truncatus* density, but not vice versa (Holst and Meikle, 2003).

The population dynamics of *P. truncatus* in rural maize stores are characterized by high infestation levels, accompanied by significant grain losses, from the fourth month of storage onwards (e.g. Henckes, 1992; Borgemeister *et al.*, 1994). The population growth rate of *P. truncatus* decreases with increasing weight loss, but only becomes negative above 50% weight loss (Holst and Meikle, 2003). The onset of a *P. truncatus* infestation in a maize store is difficult to determine. During the first months of storage, *P. truncatus* densities are not only very low, but the beetles tend to be highly aggregated in the stores (Meikle *et al.*, 2000). *P. truncatus* (and *S. zeamais*) will infest maize both in-field and in-store, although the level of *P. truncatus* field infestation is highly variable (see Borgemeister *et al.*, 1994 and references cited therein), but even a single pheromone-emitting male can attract enough conspecifics to cause a serious pest infestation in a store (Scholz *et al.*, 1997b). Results of a population simulation model suggest that *P. truncatus* densities as low as one beetle per 10 kg grain at harvest could result in a devastating larger grain borer outbreak during the storage season (Meikle *et al.*, 1998a). Detecting such low infestation levels at harvest necessitates extremely large sample sizes, often exceeding 1000 cobs, which are not practical.

Initial Control Strategies against *Prostephanus truncatus*

Initial attempts to combat the beetle in Africa mainly focused on conventional chemical control strategies. Organophosphates like pirimiphos methyl or fenitrothion, although highly effective against the usual storage pests such as *S. zeamais* and *S. cereallela*, proved rather ineffective against *P. truncatus*. However, larger grain borer is highly susceptible to synthetic pyrethroids like permethrin, deltamethrin or fenvalerate that have a relatively low toxicity to the other pests. Therefore, insecticide mixtures such as permethrin and pirimiphos methyl were developed to control *P. truncatus* and other important postharvest maize pests. In addition to insecticides, *P. truncatus* is also highly susceptible to fumigants like phostoxin or methyl bromide, which can be used in large warehouses, but not normally for the disinfestation of smallholder stores.

In the first campaign to control larger grain borer in Tanzania, farmers were encouraged to shell their maize cobs and admix a suitable dilute-dust insecticide mixture. In Togo, where the initial introduction was apparently restricted to an area close to Lomé, it was thought that eradication might prove possible by fumigating smallholder stores. However, these fumigations proved ineffective due to the high gas leakage rate through earthen floors. In any case, this approach would probably not have succeeded because there had already been sufficient time for *P. truncatus* to establish itself in its natural forest environment from where it could reinvade the disinfested stores. The indicated insecticide mixtures have been used very successfully in several East African countries, and widespread adoption has led to a considerable reduction in postharvest losses in maize. In West Africa though, adoption rates have been much lower, mainly due to a smaller government subsidy of pesticides, socio-economic constraints faced by farmers and marketing problems with the products (Agbaka, 1996).

Biological Control of *Prostephanus truncatus*

Classical biological control was believed to be a control option for *P. truncatus* in Africa (Markham *et al.*, 1991), because of it being an outbreak pest and causing much greater damage to stored maize and cassava in its new area of distribution than in its native home in meso-America. Hence, large surveys on the natural enemy fauna of *P. truncatus* in Mexico and several Central American countries were undertaken in the early 1980s. However, the natural enemy complex associated with *P. truncatus* in its area of origin turned out to be rather limited in diversity. Two pteromalid parasitoids, *Anisopteromalus calandrae* Howard and *Theocolax elegans* Westwood, were able to slow down *P. truncatus* development in laboratory and cage trials in Costa Rica (Böye, 1990). However, parasitoids could not control the build-up of larger grain borer populations in traditional maize stores in Togo (Helbig, 1999a). Moreover, both *A. calandrae* and *T. elegans* possess a broad host spectrum, attacking various storage pests, are of cosmopolitan distribution and also occurred in Africa, where they apparently had limited effect on the population dynamics of *P. truncatus*.

In a number of studies, the potential impact of entomopathogens on *P. truncatus* was investigated. Several protozoa were isolated from *P. truncatus* in Central America (Burde, 1988) and Tanzania (Purrini and Keil, 1989). The most promising isolate, a *Mattesia* sp. (Neogregarinida, Ophryocystinae), killed up to 90% of first instar *P. truncatus* in the laboratory, but in subsequent field trials in Togo only insufficient levels of control were recorded (Henning, 1993). Burde (1988) isolated a number of strains of entomopathogenic fungi in Central America that showed moderate levels of pathogenicity to *P. truncatus* under laboratory conditions. Yet, the most efficient fungi turned out to be *Aspergillus* spp., which due to their highly toxic metabolites were excluded from further testing. More recently, the potential of *Beauveria bassiana* (Bals.) and *Metarhizium anisopliae* (Metsch.) (both *Hyphomycetes*) for *P. truncatus* control were investigated. Surveys in farmers' maize stores in Kenya revealed only very low levels of infection of *S. zeamais* and *P. truncatus* by *B. bassiana* (Oduor *et al.*, 2000). By contrast, in laboratory trials, three strains of *B. bassiana* and one *M. anisopliae* strain caused high mortalities in *P. truncatus* (Bourassa *et al.*, 2001). In storage trials, the application of the most promising *B. bassiana* strain significantly reduced *P. truncatus* numbers, but pest densities and grain losses still reached unacceptable levels (Meikle *et al.*, 2001).

The anthocorid predator *Calliodis* spp. readily preyed on, but was poorly synchronized with, *P. truncatus*, occurring too late during the storage season to successfully control the pest (Böye, 1990). Likewise, the indigenous anthocorid *Xylocoris flavipes* (Reuter), which is often found in the storage environment, proved to be a rather inefficient predator of *P. truncatus* in store trials in Togo (Helbig, 1999b).

The only specialized natural enemy of *P. truncatus* that has been identified in Central America is the histerid predator *Teretrius* (formerly *Teretriosoma*) *nigrescens* (Lewis) (Plate 60), identified from Mexican specimens sent to the Natural Resources Institute, UK. The close association between these two species was confirmed during exploratory surveys for natural enemies of *P. truncatus* in Mexico and Costa Rica where the histerid predator was found in all maize stores infested by *P. truncatus* (Böye, 1990). Interestingly, the male aggregation pheromone of *P. truncatus* is a kairomone for *T. nigrescens* (Rees *et al.*, 1990; Böye *et al.*, 1992a), so that as soon as pheromone traps were used to monitor the pest in the neo-tropics, *T. nigrescens* was also captured.

The basic biology and the ability of the predator to suppress *P. truncatus* populations were investigated in the laboratory in meso-America and under third-party-quarantine conditions in Europe (Rees, 1985). Both *T. nigrescens* adults and larvae prey on larger grain borer eggs and larvae, although the larvae are far more voracious than adult beetles. With 30 days under optimal conditions (30°C and 75% r.h.), the egg to adult developmental time in *T. nigrescens* is very similar to that of *P. truncatus* (Oussou *et al.*, 1998). However, due to its comparatively low fecundity, the intrinsic rate of increase of *T. nigrescens* is only two-thirds or less than that of *P. truncatus* (Holst and Meikle, 2003). Adult *T. nigrescens* are strong flyers and live up to 2 years. Moreover, adult beetles can survive starvation periods of 60 days.

Prior to the first releases of *T. nigrescens* in Africa, the prey spectrum and prey preferences of the predator were investigated under third-party-quarantine conditions. These studies showed that *T. nigrescens* readily takes other prey, especially the larvae of other storage Coleoptera, but given a choice prefers *P. truncatus* larvae (summarized in Böye *et al.*, 1992b). Moreover, adult *T. nigrescens* can survive, but not reproduce, when maintained only on maize flour. This led to concerns regarding the potential ecological risks of introducing *T. nigrescens* into Africa. Histerids are generally not species-specific predators and have thus rarely been used in classical biological control projects. In a comparative study after the first releases of *T. nigrescens* in Africa, the prey spectrum and prey composition of adults and larvae, collected in rural grain stores, and of beetles trapped with the pheromone, were investigated by electrophoresis of gut contents in meso-America and West Africa (Camara, 1996). Results of this study revealed that *T. nigrescens* adults and larvae occasionally preyed upon pests other than *P. truncatus*, but the large majority of the analysed specimens (>90%) had previously fed on larger grain borer, indicating that *T. nigrescens* has a strong prey preference.

The predator was first released in Togo (Biliwa *et al.*, 1992), followed by releases in Kenya (Giles *et al.*, 1996), Bénin (1992), Ghana (J.A.F. Compton and A. Ofosu, Accra, 1994, unpublished results), and – less well documented – in Tanzania, Guinea-Conakry, Zambia and Malawi. In all release countries, *T. nigrescens* became established, but data on impact are available only for Togo, Bénin, Ghana and Kenya. Following the releases in southern Togo, both Mutlu (1994) and Richter *et al.* (1998) recorded significantly reduced *P. truncatus* densities and grain weight losses in rural maize stores in release villages compared with villages and/or regions where *T. nigrescens* had not been released. Yet in Mutlu's study, only in one out of two seasons were *P. truncatus* densities reduced by the presence of the predator. The impact assessment studies in Togo were considerably hampered by the political instabilities from 1992 onwards. More data are available for southern and central Bénin. Prior to the first releases in Bénin, Borgemeister *et al.* (1997) found *T. nigrescens* in pheromone traps in August 1992 in the Mono province of south-western Bénin in locations close to the border of neighbouring Togo. Most likely the predator had dispersed from release sites in southern Togo into Bénin. In the following years the beetle rapidly spread eastward, with high pheromone trap catches of *T. nigrescens* in sites of up to 70 km distance from the Togo border. Continuous pheromone trapping data from seven to nine sites in the Mono province between 1992 and 1997 revealed increasing numbers of *T. nigrescens*, particularly in 1994 and 1995, and a sharp decline in *P. truncatus* trap catches. Finally, in three consecutive on-farm storage trials in the Mono province, *P. truncatus* numbers and grain weight losses decreased tremendously between 1992 and 1995, reaching levels of grain losses that were comparable to those recorded in southern Togo prior to the accidental introduction of *P. truncatus*. In a larger study, Schneider (1999) recorded pheromone trap data of *P. truncatus* and *T. nigrescens* in 124 trapping sites in southern Togo and the whole of Bénin between 1995 and 1997. These data were coupled with samples taken in nearby rural maize stores. As in the previous study, a sharp decline of *P. truncatus* numbers in traps and maize stores were observed in southern Togo and the south-west of Bénin, accompanied by increasing records of *T. nigrescens* in traps and store samples. In addition, a rapid west–east spread of the predator was observed in southern Bénin, as indicated by rising trap catches of *T. nigrescens* during the study period. It is possible that the fast spread of *T. nigrescens* prevented *P. truncatus* from developing damaging populations in the Ouémé department of south-eastern Bénin (Fig. 20.2). Trap catches in central and especially the northern parts of Bénin were considerably lower, although the *P. truncatus* distribution in these areas was highly aggregated, with outbreaks of larger grain borer mostly confined to sites near major market places. However, in 1998 and 1999, Nansen (2000) recorded rising trap catches of *P. truncatus* in the Mono province of Bénin. In the 1999/2000-storage season, Meikle *et al.* (2002) found half of the investigated farmer maize stores in southern and central Bénin infested with *P. truncatus*. While the data could not resolve whether *T. nigrescens* had slowed the attack of *P. truncatus*, they documented several cases where the predator did not prevent serious losses by *P. truncatus*.

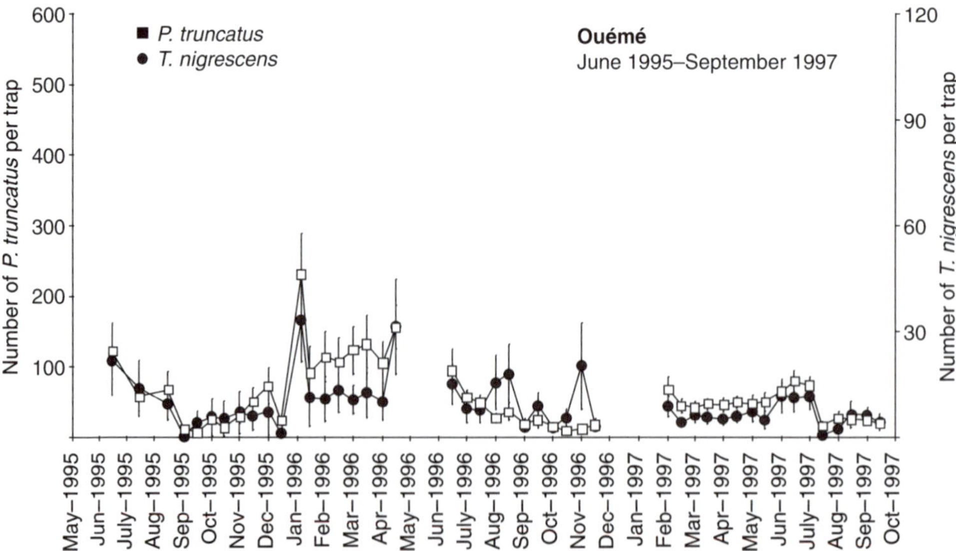

Fig. 20.2. Flight activity of *Prostephanus truncatus* and *Teretrius nigrescens* in the Ouémé department of south-eastern Bénin. Means and SE of 16 pheromone traps, changed at 2-week intervals (from Schneider, 1999).

In the Volta Region of Ghana, *T. nigrescens* was released in several sites in 1994. During the same period, the predator was also recorded entering the Volta Region from Togo, being captured in traps along the Ghana/Togo border. Attempts to demonstrate impact of the predator gave inconclusive results. However, pheromone trap monitoring of the two insects over extended periods showed that even though *T. nigrescens* was well established, it was not able to prevent *P. truncatus* from reaching massive peaks in years when the climate was particularly favourable to the pest, such as 1998/1999 and 1999/2000. Strangely, the *T. nigrescens* catch remained little changed from year to year despite large difference in prey counts. It has been shown that the likelihood of an open store, such as the maize barns used in West Africa, becoming infested is related to the numbers of actively dispersing *P. truncatus* individuals in the area (Birkinshaw *et al.*, 2002). Thus, despite being well established in south-eastern Ghana, *T. nigrescens* was not able to limit *P. truncatus* densities to below damage levels.

While assessing the impact of the initial *T. nigrescens* releases in Kenya in 1991, Giles *et al.* (1996) found no benefits due to biological control in maize stores at two study locations. Average temperatures at these locations were 17.7 and 22.2°C, which are close to the lower threshold for *T. nigrescens* development (Holst and Meikle, 2003). However, in Kenya most of the predators were released in a warmer, deciduous bush-land area. In the subsequent impact assessment, pheromone trap catches in release transects were compared with those from non-release transects (Nang'ayo, 1996). Over a 3-year period, *P. truncatus* numbers in the release transects declined sharply – though overall numbers were low – while those in the non-release transects remained relatively unchanged. These results indicate that *T. nigrescens* is also capable of having an impact on *P. truncatus* in its natural forest habitat. No data from post-release studies from Guinea-Conakry, Tanzania and Malawi have been published.

In laboratory and field trials in southern Bénin, Schneider (1999) studied the predator–prey system in considerable detail. He showed that high *P. truncatus* densities in stores can nullify the impact of the predator, mainly because the rate of increase of *P. truncatus* is higher than that of *T. nigrescens*, eventually leading to an oversupply of prey and satiation of the predator. The main effect of *T. nigrescens* on the population dynamics of *P. truncatus* is the early curtailment of population increase of the prey. Yet this mechanism is only effective up to a certain threshold in prey density, beyond which prey population will escape predator control. The best control of *P. truncatus* populations in maize stores may be achieved under the following conditions: (i) early synchronization of prey and predator; (ii) clumped distribution of *P. truncatus* on a few infested cobs in the store; and (iii) very few *P. truncatus* adults per cob when *T. nigrescens* arrives. In the natural forest habitat of *P. truncatus* and *T. nigrescens*, the conditions for effective control of larger grain borer are probably much better, because here *P. truncatus* generally occurs at much lower densities due to the comparatively lower nutritive value of woody substrates and the consequently lower reproduction rate of *P. truncatus* on wood compared with starch-rich substrates like maize and/or cassava. This could also explain the documented impact of *T. nigrescens* following the releases in Kenya's forest environments.

Two alternative, not mutually exclusive, modes of action are possible for biological control through *T. nigrescens*. First, regional control, in which there is a general depression of the pest population in the wild habitat where the pest, subsisting on poor substrates, could more readily be overcome by the predator than in the rich substrate of stored produce (Markham *et al.*, 1991; Mutlu, 1994). Second, local control, in which the predator seeks out and overcomes the pest inside the store before serious damage accrues (Richter *et al.*, 1998). The field studies cited above were designed to assess natural enemy impact acting via the first mode. In a modelling study, based on laboratory data and field data from southern Bénin, Holst and Meikle (2003) assessed the prospect of control via the second mode. They showed that the predator has a strong negative effect on its own population build-up (negative density-dependence) prohibiting an effective numerical response; moreover, the predator was unable to push the population growth rate of its prey down to zero, except at very high predator densities (above ~60 predator beetles per kg grain). It was concluded that a store infested early in the season with *P. truncatus* would ultimately suffer serious losses, irrespective of the density of *T. nigrescens* in the store, given climatic conditions conducive to the proliferation of the pest.

Thus, it is too early for a final assessment on the potential impact of *T. nigrescens* as a classical biological control agent. Only a few years after the release and establishment of *T. nigrescens*, the ecology of the predator–prey system in release countries in Africa may still be in an unstable state. It seems that the predator can successfully control *P. truncatus* in their natural forest habitat, as shown in Kenya. Moreover, the faster spread of *T. nigrescens* may prevent larger grain borer outbreaks in regions where *P. truncatus* has not yet reached damaging levels (e.g. south-eastern Bénin). The situation in the storage environment is less conclusive. Some studies from West Africa suggest that *T. nigrescens* can successfully suppress *P. truncatus* populations in maize stores, while in others the predator was unable to sufficiently control larger grain borer densities. The predator is most likely incapable of controlling *P. truncatus* once prey densities have surpassed a certain threshold. This may happen on a local scale (i.e. in a single store) or on a regional scale. Moreover, *T. nigrescens* seems to have less impact under drier conditions. Therefore, biological control of the larger grain borer needs to be accompanied by compatible intervention techniques within the framework of IPM.

IPM

Early efforts to keep losses in grain stores to levels acceptable to smallholder farmers followed a rigid pattern of shelling maize cobs and admixing a suitable binary insecticide. This approach was successful while supported by an intensive extension effort and external financial support, but is unsustainable in the long-term for several reasons. In many places, shelling cobs soon after harvest is not consistent with farming practices. The use of synthetic pesticides is generally undesirable on environmental and health grounds, and in many cases may not be an option for farmers because insecticides are either too expensive or locally unavailable. Attack by *P. truncatus* is sporadic. Pest incidence may be insignificant for several years and then suddenly increase in a 'bad' year. Furthermore, great variability exists in the severity of infestations, even among close-standing stores. Given this scenario, it is difficult to introduce practices that require significant investment by poor farmers.

It is in this context that IPM approaches may help smallholder farmers. It is inevitable that IPM for the protection of smallholder's farm stores will be more limited than its relatively sophisticated counterpart for the protection of large-scale crop production. Smallholder farmers generally lack the resources and flexibility to alter pest management at will, but if offered acceptable options and a means of decision making so that the proposed action is cost effective in relation to the risk of losses, then a more sustainable and cost-effective approach is possible.

Detecting the problem

Determining the risk of a serious grain loss due to *P. truncatus* is an important element in any IPM approach and a sequential sampling system has been developed that can be used to determine whether *P. truncatus* is present in stores in small numbers (Meikle *et al.*, 2000). An alternative, or complementary, approach would be to predict those years when *P. truncatus* attack in store is likely (i.e. predict the 'bad years'). Examples of 'bad years' were recently experienced around Morogoro in Tanzania in 1998, northern Ghana 1999/2000 and eastern Kenya 2001/2002, when devastating *P. truncatus* attacks resulted in high grain losses. Such years are ones in which there are very high numbers of *P. truncatus* flying around searching for food. The likelihood of a store becoming infested is related to the cumulative beetle catch in pheromone traps over the period of storage (Birkinshaw *et al.*, 2002), but determining risk on a routine basis using pheromone traps would be expensive and unsustainable. For this reason, models based on climatic data have been developed to predict trap catches for Bénin (Nansen *et al.*, 2002) and Ghana (Hodges *et al.*, 2003). The model developed in Ghana showed very good prediction of the cumulative trap catch of *P. truncatus* for the major and minor season harvests over 3 years. Therefore, using a predictive model, farmers can be warned several months in advance of impending *P. truncatus* problems.

Range of options for the smallholder farmer

There may be several options open to the farmer according to existing circumstances, but in all cases essential store hygiene considerations are very important. Old store residues must be removed from the store before the new stock is loaded. Before loading the new harvest, the maize must be carefully inspected and any cobs showing signs of damage must be rejected. When storing maize on the cob, traditional varieties with good husk cover are much less likely to be attacked (Meikle *et al.*, 1998b), so farmers who store these can safeguard their stock by rejecting any cobs with damaged or open sheathing leaves. For the future, plant breeding to improve the husk cover of high-yielding varieties would certainly reduce pest problems, as would the introduction of varieties with grain that has a low susceptibility to *P. truncatus* (Kumar, 2002).

According to the extent of the risk, farmers can: (i) sell the maize stock within 3 months so that specific pest management measures are not needed. (ii) Treat the whole stock with pesticide. Maize cobs can be sprayed with, or dipped in, an emulsion or dusted with dilute dust layer by layer. Shelled grain would be treated with dilute dust. (iii) Treat only a portion of the stock with pesticide. If some of the grain is to be stored for less than 3 months, then this may not require any pesticide treatment. The grain to be treated should be placed at the base of the store where insect infestation pressure is greater (Hodges *et al.*, 1999b). Untreated grain should remain at the top and may be easily removed for consumption or sale. (iv) Treat the stock with a local, botanical pesticide. Farmers use a wide variety of plants in this way. These vary in their efficacy according to their mode of application, time of year of collection and area of collection. They may be recommended depending on the risk of attack and the circumstances of the farmer (e.g. availability of botanicals, ability to afford more effective alternatives, etc.). (v) Treat stocks with inert dusts. A thick layer of paddy husk ash covering the stock is effective in preventing attack. Commercial preparations of diatomaceous earths are effective in dry areas against several pest species and are currently being tested for their efficacy against *P. truncatus*. (vi) Adopt a sealed storage system, such as mud silos or the mudding of traditionally unmudded structures. Such sealed storage can provide a very effective barrier to pest attack and can be adopted in situations where the stock is sufficiently dry so that ventilation is not required. This has been achieved with several communities in northern Ghana that do not traditionally use mudded structures. Use of mudded structures may be combined with the use of synthetic pesticides or botanicals according to risk.

Deciding what to do

The circumstances and aspirations of individual farmers will vary. For this reason, simple blanket recommendations for pest management are of limited value. In view of the range of options available for pest management in smallholder stores, a decision-tree approach such as that described by Farrell *et al.* (1996) is particularly appropriate (Fig. 20.3). Both in East and West Africa, farmers tend to leave their maize in stores for extended periods (sometimes exceeding 8 months) because maize prices on rural and urban markets are lowest immediately after the harvest and highest into the 'lean' season when there is little maize available. Because the extent of *P. truncatus* infestation during the first 3 months of storage are generally low, this has led to recommendations to farmers to divide their maize harvest into two portions. One portion is destined for consumption by their families and is usually kept for no longer than 3 months in the store. The other portion is intended to be sold later in markets and is thus stored for longer periods. Fieldwork in both Tanzania and Kenya

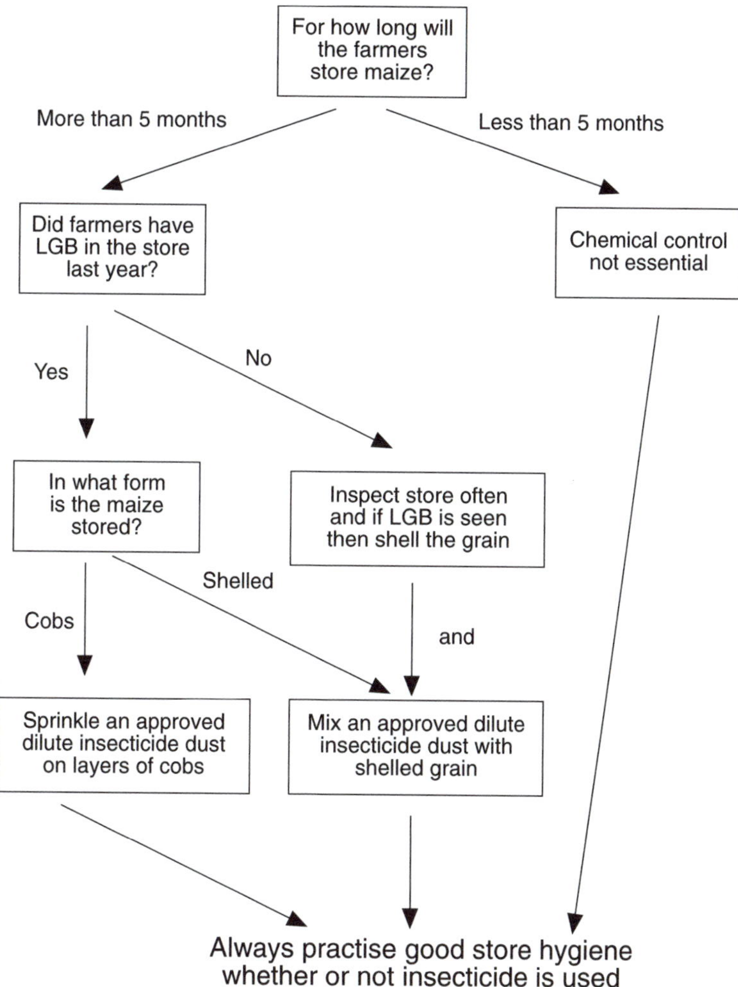

Fig. 20.3. A decision tree to help extension services advise farmers on the protection of their maize against *Prostephanus truncatus* (from Farrell *et al.* 1996).

indicated that if maize was to be stored for less than 5 months in either country then pesticide use was probably not justified (Henckes, 1992). This conclusion was based on the break-even point between the value of losses and the costs of treating grain with a dilute dust insecticide. As the value of maize and cost of pesticide treatment changes from year to year, the break-even point for pesticide treatment needs to be recalculated from time to time. This would be a job for the extension personnel who advise subsistence farmers.

Looking at the decision tree (Fig. 20.3), it can be seen that if the storage period was to be greater than 5 months then knowledge of the previous history of *P. truncatus* infestation in a store is important. If there was *P. truncatus* in the store during the previous year, then the risk of renewed infestation in the current year is higher and the admixture of a dilute dust insecticide is recommended. However, if no *P. truncatus* were observed last year then regular inspection is required. If the pest is found subsequently, then grain shelling and insecticide treatment are required. The decision tree could be developed much further to include more of the options listed above, and, where appropriate, links to predictions on the likelihood of it being a 'bad' year for *P. truncatus* so that the safest options are selected.

Conclusion: Future Role of Biological Control in IPM for Larger Grain Borer

So far, biological control has not solved the larger grain borer problem consistently over large areas for long time spans. Thus, any particular farmer's maize or cassava stores remain at risk of serious damage caused by *P. truncatus*. While this status may change over the years, it is currently clear that farmers need to learn and practise IPM procedures such as those described above. A huge task lies ahead, both in implementing those methods already developed and tested and in further research and development.

Because *T. nigrescens* is inexpensive to rear and release, it should be introduced in the wake of the pest as it spreads. Although the predator works unpredictably, when present in a region it may reduce the risk of any particular store being infested with *P. truncatus*. *T. nigrescens* works (or not) independently of the IPM measures suggested above. In the case of heavy infestations, *T. nigrescens* cannot prevent serious losses; hence the application of insecticides or other actions to be taken is warranted, even if the action decimates the natural enemy population in the store. Thus, biological control can serve as a background, regional control strategy augmented by those IPM activities that farmers may decide to implement.

References

Agbaka, A. (1996) Etude biologique et possibilité de lutte intégrée contre *Prostephanus truncatus* (Horn) (Coleoptera: Bostrichidae) ravageur des stocks de maïs dans les milieux paysans en République du Bénin. PhD thesis, Université National de Côte d'Ivoire, Abidjan, Côte d'Ivoire.

Barrer, P.A. (1983) A field demonstration of odour-base, host-food finding behaviour in several species of stored grain insects. *Journal of Stored Products Research* 19, 105–110.

Bell, R.J. and Watters, F.L. (1982) Environmental factors influencing the development and rate of increase of *Prostephanus truncatus* (Horn) (Coleoptera: Bostrichidae) on stored maize. *Journal of Stored Products Research* 18, 131–142.

Biliwa, A., Böye, J., Fischer, H.U. and Helbig, J. (1992) Stratégie de lâchers et études de suivi de *Teretriosoma nigrescens* au Togo. In: Boeye, J., Wright, M. and Laborius, G.-A. (eds) *Implementation of and Further Research on Biological Control of the Larger Grain Borer. Proceedings of an FAO/GTZ Co-ordination Meeting*. FAO, Rome, Italy and GTZ, Eschborn, Germany, pp. 138–142.

Birkinshaw, L.A., Hodges, R.J., Addo, S. and Riwa, W. (2002) Can 'bad' years for damage by *Prostephanus truncatus* be predicted? *Crop Protection* 21, 783–791.

Borgemeister, C., Adda, C., Djomamou, B., Degbey, P., Agbaka, A., Djossou, F., Meikle, W.G. and Markham, R.H. (1994) The effect of maize cob selection and the impact of field infestation on stored maize losses by larger grain borer (*Prostephanus truncatus* [Horn] Col., Bostrichidae) and associated storage pests. In: Highley, E., Wright, E.J., Banks, H.J. and Champ, B.R. (eds) *Stored Product Protection. Proceedings of the 6th International Working Conference on Stored-product Protection*, Vol. 2. CAB International, Wallingford, UK, pp. 906–909.

Borgemeister, C., Djossou, F., Adda, C., Schneider, H., Djomamou, B., Azoma, K. and Markham, R.H. (1997) Establishment, spread and impact of *Teretriosoma nigrescens* Lewis (Coleoptera: Histeridae), an exotic predator of the larger grain borer *Prostephanus truncatus* (Horn) (Coleoptera: Bostrichidae), in south-western Bénin. *Environmental Entomology* 26, 1405–1415.

Borgemeister, C., Tchabi, A. and Scholz, D. (1998a) Trees or stores? The origin of migrating *Prostephanus truncatus* collected in different ecological habitats in southern Bénin. *Entomologia Experimentalis et Applicata* 87, 285–294.

Borgemeister, C., Goergen, G., Tchabi, A., Awande, S., Markham, R.H. and Scholz, D. (1998b) Exploitation of a woody host plant and cerambycid associated volatiles as host finding cues by the larger grain borer, *Prostephanus truncatus* (Horn) (Coleoptera: Bostrichidae). *Annals of the Entomological Society of America* 91, 741–747.

Bourassa, C., Vincent, C., Lomer, C.J., Borgemeister, C. and Mauffette, Y. (2001) Effects of entomopathogenic Hyphomycetes against the larger grain borer, *Prostephanus truncatus* (Horn) (Coleoptera: Bostrichidae) and its predator, *Teretriosoma nigrescens* Lewis (Coleoptera: Histeridae). *Journal of Invertebrate Pathology* 77, 75–77.

Böye, J. (1990) Autökologische Untersuchungen zum Verhalten des großen Kornbohrers *Prostephanus truncatus* (Horn) (Colepotera: Bostrichidae). PhD thesis, Christian-Albrechts-University, Kiel, Germany.

Böye, J., Laborius, G.A. and Schulz, F.A. (1992a) The response of *Teretriosoma nigrescens* Lewis (Col., Histeridae) to the pheromone of *Prostephanus truncatus* (Horn) (Col., Bostrichidae). *Anzeiger für Schädlingskunde Pflanzenschutz Umweltschutz* 65, 153–157.

Böye, J., Wright, M. and Laborius, G.A. (eds) (1992b) *Implementation of and Further Research on Biological Control of the Larger Grain Borer. Proceedings of an FAO/GTZ Co-ordination Meeting*. FAO, Rome, Italy and GTZ, Eschborn, Germany.

Burde, S. (1988) Mikrobielle Antagonisten von *Prostephanus truncatus* (Horn) (Coleoptera: Bostrichidae) – Grundlagen für eine Biotherapie im tropischen Vorratsschutz. GTZ, Eschborn, Germany.

Camara, M. (1996) Elektrophoretische Untersuchung des Beutespektrums von *Teretriosoma nigrescens* Lewis (Coleoptera: Histeridae), ein natürlicher Feind des Großen Kornbohrers *Prostephanus truncatus* (Horn) (Coleoptera: Bostrichidae). PhD Thesis, University of Georg-August-University, Göttingen, Germany.

Chittenden, F.H. (1911) Papers on insects affecting stored products. The larger grain borer. *U.S. Department of Agriculture, Division of Entomology, Bulletin* 96, 48–52.

Cork, A., Hall, D.R., Hodges, R.J. and Pickett, J.A. (1991) Identification of major component of male-produced aggregation pheromone of larger grain borer, *Prostephanus truncatus* (Horn) (Coleoptera: Bostrichidae). *Journal of Chemical Ecology* 17, 789–803.

Dick, K. (1988) A review of insect infestation of maize in farm storage in Africa with special reference to the ecology and control of *Prostephanus truncatus*. Natural Resources Institute Bulletin No. 18. Natural Resources Institute, Chatham, UK.

Farrell, G. (2000) Dispersal, phenology and predicted abundance of the larger grain borer in different environments. *African Crop Science Journal* 8, 337–343.

Farrell, G., Hodges, R.J. and Golob, P. (1996) Integration of control methods for stored products pests in East Africa. In: Farrell, G., Greathead, A.H., Hill, M.G. and Kibata, G.N. (eds) *Management of Farm Storage Pests in East and Central Africa. Proceedings of the East and Central Africa Storage Pest Management Workshop*, Naivasha, Kenya, 14–19 April 1996. IIBC, Ascot, pp. 57–67.

Giga, D.P. and Canhao, Sr.J. (1993) Competition between *Prostephanus truncatus* (Horn) and *Sitophilus zeamais* (Motsch.) in maize at two temperatures. *Journal of Stored Products Research* 29, 63–70.

Giles, P.H., Hill, M.G., Nang'ayo, F.L.O., Farrell, G. and Kibata, G.N. (1996) Release and establishment of the predator *Teretriosoma nigrescens* Lewis for the biological control of *Prostephanus truncatus* (Horn) in Kenya. *African Crop Science Journal* 4, 325–337.

Helbig, J. (1999a) Ability of naturally occurring parasitoids to suppress the introduced pest *Prostephanus truncatus* (Horn) (Coleoptera, Bostrichidae) in traditional maize stores in Togo. *Journal of Stored Products Research* 34, 287–295.

Helbig, J. (1999b) Efficacy of *Xylocoris flavipes* (Reuter) (Het., Anthocoridae) to suppress *Prostephanus truncatus* (Horn) (Col., Bostrichidae) in traditional maize stores in southern Togo. *Journal of Applied Entomology* 123, 503–508.

Henckes, C. (1992) Investigations into insect population dynamics, damage and losses of stored maize – an approach to IPM on small farms in Tanzania with special reference to *Prostephanus truncatus* (Horn). GTZ, Eschborn, Germany.

Henning, S.M.I. (1993) Untersuchungen zur Pathogenität von *Matthesia* sp. (Neogregrinidia: Ophryocystidae) und *Nosema* sp. (Microsporida: Nosematidae) an *Prostephanus truncatus* (Horn) (Coleoptera: Bostrichidae) zum Einsatz der Protozoen in traditionellen Maislagern Togos. PhD thesis, Technical-University-Berlin, Berlin, Germany.

Hodges, R.J. (1986) The biology and control of *Prostephanus truncatus* – a destructive pest with an increasing range. *Journal of Stored Products Research*, 22, 1–14.

Hodges, R.J. (1994) Recent advances in the biology and control of *Prostephanus truncatus*. In: Highley, E., Wright, E.J., Banks, H.J. and Champ, B.R. (eds) *Stored Product Protection. Proceedings of the 6th International Working Conference on Stored-product Protection*, Vol. 2. CAB International, Wallingford, UK, pp. 929–934.

Hodges, R.J. and Meik, J. (1984) Infestation of maize cobs by *Prostephanus truncatus* – aspects of biology and control. *Journal of Stored Products Research* 20, 205–213.

Hodges, R.J., Dunstan, W.R., Magazini, I. and Golob P. (1983) An outbreak of *Prostephanus truncatus* (Horn) (Coleoptera: Bostrichidae) in East Africa. *Protection Ecology* 5, 183–194.

Hodges, R.J., Birkinshaw, L.A. and Smith, R.H. (1999a) Host selection or mate selection? Lessons from *Prostephanus truncatus*, a pest poorly adapted to stored products. In: *Stored Product Protection. Proceedings of the 7th International Working Conference on Stored-Product Protection*, 14–19 October 1998, Beijing, China, Vol. 2, pp 1788–1794.

Hodges, R.J., Carr, P. and Hussein, A.I. (1999b) Limiting the amount of pesticide applied to small bulks of maize in rural stores. In: *Stored Product Protection. Proceedings of the 7th International Working Conference on Stored-Product Protection*, 14–19 October 1998, Beijing, China, Vol. 1, pp 956–963.

Hodges, R.J., Addo, S. and Birkinshaw, L.A. (2003) Can observations of climatic variables be used to predict the flight dispersal rates of *Prostephanus truncatus*? *Agricultural and Forest Entomology* (in press).

Holst, N. and Meikle, W.G. (2003) *Teretrius nigrescens* against larger grain borer *Prostephanus truncatus* in African maize stores: biological control at work. *Journal of Applied Ecology* 40(2) (in press).

Kumar, H. (2002) Resistance in maize to the larger grain borer, *Prostephanus truncatus* (Horn) (Coleoptera: Bostrichidae). *Journal of Stored Products Research* 38, 267–280.

Li, L. (1988) Behavioural ecology and life history evolution in the larger grain borer *Prostephanus truncatus* (Horn). PhD thesis, University of Reading, Reading, UK.

Markham, R.H., Wright, V.F. and Ríos Ibarra, R.M. (1991) A selective review of research on *Prostephanus truncatus* (Col.: Bostrichidae) with an annotated and updated bibliography. *Ceiba* 32, 1–90.

Meikle, W.G., Holst, N., Scholz, D. and Markham, R.H. (1998a) A simulation model of *Prostephanus truncatus* (Horn) (Coleoptera: Bostrichidae) in rural maize stores in the Republic of Bénin. *Environmental Entomology* 27, 59–69.

Meikle, W.G., Adda, C., Azoma, K., Borgemeister, C., Degbey, P., Djomamou, B. and Markham, R.H. (1998b) The effects of maize variety on the density of *Prostephanus truncatus* (Coleoptera: Bostrichidae) and *Sitophilus zeamais* (Coleoptera: Curculionidae) in postharvest stores in Bénin Republic. *Journal of Stored Products Research* 34, 45–58.

Meikle, W.G., Holst, N., Degbey, P. and Oussou, R. (2000) Evaluation of sequential sampling plans for the larger grain borer (Coleoptera: Bostrichidae) and the maize weevil (Coleoptera: Curculionidae) and of visual grain assessment in West Africa. *Journal of Economic Entomology* 93, 1822–1831.

Meikle, W.G., Cherry, A.J., Holst, N., Hounna, B. and Markham, R.H. (2001) The effects of an entomopathogenic fungus, *Beauveria bassiana* (Balsamo) Vuillemin (Hyphomycetes), on *Prostephanus truncatus* (Horn) (Col.: Bostrichidae), *Sitophilus zeamais* Motschulsky (Col.: Curculionidae), and grain losses in stored maize in the Bénin Republic. *Journal of Invertebrate Pathology* 77, 198–205.

Meikle, W.G., Markham, R.H., Nansen, C., Holst, N., Degby, P., Azoma, K. and Korie, S. (2002) Pest management in traditional maize stores in West Africa: A farmer's perspective. *Journal of Economic Entomology* 95, 1088–1097.

Mutlu, P. (1994) Ability of the predator *Teretriosoma nigrescens* Lewis (Col.: Histeridae) to control the larger grain borer *Prostephanus truncatus* (Horn) (Col.: Bostichidae) under rural storage conditions in the southern region of Togo. In: Highley, E., Wright, E.J., Banks, H.J. and Champ, B.R. (eds) *Stored Product Protection. Proceedings of the 6th International Working Conference on Stored-product Protection*, Vol. 2. CAB International, Wallingford, UK, pp. 1116–1121.

Nang'ayo, F.L.O. (1996) Ecological studies on larger grain borer in savanna woodlands of Kenya. PhD thesis, Imperial College, London, UK.

Nang'ayo, F.L.O., Hill, M.G., Chandi, E.A., Nzeve, N.V. and Obiero, J. (1993) The natural environment as a reservoir for the larger grain borer *Prostephanus truncatus* (Horn) (Coleoptera: Bostrichidae) in Kenya. *African Crop Science Journal* 1, 39–47.

Nansen, C. (2000) Spatial distribution and potential hosts of the larger grain borer *Prostephanus truncatus* (Horn) (Coleoptera: Bostrichidae), in a forest in Bénin, West Africa. PhD thesis, Royal Veterinary and Agricultural University, Copenhagen, Denmark.

Nansen, C., Meikle, W.G. and Korie, S. (2002) Spatial analysis of *Prostephanus truncatus* (Bostrichidae: Coleoptera) flight activity near maize stores and in different forest types in southern Bénin, West Africa. *Annals of the Entomological Society of America* 95, 66–74.

Oduor, G.I., Smith, S.M., Chandi, E.A., Karanja, L.W., Agano, J.O. and Moore, D. (2000) Occurrence of *Beauveria bassiana* on insect pests of stored maize in Kenya. *Journal of Stored Products Research* 36, 177–185.

Oussou, R.D., Meikle, W.G. and Markham, R.H. (1998) Factors affecting larval survivorship and development rate of *Teretriosoma nigrescens* Lewis. *Insect Science and its Application* 18, 53–58.

Pantenius, C. (1987) Verlustanalyse in kleinbäuerlichen Maislagerungssystemen der Tropen, dargestellt am Beispiel von Togo. PhD thesis, Christian-Albrechts-University, Kiel, Germany.

Pike, V. (1993) Development of pheromone-baited traps and their uses for the larger grain borer. *IOBC Bulletin* 16, 64–70.

Purrini, K. and Keil, H. (1989) *Ascogregarina bostrichidorum* n. sp. (Lecudinidae, Eugregarinida), a new gregarine parasitising the larger grain borer, *Prostephanus truncatus* (Horn) (Coleoptera: Bostrichidae). *Archiv für Protistenkunde* 137, 165–171.

Ramírez-Martinez, M., de Alba-Avila, A. and Ramírez-Zuriba, R. (1994) Discovery of the larger grain borer in a tropical deciduous forest in Mexico. *Journal of Applied Entomology* 118, 354–360.

Rees, D.P. (1985) Life history of *Teretriosoma nigrescens* Lewis (Coleoptera: Histeridae) and its ability to suppress populations of *Prostephanus truncatus* (Horn) (Coleoptera: Bostrichidae). *Journal of Stored Products Research* 21, 115–118.

Rees, D.P., Rodriguez Rivera, R. and Herrera Rodriguez, F.J. (1990) Observations of the ecology of *Teretriosoma nigrescens* Lewis (Col.: Histeridae) and its prey *Prostephanus truncatus* (Horn) (Coleoptera: Bostrichidae) in the Yucatan peninsula, Mexico. *Tropical Science* 30, 153–165.

Richter, J., Biliwa, A., Helbig, J. and Henning-Helbig, S. (1998) First release of *Teretriosoma nigrescens* Lewis (Col., Histeridae), the predator of *Prostephanus truncatus* (Horn) (Col.: Bostrichidae), and follow-up investigations in southern Togo. *Journal of Applied Entomology* 122, 383–387.

Schneider, H. (1999) Impact assessment of *Teretriosoma nigrescens* Lewis (Coleoptera: Histeridae), a predator of the larger grain borer *Prostephanus truncatus* (Horn) (Coleoptera: Bostrichidae). PhD thesis, University of Hannover, Hannover, Germany.

Scholz, D., Tchabi, A., Borgemeister, C., Markham, R.H., Poehling, H.-M. and Lawson, A. (1997a) Host-finding behaviour of *Prostephanus truncatus* (Horn) (Col., Bostrichidae): primary attraction or random attack? *Journal of Applied Entomology* 121, 261–269.

Scholz, D., Borgemeister, C., Meikle, W.G., Markham, R.H. and Poehling, H.-M. (1997b) Infestation of maize by *Prostephanus truncatus* initiated by male-produced pheromone. *Entomologia Experimentalis et Applicata* 83, 53–61.

Shires, S.W. (1980) Life history of *Prostephanus truncatus* (Horn) (Coleoptera: Bostrichidae) at optimum conditions of temperature and humidity. *Journal of Stored Products Research* 16, 147–150.

Subramanyam, Bh., Cutkomp, L.K. and Darveaux, B.A. (1985) A new character for identifying larval instars of *Prostephanus truncatus* (Horn) (Coleoptera: Bostrichidae). *Journal of Stored Products Research* 21, 101–104.

Vowotor, K.A., Meikle, W.G., Ayertey, J.N., Borgemeister, C. and Markham, R.H. (1998) Intraspecific competition in larvae of the larger grain borer, *Prostephanus truncatus* (Horn) within maize grains. *Insect Science and its Application* 18, 171–175.

21

Biological Control of *Helicoverpa armigera* in Africa

Andy Cherry,[1] Matthew Cock,[2] Henk van den Berg[3] and Rami Kfir[4]

[1]*International Institute of Tropical Agriculture, Cotonou, Bénin;* [2]*CABI Bioscience Switzerland, Delémont, Switzerland;* [3]*Laboratory of Entomology, Department of Plant Sciences, Wageningen University, Wageningen, The Netherlands;* [4]*Agricultural Research Council – Plant Protection Research Institute, Division of Insect Ecology – Biological Control, Pretoria, South Africa*

Introduction – the African Bollworm as a Pest

The African bollworm, *Helicoverpa armigera* (*Heliothis armigera*) (Hübner) (Lepidoptera, Noctuidae), is an indigenous species considered to be a major constraint to food, fibre and horticultural crop production in Africa. Its perception as a particularly serious pest derives from its polyphagy, high fecundity and short generation time (often being multivoltine), high mobility, preference for the harvestable fruiting parts of its host plant, and its propensity to develop resistance to chemical insecticides. The severity of *H. armigera* attack varies not only between crops and regions, but also on a temporal scale. Moreover, due also to its dispersive and migrational attributes, incidence of *H. armigera* is unpredictable. Four heliothine species are reported as of economic importance in Africa: *Helicoverpa armigera*, *Helicoverpa assulta afra* (Hardwick), *Helicoverpa fletcheri* (Hardwick) and *Helicoverpa peltigera* (Schiffermüller), but *H. armigera* is the only species of major economic importance and the published literature is overwhelmingly concerned with this species (Greathead and Girling, 1989). This chapter confines itself to *H. armigera* and we focus attention principally on the two best studied areas, southern and East Africa.

Greathead and Girling (1989) list 35 crop hosts of *H. armigera* plus 25 wild host plants in eastern and southern Africa. In East Africa it attacks various crops including cotton, legumes, maize, sorghum, sunflower, tobacco and tomato. In South Africa crops attacked include peas, beans, wheat, cotton, maize, grain sorghum, oats, barley, sunflower, tobacco, citrus, cucurbits, potato, tomato, lucerne, sunnhemp, cape gooseberry, chickpea and groundnuts (Annecke and Moran, 1982). To this list may be added a large variety of garden ornamentals and wild host plants. In South Africa it is regarded as one of the most serious citrus pests (Bedford, 1968) and is the key pest on cotton (van Hamburg and Guest, 1997).

Low economic damage thresholds in high value crops like cotton, tomato, pulses and tobacco require a high level of control that leads to reliance on heavy and frequent use of synthetic insecticides. However, regional and even relatively local differences in host preference can give rise to differences in pest status on particular crops (*Crop Protection Compendium*, 2001), and may also lead to exaggerated perceptions of the species as a pest. Integrated control programmes exist for *H. armigera* that seek to minimize pesticide inputs and maximize the impact of natural enemies, but a major constraint to their development, particularly on cotton, has been the need to deal with a complex of pests where control needs may conflict (e.g. Kuklinski and Borgemeister, 2002; Mensah, 2002).

Indigenous Natural Enemy Complexes

Most studies on the natural enemies of the African bollworm have been carried out in East and southern Africa and have focused on parasitoids that attack bollworm eggs and larvae, often in cotton cropping systems. Predators and naturally occurring pathogens have been less well studied. Natural enemy data are largely restricted to records only, with few studies providing quantitative information on percentage parasitism or infection. Data on impact in the context of pest life tables were generally lacking until recent studies in East Africa by van den Berg and Cock (1993a,b, 1995a,b), and van den Berg *et al.* (1993, 1997).

Parasitoids

The natural enemies of *H. armigera* in Africa have been reviewed and catalogued by van den Berg *et al.* (1988). A total of 83 identified and 93 partially identified species of parasitoids, some of which are important biological control agents, have been recorded from *H. armigera* (van den Berg *et al.* 1988). Most records are from southern and East Africa and concern larval parasitoids in the families Ichneumonidae, Braconidae and Tachinidae.

Most parasitoids recorded from *H. armigera* in Africa attack a range of host species. A minority of parasitoids recorded are host specific, notably five members of the braconid genus *Cardiochiles* that have only been recorded from the African bollworm. Among egg parasitoids, the scelionid wasp *Telenomus ullyetti* Nixon (Hymenoptera, Scelionidae) is specific to African bollworm whereas trichogrammatid wasp species will parasitize a wide range of lepidopteran eggs in a specific habitat.

Records suggest that geographical and temporal distribution of parasitoids and their importance within the local natural enemy complex vary considerably within and between East and southern Africa. For example in Kenya, van den Berg *et al.* (1993) found that in smallholder crops (sunflower, maize, sorghum and cotton) in several agroecological zones, occurrence of parasitoids varied greatly between seasons and sites. *Trichogrammatoidea* spp. (Hymenoptera, Trichogrammatidae) egg parasitoids, and *Linnaemya longirostris* (Macquart)

(Diptera, Tachinidae), a late-larval parasitoid, were the most common parasitoid species, but mean percentage parasitism was generally rather low (<5%), even though egg parasitism by *Trichogramma* spp. and late larval parasitism by *L. longirostris* occasionally reached up to 20%. Levels of parasitism were lower and species diversity poorer during this Kenyan study than in a similar study in western Tanzania (Nyambo, 1990).

One of the few quantitative studies of parasitism of *H. armigera* in East Africa was conducted in the western cotton growing area of Tanzania over the period 1981–1985 (Nyambo, 1990) on five major crops; cotton, tomato, chickpea, sorghum and maize, and one weed *Cleome* sp. (*Capparidaceae*). The study found variation within and between seasons, and between crops in the levels of both mortality factors. Twelve parasitoid species are listed belonging to the families Braconidae, Ichneumonidae and Tachinidae. Peak parasitism levels were much greater on the weed host *Cleome* sp. than on any of the crops. Except in sorghum, disease caused more mortality than parasitism, and mean peak parasitism did not exceed 24.5% in any of the five crops. Across all seasons and all crops, however, mean percentage parasitism did not rise above 2% for any of the 12 parasitoid species listed.

In East Africa in general, hymenopteran larval parasitoids are the most commonly recorded parasitoids, whereas in southern Africa dipterans are more frequently recorded (van den Berg, 1993). The braconid *Apanteles diparopsidis* Lyle is an important parasitoid of *H. armigera* in Tanzania, where parasitism of up to 26% has been recorded. In southern Africa, the same parasitoid is only found on the red bollworm *Diparopsis castanea* Hampson (Lepidoptera, Noctuidae) and spiny bollworms *Earias* spp. (Lepidoptera, Noctuidae). Records show high levels of egg parasitism in southern Africa with figures of up to 60–70% in the field in maize, cotton and other crops. In eastern Africa, egg parasitoid records are rare and in sunflower, maize and sorghum egg parasitism was between 0% and 22% (van den Berg and Cock, 1993b), while in Malagasy cotton, Kuklinski and Borgemeister (2002) recorded a peak of 32% egg parasitism by *Trichogramma evanescens* Westwood. In a 3-year study in sunflowers in South Africa, von Maltitz (E.F. von Maltitz, 1992, Grahamstown, unpublished results) recorded an average of 19% egg parasitism and 27% larval parasitism. The most abundant parasitoid was *Palexorista laxa* (Curran) (Diptera, Tachinidae), which emerged from 44% of parasitized larvae. Also in South Africa, *Paradrino halli* Curran (Diptera, Tachinidae) was recorded as the first tachinid species to emerge in spring and was responsible for up to 24.8% parasitism in citrus (Parry-Jones, 1938). Only three species of parasitoids are important in both areas: the braconid *Chelonus curvimaculatus* Cameron and the tachinids *P. laxa* and *P. halli*. All other species are important in only one of the areas, although they may be present in both.

Parry-Jones (1937) noted temporal variation in the bionomics of two egg parasitoids in citrus in Zimbabwe: *Trichogrammatoidea lutea* Girault was most active during the summer months (December–March) when *H. armigera* was most abundant on alternative host plants, whereas *T. ullyetti* was more abundant in the spring months. Similarly, Parsons and Ullyett (1934, 1936) showed that in South Africa, *T. ullyetti* was abundant in winter on irrigated vegetable crops and citrus and became scarce in summer. On the other hand, *T. lutea* was inactive during winter, but from spring until the end of summer, it predominated on rain grown cotton, maize and other summer crops (Parsons, 1940).

One of the most interesting findings in African studies on bollworm natural enemies is the huge variation in occurrence and impact in different host-plant associations. This is especially relevant for African smallholder farmers who typically cultivate a mixture of different crops within a small area. For example in western Tanzania, a parasitoid guild comprising the tachinid fly *P. laxa,* and the braconids *C. curvimaculatus* and *A. diparopsidis*, inflicts heavy parasitism on *H. armigera* in sorghum (van den Berg *et al.* 1990), but not to a significant degree on other crops. In contrast, *Cardiochiles* spp. were associated with cotton and cleome, and were rare on sorghum and maize (Nyambo, 1986). The ichneumonid *Charops* sp. was the only parasitoid species to occur on all crops and was the most frequent species found. *A. diparopsidis* was frequent on sorghum in two seasons, and *Cardiochiles* spp. were the dominant parasitoids in each season on cleome (Nyambo, 1986). Cleome itself was associated with a more diverse population of parasitoids than any of the other host plants (Nyambo, 1990). In the few studies that have compared natural enemy numbers in different crops, sorghum and maize seem to attract more predators and parasitoids as bollworm moths prefer to lay their eggs on these plants. Caution is required with crop-specific interpretations since parasitoids tend to have habitat-specific, rather than crop-specific host searching behaviour. Furthermore, inter-variety variation in crop plants also plays a role, and hairiness of plants in particular has been associated with natural enemy-mediated differences in pest incidence (Bottrell *et al.*, 1998).

Predators

van den Berg *et al.* (1988) list predator records from Africa available at that time including Anthocoridae (five or more species), Reduviidae (eight species), Carabidae (four species), Staphylinidae (one species), Coccinellidae (two species), Asilidae (one species), Vespidae (one species), Eumenidae (two species), Sphecidae (two species), and Formicidae (four species). However, as subsequent studies in Kenya have shown (e.g. van den Berg and Cock, 1993a), this list just scratches the surface of the generalist predators associated with *H. armigera* in different crops. More recent studies in cotton fields in South Africa considerably expand this list of predators and show even the mouse *Mastomys natalensis* (Smith) to be an important predator of *H. armigera* pupae (Watmough, 1991; Watmough and Kfir, 1995; van Hamburg and Guest, 1997).

The impact of predation in the field is more difficult to assess than that of parasitism. An extensive investigation conducted in western, central and coastal Kenya over the period 1987–1991 focused on the impact of indigenous predators, as well as parasitoids and pathogens, in the population dynamics of *H. armigera* in smallholder crops, including cotton and sunflower, that suffer yield loss from *H. armigera*, and secondary crops, including maize and sorghum, that generally tolerate *H. armigera* infestation. Studies were also carried out on several important local crops known to be attacked by *H. armigera*, although detailed studies subsequently focused on sunflower and cotton. The incidence of *H. armigera* and natural enemies was monitored over several seasons (van den Berg *et al.*, 1993). Throughout the study period, the incidence of *H. armigera* on unsprayed experimental crops was low.

Accumulated data were used to construct partial life tables for each crop and each season (van den Berg and Cock 1993b). These showed that mortality due to parasitoids and diseases was not significant, but that mortality due to unknown factors including predation was very large. Thus, partial lifetables showed that mortality was highest on maize, where between the egg stage and the second larval stage more than 95% died. On cotton, sunflower and sorghum, this figure was 78–85%. In the latter crops, second- to sixth-instar larvae showed high mortality. Total mortality until the sixth instar was around 99% on maize, sorghum and cotton, and 95% on sunflower. The most important mortality factors were predation and/or unknown mortality, which includes disappearence due to abiotic factors, during the egg and young larval stages, respectively.

Data from regular sampling of the experimental plots were used to examine the degree of temporal overlap among the common groups of predators known to attack *H. armigera* (van den Berg *et al.*, 1993). The data were further analysed by location within the plant of both *H. armigera* and the major groups of predators, to establish the degree of spatial overlap between the predators and their potential prey (van den Berg and Cock, 1995b).

Of the large complex of predators recorded in the same crops in several agroecological zones across Kenya, only anthocorids (*Orius* spp. (Heteroptera, Anthocoridae)) and ants (predominantly *Pheidole* spp., *Myrmicaria* spp. and *Camponotus* spp. (Hymenoptera, Formicidae)) were sufficiently common and widespread to be of importance in suppressing *H. armigera* (van den Berg *et al.*, 1993; van den Berg and Cock, 1993b, 1995a).

The presence of anthocorids was associated with high mortality of *H. armigera* eggs, which suggests that anthocorids can suppress *H. armigera* when they are common concurrently with the egg stage. However, correlative field data on predators and prey demonstrated in the Kenyan study that anthocorids were generally poorly associated with eggs on sunflower and sorghum. This may explain the relative high survival of young stages on sunflower. On maize, the association of anthocorids with eggs and larvae of *H. armigera* was stronger.

Ants showed differences in behaviour between crops. For instance, *Myrmicaria* spp., a common predacious ant in western Kenya, visited sunflower plants more often than maize or sorghum, even though it was equally common in pitfall traps in the three crops. Ant foraging activity in the canopy may be directed by the availability of alternative food sources, such as plant exudates or honeydew-producing Homoptera. *Myrmicaria* spp. was often found feeding on exudates on sunflower (van den Berg *et al.*, 1997). The degree to which ants visit vegetation may have important implications for the role of ants in suppressing *H. armigera*, particularly in sunflower, where ants may have contributed to the relatively high late-larval mortality. Furthermore, ants were most common on the sunflower plants with the highest number of larvae, indicating new recruitment of workers in response to *H. armigera* density (van den Berg and Cock, 1995b).

Subsequent experiments on cotton and sunflower were aimed at assessing the relative importance of these two main groups of predators. Manipulative experiments in which crawling predators, dominated by *Pheidole* spp. ants, were excluded from sunflower plots showed the role of ants to vary considerably. In one location, *H. armigera* levels were almost seven times greater in the absence of ants than in control plots where predators were not excluded (van den Berg *et al.*, 1997). Exclusion of flying predators, dominated by anthocorids, had little impact on *H. armigera*. Exclusion trials conducted on cotton and sunflower showed no irreplaceable mortality by predation in either crop. In sunflower, this was attributable to the lack of predators during the trial; in cotton, an extremely high background mortality had masked the effect of predation (van den Berg and Cock, 1995a; van den Berg *et al.*, 1997).

The combined effect of predators and parasitoids was measured using a cage exclusion experiment with artificially enhanced prey density on cotton (van den Berg and Cock, 1993c). The results clearly indicated the ability of natural enemies to reduce pest numbers; in the absence of natural enemies, *H. armigera* were four to six times as numerous and there was a corresponding increase in damaged plant parts. Background mortality was again high, but egg numbers at inoculation were sufficient to measure significant differences in larval numbers.

The effectiveness of anthocorids as predators of eggs was studied in a separate experiment where cohorts of eggs of *H. armigera* were laid on experimental plants in the field and monitored for 48 h. The fate of the eggs (present, disappeared, sucked) was recorded, showing that the percentage of predation by sucking increased from 12% early in the season to 65% late in the season, contributing 23–83% of the total recorded mortality. Other mortalities included 15% eggs that were lost and 6% that were parasitized.

A field experiment on natural populations of *H. armigera* attempted to distinguish the different categories of predation and their effect upon the *H. armigera* population on cotton (van den Berg and Cock, 1995a) by excluding either walking predators or all natural enemies. Although the incidence of *H. armigera* was low, there were strong indications of an effect due to predators. When anthocorids and ants were both present, there were 71% fewer *H. armigera*. When crawling predators (i.e. ants) were excluded, but not flying predators such as anthocorids, there were again 71% fewer *H. armigera*. The implication being that in this experiment nearly all the mortality was due to anthocorids, and not due to ants. A parallel experiment exposing eggs on cotton plants showed 20% were sucked by anthocorids over 96 h.

Thus, although it was demonstrated that predators are the most important group of natural enemies, no generalizations could be made about the impact of predation on field populations of *H. armigera* in Kenya (van den Berg, 1993). The variable mortality due to predation could be obscured by other, larger mortality factors. In some instances, high background mortality suppressed the pest, in other instances survival was better and the role of predation became more obvious. For example, where anthocorid bugs are common *H. armigera* populations are always low; where anthocorid populations are low *H. armigera* populations are sometimes low and sometimes high. This shows that anthocorids are generally poorly associated with the pest, mainly because they arrive too late in the season.

Because populations of *H. armigera* almost never reached economically damaging levels during this study, it is difficult to say what effect these natural enemies might have had on a high incidence of *H. armigera*. The implication is that they would have contributed substantial mortality, which would have been easily disrupted by the use of broad spectrum insecticides. It is thus important that any IPM strategy for cotton in this region should take this natural mortality due to predators into consideration, and any control strategies should be based on conserving and encouraging these predators.

Enhancement of indigenous natural enemies

H. armigera has a large array of indigenous natural enemies that are not always able to prevent the pest from causing economic damage. Apart from augmentative and inundative releases of indigenous natural enemies to enhance the efficacy, there have been considerable advances in the understanding of tritrophic natural enemy–pest–crop interactions and there is evidence that environmental manipulations can improve the contribution of natural enemies to *H. armigera* control. Ecological theory predicts that pests find plants more easily if concentrated in a monoculture than plants grown in a polyculture (Root, 1973) and increasing crop diversity often reduces pest infestation (Risch, 1983; Andow, 1991). An increased abundance and action of natural enemies in polycultures may be responsible for reduced pest levels (Russell, 1989). Environmental manipulation to enhance natural enemies is intimately associated with cultural control techniques. Greathead and Girling (1989) concluded that improvements to the cultural controls employed in traditional farming to conserve and enhance the impact of natural enemies represents the best prospect for biological control of *H. armigera* in Africa. Data from East Africa suggest that there are prospects to improve the impact of anthocorids, ants and parasitoids (van den Berg *et al.*, 1990, 1997; van den Berg and Cock, 1995b) through intercropping or adjacent planting of susceptible host crops with crops that are attractive to these natural enemies. For example, the BioRe project in Tanzania successfully uses sunflowers as a trap crop in and around organic cotton fields. Cannibalism and predation by ants (*Pheidole* spp.) on these sunflowers induce high mortality among *H. armigera* larvae (Saro Ratter, Consultant, Germany, 2002, personal communication).

With the advent of readily available broad-spectrum pesticides, it became the norm in South Africa for farmers to spray cotton against *H. armigera*, mainly for preventative control purposes with up to 15 treatments per season (van Hamburg and Guest, 1997). As a result, the pest populations became resistant to a number of insecticides, which led to more frequent spraying and ever escalating costs (Whitlock, 1973; Anon., 1992). Since 1975, a spray programme against *H. armigera,* based on scouting for eggs, was developed. This led to a reduction in the average number of insecticide applications, from 15 for preventative control to eight, when sprayed according to egg density counts (van Hamburg and Kfir, 1982). Later it was learned that the egg population was a poor indicator of the damaging larval populations, due to loss from parasitism and predation (van Hamburg, 1981). A new scouting method based on larval counts was developed, whereby an average of only two to three sprays per season were required, without substantial decline in yield (Kfir and van Hamburg, 1983). This new system resulted in a 60% reduction in pest control costs for cotton growers. Since the importance of natural enemies is now better appreciated, to protect and conserve them restrictions were imposed on the use of certain harmful insecticides including a ban on the use of synthetic pyrethroids on cotton less than 12 weeks old (Charleston *et al.*, 2003). This approach has favoured more sustainable biological control in cotton. Despite the recent arrival and rapid expansion of the area under transgenic *Bt* cotton, this system is still relevant today.

Classical, Inundative and Augmentative Biological Control

The apparent variation in geographical and temporal distribution of indigenous natural enemies, as well as in host-plant association, offers the possibility of natural enemy introduction or redistribution. Gaps in parasitoids guilds indicate which groups of species may be of value to particular ecological niches.

Egg parasitoids

In 1930, Parsons and Ullyett (1936) were the first to consider inundative biological control against *H. armigera* in South Africa and undertook investigations on the mass production and releases of *T. lutea.* Up to 1 million parasitoids day^{-1} were produced and released on maize, cotton, and citrus crops. Although releases of up to 10,000 parasitoids acre^{-1} week^{-1} were made with egg parasitism ranging from 21% to 82%, no reduction in the larval pest population was achieved.

Attempts at classical biological control with egg parasitoids were made in South Africa in the 1970s and 1980s, when *Trichogramma chilonis* Ishii (from Columbia and Taiwan), *Trichogramma perkinsi* Girault and *Trichogramma semifumatum* Perkins (from Columbia), *Trichogramma pretiosum* Riley (from the USA), *Trichogramma ostrinia* Pang and Chen (from Taiwan) and *Trichogrammatoidea brasiliensis* (Ashmead) (from Colombia and France) were released in cotton and maize fields, but did not become permanently established (van Hamburg, 1980; Kfir, 1981, 1982, 1994).

Classical biocontrol attempts with egg parasitoids are also reported from elsewhere in Africa. Bournier and Peyrelongue (1973) were able to control *H. armigera* in Madagascar with regular inundative releases of *T. brasiliensis.* Two million adults were released in 2 ha of cotton over a 2-month period. This schedule permitted a delay in the start of insecticide treatments and a reduction by five in the number of treatments from the usual ten to 12. In the Sudan, augmentative releases of *T. pretiosum* on cotton were reported to be promising (Abdelrahman and Munir, 1989). In 1986 and 1987, *T. chilonis* was released in rather small numbers on the island of Santiago, Cape Verde and is reported to be possibly established, but there is no discussion of impact (van Harten *et al.*, 1990).

Larval parasitoids

In western Kenya, parasitoids of young *H. armigera* larvae were almost absent from experimental sites (van den Berg *et al.*, 1993), yet outside of Africa parasitoids of young larvae such as *Campoletis chlorideae* Uchida (in India), *Hyposoter didymator* (Thunberg) (Hymenoptera, Ichneumonidae) (a European species with a wide host range), *Glabromicroplitis croceipes* (Cresson) (a North American species) and *Cotesia kazak* Telenga (Hyenoptera, Braconidae) (a European species), often have a substantial impact on *H. armigera* or related Heliothinae (Messenger, 1974; King *et al.*, 1985; Carl, 1989; Mohyuddin, 1989). Greathead and Girling (1982) proposed that the introduction of exotic larval parasitoids into East Africa might improve the overall level and reliability of biological control. It is, however, worth noting that in India, where *H. armigera* causes substantial crop losses, the many introductions of exotic egg, egg-larval and larval parasitoids have met with limited success only (Romeis and Shanower, 1996).

There have been several attempts at classical and augmentative biological control with larval parasitoids of *H. armigera* in South Africa. Taylor (1932) experimented with the indigenous parasitoid *Habrobracon brevicornis* (Wesmael) (Hymenoptera, Braconidae). Ullyett (1933) and Parsons and Ullyett (1934) also worked with the parasitoid, but discontinued when no practical results were obtained. Between 1944 and 1949 another attempt to control *H. armigera* was made in South Africa, when *Chelonus texanus* Cresson (Hymenoptera, Braconidae) was imported from the USA (originally against the Karoo caterpillar *Loxostege frustralis* Zeller (Lepidoptera, Pyralidae)). Altogether 1.4 million parasitoids were released on vegetables and citrus, but no subsequent recoveries were made (Bedford, 1954). In the 1980s, the braconid *C. kazak* was introduced from New Zealand to cotton fields in South Africa (Anon., 1992) but, like the egg parasitoids, did not became permanently established. *C. kazak* was also introduced to Cape Verde between 1983 and 1987 (van Harten *et al.*, 1990) along with *H. didymator* and *Cotesia ruficrus* (Haliday). However, none of these three species were subsequently recovered from the field.

Classical, inundative and augmentative releases of exotic and indigenous egg parasitoids can lead to high levels of parasitism and serve to reduce or avoid early-season insecticide applications, but while this approach has been widely employed, it has not worked everywhere. Furthermore, against the background of very high early natural mortality, the significance of even high levels of egg parasitism is unclear. Attempts at biological control of *H. armigera* through introduction and augmentation of exotic and indigenous larval parasitoids were not successful in South Africa. The recommendation to introduce exotic larval parasitoids to East Africa (Greathead and Girling, 1982) appears not to have been followed, while results from elsewhere are mixed.

There are no records of introductions of exotic predators of *H. armigera* to Africa. Moreover, as predators are not normally host specific, the increasing difficulty of introducing exotic species that have a wide host range could constrain any such introductions in the future. From the examples given above the potential value of classical or inoculative and augmentative biological control to the integrated management of *H. armigera* appears to be limited. Indeed, Greathead and Girling (1989), in an apparent revision of their earlier proposal, concluded that since *H. armigera* is highly polyphagous, moves between crops and weeds as they flower, and is capable of migrating long distances, it does not seem to be a promising target for classical biological control.

Experiences with Microbial Control

Of the four major groups of entomopathogens, viruses and bacteria have been most used against *H. armigera* in Africa. While fungi and protozoa have been isolated on many occasions, they have received little attention. Although the most extensive literature relates to the use of baculoviruses against *H. armigera*, in practical control terms *Bacillus thuringiensis* Berliner (*Bt*) has undoubtedly been more widely used.

H. armigera is possibly the most widely targeted of African field crop pests for control by viruses; it is susceptible to its homologous nucleopolyhedrovirus (*Ha*NPV) (*Baculoviridae*) (Plate 61a,b) and to those of *Mamestra brassicae* (L.) and *Helicoverpa zea* (Boddie) (Lepidoptera, Noctuidae). A granulovirus (GV) (*Baculoviridae*) of *H. armigera* is known from South Africa, but may not be suitable for field control because of its effect on larval maturation. Use of baculoviruses against *H. armigera* in Africa was reviewed by Kunjeku *et al.* (1998). A cypovirus (CPV) (*Reoviridae*) is also known from *H. armigera* in Africa, but has not been seriously considered as a pest management option.

Although the NPV of *H. armigera* is widespread, natural control by the virus is usually inadequate. For example, of the diseases infecting *H. armigera* in Tanzania on maize, sorghum, cotton, chickpea, tomato and cleome, NPV was the most frequent, particularly on maize and cleome, but it was not able to prevent the pest from causing economic damage on the crops (Nyambo, 1990). von Maltitz (E.F. von Maltitz, 1992, Grahamstown, unpublished results) recorded up to 40% larval mortality from a polyhedral virus in sunflower in South Africa. Epizootics of baculoviruses in insect pests of short-term crops are rather rare; pest populations are not normally allowed to reach critical thresholds, rates of transmission remain low, and the reservoir of virus does not accumulate.

The experimental use of baculoviruses against *H. armigera* has been most extensive on cotton and dates back to the late 1950s, when Coaker (1958) tested the efficacy of a local isolate of *Ha*NPV against *H. armigera* on cotton in Uganda. While the virus was capable of killing larvae on treated foliage in cages, it was not recommended for use in Uganda where *H. armigera* populations were low and the virus endemic. The greatest concentration of experimental use of baculoviruses against *H. armigera* on cotton has been in West and Central Africa. Over the period spanning the 1960s to the 1980s, numerous trials conducted in Cameroon, Togo, Côte d'Ivoire and Chad focused on the use of indigenous isolates of *H. armigera* NPV and two imported species *M. brassicae* NPV and *H. zea* NPV (as the commercial formulations Biotrol VHF from Nutrilite Products Inc., Viron-H from International Minerals Corp. and Elcar from Sandoz). All three viruses were tested either alone or in combination with reduced doses of chemical insecticides or *Bt*. Results have been variable, but indigenous isolates of *H. armigera* NPV generally performed better than imported *H. zea* NPV (Atger, 1969; Angelini and Couilloud, 1972). Use of *M. brassicae* NPV against *H. armigera* has the advantage of also controlling *Diparopsis watersi* (Rothschild) (Lepidoptera, Noctuidae). Yields with *H. armigera* NPV and *M. brassicae* NPV have matched those achieved with standard chemical insecticides, but high dose rates of up to 1×10^{13} occlusion bodies (OB) ha^{-1} are usually required. Additionally, in combination with chemical insecticides at reduced dose and a phagostimulant, all three baculoviruses have given control that is often as good as control achieved with chemical insecticides alone (Angelini and Couilloud, 1972; Cadou and Soubrier, 1974; Montaldo, 1991; Silvie *et al.*, 1993; Vaissayre *et al.*, 1995).

On other crops, baculoviruses have had mixed success. In Botswana, Roome (1971) reported promising results with *Ha*NPV on sorghum, cowpeas and maize. In later trials in sorghum, neither *Hz*NPV nor *Bt* were as effective as a local *Ha*NPV, which itself could be as effective as the chemical standard if used at high doses. According to Kunjeku *et al.* (1998), *H. armigera* control with NPV on sorghum has been exceptionally successful and should be pressed forward. In Cape Verde, Elcar (*H. zea* NPV) was used successfully in tomatoes as part of an IPM programme. Withdrawal of Elcar from the market led to substitution with *Bt* that was reported to be less efficient (van Harten and Viereck, 1986). Lutwama and Matanmi (1988) used Elcar at 1 and 2×10^{12} OB ha^{-1} and *Bt* against *H. armigera* on tomatoes in southern Nigeria. At these rates, Elcar was not significantly better than *Bt* or Carbaryl.

Despite extensive testing in Africa, there is no evidence that baculoviruses have ever been used on anything other than an experimental scale and nothing to indicate that they are being widely used today. At least two foreign companies have, at one time or another, invested in either the development and/or testing of baculoviruses against *H. armigera* in West Africa. Regrettably, progress was inadequate, investment was halted and today neither company is present on the ground. Two particular problems with NPV for control of *H. armigera* in Africa are highlighted by Kunjeku *et al.* (1998). The first is the need for better formulations and the use of frequent application of fairly low doses, seen in North America as one of the main keys to successful *H. zea* control. The taxonomic diversity of cotton bollworm species in Africa is the second problem faced by the use of highly specific baculoviruses.

B. thuringiensis is widely used against lepidopteran pests in Africa and several subspecies/serovars are effective against *H. armigera* (Glare and O'Callaghan, 2000). Although *Bt* has been tested against *H. armigera* for many years and was included alongside NPV in many of the trials mentioned above in cotton, literature on the use of *Bt* against *H. armigera* in Africa is scant. In the coastal countries of West Africa, *H. armigera* is an important pest of vegetables and *Bt* has become one of the most important insecticides used against lepidopteran larvae in peri-urban vegetable crops. This has been more in response to farmer concern over chemical insecticide efficacy than any concern for the environment (A. Cherry, Cotonou, 2002, unpublished observations).

Conclusions

There are rich and varied indigenous natural enemy complexes attacking *H. armigera* in both East and southern Africa. These natural enemies vary geographically, temporally and in their host-plant associations. Generalist predators tend to be more important than parasitoids and this makes the situation more difficult to grasp, since their effectiveness depends on the presence of other concurrent food sources. The combined impact of indigenous natural enemies is not negligible, they can cause major mortality and contribute to pest suppression. Yet, they have no clear cut role and the consensus of opinion is that alone they do not always prevent economic damage. This seems more true in high-value crops with their low economic-damage thresholds. The absence of consistent effect dictates the need for locally adapted pest control strategies that take account of regional variation.

Nevertheless, the potential to enhance the impact of natural enemies exists through a variety of mechanisms. Classical biological control exchanges between regions, the use of new associations, and augmentative and inoculative releases of indigenous and exotic natural enemies have been tested. These options have, however, generally failed to achieve substantial or permanent gains in control, and it seems that particularly with high-value crops, these approaches alone will not solve the *H. armigera* problem.

Manipulation and diversification of the crop environment through cultural techniques to conserve and enhance indigenous natural enemy impact may hold a key role, and was seen by Greathead and Girling (1989) as among the best prospects for biological control in Africa. In this regard, Africa already has an advantage in that much of Africa's agriculture is practised in smallholdings where a diversity of crops are grown in relatively small plots and where pesticide inputs are relatively low. This mosaic, which on the one hand provides a continuous supply of food for *H. armigera*, on the other hand conserves or even promotes natural enemies.

Farmers' pesticide spraying practices can negatively affect natural enemy populations, particularly when applied early in the season. Natural enemies that might otherwise have built up and suppressed the pest are killed and more insecticides are required. Reductions in the use of hazardous insecticides and early season substitution of broad spectrum insecticides with softer biological alternatives such as *Bt*, NPV and botanical insecticides may permit early establishment of natural enemies and contribute to pest suppression. However, as Landis *et al.* (2000) notes, while eliminating a pesticide treatment within a field may permit the establishment or persistence of a natural enemy population, if viable meta-populations do not exist at the landscape level to provide immigrants, the within-field effort may be ineffective.

Use of microbial control agents against *H. armigera* was rejected by Greathead and Girling (1989) because of high costs, short shelf-life and the difficulty of finding suitable formulations. Increased availability and improvements to the formulations of commercially available products, together with increasing resistance to chemical insecticides will make microbials a more favourable option. Nevertheless, the specificity of some microbials, and the frequent occurrence of *H. armigera* as part of a pest complex, means that there will continue to be a need for chemical pesticides.

Effective mating disruption by pheromones has been demonstrated in small plots outside Africa (Kehat *et al.*, 1998; Chamberlain *et al.*, 2000), although it is unlikely that this approach will provide cost-effective control because of the mobility and polyphagy of the pest. In Africa, use of the pheromone has been very limited. Pheromone traps were used for monitoring *H. armigera* as well as the other bollworms on cotton throughout Egypt during the 1980s and 1990s (Critchley, 1991). Lures have been provided to Kenya, Tanzania, Malawi, Sudan and Zimbabwe for evaluation. Synthetic sex pheromone traps for *H. armigera* were also tested in South Africa with the idea of establishing economic threshold levels for chemical control of the pest. The study was terminated because no correlation was found between larval infestations in the field and moth catches. Bourdouxhe (1982) reported use of pheromone traps for monitoring the flight periods of *H. armigera* in Senegal and Nyambo (1989) reported similar work in Tanzania.

Progress in biological control and IPM of *H. armigera* requires that situation-adapted strategies employing combinations of the options discussed above be adopted by farmers on a regional scale. Few examples exist and even though the results of the research conducted in Kenya (van den Berg, 1993) had important implications for the way farmers grow and manage their crops, the results remain unutilized at the field level. Recent developments in participatory IPM and action research in the region suggest ways in which research findings can be adopted and expanded by farmer communities in a process which is continuous and which includes direct stakeholders as well as researchers (Bruin and Meerman, 2001).

One way forward was exemplified by the cotton IPM programme in South Africa, driven not by a demand for biological control, but by the need to reduce pest management costs that had escalated as a result of the increasing frequency of pesticide applications. An example of how this approach might be extended can be seen in Australia. There, high cotton yields are maintained by intensive synthetic insecticide dependence, but in one study indigenous predators were used as the basic component of an IPM programme supported by interplanting with lucerne, the use of supplementary food sprays, biopesticides based on *Bt* and NPV, and judicious use of synthetic insecticides. The project demonstrated yields and economic returns that are equivalent to, or better than, the conventional system (Mensah, 2002).

The advent of *Bt*-transgenic plant varieties to Africa offers the promise of reduced insecticide inputs and the knock-on effect of conserved natural enemies. *Bt*-transgenic cotton and several *Bt*-transgenic maize hybrids are now commercially available in South Africa. During the 1998/1999 growing season, approximately 50,000 ha of irrigated land was planted with *Bt*-transgenic maize hybrids. In 2001, about 70% of cotton planted in South Africa was *Bt* cotton, but in some regions the figure was above 80%. South African legislation requires cotton farmers to plant at least 20% of each field with non-transgenic varieties as refugia to slow the potential for accelerated development of resistance to *Bt*. Only the smooth-leaved *Bt* cotton variety is available and, although very efficient against *H. armigera* and other bollworms, it is susceptible to leafhoppers (Cicadellidae). As a consequence, transgenic cotton usually still has to be sprayed against jassids. However, two recent reports show that both *Bt* cotton and *Bt* maize do enhance biodiversity through reduced insecticide application (Carpenter *et al.*, 2002; Gianessi *et al.*, 2002).

Finally, prospects for improved implementation of biological control and IPM in the future could be driven by external factors such as changes to market structure, consumer requirements and pest management economics. For instance, increasingly strict pesticide residue tolerance limits imposed by foreign markets on export products will fuel the demand for pest management approaches with reduced synthetic pesticide inputs. Such developments are already visible among those countries with a significant horticultural export sector. In the domestic market, increasing consumer awareness of pesticide residue hazards will drive a demand for improved control of residue limits, and an extension of this is the development and expansion of the organic products market whose consumers are willing to pay a green premium for pesticide-free products. The propensity of *H. armigera* to develop resistance will continue to escalate the costs of conventional insecticide-dependent pest management. At the same time, it will fuel the search for more economic and environmentally sound alternatives, but also alter the cost–benefit ratio of currently uneconomic or unattractive options.

References

Abdelrahman, A.A. and Munir, B. (1989) Sudanese experience in integrated pest management of cotton. *Insect Science and its Application* 10, 787–794.

Andow, D.A. (1991) Vegetational diversity and arthropod population response. *Annual Review of Entomology* 36, 561–586.

Angelini, A. and Couilloud, R. (1972) Les moyens de lutte biologique contre certains ravageurs du cotonnier et une perspective sur la lutte intégrée en Côte d'Ivoire. *Coton et Fibres Tropicales* 27, 283–289.

Annecke, D.P. and Moran, V.C. (1982) *Insects and Mites of Cultivated Plants in South Africa*. Butterworths, Pretoria, South Africa.

Anonymous (1992) Integrated pest management of cotton pests: a compromise between biological and chemical control. *Plant Protection News*, Bulletin of the Plant Protection Research Institute, Pretoria, South Africa, 29, 6–7.

Atger, P. (1969) Observations sur la polyhedrose nucléaire d'*Heliothis armigera* (Hbn.) au Tchad. *Coton et Fibres Tropicales* 24, 243–244.

Bedford, E.C.G. (1954) Summary of the biological control of insect pests in South Africa 1895–1953. Department of Agriculture and Technical Services, Pretoria, South Africa. Special Report 2/53–54.

Bedford, E.C.G. (1968) Biological and chemical control of citrus pests in the Western Transvaal. An integrated spray programme. *South African Citrus Journal* 417, 21–28.

Bottrell, D.G., Barbosa, P. and Gould, F. (1998) Manipulating natural enemies by plant variety selection and modification: a realistic strategy? *Annual Review of Entomology* 43, 347–367.

Bournier, J.P. and Peyrelongue, J.Y. (1973) Introduction, rearing and releases of *Trichogramma brasiliensis* Ashm. (Hym. Chalcididae) with a view to controlling *Heliothis armigera* Hbn. (Lep. Noctuidae) in Madagascar. *Coton et Fibres Tropicales* 28, 231–237.

Bourdouxhe, L. (1982) Comparaison de deux types de pièges pour le piégeage sexuel de *Heliothis armigera* au Sénégal. *Plant Protection Bulletin FAO* 30, 131–136.

Bruin, G.C.A. and Meerman, F. (2001) *New Ways of Developing Agricultural Technologies: the Zanzibar Experience with Participatory Integrated Pest Management.* Co-publication Wageningen University and Research Centre and CTA (CTA number 1047), Wageningen, The Netherlands.

Cadou, J. and Soubrier, G. (1974) Utilisation d'une polyédrose nucléaire dans la lutte contre *Heliothis armigera* (Hb.) (Lep. Noct.) en culture cotonnière au Tchad. *Coton et Fibres Tropicales* 29, 357–365.

Carl, K.P. (1989) Attributes of effective natural enemies, including identification of natural enemies for introduction purposes. In: King, E.G. and Jackson, R.D. (eds) *Proceedings of the Workshop on Biological Control of* Heliothis*: Increasing the Effectiveness of Natural Enemies. November 11–15, 1985, New Delhi, India.* Far Eastern Regional Research Office, U.S. Department of Agriculture, New Delhi, India, pp. 351–361.

Carpenter, J., Felsot, A., Goode, T., Hamig, M., Onstad, D. and Sankula, S. (2002) *Comparative Environmental Impacts of Biotechnology-derived and Traditional Soybean, Corn and Cotton Crops.* Council for Agricultural Science and Technology, Ames, Iowa, USA. www.cast-science.org

Chamberlain, D.J., Brown, N.J., Jones, O.T. and Casagrande, E. (2000) Field evaluation of a slow release pheromone formulation to control the American bollworm, *Helicoverpa armigera* (Lepidoptera: Noctuidae) in Pakistan. *Bulletin of Entomological Research* 90, 183–190.

Charleston, D.S., Kfir, R., van Rensburg, N.J., Barnes, B.N., Hattingh, V., Conlong, D.E., Visser, D. and Prinsloo, G.J. (2003) Integrated pest management in South Africa. In: Maredia, K., Dakouo, D. and Mota-Sanchez, D. (eds) *Integrated Pest Management in the Global Arena.* CAB International, Wallingford, UK.

Coaker, T.H. (1958) Experiments with a virus disease of the cotton bollworm *Heliothis armigera* (Hbn). *Annals of Applied Biology* 46, 536–541.

Critchley, B.R. (1991) Commercial use of pink bollworm pheromone formulations in Egyptian cotton pest management. *Pesticide Outlook* 2, 9–13.

Crop Protection Compendium (2001) CD ROM. CAB International, Wallingford, UK.

Gianessi, L.P., Silvers, C.S., Sankula, S. and Carpenter, J.E. (2002) *Plant Biotechnology: Current and Potential Impact for Improving Pest Management in US Agriculture.* National Centre for Food and Agricultural Policy. www.ncfap.org

Glare, T.R. and O'Callaghan, M. (2000) *Bacillus thuringiensis: Biology, Ecology and Safety.* John Wiley & Sons, Chichester, UK.

Greathead, D.J. and Girling, D.J. (1982) Possibilities for natural enemies in *Heliothis* management and the contribution of the Commonwealth Institute of Biological Control. In: *Proceedings of the International Workshop on* Heliothis *Management,* November 15–20, 1981, Patancheru, A.P., India. International Crops Research Institute for the Semi-Arid Tropics, Patancheru, India, pp. 147–158.

Greathead, D.J. and Girling, D.J. (1989) Distribution and economic importance of *Heliothis* and of their natural enemies and host plants in southern and eastern Africa. In: King, E.G. and Jackson, R.D. (eds) *Proceedings of the Workshop on Biological Control of* Heliothis*: Increasing the Effectiveness of Natural Enemies.* November 11–15, 1985, New Delhi, India. Far Eastern Regional Research Office, United States Department of Agriculture, New Delhi, India, pp. 329–345.

Kehat, M., Anshelevich, L., Gordon, D., Harel, M. and Dunkelblum, E. (1998) Evaluation of Shin-Etsu twist-tie rope dispensers by the mating table technique for disrupting mating of the cotton bollworm, *Helicoverpa armigera* (Lepidoptera: Noctuidae), and the pink bollworm, *Pectinophora gossypiella* (Lepidoptera: Gelechiidae). *Bulletin of Entomological Research* 88, 141–148.

Kfir, R. (1981) Effect of host and parasite density on the egg parasite *Trichogramma pretiosum* (Hym.: Trichogrammatidae). *Entomophaga* 26, 445–451.

Kfir, R. (1982) Reproduction characteristics of *Trichogramma brasiliensis* and *T. lutea*, parasitising eggs of *Heliothis armigera. Entomologia Experimentalis et Applicata* 32, 249–255.

Kfir, R. (1994) Attempts at biological control of the stem borer *Chilo partellus* (Swinhoe) (Lepidoptera: Pyralidae) in South Africa. *African Entomology* 2, 67–68.

Kfir, R. and van Hamburg, H. (1983) Further tests of threshold levels for control of cotton bollworms (mainly *Heliothis armigera*). *Journal of the Entomological Society of Southern Africa* 46, 49–58.

King, E.G., Powell, J.E. and Coleman, R. (1985) A high incidence of parasitism of *Heliothis* spp. (Lepidoptera: Noctuidae) larvae in cotton in southeastern Arkansas. *Entomophaga* 30, 419–426.

Kuklinski, F. and Borgemeister, C. (2002) Cotton pests and their natural enemies in Madagascar. *Journal of Applied Entomology* 126, 55–65.

Kunjeku, E., Jones, K.A. and Moawad, G.M. (1998) Africa, the Near and Middle East. In: Hunter Fujita, F.R., Entwistle, P.F., Evans, H.F. and Crook, N.E. (eds) *Insect Viruses and Pest Management*. John Wiley & Sons, Chichester, UK, pp. 280–302.

Landis, D.A., Wratten, S.D. and Gurr, G.M. (2000) Habitat management to conserve natural enemies of arthropod pests in agriculture. *Annual Review of Entomology* 45, 175–201.

Lutwama, J.J. and Matanmi, B.A. (1988) Efficacy of *Bacillus thuringiensis* subsp. *kurstaki* and *Baculovirus heliothis* foliar applications for suppression of *Heliothis armigera* (Hübner). *Bulletin of Entomological Research* 78, 173–179.

Mensah, R.K. (2002) Development of an integrated pest management programme for cotton. Part 2: Integration of a lucerne/cotton interplant system, food supplement sprays with biological and synthetic insecticides. *International Journal of Pest Management* 48, 95–105.

Messenger, P.S. (1974) Procedures for conducting a biological control programme against *Heliothis armigera. Plant Protection Service Technical Bulletin,* Department of Agriculture, Thailand 19.

Mohyuddin, A.I. (1989) Distribution and economic importance of *Heliothis* spp. in Pakistan and their natural enemies and host plants. In: King, E.G. and Jackson, R.D. (eds) *Proceedings of the Workshop on Biological Control of* Heliothis*: Increasing the Effectiveness of Natural Enemies.* November 11–15, 1985, New Delhi, India. Far Eastern Regional Research Office, United States Department of Agriculture, New Delhi, India, pp. 229–240.

Montaldo, T. (1991) Microbial control in cotton crops in North Cameroon: overview of experiments conducted from 1979 to 1988. *Coton et Fibres Tropicales* 46, 230–240.

Nyambo, B.T. (1986) Studies in the bollworm, *Heliothis armigera* Hübner, the key cotton pest in Tanzania, as a basis for improved integrated pest management. PhD thesis, University of London, UK.

Nyambo, B.T. (1989) Assessment of pheromone traps for monitoring and early warning of *Heliothis armigera* Hübner (Lepidoptera, Noctuidae) in the western cotton-growing areas of Tanzania. *Crop Protection* 8, 188–192.

Nyambo, B.T. (1990) Effect of natural enemies on the cotton bollworm, *Heliothis armigera* Hübner (Lepidoptera: Noctuidae) in Western Tanzania. *Tropical Pest Management* 36, 50–58.

Parry-Jones, E. (1937) The egg parasites of the cotton bollworm, *Heliothis armigera* Hübner (*obsuleta* Fabr.), in southern Rhodesia. *Publication, Mazoe Citrus Experimental Station* 6, 37–105.

Parry-Jones, E. (1938) The biology of a tachinid parasite (*Sturmia rhodesiensis*, sp. n.) of the cotton bollworm (*Heliothis armigera* Hübner) in southern Rhodesia. *Publication, Mazoe Citrus Experimental Station* 7, 11–34.

Parsons, F.S. (1940) Investigations on the cotton bollworm *Heliothis armigera* Hübn. (*obsoleta* Fabr.). Part II. The incidence of parasites in quantitative relation to bollworm populations in South Africa. *Bulletin of Entomological Research* 31, 89–109.

Parsons, F.S. and Ullyett, G.C. (1934) Investigations on the control of the American and red bollworms of cotton in South Africa. *Bulletin of Entomological Research* 25, 349–381.

Parsons, F.S. and Ullyett, G.C. (1936) Investigations of *Trichogramma lutea*, Gir., as a parasite of the cotton bollworm, *Heliothis obsoleta* Fabr. *Bulletin of Entomological Research* 27, 219–235.

Risch, S.J. (1983) Intercropping as cultural pest control: prospects and limitations. *Environmental Management* 7, 9–14.

Romeis, J. and Shanower, T.G. (1996) Arthropod natural enemies of *Helicoverpa armigera* (Hübner) (Lepidoptera: Noctuidae) in India. *Biocontrol Science and Technology* 6, 481–508.

Roome, R.E. (1971) A note on the use of biological insecticides against *Heliothis armigera* (Hbn.) in Botswana. *Proceedings of the Cotton Insect Control Conference*, Blantyre, Malawi, pp. 221–238.

Root, R.B. (1973) Organization of a plant-arthropod association in simple and diverse habitats: the fauna of collards (*Brassica oleracea*). *Ecological Monographs* 43, 95–124.

Russell, E.P. (1989) Enemies hypothesis: a review of the effect of vegetational diversity on predatory insects and parasitoids. *Environmental Entomology* 18, 590–599.

Silvie, P., Le Gall, P. and Sognigbé, B. (1993) Evaluation of a virus-insecticide combination for cotton pest control in Togo. *Crop Protection* 12, 591–596.

Taylor, J.S. (1932) Report on cotton insect and disease investigations part II. Notes on the American bollworm (*Heliothis obsoleta* Fabr.) on cotton, and its parasite, *Microbracon brevicornis* Wesm. *Science Bulletin,* Department of Agriculture, Union of South Africa 113, 1–18.

Ullyett, G.C. (1933) The mass-rearing of *Bracon brevicornis* Wesm. *South African Journal of Science* 30, 426–432.

Vaissayre, M., Cauquil, J. and Sylvie, P. (1995) Protection phytosanitaire du cotonnier en Afrique tropicale. *Agriculture et Développement* 8, 3–23.

van den Berg, H. (1993) Natural control of *Helicoverpa armigera* in smallholder crops in East Africa. PhD thesis, Wageningen University, The Netherlands.

van den Berg, H. and Cock, M.J.W. (1993a) *African Bollworm and its Natural Enemies in Kenya*. CAB International, Wallingford, UK.

van den Berg, H. and Cock, M.J.W. (1993b) Stage-specific mortality of *Helicoverpa armigera* in three smallholder crops in Kenya. *Journal of Applied Ecology* 30, 640–653.

van den Berg, H. and Cock, M.J.W. (1993c) Exclusion cage studies on the impact of predation on *Helicoverpa armigera* in cotton. *Biocontrol Science and Technology* 3, 491–497.

van den Berg, H. and Cock, M.J.W. (1995a) Natural control of *Helicoverpa armigera* in cotton: assessment of the role of predation. *Biocontrol Science and Technology* 5, 453–463.

van den Berg, H. and Cock, M.J.W. (1995b) Spatial association between *Helicoverpa armigera* and its predators in smallholder crops in Kenya. *Journal of Applied Ecology* 32, 242–252.

van den Berg, H., Waage, J.K. and Cock, M.J.W. (1988) *Natural Enemies of* Helicoverpa armigera *in Africa: a Review.* CAB International and Institute of Biological Control, Ascot, UK.

van den Berg, H., Nyambo, B.T. and Waage, J.K. (1990) Parasitism of *Helicoverpa armigera* (Lepidoptera: Noctuidae) in Tanzania: analysis of parasitoid-crop associations. *Environmental Entomology* 19, 1141–1145.

van den Berg, H., Cock, M.J.W., Oduor, G.I. and Onsongo, E.K. (1993) Incidence of *Helicoverpa armigera* (Lepidoptera: Noctuidae) and its natural enemies on smallholder crops in Kenya. *Bulletin of Entomological Research* 83, 321–328.

van den Berg, H., Cock, M.J.W. and Oduor, G.I. (1997) Natural control of *Helicoverpa armigera* in sunflower: assessment of the role of predation. *Biocontrol Science and Technology* 7, 613–629.

van Hamburg, H. (1980) Biological control efforts of *Heliothis armigera* in South Africa. In: *Workshop on Biological Control of* Heliothis *spp.* Queensland Department of Primary Industries, 25 September 1980. Toowomba, Australia.

van Hamburg, H. (1981) The inadequacy of egg counts as indicator of threshold levels for control of *Heliothis armigera* on cotton. *Journal of the Entomological Society of Southern Africa* 44, 289–295.

van Hamburg, H. and Kfir, R. (1982) Tests of threshold levels based on larval counts for chemical control of cotton bollworms. *Journal of the Entomological Society of Southern Africa* 45, 109–121.

van Hamburg, H. and Guest, P.J. (1997) The impact of insecticides on beneficial arthropods in cotton agroecosystems in South Africa. *Archives of Environmental Contamination and Toxicology* 32, 63–68.

van Harten, A. and Viereck, A. (1986) Umweltfreudlichere Schädlingsbekämpfung durch integrierten Pflanzenschutz in Kapverden. *Entwicklung and Ländlicher Raum* 4, 22–24

van Harten, A., Neves, A.M. and Brito, J.M. (1990) Importaçáo, criaçáo, libertaçáo e recuperaçáo de parasitas e predatores de pragas em Cabo Verde no período Abril 1983 – Dezembro 1987. *Investigaçáo agrária (INIA) Cabo Verde* 3, 39–49.

Watmough, R.H. (1991) Predation, an important cause of mortality in the pupae of the American bollworm. *Plant Protection News,* Bulletin of the Plant Protection Research Institute, Pretoria, South Africa, 26, 4.

Watmough, R.H. and Kfir, R. (1995) Predation on pupae of *Helicoverpa armigera* Hbn. (Lep., Noctuidae) and its relation to stem borer numbers in summer grain crops. *Journal of Applied Entomology* 119, 679–688.

Whitlock, V.H. (1973) Studies on insecticidal resistance in the bollworm *Heliothis armigera* Hübner (Lepidoptera: Noctuidae). *Phytophylactica* 5, 71–74.

22

Biological Control in IPM for Coffee

George I. Oduor and Sarah A. Simons
CAB International – Africa Regional Centre, Nairobi, Kenya

Introduction

Coffee (*Coffea* spp., *Rubiaceae*) is the world's most important traded commodity in terms of monetary value, after oil, and is the primary export of many developing countries. The genus *Coffea* consists of about 90 species, all of which are endemic to tropical Africa and the Mascerenes (Mabberly, 1997). There are three species of economic importance: *Coffea arabica* L. (Arabica), *Coffea canephora* Pierre ex Froehner (Robusta) and *Coffea liberica* Bull ex Hiern (Liberica). *C. arabica* and *C. canephora* originated in the forests of East and Central Africa, whereas *C. liberica* originated in Liberia.

In sub-Saharan Africa, coffee is a mainstay of the economies of more than 20 countries, and central to the livelihoods of more than 20 million rural families. While it is still grown in large estates, during the past 30 years it has also become a major source of income for millions of smallholder coffee growers, who are now responsible for an estimated 80% of coffee production in Africa. Coffee is the most important cash crop for Africa as a whole, contributing some US$2.5 billion annum^{-1} (10% of the total) to the foreign exchange earnings in the continent (FAO, 2000). A number of coffee-producing countries in sub-Saharan Africa, including Uganda, Ethiopia, Rwanda and Burundi, depend on the export of this commodity for more than half of their foreign exchange earnings.

Global coffee production in 2001 was 115.4 million bags of which 60% was Arabica and 39% Robusta (ACPC, 2001), but whereas production has nearly doubled since 1970, Africa's share of total production declined from 30% to less than 15% (FAO, 2000) despite accounting for one-third of the world's coffee hectareage. Average yields in Africa are generally low and declining, ranging from 0.3 to 0.38 t ha^{-1}, half that achieved in Latin America and almost a third of Asia's. This is partly due to history: some 40% of African plantations date from the pre-independence era and have not been renewed since. The continued reliance on outdated and often unproductive varieties in

(eds P. Neuenschwander, C. Borgemeister and J. Langewald)

the face of widespread prevalence of pests and diseases such as coffee berry borer antesia bugs, coffee berry disease (CBD), coffee rust and more recently *Fusarium* wilt disease, has seriously undermined coffee production in Africa.

Globally, coffee pests are estimated to cause losses of about 13% (Bardner, 1978) but in Africa, yield losses can be much higher, e.g. losses of up to 96% have been reportedly caused by coffee berry borer in Tanzania (Waterhouse and Norris, 1989). In addition to their impact on yield, direct pests such as antestia bugs and coffee berry borer also lower the bean and liquor qualities of coffee, particularly on Arabica coffee. Until recently, recommended pest control measures relied heavily on the use of synthetic pesticides. However, as global coffee prices have fallen, the costs of inputs, particularly pesticides, have become prohibitive for most farmers in Africa, and this, coupled with increasing concern about pesticide residue risks to the environment and the increasing reports of insect resistance has resulted in the need to develop alternative pest management strategies. IPM strategies include a combination of resistant/tolerant cultivars, cultural practices, biological control and, only where absolutely necessarily, synthetic pesticides. In Africa, the diverse agroecological zones and farming systems in which coffee is grown support a wide range of natural enemies of coffee pests. Thus, biological control could potentially provide the basis for an IPM approach to the management of pests in coffee production systems in Africa.

Pests of Coffee

Arthropod pests

Coffee is indigenous to Africa and so are most of its major pests. However, fewer than 20 of the arthropod pests constitute major constraints to coffee production and productivity. The pest status of herbivorous arthropods varies in different parts of Africa. In terms of economic impact, there are five arthropod pests that are particularly important on coffee in Africa.

Antestia bugs (shield bug), *Antestiopsis* spp. (Heteroptera, Pentatomidae) (Plate 62), consist of a species complex, which includes *Antestiopsis orbitalis orbitalis* (Westw.), *Antestiopsis orbitalis bechuana* (Kirk.), *Antestiopsis orbitalis intricata* (Ghesquire and Carayon) and *Antestiopsis facetoides* Greath. Of these, the two most important are *A. o. intricata* in west and central Africa, and *A. o. orbitalis* in east and southern Africa. Antestia bugs are the most destructive pests of Arabica coffee, and are found throughout Africa. They are responsible for the continuing use of large quantities of potentially hazardous insecticides (e.g. Parathion and Aldicarb) (Le Pelley, 1968). The nymphs and adults pierce and suck green coffee berries, cause premature loss of flowers and berries, and sometimes affect the growth of the plant (Le Pelley, 1968). *Antestia* spp. have low economic thresholds, i.e. two antestia bugs per tree (Hill, 1983). One of the reasons for this is that the pests serve as vectors of the fungi, *Nematospora* spp., which causes rotting of the coffee endosperm (Le Pelley, 1968). Infestation often leads to berry drop, with surviving berries producing cracked beans on ripening (zebra or ragged beans).

Coffee leafminers, *Leucoptera* spp. (Lepidoptera, Gracillariidae) are also major pests of coffee in Africa, with *Leucoptera meyricki* Ghesq. as the most common species in unshaded coffee, whereas *Leucoptera caffeina* Wash. predominates in shaded coffee. Their larvae mine coffee leaves leading to the formation of brown blotches on the upper surfaces of leaves. Severe attacks lead to shedding of leaves. Outbreaks of leafminers have been associated with the use of pesticides (Abasa, 1975).

White coffee stem borer, *Monochamus leuconotus* (Pascoe) (Coleoptera, Cerambycidae) (Plate 63), is endemic to Africa. It mainly attacks Arabica coffee grown at altitudes of below 1700 m and is a particularly serious pest in southern parts of Africa. The larvae feed on the bark and finally bore into the coffee stem, weakening the plant and causing yellowing of the foliage. Infested trees that are less than 2 years old are inevitably killed, and a high percentage of older trees also succumb. Routine crop losses of >5% have been attributed to stem borers in Africa, although cumulative yield losses of up to 25% in South Africa (P.S. Schoeman, 1994, South Africa, unpublished results) and on smallholder farms in northern Malawi incidences of up to 80% have been recorded (Oduor and Simons, 1999).

Coffee berry borer, *Hypothenemus hampei* (Ferrari) (Coleoptera, Scolytidae), is endemic to East and Central Africa where it is also found attacking wild coffee in the natural forests and has spread to become a major pest of coffee throughout the world (Le Pelley, 1968). It is a serious pest of Robusta and low-altitude Arabica coffee. The adults and larvae bore into the bean and feed on the endosperm, giving the bean a characteristic blue-green stain. Heavy attacks usually lead to berry drop. *H. hampei* may be a vector of the fungus *Aspergillus ochraceus*, which produces ochratoxin A.

Four main species of the coffee green scale, *Coccus* spp. (Homoptera, Coccidae), attack coffee in Africa. *Coccus viridis* (Green) is a pest of coffee grown at low altitudes, i.e. below 1300 m, and *Coccus alpinus* De Lotto at altitudes above 1300 m. Other species include *Coccus celatus* De Lotto and *C. viridulus* De Lotto. Damage to coffee plants is either directly through sap sucking or indirectly by sooty mould growing on honeydew deposited on the upper surfaces of leaves of infested plants.

Other important pests include the giant looper, *Ascotis selenaria reciprocaria* (Wlk) (Lepidoptera, Gelechiidae), yellow headed borer, *Dirphya nigricornis* (Ol.) (Coleoptera, Cerambycidae), and coffee berry moth, *Prophantis smaragdina* (Battl.) (Lepidoptera, Pyralidae).

Weeds

Many different weed species have been reported as pests of coffee in Africa (Njoroge, 1994). These include spanish needle, *Bidens pilosa* L. (*Asteraceae*); chromolaena, *Chromolaena odorata* (L.) King & Rob (*Asteraceae*); wandering Jew, *Commelina benghalensis* L., (*Commelinaceae*); nutgrass, *Cyperus rotundus* L., (*Cyperaceae*); lantana, *Lantana camara* L., (*Verbenaceae*); sorrel,

Oxalis latifolia H.B.K. (*Oxalidaceae*); parthenium weed, *Parthenium hysterophorus* L. (*Asteraceae*); and couch grass, *Digitaria scalarum* (Schweinf.) Chiov. (*Poaceae*). In terms of economic importance, *C. odorata* is problematic in coffee plantations throughout Africa. It is particularly severe in Côte d'Ivoire, where it is considered to be a pest on coffee farms in all but a small part of the north east of the country (Zebeyou, 1991).

Pest Control Methods

Apart from the costs, routine spraying of synthetic pesticides is becoming less acceptable because of increasing concerns about their detrimental effects on human/animal health and the environment, in particular their negative impact on natural enemies. The increasing use of synthetic insecticides and copper-based fungicides in Africa has resulted in the emergence of leaf miner (*Leucoptera* spp.) as a major pest of coffee (Crowe, 1964; Abasa, 1975), the development of resistance to pesticides by coffee berry borer *H. hampei* (Brun *et al.*, 1989), and tolerance of the invasive weeds *B. pilosa* and *P. hysterophrus* towards some common herbicides (Njoroge, 1991).

Over the years, a variety of pest management strategies have been developed and used in the management of individual coffee pests (e.g. Le Pelley, 1968; Abasa, 1975; Walyaro *et al.*, 1984; Oduor and Simons, 1999). In some parts of Africa like Kenya, such methods are now being combined, and an effective but simple IPM strategy for coffee production is evolving. There are inevitably some pest constraints that remain intractable, and further research will be required to optimize the IPM strategy and adapt it for other coffee-producing regions. However, the system is ideally suited to areas where labour is cheap but technology is expensive (Bardner, 1978).

Natural Enemies of Coffee Pests

Despite the large numbers of arthropods found on coffee (>1420 species), relatively few attain the status of major pests, and only rarely do pest populations exceed economic thresholds because they are held in check for most of the time by their natural enemies, i.e. parasitoids, predators and pathogens (e.g. Bardner, 1978). More than 750 species of natural enemies occur in coffee ecosystems, the majority of which are associated with arthropods endemic to Africa. Following the discovery of parasitoids of leafminers (*Leucoptera* spp.) in Réunion in 1898, i.e. *Eulophus borbonicus* Giard. (Hymenoptera, Eulophidae) and *Apanteles bordagei* Giard (Hymenoptra, Braconidae) (Giard, 1898), hundreds of natural enemies have been collected from coffee pests.

Besides arthropods and entomopathogens, there are other natural enemies of coffee pests. These include the predatory woodpeckers against stem borers (Tapley, 1960) and praying mantis against the coffee giant looper (Abasa, 1971).

Case Studies in Biological Control

Literature on biological control of coffee pests is immense but tends to be scattered in non peer-reviewed scientific journals, monthly reports and annual reports of Departments of Agriculture/Coffee Research Institutes. Much of this research was carried out during the pre-independence period and pre-dated most countries' quarantine regulations. Experimental designs in much of this early work were poor or non-existent, which meant that it was difficult to accurately quantify the impact of natural enemy releases (see review of Greathead, 1971). Biological control of coffee pests is also complicated because being an indigenous crop, most of the pests of coffee are also indigenous, severely limiting the chances of using exotic natural enemies to manage these pests within the framework of classical biological control programmes.

To date, attempts at biological control of coffee pests have been relatively unsuccessful. This can be attributed to a poor understanding of the total pest complex and the role of farming systems in pest control strategies, particularly where chemical pesticides may be used on other crops in the system. Case studies of biological control attempts of five major coffee pests in Africa, i.e. mealybugs, antestia bugs, coffee berry borer, coffee green scale and coffee giant looper, are given below, while the invasive plant *C. odorata* is discussed in Timbilla *et al.* in this volume.

Kenya mealybug

Planococcus kenyae (Le Pelley) (Homoptera, Pseudococcidae) was imported into Kenya from Uganda on infested passion fruit in 1923. However, a serious outbreak of the pest on coffee was reported in Central Province of Kenya only in 1926. Chemical control proved ineffective, and efforts were directed towards biological control. Coccinellid predators like the exotic *Cryptolaemus montrouzieri* Mulsant and two indigenous *Chilocorus angolensis* Crotch and *Scymnus guttulatus* Lec. failed to control the mealybugs. However, banding of coffee trees to keep away the ant *Acantholepis capensis* Mayr (Hymenoptera, Formicidae) was shown to encourage the control of mealybugs by the predators *Hyperaspis senegalensis* Muls. (Coleoptera, Coccinellidae) and *Leucopis africana* Malloch (Diptera, Chamaemyiidae), although repeated outbreaks of the pests were still frequently observed. It was subsequently found that the recurrence of the pest was due to heavy parasitism of the predators by *Xenocrepis secundus* Crawf. (Hymenoptera, Pteromalidae). There was, therefore, a need for the use of other natural enemies to supplement the efforts of the predators.

Attempts to control the mealybugs by use of parasitoids were severely delayed following the incorrect identification of the pest as *Planococcus lilacinus* Ckll., which is indigenous to South-East Asia. After searches for parasitoids in Asia, *Anagyrus subalbipes* Ishii and *Pseudaphycus* sp. (both Hymenoptera, Encyrtidae) as well as other parasitoids of *P. lilacinus* were imported into Kenya, but did not successfully breed on the mealybug. Later studies revealed that the *P. kenyae* originated in Uganda, not Asia, and that the mealybug had become a major pest in Kenya because it had been released from its natural control by the complex of natural enemies which occurred in Uganda. Searches for these natural enemies in Uganda led to the release in Kenya of many species of *Anagyrus*, including *Anagyrus* sp. near *kivuensis* Compere, *Anagyrus beneficians* Compere, *Anagyrus* sp., and also *Leptomastix bifasciatus* Compere, *Pauridia peregrina* Timb and *Pseudaphycus* sp. (all Hymenoptera, Encyrtidae) (Le Pelley, 1954). By 1940, control of the mealybug was widespread, with most credit going to *A.* sp. near *kivuensis* and *A. beneficians*. The former managed to establish even where the mealybug population was extremely low. Active dispersal and distribution of parasitized mealybugs by farmers also assisted the spread of the parasitoids. Biological control of *P. kenyae* is one of the most spectacular success stories of biological control in Africa and is estimated to have saved the coffee industry in Kenya more than £10 million (Melville, 1959). New mealybug outbreaks remain a possibility, particularly where broad-spectrum insecticides, which may destabilize the ecological balance between the mealybug and its parasitoids, continue to be used.

Antestia bugs

Antestia spp. is a serious direct pest that is indigenous to East Africa. All stages of the pest are attacked by parasitoids. A number of egg parasitoids have been reported including *Telenomus* (=*Asolcus*) *seychellensis* Dodd, *Telenomus mopsus* (Nixon), *Telenomus suranus* (Nixon), *Hadronotus antestiae* Dodd (all Hymenoptera, Scelioniidae), *Anastatus antestiae* Ferrière (Hymenoptera, Eupelmidae) and *Acroclisoides africanus* Ferrière (Hymenoptera, Pteromalidae) (Le Pelley, 1959; Abebe, 1987). Of these, *T. seychellensis* is the most common species and is considered to be the most important. Taylor (1945) released 68 parasitized antestia adults containing immature stages of the parasitoid, *Corioxenos antestiae* Blair (Strepsiptera) in a coffee field in Toro District, Uganda. One year later, 38% of the sampled antestia collected from the release site were parasitized, although dispersal of the parasitoid was very limited. Greathead and Bigger (1967) released 30 adult *Bogosia rubens* (Villen.) (Diptera, Tachinidiae) from Uganda into a coffee plantation in Lyamungu, Tanzania. Assessment 1 year later showed that 3% of the collected antestia were parasitized. By exposing different numbers of antestia eggs to a fixed number of *T. seychellensis*, Mugo and Ndoiru (1997) showed that 100% parasitism is attained when the parasitoid:pest-egg ratio reaches 1:6.

Coffee berry borer

Coffee berry borer was first reported in Uganda in 1908, and subsequently recorded in neighbouring Kenya in 1928, where its biological control was initiated. The parasitoids, *Prorops nasuta* (Waterston) (Hymenoptera, Bethylidae) and *Heterospila caffeicola* Schmiedeknecht (Hymenoptera,

Braconidae) were imported from Uganda and released into Kenya where they both failed to provide economic control of the coffee berry borer (Greathead, 1971). Murphy *et al.* (1986) conducted quantitative studies on the population dynamics of *H. hampei* and *P. nasuta* on coffee farms in western Kenya and reported that these parasitoids did not have a major regulatory impact on the population of this pest.

Besides *P. nasuta* and *H. caffeicola*, a number of other parasitoids of *H. hampei* have been collected from Africa including *Cephalonomia stephanoderis* Betrem, *Sclerodermus cadavericus* Benoit (both Hymenoptera, Bethylidae) and *Phymastichus coffeae* LaSalle (Hymenoptera, Eulophidae) (Benoit, 1957; Ticheler, 1961; Borbon-Martinez, 1989). However, attempts at biological control using these parasitoids have been targeted mainly at *H. hampei* in Latin America (Barrera *et al.*, 1990; Murphy and Moore, 1990; Murphy and Rangi, 1991). Interestingly, *S. cadavericus* has not been used in biological control because its bite causes dermatitis in man. Quintero (1997) quantified the parasitism of *H. hampei* by both *C. stephanoderis* and *P. nasuta* 2 years after the release of the parasitoids in Colombia. The mean parasitism attained was less than 5%, which led him to conclude that successful classical biological control of *H. hampei* by these bethylid parasitoids was unlikely. One of the major problems was that the populations of the parasitoids fell drastically when the coffee crop was harvested. Furthermore, the low natural level of parasitism in the field, low fecundity and the apparent inability of the parasitoid to prevent coffee parchment damage all mitigate against their use in augmentative biological control (Baker, 1999).

Entomopathogens have also been isolated from *H. hampei* populations collected in Africa. One of the most promising pathogens is *Beauveria bassiana* (Balsamo) Vuill., and its potential as a biopesticide against *H. hampei* was discussed by Murphy and Moore (1990). Most of the field evaluations of this pathogen have been conducted in Latin America, e.g. in Columbia, where *B. bassiana* is applied against *H. hampei* over an area of 50,000 ha (Sosa-Gomez and Lanteri, 1999). Studies show that application using a lever-operated knapsack sprayer requires a concentration of 10^{10} spores per tree to achieve mortality equal to that achieved by synthetic pesticides (i.e. 80%) (Baker, 1999). This is prohibitively expensive and more research needs to be done to identify more pathogenic strains, improve the formulation (to ensure extended spore viability) and refine application methods (to optimize coverage of infested berries by the spores).

Biological Control as a Component of IPM in Coffee Farming Systems

In Africa, coffee is typically grown as an intercrop with either food crops (to increase food security) or agro-forestry trees (to provide shade and/or serve as windbreaks). Both of these practices are likely to continue because of increasing competition for land, food shortages and because shade is generally considered to improve coffee production (Nyambo *et al.*, 1997). Thus, any IPM strategy that is being developed for coffee pests, particularly one based on natural biological control systems, must take cognizance of the typical coffee production systems in Africa. The existence and/or effect of biological control systems often only become apparent when they are disturbed or removed, especially through the indiscriminate use of pesticides. In developing an effective IPM strategy that is based on biological control, it is essential that the nature of the tri-trophic interactions between coffee, coffee pests and their natural enemies, and how these are influenced by the environment, are well understood.

Biological control in relation to pesticides

The use of synthetic pesticides will continue as a pest management option in the short term, but where they are used as a component of an IPM strategy, improvements are required in terms of better targeting (e.g. use of granular formulation of systemic pesticides), use of the lowest volume possible (e.g. ULV sprays), better timing and target specificity. In this context, 'scouting' of coffee trees by farmers should be encouraged to ensure that pesticides are applied only after the pest populations exceed specified thresholds. Scouting should result in a reduction in the quantity of pesticides applied, and in turn, this should lessen the impact of the pesticides on natural enemies. The Coffee Research Foundation in Kenya has developed pest population thresholds for some of the major pests of coffee, above which spraying with specific pesticides is recommended (CRF, 1993). In terms of compatibility with natural enemies, Mejia *et al.* (2000) conducted field studies on the effect of commonly used chemical pesticides (i.e. chlorpyriphos, endosulfan, pirimiphos-methyl, fenitrothion, as well as *B. bassiana*) on adults of *P. nasuta*, an exotic parasitoid introduced to manage *H. hampei*, in order to determine appropriate time lapses between applications of these two methods of control. The results were that all of the synthetic pesticides were toxic to *P. nasuta*, although mortality decreased with increasing delay of pesticide applications. They further reported that when the parasitoids were released first, it was advisable to wait 9 days before spraying *B. bassiana* and 20 days before spraying any of the synthetic pesticides, but when pesticides were sprayed prior to the release of *P. nasuta*, the recommended minimum time interval before the release of the parasitoids was 22 days.

Selective pesticides, particularly bio-pesticides, serve to control the pests while conserving their natural enemies, and thus form an integral component of an IPM strategy. Gusmao *et al.* (2000) tested the selectivity of six synthetic pesticides commonly used to control the coffee leafminer *Perileucoptera coffeella* Guerin-Menv. (Lepidoptera, Lyonetidae) to predatory wasps which are natural enemies of this leafminer. Deltamethrin was found to be selective in favour of the *Polistes versicolor versicolor* Olivier and *Apoica pallens* Fab., whereas ethion showed medium selectivity to the same predators and good selectivity to *Brachygastra lecheguana* Latreille (all Hymenoptera, Vespidae).

Judicious use of either deltamethrin or ethion could therefore supplement the pest control effort of the aforementioned predators. Where selective pesticides are not available, an alternative strategy would be the application of broad-spectrum pesticides with a short persistence. In this way, applications could be timed such that they had minimal impact on natural enemies, e.g. application when adult parasitoids are inactive.

Given the current concern surrounding the use of pesticides on food crops, it seems likely that bio-pesticides will play an increasingly important role in the management of coffee pests. Currently the most likely candidate entomopathogens are *B. bassiana* and *Metarhizium anisopliae* (Metsch.) Sorokin. However, in coffee production systems most of the research to date has been directed at using *B. bassiana* against *H. hampei* in Latin America (Baker, 1999; Sosa-Gomez and Lanteri, 1999; Edington *et al.*, 2000; Rosa *et al.*, 2000). As noted earlier, the current application rate of *B. bassiana* required to provide satisfactory levels of control, i.e. comparable to that of synthetic pesticides, is not cost-effective (Baker, 1999). Cheaper methods of producing high-quality spores now need to be developed. Attempts have been made to substitute rice with coffee husks as a substratum for mass-producing spores (Borges *et al.*, 1997), which is certainly a step in the right direction. Improved formulations using photo-protectants would also reduce mortality in the field as a result of spores being exposed to lethal doses of UV from the sun. Edington *et al.* (2000) evaluated 22 substances for their ability to protect spores of *B. bassiana* from sunlight and found that a mixture of 3% (w/v) albumen and 4% (w/v) milk powder gave the highest degree of spore protection per unit cost. Timing of sprays to coincide with the dry season when most berries are more susceptible to attack by the borer and spraying shaded coffee also need to be evaluated in terms of enhancing the efficacy of *B. bassiana*.

Biological control in relation to cultural practices

Cultural practices recommended for the improvement of coffee productivity may also influence the success of biological control of coffee pests. Pruning has been shown to influence the natural enemies in a number of ways. In unpruned coffee trees, temperature and light decrease, whereas humidity increases under the tree canopy. Such environmental conditions have been known to induce sporulation by fungal entomopathogens, thereby encouraging fungal epizootics (Gottwald and Tedders, 1982). A good example is the entomopathogenic fungus *Verticillium lecanii* (Zim.), which is reported to develop more rapidly amongst populations of its host *C. viridis* in shaded conditions and on the underside of coffee leaves (Kohler, 1980). The same fungus was also reported to exert only limited impact on populations of *C. viridis* in parts of Tanzania due to prevailing low humidity (Ritchie, 1935). Similarly, Wanjala (1978) studied the effect of capping (a type of pruning) the canopies of coffee trees on the abundance of parasitoids of *L. meyricki*. A total of ten species of parasitoids were reared from the pest and uncapped canopies were found to provide more stable ecosystems and support higher levels of parasitism than capped canopies. Conversely, the impact of parasitoids on antestia bugs and white stem borer has been shown to be higher on pruned coffee (Taylor, 1945; Oduor and Simons, 1999).

Mulching is frequently recommended for the management of weeds and conservation of soil moisture in coffee production systems (Abasa, 1975). However, mulching has also been shown to encourage the population build-up of *Leucoptera* spp. by lessening the impact of predatory ants on the larvae of this pest (Crowe, 1964).

A number of agroforestry trees are grown to provide shade to coffee trees. In some respects, shade trees provide a similar environment to that created when coffee trees are left unpruned (see above). However, unlike unpruned coffee trees, in some instances, shade trees reportedly increase damage to coffee by pests such as *A. o. intricata* (Mbondji, 1999) and *L. caffeina* (Hill, 1983), by destabilizing the natural regulatory effects of their natural enemies.

Biological control in relation to host resistance

Breeding for resistant, high-yielding (in terms of quantity and cup-quality) coffee germplasm is potentially an ideal pest management strategy for coffee farmers. However, a coffee variety with all the required traits remains to be bred. A relevant example is Ruiru 11, a coffee variety which was bred and released by the Coffee Research Foundation in Kenya, and is resistant to both coffee leaf rust and coffee berry disease (Walyaro *et al.*, 1984). The use of coffee varieties resistant to specific diseases and pests means that the farmer will not have to apply pesticides aimed at managing these diseases and pests. Abstaining from the use of pesticides encourages the natural pest control and is therefore compatible with biological control. Inability to raise enough seedlings of improved coffee varieties to satisfy the demand of the farmers continues to be a problem.

Biological control with entomopathogens in relation to other natural enemies

The possibility of combining the use of fungal entomopathogens (*B. bassiana* or *M. anisopliae*) with parasitoids like *P. nasuta* for the management of *H. hampei* has been evaluated in the laboratory (Rosa *et al.*, 2000). Spore suspensions of three isolates of each of the entomopathogens were directly applied to adults of *P. nasuta*. The LC_{50} of the isolates that caused the lowest mortality were 8.3×10^6 spores ml^{-1} for *M. anisopliae* and 4.0×10^6 spores ml^{-1} for *B. bassiana*. It is encouraging to note that despite these levels of pathogenicity, treating borer-infested berries with either pathogen did not significantly affect the capacity of the parasitoid on the borers therein. These preliminary data suggest that the isolates of either pathogen may be compatible with the action of *P. nasuta* under field conditions, provided that pathogen applications and parasitoid liberations are timed not to coincide. Field trials are now required in order to confirm their compatibility as components of an IPM strategy in the field.

Enhancing biological control

A number of approaches could be used to increase the impact of natural enemies on their hosts (pests). Successful biological control of coffee pests depends both on the number and also types of natural enemies released, as well as the production system into which they are released. Laboratory studies have shown that relying on several different natural enemies which attack different stages of the same pest or have different searching capacities is significantly more effective than relying on only one natural enemy. An evaluation of the potential competition between *C. stephanoderis* and *P. nasuta* over their host *H. hampei* in infested coffee berries revealed that both parasitoids avoided ovipositing on previously parasitized hosts (Infante *et al.*, 2001). Furthermore, neither aggressive behaviour nor interference between the two parasitoids was observed outside the coffee berries. These results suggest that the parasitoids would complement each other in the field. Similarly, in the control of *C. viridis*, it has been shown that mixtures of *V. lecanii* and 0.1% solutions of the pesticides fenthion and phosphamidon were found to be more effective in combination than either the fungus or the pesticide applied alone (Easwaramoorthy *et al.*, 1978; Jayaraj, 1989).

Natural enemies of pests that bore into the coffee berry are inevitably removed when coffee is harvested. Baker (1999) attributes the slow pace of biological control of *H. hampei* using parasitoids in Colombia to the repeated removal of these natural enemies in the picked berries. The loss of such natural enemies could be minimized through strip harvesting and the conservation of berries that have dropped to the ground (*mbuni*). The unharvested infested berries, although harbouring the pest, would serve as reservoirs for its natural enemies as well.

It has been demonstrated that coffee-farming systems can be modified in favour of natural enemies by intercropping coffee with plants that are attacked by similar pests. The aim is to ensure a continuous supply of pests (hosts) to their natural enemies at times when the pests' populations on the coffee plants are low. Adult parasitoids feed on pollen and nectar from flowers, although their immature stages develop, and in the process kill the target pests. Therefore, growing flowering plants within or around coffee plantations can serve to restrict these parasitoids to the coffee farms. This will serve to direct the natural enemies to their hosts (pests).

It is reported that one of the advantages of biological control is that it is free to, and does not need the full participation of, the farmers. However, it has also been recognized that failure to involve the farmer in identifying pest problems, and the poor linkages between researchers, extension workers and farmers are the main reasons for the limited adoption of biological control strategies. Involving farmers in the implementation of biological control can improve the farmers' understanding and enhance success of biological control. Using discovery-learning in the Farmer Field School (FFS) approach, Nyambo *et al.* (1997) introduced the IPM concept to farmers who were initially using excessive amounts of pesticides to grow coffee and vegetables in Kenya. From these FFS, which generated a lot of enthusiasm, farmers were able to identify and understand the role of predators and parasitoids in controlling pests. The farmers were able to confirm higher numbers of parasitized eggs of antestia bugs in pruned than in unpruned coffee plots. It was then demonstrated that high coffee and vegetable yields could be attained with limited use of pesticides.

The Future and Conclusions

Coffee in sub-Saharan Africa is rapidly losing its competitive edge with the rest of the world, and one of the major reasons for this is poor yields. None the less, this decline could be reversed if fundamental interventions were made in two key areas of coffee production, i.e. regeneration of old, unproductive coffee, and the development of a sustainable approach towards pest management.

A number of coffee replanting programmes are currently underway in African countries aimed at replacing many of the present coffee trees which are considered to be too old and unproductive. New hybrids, which are bred to produce high yields, are preferred over the old varieties. However, the susceptibility of these new hybrids to existing or potential pests has not yet been established. Increased movement of plant material (intended or otherwise) further increases the chances of introducing new pests into the main coffee-producing areas.

Coffee farmers have tended to rely on pesticides as their preferred method for managing pests of coffee. However, increasing concern about the harmful effects of pesticides both to humans/animals and the environment has resulted in the need to identify and develop alternative, sustainable IPM strategies for the management of coffee pests. Biological control offers a sustainable, environmentally friendly and cost-effective alternative to synthetic pesticides, which can be used as the basis for an IPM strategy. To date, classical biological control of coffee pests in Africa has not been particularly successful. This is partly because coffee, together with most of its pests, is indigenous to Africa, thereby limiting the chances of controlling pest outbreaks through the use of exotic natural enemies. However, it is also worth noting that many of the earlier attempts at biological control were undertaken without a proper understanding of the biology and ecology of either the targeted pests or their natural enemies. There is now an urgent need to repeat some of the earlier (failed) attempts at biological control, using improved scientific approaches.

An important factor in successful biological control is the need for accurate taxonomy of the target pest and its natural enemies. There is no better example than the case of *P. kenyae* (Abasa, 1975), described previously, which shows how much time and resources can be wasted when a pest species is identified incorrectly. The need for correct identification of new pests before biological control is attempted is reiterated by Rosen (1978).

The failure of parasitoids to control the leafminer *P. coffeella* was reported by Reis *et al.* (2000) to be due to parasitized pest larvae being predated upon by wasps. In so doing, predatory wasps indirectly killed parasitoids, thereby undermining the impact of the parasitoid on the leafminer population. Similar detailed studies on the biology as well as the ecological requirements of other potential biological control agents are needed. This is highlighted in an isolated report, which suggests that arthropod natural enemies are responsible for spreading major coffee diseases, thereby posing a phytosanitary risk. Nemeye *et al.* (1990) provide some evidence that suggests that *H. coffeicolla*, a potential biological control agent of *H. hampei*, could transmit coffee berry disease. Whilst this may be a relatively minor issue, further work is required in order to confirm and, where necessary, quantify the risk posed.

Biological control will undoubtedly provide the fundamental component of a sustainable IPM strategy for coffee pests in Africa in the future. Despite the limited success using biological control to date, the rich diversity of natural enemies of coffee pests in Africa should enable the potential of biological control to be realized. Since classical biological control has not been very successful, more emphasis should be placed on inundative, augmentative or other biological control strategies. This should involve the mass-rearing and release, or enhancement, of promising natural enemies in an integrated approach to maximize their impact on the pest complex in coffee production systems in Africa.

References

Abasa, R.O. (1971) The preying mantis, *Sphodromantis* sp., predator of the giant looper, *Ascotis selenaria reciprocaria* Wlk. in Kenya coffee. *East Africa Agricultural and Forestry Journal* 37, 177–180.

Abasa, R.O. (1975) A review of biological control of coffee pests in Kenya. *East African Agricultural and Forestry Journal* 40, 292–299.

Abebe, M. (1987) Insect pests of coffee with special emphasis on Antestia, *Antestiopsis intricata* in Ethiopia. *Insect Science and its Applications* 8, 977–980.

ACPC (2001). *Coffee Market Report,* No. 22, September, London, UK.

Baker, P.S. (1999) *The Coffee Berry Borer in Colombia.* Final Report of the DFID-Cenicafe-CABI Bioscience IPM of coffee project, Chinchina, Colombia 154pp.

Bardner, R. (1978) Pest control in coffee. *Pesticide Science* 9, 458–464.

Barrera, J.F., Baker, P.S., Valenzuela, J.E. and Schwartz, A. (1990) Introduccion de dos especies de parasitoids Africanos a Mexico para el controle biologico de la broca del café, *Hampothenemus hampei* (Ferrari) (Coleoptera: Scolytidae). *Folia Entomologica Mexicana* 79, 245–247.

Benoit, P.L.G. (1957) Un nouveau *Sclerodermus* vulnerant pour l'homme en Afrique Central. *Bulletin et Annales de la Societé Royale d'Entomologie de Belgique* 93, 42–46.

Borbon-Martinez, O. (1989) Bioécologie d'un ravageur des baies de caféier, *Hypothenemus hampei* Ferr. (Coleoptera: Scolytidae) et de ses parasitoids au Togo. PhD thesis, l'Université Paul-Sabatier de Toulouse.

Borges, M., Garcia, C., Fonseca, R. and Levya, M. (1997) Suplementacion del mosto de destileria al 40% para su aprovechamieto mas eficiente en la reproduccion masiva de *Beauveria bassiana* (Bals.) Vuill., *Paecilomyces lilacinus* (Thorn) Samson y *Verticillium lecanii* (Zim.) Viegas. *Revista de Proteccion Vegetal* 12, 27–31.

Brun, L.O., Marcillaud, C., Gaudichon, V. and Suckling, D.M. (1989) Endosulfan resistance in *Hypothenemus hampei. Journal of Economic Entomology* 82, 1311–1316.

CRF (1993) Coffee insect pest control. *Kenya Coffee* 58, 1577–1578.

Crowe, T.J. (1964) Coffee leaf-miner in Kenya. II. Causes of outbreaks. *Kenya Coffee* 29, 223–231.

Easwaramoorthy, S., Regupathy, A., Santharam, G. and Jayaraj, S. (1978) The effect of sub normal concentrations of insecticides in combination with the fungal pathogen, *Cephalosporium lecanii* Zimm. in the control of coffee green scale, *Coccus viridis* Green. *Zeitschrift für Angewandte Entomologie* 86, 161–466.

Edington, S., de Segura, H., Rosa, W. and Williams, T. (2000) Photoprotection of *Beauveria bassisna:* testing simple formulations for control of the coffee berry borer. *International Journal of Pest Management* 46, 169–176.

FAO (2000) FAOSTATS statistics database. Food and Agricultural Organisation of the United Nations. http://apps.fao.org/

Giard, A. (1898) Sur l'existence du *Cemiostoma coffeella* à l'ile de la Réunion. *Bulletin Société Agricole de France*, 210–203.

Gottwald, T.R. and Tedders, W.L. (1982) Studies on the conidial release by the entomogenous fungi *Beauveria bassiana* and *Metarhizium anisopliae* (Deuteromycotina: Hyphomycetes) from adult pecan weevil (Coleoptera: Curculionidae). *Environmental Entomology* 11, 1274–1279.

Greathead, D.J. (1971) A review of biological control in the Ethiopian region. *Technical Bulletin of Commonwealth Institute of Biological Control* 5.

Greathead, D.J. and Bigger, M. (1967) Notes on the biology of *Bogosia rubens* (Diptera: Tachinidae) a parasite of *Antestiopsis* spp. (Hem.: Pentatomidae) and its introduction into Kilimanjaro, Tanzania. *Technical Bulletin of Commonwealth Institute of Biological Control* 9, 1–8.

Gusmao, M.R., Picanco, M., Gonring, A.H.R. and Moura, M.F. (2000) Physiological selectivity of insecticides to wasp predators of the leafminer. *Pesquisa Agropecuaria Brasiliera* 35, 681–686.

Hill, D. (1983) *Agricultural Insect Pests of the Tropics and their Control.* Cambridge University Press, London, UK.

Infante, F., Mumford, J., Baker, P., Barrera, J. and Fowler, S. (2001) Interspecific competition between *Cephalonomia stephanoderis* and *Prorops nasuta* (Hymenoptera: Bethylidae), parasitoids of the coffee berry borer, *Hypothenemus hampei* (Col., Bethylidae). *Journal of Applied Entomology* 125, 63–70.

Jayaraj, S. (1989) Integrated management of coffee green scale *Coccus viridis* (Green) (Homoptera, Coccidae). *Journal of Plantation Crops* 16, 195–201.

Kohler, G. (1980) Los parasitos y episitos de la guagua verde del cafeto *Coccus viridis* Green (Hemiptera, Coccoidea) en efetelas de Cuba. *Centro Agricola* 7, 75–105.

Le Pelley, R.H. (1954) *Biological Control in Kenya.* Kenya Department of Agriculture, Nairobi, Kenya.

Le Pelley, R.H. (1959) *Agricultural Insects of East Africa.* East African High Commission, Nairobi, Kenya.

Le Pelley, R.H. (1968) *Pests of Coffee.* Longmans, London, UK.

Mabberly, D.J. (1997) *The Plant Book.* Cambridge University Press, Cambridge, UK.

Mbondji, P.M. (1999) Observations eco-biologiques sur *Antestiopsis lineaticollis* intricate au Cameroun (Hemiptera: Pentatomidae). *Annales de la Société Entomologique de France* 35, 77–81.

Mejia, M.J.W., Bustillo, P.A.E., Orozco, H.J. and Chaves, C.B. (2000) Efecto de quarto insecticidas y de *Beauveria bassiana* sobre *Prorops nasuta* (Hymenoptera: Bethylidae), parasitoide de la broca del café. *Revista Colombiana de Entomologia* 3/4, 117–123.

Melville, A.R. (1959) The place of biological control in the modern science of entomology. *Kenya Coffee* 24, 81–84.

Mugo, H.M. and Ndoiru, S.K. (1997) Use of *Telenomus (Asolcus) seychellensis* (Hymenoptera:Scelionidae) in biological control of antestia bugs, *Antestiopsis* spp. (Hemiptera: Pentatomidae) in coffee. *Kenya Coffee* 62, 2455–2459.

Murphy, S.T. and Moore, D. (1990) Biological control of the coffee berry borer, *Hypothenemus hampei* (Ferrari) (Coleoptera: Scolytidae): previous programmes and possibilities for the future. *Biocontrol News and Information* 11, 107–117.

Murphy, S.T. and Rangi, D.K. (1991) The use of African wasp, *Prorops nasuta* for the control of the coffee berry borer, *Hypothenemus hampei* in Mexico and Ecuador: the introduction programme. *Insect Science and its Applications* 12, 27–34.

Murphy, S.T., O'Donell, D.J., Nangayo, F.L.O., Cross, A. and Evans, H.C. (1986) *First Report on the Coffee Berry Borer Project*. CAB International Institute of Biological Control, Silwood Park, UK.

Nemeye, P.S., Moore, D. and Prior, C. (1990) Potential of the parasitoid *Heterospilus prosopidis* (Hymenoptera: Braconidae) as a vector of plant-pathogenic *Colletotrichum* spp. *Annals of Applied Biology* 116, 11–19.

Njoroge, J.M. (1991) Tolerance of *Bidens pilosa* and *Parthenium hysterophrus* L. to paraquat (Gramoxone) in Kenya coffee. *Kenya Coffee* 56, 999–1002.

Njoroge, J.M. (1994) Weeds and weed control in coffee. *Experimental Agriculture* 30, 421–429.

Nyambo, B.T., Kimani, M. and Williamson, S. (1997) Developing an African model for effective IPM training. *ILEIA Newsletter for Low External Input and Sustainable Agriculture* 13, 29–30.

Oduor, G.I. and Simons, S.A. (1999) Integrated pest management of coffee pests on smallholder farms in Malawi. *Final Technical Report,* CAB International, Wallingford, UK.

Quintero, C.V., Bustillo, A.E., Benavides, P. and Chaves, B. (1997) Establesimiento y dispersion de *Cephalonomia* y *Prorops nasuta*, parasitoids de la broca del café, en el departamento de narinho. In: *Congreso de la Sociedad Colombiana de la Entomologia,* Pereira, 16–18 July 1997, Bogota, Colombia, pp. 22–23.

Reis, R., De Souza, O. and Vilela, E.F. (2000) Predators impairing the natural biological control of parasitoids. *Neotropical Entomology* 29, 507–517.

Ritchie, A.H. (1935) Report of the entomologist. In: *Report of the Department of Agriculture*, Dar es Salaam, Tanganyika, pp. 73–83.

Rosa, W., de la, Segura, H.R., Barrera, H.R. and William, T. (2000) Laboratory evaluation of the impact of entomopathogenic fungi on *Prorops nasuta* (Hymenoptera: Bethylidae), a parasitoid of the coffee berry borer. *Environmental Entomology* 29, 126–131.

Rosen, D. (1978) The importance of cryptic species and specific identifications as related to biological control. In: *Proceedings of Symposium on Agricultural Research: Biosystematics in Agriculture*, 8–11 May 1977, Beltsville, Maryland, Allanheld, Osmon and Co., New York, USA, pp 23–35.

Sosa-Gomez, D.R. and Lanteri, A.A. (1999) Estado actual del control biologico de plagas agricolas con hongos entomopatogenos. *Revista de la Sociedad Entomologica Argentina* 58, 295–300.

Tapley, R.C. (1960) White coffee stem borer, *Anthores leuconotus* Pasc. and its control. *Bulletin of Entomological Research* 51, 279–301.

Taylor, T.H.C. (1945) Recent investigations of *Antestia* spp. in Uganda. Part I. *East African Agricultural Journal* 10, 223–233.

Ticheler, J.H.G. (1961) Etude analytique de l'épidémiologie du scolyte des graines de café *Stephanoderes hampei* Ferr., en Côte d'Ivoire. *Mededelingen Landbouwhogeschool Wageningen, Nederland* 61, 1–49.

Wanjala, F.M.E. (1978) Relative abundance and within canopy distribution of the parasites of the coffee miner, *Leucoptera meyricki* (Lep: Lyonetidae) in Kenya. *Entomophaga* 23, 57–62.

Walyaro, D.J., van der Vossen, H.A.M., Owuor, J.B.O. and Masaba, D.M. (1984) New varieties to combat old problems in arabica coffee production in Kenya. In: Hawksworth, D.L. (ed) *Proceedings of CAB's First Scientific Conference*, 12–18 February 1984, Arusha, Tanzania. CAB International, London, UK, pp. 1146–1151.

Waterhouse, D.F. and Norris, K.R. (1989) *Biological Control: Pacific Prospects*. Supplement 1. *Hypothenemus hampei* (Ferrari). Australian Centre for International Agricultural Research, Canberra, Australia, pp. 57–75.

Zebeyou, M.G. (1991) Distribution and importance of *Chromolaena odorata* in Côte d'Ivoire. *Biotrop Special Publication* 44, 67–69.

1.

2.

3.

4.

Plate 1. *Opuntia stricta* in Kruger National Park, before control (photo by J.H. Hoffmann, University of Cape Town).
Plate 2. *Opuntia stricta* in Kruger National Park, after biological control by *Dactylopius* sp. (**inset**) (photos by H. Zimmermann, ARC – PPRI).
Plate 3. Cassava mealybug (photo by G. Goergen, IITA).
Plate 4. *Anagyrus lopezi* (photo by G. Goergen, IITA).

5.

6.

7.

8.

Plate 5. Mass rearing of biological control agents (photo by IITA).
Plate 6. *Rastrococcus invadens* (photo by G. Goergen, IITA).
Plate 7. *Gyranusoidea tebygi* (photo by G. Goergen, IITA).
Plate 8. *Anagyrus mangicola* (photo by S. Keller, FAL, Reckenholz).

9.

10.

11.

12.

Plate 9. Cassava green mite damage (photo by G. Goergen, IITA).
Plate 10. *Typhlodromalus manihoti* feeding on cassava green mite (photo by G. Goergen, IITA).
Plate 11. *Typhlodromalus aripo* inside a cassava shoot tip (photo by G. Goergen, IITA).
Plate 12. Cassava green mite mummified by the fungus *Neozygites tanajoae* (photo by G. Goergen, IITA).

13.

14A.

14B.

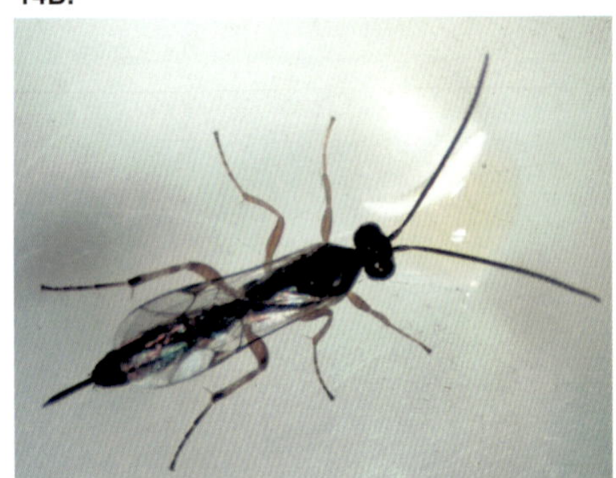

16.

15.

Plate 13. Potato tuber moth (photo by D. Visser, ARC – PPRI).
Plate 14. A Damage by potato tuber moth to potato tubers in storage (photo by D. Visser, ARC – PPRI).
B *Diadegma mollipla* (photo by R. Kfir, ARC – PPRI).
Plate 15. *Eretmocerus mundus* (photo by Bio-Bee Biological Systems, Sde Eliyahu, Israel). **Inset** Woolly whitefly, adults and eggs on citrus (photo by J. Legg, IITA and NRI).
Plate 16. *Aleurodicus dispersus* (photo by G. Goergen, IITA).

17.

18.

19.

20.

Plate 17. *Pauesia juniperorum* attacking *Cinara* sp. on cypress (photo by Holt Studios International, UK).
Plate 18. Mummies of *Cinara cronartii* with emergence holes of *Pauesia cinaravora* (photo by Rami Kfir, ARC – PPRI).
Plate 19. *Imbrasia cytherea* parasitized (photo by G.D. Tribe, ARC – PPRI).
Plate 20. *Euproctis terminalis* (photo by G.D. Tribe, ARC – PPRI).

22.

24.

Plate 21. Larva of *Sesamia calamistis* in maize (photo by K. Maes, National Museum of Kenya).
Plate 22. *Eldana saccharina* (photo by R. Copeland, Texas A&M University).
Plate 23. *Chilo partellus* (photo by R. Copeland, Texas A&M University).
Plate 24. *Xanthopimpla stemmator* (photo by Rami Kfir, ARC – PPRI).

25.

26A.

26B.

27.

28.

Plate 25. *Cotesia sesamiae* (photo by G. Goergen, IITA).
Plate 26. A *Pareuchaetes pseudoinsulata*. **B** *Pareuchaetes aurata* (photos by C. Zachariades, ARC – PPRI).
Plate 27. *Pareuchaetes insulata* larva (photo by C. Zachariades, ARC – PPRI).
Plate 28. *Actinote thalia pyrrha* adult (photo by C. Zachariades, ARC – PPRI).

29.

30.

31A.

31B.

32.

Plate 29. *Actinote thalia pyrrha* larva (photo by C. Zachariades, ARC – PPRI).
Plate 30. *Lixus aemulus* (photo by C. Zachariades, ARC – PPRI).
Plate 31. A *Neochetina* damaged leaves of water hyacinth. **B** *Neochetina bruchi* (photos by C.J. Cilliers, ARC – PPRI).
Plate 32. *Neochetina eichhorniae* (photo by C.J. Cilliers, ARC – PPRI).

33A.

33B.

36.

Plate 33. A *Niphograpta albiguttalis* larva. **B** *Niphograpta albiguttalis* adult (photo by C.J. Cilliers, ARC – PPRI).
Plate 34. *Orthogalumna terebrantis* damage. **Inset** *Orthogalumna terebrantis* (photo by C.J. Cilliers, ARC – PPRI).
Plate 35. *Eccritotarsus catarinensis* (photo by C.J. Cilliers, ARC – PPRI).
Plate 36. Water hyacinth leaves damaged by *E. catarinensis* (photo by C.J. Cilliers, ARC – PPRI).

37.

39.

38.

40.

Plate 37. Pond near Craddock, RSA, covered with *Azolla filiculoides* before control (photo by A. McConnachie, Wits University).

Plate 38. Pond near Craddock, RSA, covered with *Azolla filiculoides* after control. **Inset** *A. filiculoides* (photos by A. McConnachie, Wits University).

Plate 39. Tomato with root-knot nematodes. **Inset** AMF (arbuscular mycorrhizal fungi) endophytes in root of maize (photos by R.A. Sikora, University of Bonn).

Plate 40. Strains of *Aspergillus flavus* on maize kernels. Left: L strain with green conidial masses; right: S strain with white mycelium and small black sclerotia, with negligible conidiation (photo by K.F. Cardwell, CSREES, USDA, Washington DC and P.J. Cotty, USDA, New Orleans).

41.

42.

44.

43.

Plate 41. *Schistocerca gregaria* (photo by C. Lomer, IITA).
Plate 42. Grasshopper infected by *Metarhizium anisopliae* var. *acridum* (photo by C. Lomer, IITA).
Plate 43. Cadaver of *Zonocerus variegatus* (photo by C. Lomer, IITA).
Plate 44. Green Muscle™ (photo by C. Lomer, IITA).

45.

46.

47.

48.

Plate 45. Termite mound being treated experimentally with *Metarhizium* spores (photo by O.-K. Douro Kpindou, IITA).
Plate 46. Direct economic damage by termites (photo by O.-K. Douro Kpindou, IITA).
Plate 47. Banana weevil (photo by C. Gold, IITA).
Plate 48. Banana weevil infected with *Beauveria bassiana* (photo by C. Nankinga, IITA).

Plate 49. Sorghum attacked by *Striga hermonthica.* **Inset** Flowers of *S. hermonthica* (photos by D. Berner, IITA).
Plate 50. *Ceratitis cosyra* male (photo by R. Copeland, ICIPE).
Plate 51. *Ceratitis anonae* male (photo by R. Copeland, ICIPE).
Plate 52. *Ceratitis fasciventris* male (photo by R. Copeland, ICIPE).

53.

54.

55A.

55B.

56.

Plate 53. *Ceratitis rosa* male (photo by R. Copeland, ICIPE).
Plate 54. *Megalurothrips sjostedti* adult female (photo by G. Goergen, IITA).
Plate 55. A *Ceranisus femoratus* adult female. **B** Pupa of *C. femoratus* with the remaining host skin at the bottom (photos by G. Goergen, IITA).
Plate 56. *Trichogrammatoidea ?eldanae* adult female (photo by G. Goergen, IITA).

57.

58.

59.

60.

Plate 57. *Clavigralla tomentosicollis* adult killed by *Metarhizium anisopliae* (photo by G. Goergen, IITA).
Plate 58. *Maruca vitrata* adult male (photo by G. Goergen, IITA).
Plate 59. *Prostephanus truncatus* (photo by G. Goergen, IITA).
Plate 60. *Teretrius nigrescens* (photo by G. Goergen, IITA).

Plate 61. A and **B** *Helicoverpa armigera* larvae killed by nucleopolyhedrovirus (photos by A. Cherry, IITA).
Plate 62. *Antestiopsis* spp. (photo by S. Simons, CAB *International*).
Plate 63. White coffee stem borer (photo by S. Simons, CAB *International*).
Plate 64. Young diamondback moth larva feeding on a cabbage. **Insets** Diamondback moth and pupae of *Oomyzus sokolowskii* dissected from diamondback moth larva (photos by R. Kfir, ARC – PPRI).

23

Biological Control of the Diamondback Moth *Plutella xylostella* in Africa

Rami Kfir
Agricultural Research Council – Plant Protection Research Institute, Division of Insect Ecology – Biological Control, Pretoria, South Africa

Introduction

Diamondback moth, *Plutella xylostella* (L.) (Lepidoptera, Plutellidae) (Plate 64), is the most injurious insect pest of cabbage and other cole crops throughout the world (Talekar and Shelton, 1993). This pest is the most universally distributed of all lepidopterans (Meyrick, 1928) and occurs wherever brassica crops are cultivated (Talekar and Shelton, 1993). In the field it has developed resistance to almost every synthetic insecticide used against it including *Bacillus thuringiensis* Berliner (*Bt*) formulations (Tabashnik *et al.*, 1990; Liu *et al.*, 1995). The increasing usage of *Bt* products resulted in an increasing number of reports of field resistance by diamondback moth populations (Tabashnik, 1994). New insecticides are continuously being developed as existing insecticides become useless, but *P. xylostella* has developed resistance very quickly to many of these (Shelton *et al.*, 2000; Nisin *et al.*, 2000). Diamondback moth has also developed cross-resistance and multiple-resistance to different chemical pesticides (Shelton *et al.*, 2000).

Chemical control of *P. xylostella* worldwide is estimated to cost about US$1 billion annually (Talekar, 1992). Lack of effective natural control is considered to be the major cause of its high pest status in many parts of the world (Lim, 1986). In many countries, in addition to the development of resistance, the destruction of non-target beneficials through indiscriminate and widespread use of broad-spectrum insecticides (Ooi and Sudderuddin, 1978) is also considered responsible for this imbalance (Talekar and Shelton, 1993).

Although diamondback moth is regarded as a serious pest of crucifer crops in Africa, there is a paucity of published information about the pest and its natural enemies from this continent. Most of the published information is from South Africa. This chapter is an attempt to summarize the published information of this important pest in sub-Saharan Africa, with emphasis on its natural enemies.

The Status of Diamondback Moth in South Africa

Since heat-tolerant cultivars of cabbage have become available, cabbage, *Brassica oleracea* var. *capitata* (L.) Alef. (*Brassicaceae*), is now cultivated throughout the year in southern Africa (Hemy, 1984). In South Africa, cruciferous crops are grown commercially near markets in peri-urban areas and are a staple food crop for resource-poor farming communities. In the 1930s, Ullyett (1947) studied *P. xylostella* and its natural enemies in the Pretoria district of South Africa and recorded 19 parasitoids and hyperparasitoids, several predators, two bacteria and an entomopathogenic fungus associated with it. Ullyett concluded at the time that diamondback moth was well controlled by its natural enemies in South Africa.

Without further reports of serious damage, interest in *P. xylostella* lapsed until Annecke and Moran (1982) drew attention to its increasing economic importance in the country. Nevertheless, its pest status in South Africa is lower than in other areas with a similar climate (Kfir, 1996). Renewed interest in South Africa in *P. xylostella* started in the 1990s after farmers reported recurring outbreaks of diamondback moth in cabbage fields and also increasing difficulties with chemical control. Dennill and Pretorius (1995) demonstrated that high infestation levels by diamondback moth are a result of excessive insecticide applications. At one study site where insecticides were applied only once every 3 weeks, parasitism of diamondback moth reached 90% and the pest did not cause economic losses. In contrast, at a second study site with regular and excessive chemical applications, parasitism levels were negligible and serious outbreaks of *P. xylostella* caused total crop loss.

Because of wide and indiscriminate use of insecticides, local field populations of diamondback moth in South Africa started showing signs of resistance to synthetic pyrethroids, organophosphates and carbamates (Sereda *et al.*, 1997). It was also expected that with the recent rapid expansion of oilseed rape (*Brassica napus* L. (*Brassicaceae*)), canola cultivation, in South Africa (M.S. Mosiane, 2001, Pretoria, South Africa, unpublished results) the problem might be exacerbated (Kfir, 1997a). Of concern was that this could lead to accelerated resistance to pesticides by diamondback moth with serious consequences for the cultivation of cabbage (Kfir, 1997a).

Currently *Bt* var. *aizawai* and var. *kurstaki* products are recommended for the control of diamondback moth in South Africa (Nel *et al.*, 1999) and are widely used by commercial farmers. So far Triazine-resistant transgenic canola (resistant to the herbicide Simazine) is the only transgenic crucifer grown commercially in South Africa. There are also plans to introduce *Bt*-transgenic canola, but the implications of this resistance to diamondback moth are still unknown.

Parasitoids

Le Pelley (1959), Kibata (1997) and Oduor *et al.* (1997) recorded several parasitoids attacking *P. xylostella* in Kenya, but with relatively low rates of parasitism. Surveys in East Africa indicated that parasitism rates of diamondback moth are generally low and seldom surpass 15% (B. Löhr, ICIPE, Nairobi,

2002, personal communication). In the Cotonou area of Bénin, West Africa, only one parasitoid species of diamondback moth, namely the larval parasitoid *Cotesia plutellae* (Kurdjumov) (Hymenoptera, Braconidae), has been recorded (Goudegnon *et al.*, 2000). In contrast, Kfir (1997a,b) confirmed the presence of a rich indigenous parasitoid fauna from diamondback moth larvae and pupae in South Africa. High levels of parasitism of >90% were recorded on unsprayed cabbages in the North-West Province (Kfir, 1997a,b), in the Eastern Cape Province (Waladde *et al.*, 2001; Smith and Villet, 2003), and in Gauteng Province in unsprayed canola fields (M.S. Mosiane, 2001, Pretoria, South Africa, unpublished results). A total of 33 different species of parasitoids and hyperparasitoids associated with *P. xylostella* have been recorded in South Africa (Table 23.1) (Ullyett, 1947; Kfir, 1996, 1997b, 2002).

By far the most abundant species in all regions was *C. plutellae*. Other abundant parasitoids were *Oomyzus sokolowskii* (Kurdjumov) (Hymenoptera, Eulophidae) (Plate 64) and *Diadromus collaris* Gravenhorst (Hymenoptera, Ichneumonidae), and the hyperparasitoids *Mesochorus* sp. (Hymenoptera, Ichneumonidae) and *Pteromalus* sp. (Hymenoptera, Pteromalidae). *Apanteles halfordi* Ullyett (Hymenoptera, Braconidae), another common parasitoid, is specific to diamondback moth and is known only from South Africa (Walker and Fitton, 1992; Prinsloo, 2002). *O. sokolowskii* has been introduced into several countries including the Cape Verde Islands for biological control of diamondback moth (Lim, 1992; M.J.W. Cock, CIBC, Silwood Park, 1983, unpublished results) and is the only gregarious primary parasitoid known from diamondback moth. The hyperparasitoids *Mesochorus* sp. and *Pteromalus* sp. have emerged from cocoons of *C. plutellae*, *A. halfordi*, *Diadegma mollipla* (Holmgren) (Hymenoptera, Ichneumonidae) and occasionally *D. collaris* (Kfir, 1996, 1997b, 2002).

This rich indigenous parasitoids fauna, found attacking all diamondback moth developmental stages in South Africa, indicates a long association between parasitoids and the pest in the region (Kfir, 1998). In addition, the large number (175) of wild plant species in the *Brassicaceae* recorded from South Africa (Jordaan, 1993), on which diamondback moth might develop, caused Kfir (1998) to speculate that the pest itself might have originated in the Cape floral kingdom of South Africa. This contradicts the widely accepted view that *P. xylostella* evolved in the Mediterranean region with *Brassica* spp. of European origin (Tsunoda, 1980) and spread further with the cultivated brassicas around the world (Hardi, 1938; Harcourt, 1954). This hypothesis has serious implications for biological control because workers have focused on Europe as a source of natural enemies of diamondback moth for introductions into other continents (Hardi, 1938; Wilson, 1960). Kfir (1998) therefore suggested that more attention should be given to South Africa in this regard.

Impact of Parasitoids

In South Africa, Kfir (2003) investigated the impact of parasitoids on *P. xylostella* populations by using the insecticide check method (DeBach, 1946).

Table 23.1. Parasitoids recorded from *Plutella xylostella* in South Africa.

Parasitoid	Order, Family	Reference
Egg–larval parasitoids		
Chelonus curvimaculatus Cameron	Hymenoptera, Braconidae	Kfir, 1997b
Chelonus ritchiei Wilkinson	Hymenoptera, Braconidae	Ullyett, 1947
Chelonus sp.	Hymenoptera, Braconidae	Kfir, 1997b
Larval parasitoids		
Apanteles halfordi Ullyett (=*A. eriophyes* Nixon)	Hymenoptera, Braconidae	Kfir, 1997b
Apanteles ruficrus Haliday	Hymenoptera, Braconidae	Ullyett, 1947
Cadurcia plutellae van Emden	Diptera, Tachinidae	Ullyett, 1947
Cotesia plutellae (Kurdjumov)	Hymenoptera, Braconidae	Kfir, 1997b
Cotesia sp.	Hymenoptera, Braconidae	Kfir, 2002
Habrobracon brevicornis (Wesmael)	Hymenoptera, Braconidae	Kfir, 1997b
Microbracon hebetor Say	Hymenoptera, Braconidae	Ullyett, 1947
Peribaea sp.	Diptera, Tachinidae	Kfir, 1997b
Larval–pupal parasitoids		
Diadegma mollipla (Holmgren)	Hymenoptera, Ichneumonidae	Kfir, 2002
Itoplectis sp.	Hymenoptera, Ichneumonidae	Kfir, 1997b
Macromalon sp.	Hymenoptera, Ichneumonidae	Kfir, unpublished
Oomyzus sokolowskii (Kurdjumov)	Hymenoptera, Eulophidae	Kfir, 1997b
Pupal parasitoids		
Brachymeria sp.	Hymenoptera, Chalcididae	Kfir, 1997b
Diadromus collaris Gravenhorst	Hymenoptera, Ichneumonidae	Kfir, 1997b
Stomatoceras sp.	Hymenoptera, Chalcididae	Ullyett, 1947
Hokeria sp.	Hymenoptera, Chalcididae	Kfir, 1997b
Tetrastichus howardi (Olliff)	Hymenoptera, Eulophidae	Kfir, 1997b
Unidentified	Hymenoptera, Ichneumonidae	Kfir, 2002
Hyperparasitoids		
Aphanogmus fijiensis (Ferrière)	Hymenoptera, Ceraphronidae	Kfir 1997b
Brachymeria sp.	Hymenoptera, Chalcididae	Kfir, 1997b
Eurytoma sp.	Hymenoptera, Eurytomidae	Kfir, 1997b
Hemiteles sp.	Hymenoptera, Ichneumonidae	Ullyett 1947
Hokeria sp.	Hymenoptera, Chalcididae	Kfir, 1997b
Itoplectis sp.	Hymenoptera, Ichneumonidae	Ullyett, 1947
Perilampus sp. 1	Hymenoptera, Perilampidae	Ullyett, 1947
Perilampus sp. 2	Hymenoptera, Perilampidae	Ullyett, 1947
Mesochorus sp.	Hymenoptera, Ichneumonidae	Kfir, 1997b
Proconura sp.	Hymenoptera, Chalcididae	Kfir, 1997b
Pteromalus sp.	Hymenoptera, Pteromalidae	Kfir 1997b
Tetrastichus sp.	Hymenoptera, Eulophidae	Kfir, 1997b

This method offers a good experimental technique for evaluating the efficacy of natural enemies (Luck *et al.*, 1999) and can be conveniently applied to measure and determine the level of control achieved by parasitoids (Jones, 1982; Kenmore *et al.*, 1984; DeBach and Rosen, 1991). Previous studies indicated that the insecticide, dimethoate, an organophosphate compound with both systemic and contact action, suppressed predaceous insects, but caused no harm to Lepidoptera and was thus 'selective' (Ehler *et al.*, 1973; Eveleens *et al.*, 1973). It was also shown that dimethoate was less toxic to first instars of the large white butterfly, *Pieris brassica* L. (Lepidoptera, Pieridae), than seven other commonly used insecticides (Sinha *et al.*, 1990). The results in South Africa showed that population levels of diamondback moth and the resulting levels of pest infestations on plants treated with dimethoate were significantly higher than on the untreated plants. Conversely, the degree of parasitism of diamondback moth larvae and pupae throughout the season was significantly higher on untreated plots. In the plots sprayed with the insecticide, parasitism fluctuated around 5–10% throughout the season, whereas in the untreated plots parasitism levels increased to >90% towards the end of the season (Kfir, 2003). This clearly demonstrates that parasitoids are playing an important role in controlling *P. xylostella* populations and stresses the importance of conserving parasitoids in crucifer fields.

Entomopathogens

Ullyett (1947) recorded the fungus *Zoophthora radicans* Brefeld (Zygomycetes, Entomophthorales), recorded as *Entomophthora sphaerosperma* Fres., and two unidentified bacterial diseases associated with the diamondback moth in South Africa. Outbreaks of these entomopathogenic fungi almost invariably follow periods of prolonged soft rains. Evidently, these diseases depend for their appearance in epidemic form upon the presence of suitable weather conditions as well as high host density.

A *P. xylostella* granulovirus (*Plxy*GV) endemic to Kenya was first isolated from *P. xylostella* in 1997. Fourteen genetically different isolates were collected and evaluated in the laboratory (Parnell *et al.*, 2002). Field trials on kale and cabbage in Kenya showed that *Plxy*GV could control diamondback moth better than lambda-cyhalothrin, the insecticide most widely used for *P. xylostella* control in Kenya, (M.A. Parnell, 1999, Greenwich, UK, unpublished results; Grzywacz *et al.*, 2003). Work on mass production of the *Plxy*GV has improved the productivity fourfold by using artificial diet-based rearing (Grzywacz *et al.*, 2003). A plan to produce *Plxy*GV is now being established in Kenya with a commercial partner. This will be the first viral insecticide to go into commercial production in sub-Saharan Africa. The product will be promoted specifically to small-scale farmers to provide a safe, effective alternative to the increasingly expensive and unreliable synthetic pesticides.

Biological Control of *Plutella xylostella*

Inoculative biological control of diamondback moth has been attempted only on a few occasions and mainly with parasitoids. The earliest attempt took place in New Zealand where only 7% parasitism was achieved by *Diadegma* sp., the only indigenous parasitoid recorded (Muggeridge, 1930). Introductions of *Diadegma semiclausum* Hellin and *D. collaris* from England were made and effective suppression of diamondback moth was achieved (Lim, 1986). In Australia, *P. xylostella* was a serious pest until several parasitoid species were introduced (Wilson, 1960). *D. semiclausum* and *D. collaris* were introduced from England and The Netherlands via New Zealand, and *C. plutellae* from Italy. The parasitoids became established and caused heavy parasitism with marked reduction in the abundance of *P. xylostella* (Wilson, 1960; Goodwin, 1979; Waterhouse and Sands, 2001). *D. semiclausum* has been successfully introduced from New Zealand to Indonesia (Vos, 1953) and then to Malaysia (Lim, 1986), Taiwan (Talekar, 1996), the Philippines (Amed *et al.*, 1994) and other countries in South-East Asia. However, good control has been achieved only in cooler high-elevation regions, but not in warmer lowland areas (Talekar, 1996).

Elsewhere, parasitoid introductions have not resulted in control success. *C. plutellae* from India was released on many Caribbean islands, but full biological control was never achieved (Bennett and Yaseen, 1972; Lim, 1986). *C. plutellae* and *O. sokolowskii* were introduced into Trinidad and became well established, but adequate control was not achieved (Yaseen, 1978). Although over 124,000 *C. plutellae* were released in cabbage fields in Florida over a 2-year period, the parasitoid was not yet found to be established (Mitchell *et al.*, 1997). In 1999, a selected population of *Diadegma insulare* (Cresson), which is native to North, Central and northern parts of South America, and one of the most important parasitoids of diamondback moth in these regions, was introduced from Florida to the Philippines and Vietnam for release in lowland areas (M. Whitten, FAO, Manila, 1999, personal communication). Unfortunately, results from this project are still unknown.

Biological Control of Diamondback Moth in the African Region

Three attempts at biological control of diamondback moth have taken place on the African continent and two on adjacent islands. The first attempt took place in 1936 when *D. semiclausum* (recorded as *Angita cerophaga* Gravenhorst) was introduced into South Africa from England (Evans, 1939). About 950 cocoons were supplied and a total of 326 parasitoids were released near Pretoria. In 1937 wasps were recovered and *D. semiclausum* was thus considered to have established (Greathead, 1971). However, since then the species was never again recovered and it is doubtful if *D. semiclausum* is really established in South Africa.

C. plutellae and *D. collaris* were introduced into Zambia. Both species established, and together with the indigenous *O. sokolowskii* contributed to 80% reduction in damage by diamondback moth (Yaseen, 1978). Unfortunately the source of the introduced parasitoids is not specified.

Until 1981 diamondback moth was considered the most important pest of cabbage in Cape Verde Islands. In 1981 and 1982, three braconids, *C. plutellae, O. sokolowskii* and *Microplitis plutellae* Muesebeck (Hymenoptera, Braconidae), were imported from Pakistan and released. Of these *C. plutellae* and *O. sokolowskii* established and subsequently led to a full control of the pest (Cock, 1983). In a survey of cabbage plots on Cape Verde in 1992, over 20 plots had negligible infestations of diamondback moth, whereas only one plot had severe infestation that was triggered by heavy use of insecticides in adjacent other crops. *O. sokolowskii* was the only parasitoid recovered in this survey (Karl, 1992).

In a cooperative project between IIBC, the Gesellschaft für Technische Zusammenarbeit (GTZ) and the Phytosanitary Services of Vegetables (SPV) of Togo, *C. plutellae* and *D. semiclausum* obtained from Taiwan were released in 1991 in heavily infested plots in Lome, Togo (Karl, 1992). By the end of 1991, the project was suspended for political reasons and unfortunately no further information is available.

On St Helena Island, where *P. xylostella* is also a serious pest, farmers were heavily dependent on chemical control. Furthermore, where recommended doses failed to produce the required control, they often overdosed or mixed cocktails containing several different insecticides. The only parasitoid of *P. xylostella* found on St Helena was *D. mollipla,* which on its own was unable to reduce pest populations below the economic damage levels (Kfir and Thomas, 2001). Since there is no airport on St Helena, during 1999 two consignments of *C. plutellae* and *D. collaris* were dispatched via boat. The assumption made was that if these larval and pupal parasitoids, respectively, were to be successful they should not compete with the resident *D. mollipla*, which is a larval–pupal parasitoid. Before release of the parasitoids farmers were also encouraged to switch to more selective *Bt* sprays, so as to give the parasitoids the best possible chance of survival. A total of 17,500 *C. plutellae* and 23,500 *D. collaris* were then released at ten different farm sites across the island (Kfir and Thomas, 2001). An initial follow-up survey of 19 farms, conducted in 2000, found that both parasitoids were well established even on farms where none had been released. Up to 80% diamondback moth larval parasitism by *C. plutellae* and 55% pupal parasitism by *D. collaris* were recorded. A survey conducted in 2001 revealed extremely low levels of diamondback moth infestations. Moreover, cocoons of *C. plutellae* were reportedly found to be present on most farms, which indicated that parasitoids had been responsible for the decline in *P. xylostella* populations. On St Helena diamondback moth outbreaks usually occur in spring (September–October). However, in 2001 farmers reported that infestations remained so low that no chemical control was needed, which augurs well for the success of biological control of *P. xylostella* on St Helena (Kfir and Thomas, 2001).

In 2001, *D. semiclausum* was also introduced into Kenya from Taiwan and field releases started in 2002. It is not yet known if the parasitoid has become established but additional releases are planned in Kenya, Tanzania, Uganda and Ethiopia (B. Löhr, ICIPE, Nairobi, 2002, personal communication).

IPM of *Plutella xylostella*

In addition to biological control, several other alternative control methods against diamondback moth have been researched in Africa, such as pheromone trapping, trap crops and botanical pesticides.

Five years of continuous monitoring of *P. xylostella* larvae and pupae in crucifer crops, and of moth catches by means of synthetic sex pheromone trapping indicated that moths are active throughout the year in South Africa and that peak flights coincide with peaks of larval infestations on the crops (Kfir, 1997a, 2002; M.S. Mosiane, 2001, Pretoria, South Africa, unpublished results). These results open up the possibility of using pheromone traps for monitoring flight peaks and for determining economic threshold levels for timing of spray applications against diamondback moth. The mating disruption technique with pheromones has also been tried in cabbage fields in South Africa, but without any conclusive results (G. Booysen, ISSA, Tzaneen, South Africa, 2002, personal communication).

The use of trap crops has special significance for subsistence farmers in developing countries, offering the possibility to reduce reliance on pesticides and production costs (Hokkanen, 1991). In most parts of Africa where subsistence farming is practised, the development of trap crops to lower diamondback moth infestations could prove beneficial. Laboratory and field experiments in South Africa indicated that female moths favour Indian mustard (*Brassica juncea* (L.) Csern. (*Brassicaceae*)) for oviposition over cabbage, cauliflower, broccoli or Chinese cabbage. Larval survival has also been shown to be lower on Indian mustard than on the other crops, indicating its potential use as a trap crop for *P. xylostella* control (Charleston and Kfir, 2000).

Several species from the mahogany (*Meliaceae*) family are known for their toxicity to insects. Trees from this family such as the neem tree (*Azadirachta indica* A. Juss.) and the syringa tree (*Melia azedarach* L.) were introduced into Africa and have become acclimatized and common in many countries. Botanical extracts from syringa and neem trees have been tested in South Africa. The extracts significantly affected both adults and larvae of diamondback moth and were reported to have no direct negative impact on the parasitoids *C. plutellae* and *D. collaris* (Charleston *et al.*, 2003). In Kenya, applications of neem prevented the development of diamondback moth for 1 week (S.O. Okoth, 1998, Ruiru, Kenya, unpublished results) and water extracts and oil-based neem formulations proved to be harmless to *D. mollipla* and *C. plutellae* (Akol, 2002). In field trials in Bénin, neem extracts not only controlled *P. xylostella* populations better than the synthetic pyrethroid deltamethrin, but also caused no detrimental effects to *C. plutellae* (Goudegnon *et al.*, 2000). These results indicate that an IPM approach combining biological control with the use of botanical insecticides has potential for diamondback moth control in Africa.

Conclusion

There is no doubt that the high pest status of the diamondback moth is manmade. The severity of its infestations in Africa and in other parts of the world has been directly related to the excessive use of insecticides. Before the mid-1960s, very few insecticides were used in Taiwan and *P. xylostella* was not an economic problem (Talekar, 1996). Similarly, diamondback moth was not considered to be a severe pest in South Africa until the 1990s, when populations started to show resistance to insecticides. Its pest status increased gradually as the number of insecticides used for its control increased and resistance developed rapidly. The increase in area under crucifer production, and the development of heat-tolerant cultivars that allowed cultivation of cabbage all year round in Africa, are additional reasons for the exacerbation of the diamondback moth problem. A no-crucifer period is important for keeping diamondback moth populations low and for resistance management. One of the reasons for the low pest status of *P. xylostella* in Europe is that crucifer production is limited to spring and summer. Introduction of new synthetic insecticides will only provide a temporary relief but in the long run it will lead to further exacerbation of the problem with additional environmental hazards. As evident in many countries, the 'pesticide treadmill' that has been the control strategy against the diamondback moth is not sustainable. Therefore, development and implementation of environmentally safe and sustainable control measures such as biological control or biological control-based IPM is urgently needed.

It is obvious that in most past biological control projects against *P. xylostella* climate matching between the parasitoids' source area and target release sites was not taken into consideration. It is understandable why the introductions of *D. semiclausum* and *D. collaris* from Europe was successful in temperate New Zealand, but in the tropics only under cooler highland conditions, and why the parasitoids failed to give any control in hotter lowland areas. In fact, sometimes the parasitoids can be found in lowlands in Taiwan, but only during cool dry periods and with negligible effects (Talekar, 1996).

South Africa is a more logical source than Europe for parasitoids against *P. xylostella* in other tropical regions of the world. Studies at different climatic regions of the country indicated a rich source of efficient, indigenous parasitoids that control the pest by normally achieving >90% parasitism. It is suggested that other African countries where diamondback moth is a severe pest and the fauna of the pest's natural enemies is poor consider introductions of parasitoids from South Africa.

References

Akol, A. (2002) Effects of two neem insecticide formulations on the trophic interactions between the diamondback moth, *Plutella xylostella* (L.) (Lepidoptera: Plutellidae) and its parasitoids. PhD thesis, Kenyatta University, Nairobi, Kenya.

Amed, J., Mangali, T. and Emaguin, D. (1994) Establishment of *Diadegma semiclausum* (Hellen) (Hymenoptera: Ichneumonidae) in the Philippine Cordillera: status in 1992 and 1993. *Philippine Journal of Plant Industry* 59, 103–108.

Annecke, D.P. and Moran, V.C. (1982) *Insects and Mites of Cultivated Plants in South Africa*. Butterworths, Durban, South Africa.

Bennett, F.D. and Yaseen, M. (1972) Parasite introduction for the biological control of three insect pests in the Lesser Antilles and British Honduras. *Pest Articles and News Summaries* 18, 468–474.

Charleston, D.S. and Kfir, R. (2000) The possibility of using Indian mustard, *Brassica juncea*, as a trap crop for the diamondback moth, *Plutella xylostella*, in South Africa. *Crop Protection* 19, 455–460.

Charleston, D.S., Dicke, M., Vet, L.E.M. and Kfir, R. (2003) Integration of biological control and botanical pesticides – evaluation in a tritrophic context. In: Endersby, N.M. and Ridland, P.M. (eds) *Proceedings of the 4th International Workshop on the Management of Diamondback Moth and Other Crucifer Pests*. Department of Natural Resources and Environment, Melbourne, Australia (in press).

Cock, M.J.W. (1983) Introduction of parasites of *Plutella xylostella* (L.) into Cape Verde Islands. *Project Report*, Commonwealth Institute of Biological Control.

DeBach, P. (1946) An insecticidal check method for measuring the efficacy of entomophagous insects. *Journal of Economic Entomology* 39, 695–697.

DeBach, P. and Rosen, D. (1991) *Biological Control by Natural Enemies*. Cambridge University Press, New York, USA.

Dennill, G.B. and Pretorius, W.L. (1995) The status of diamondback moth, *Plutella xylostella* (Linnaeus) (Lepidoptera: Plutellidae), and its parasitoids on cabbages in South Africa. *African Entomology* 3, 65–71.

Ehler, L.E., Eveleens, K.G. and van den Bosch, R. (1973) An evaluation of some natural enemies of cabbage looper on cotton in California. *Environmental Entomology* 2, 1009–1015.

Evans, I.B.P. (1939) Solving the Union's pasture, crop and insect problems. Annual Report of the Division of Plant Industry. *Farming in South Africa* The Government Printer, Pretoria, South Africa 117.

Eveleens, K.G., van den Bosch, R. and Ehler, L.E. (1973) Secondary outbreak induction of beet armyworm by experimental insecticide application in cotton in California. *Environmental Entomology* 2, 497–503.

Goodwin, S. (1979) Changes in numbers in the parasitoid complex associated with the diamondback moth, *Plutella xylostella* (L.) (Lepidoptera), in Victoria. *Australian Journal of Zoology* 27, 981–989.

Goudegnon, E.A., Kirk, A.A., Schiffers, B. and Bordat, D. (2000) Comparative effects of deltamethrin and neem kernel solution treatments on diamondback moth and *Cotesia plutellae* (Hym., Braconidae) parasitoid populations in the Cotonou periurban area in Bénin. *Journal of Applied Entomology* 124, 141–144.

Greathead, D.J. (1971) A review of biological control in the Ethiopian Region. *Technical Communication* 5. Commonwealth Institute of Biological Control, Wallingford, UK.

Grzywacz, D., Parnell, D., Kibata, G., Oduor, G., Ogutu, W.O., Miano, D. and Winstanley, D. (2003) The development of endemic baculoviruses of *Plutella xylostella* (diamondback moth, DBM) for control of DBM in East Africa. In: Endersby, N.M. and Ridland, P.M. (eds) *Proceedings of the 4th International Workshop on the Management of Diamondback Moth and Other Crucifer Pests*. Department of Natural Resources and Environment, Melbourne, Australia (in press).

Harcourt, D.G. (1954) The biology and ecology of the diamondback moth, *Plutella maculipennis*, Curtis, in Eastern Ontario. PhD thesis, Cornell University, Ithaca, New York, USA.

Hardi, J. (1938) *Plutella maculipennis* Curt., its natural and biological control in England. *Bulletin of Entomological Research* 29, 343–372.

Hemy, C. (1984) *Growing Vegetables in South Africa*. Macmillan, Johannesburg, South Africa.

Hokkanen, H.M. (1991) Trap cropping in pest management. *Annual Review of Entomology* 36, 119–138.

Jones, D. (1982) Predators and parasites of temporary row crop pests: agents of irreplaceable mortality or scavengers acting prior to other mortality factors? *Entomophaga* 27, 245–265.

Jordaan, M. (1993) Brassicaceae. In: Arnold, T.H. and De Wet, D.C. (eds) Plants of Southern Africa: Names and Distribution. *Memoirs of the Botanical Survey of South Africa* 62, 313–322.

Karl, K. (1992) DBM biocontrol project in Togo, Bénin and Cape Verde. Newsletter 1991. IOBC Global Working Group on Biological Control of *Plutella*. Asian Vegetable Research and Development Center (AVRDC), Tainan, Taiwan, Republic of China.

Kenmore, P.E., Carino, F.D., Perez, C.A., Dyck, V.A. and Guitierez, A.P. (1984) Population regulation of the rice brown planthopper (*Nilaparvata lugens* Stål) within rice fields in the Philippines. *Journal of Plant Protection of Tropics* 1, 19–37.

Kfir, R. (1996) Diamondback moth: natural enemies in South Africa. *Plant Protection News. Bulletin of the Plant Protection Research Institute, Pretoria* 43, 20–21.

Kfir, R. (1997a) The diamondback moth with special reference to its parasitoids in South Africa. In: Sivapragasam, A., Loke, W.H., Hussan, A.K. and Lim, G.S. (eds) *Proceedings of the 3rd International Workshop on the Management of Diamondback Moth and Other Crucifer Pests*. MARDI, Kuala Lumpur, Malaysia, pp. 54–60.

Kfir, R. (1997b) Parasitoids of diamondback moth, *Plutella xylostella* (L.) (Lepidoptera: Yponomeutidae), in South Africa: an annotated list. *Entomophaga* 42, 517–523.

Kfir, R. (1998) Origin of diamondback moth (Lepidoptera: Plutellidae). *Annals of the Entomological Society of America* 91, 164–167.

Kfir, R. (2002) The diamondback moth, *Plutella xylostella* (L.) (Lepidoptera: Plutellidae), and its parasitoids in South Africa. In: Melika, G. and Thuroczy, Cs. (eds) *Parasitic Wasps: Evolution, Systematics and Biological Control*. Agroinform, Budapest, pp. 418–424.

Kfir, R. (2003) Effect of parasitoid elimination on populations of diamondback moth on cabbage. In: Endersby, N.M. and Ridland, P.M. (eds) *Proceedings of the 4th International Workshop on the Management of Diamondback Moth and Other Crucifer Pests*. Department of Natural Resources and Environment, Melbourne, Australia (in press).

Kfir, R. and Thomas, J. (2001) Biological control of diamondback moth in the island of St Helena. *Biocontrol News and Information* 22, 76N.

Kibata, G.N. (1997) The diamondback moth: a problem pest of brassica crops in Kenya. In: Sivapragasam, A., Loke, W.H., Hussan, A.K. and Lim, G.S. (eds) *The Management of Diamondback Moth and Other Crucifer Pests*. Proceedings of the Third International Workshop, Kuala Lumpur, Malaysia, pp. 47–53.

Le Pelley, R.H. (1959) *Agricultural Insects of East Africa*. East Africa High Commission, Nairobi, Kenya.

Lim, G.S. (1986) Biological control of diamondback moth. In: Talekar, N.S. and Griggs, T.D. (eds) *Diamondback Moth Management*. Proceedings of the 1st International Workshop. Asian Vegetable Research and Development Center, Shanhua, Taiwan, pp. 159–171.

Lim, G.S. (1992) Integrated pest management of diamondback moth: practical realities. In: Talekar, N.S. (ed.) *Management of Diamondback Moth and Other Crucifer Pests. Proceedings of the Second International Workshop*. Asian Vegetable Research and Development Center, Shanhua, Taiwan, pp. 565–576.

Liu, Y.B., Tabashnik, B.E. and Johnson, M.W. (1995) Larval age affects resistance to *Bacillus thuringiensis* in diamondback moth (Lepidoptera: Plutellidae). *Journal of Economic Entomology* 88, 788–792.

Luck, R.F., Shepard, B.M. and Kenmore, P.E. (1999) Evaluation of biological control with experimental methods. In: Bellows, T.S. and Fisher, T.W. (eds) *Handbook of Biological Control*. Academic Press, San Diego, USA, pp. 225–242.

Meyrick, E. (1928) A Revised Handbook of British Lepidoptera. Watkins and Doncaster, London, UK.

Mitchell, E.R., Tingle, F.C., Navasero-Ward, R.C. and Kehat, M. (1997) Diamondback moth (Lepidoptera: Plutellidae): parasitism by *Cotesia plutellae* (Hymenoptera: Braconidae) in cabbage. *Florida Entomologist* 80, 477–489.

Muggeridge, J. (1930) The diamondback moth. Its occurrence and control in New Zealand. *New Zealand Journal of Agriculture* 41, 253–264.

Nel, A., Krause, M., Ramautar, N. and van Zyl, K. (1999) *A Guide for the Control of Plant Pests*, 38th edn. National Department of Agriculture, Republic of South Africa.

Nisin, K.D., Mo, J. and Miyata, T. (2000) Decreased susceptibilities of four field populations of the diamondback moth, *Plutella xylostella* (L.) (Lepidoptera: Yponomeutidae), to acetamiprid. *Applied Entomology and Zoology* 35, 591–595.

Oduor, G.I., Löhr, B. and Seif, A.A. (1997) Seasonality of major cabbage pests and incidence of their natural enemies in central Kenya. In: Sivapragasam, A., Loke, W.H., Hussan, A.K. and Lim, G.S. (eds) *The Management of Diamondback Moth and Other Crucifer Pests. Proceedings of the Third International Workshop*. Kuala Lumpur, Malaysia, pp. 37–43.

Ooi, P.A.C. and Sudderuddin, K.I. (1978) Control of diamondback moth in Cameron Highlands, Malaysia. In: Amin, L.L. (ed.) *Proceedings of Plant Protection Conference*. Rubber Research Institute of Malaysia, Kuala Lumpur, Malaysia, pp. 214–227.

Parnell, M., Oduor, G., Ong'aro, J., Grzywacz, D., Jones, K.A. and Brown, M. (2002) The strain variation and virulence of granulovirus of diamondback moth (*Plutella xylostella*) isolated in Kenya. *Journal of Invertebrate Pathology* 79, 192–196.

Prinsloo, G.L. (2002) The identity of *Apanteles halfordi* Ullyett (Hymenoptera: Braconidae), a parasitoid of the diamondback moth, *Plutella xylostella* (linnaeus) (Lepidoptera: Plutellidae). *African Entomology* 10, 351–353.

Sereda, B., Basson, N.C.J. and Marais, P. (1997) Bioassay of insecticide resistance in *Plutella xylostella* (L.) in South Africa. *African Plant Protection* 3, 67–72.

Shelton, A.M., Sances, F.V., Hawley, J., Tang, J.D., Boune, M., Jungers, D., Collins, H.L. and Farias, J. (2000) Assessment of insecticide resistance after the outbreak of diamondback moth (Lepidoptera: Plutellidae) in California in 1997. *Journal of Economic Entomology* 93, 931–936.

Sinha, S.N., Lakhani, K.H. and Davis, B.N.K. (1990) Studies of the toxicity of insecticidal drift to the first instar larvae of the large white butterfly *Pieris brassicae* (Lepidoptera: Pieridae). *Annals of Applied Biology* 116, 27–41.

Smith, T.J. and Villet, M.H. (2003) Parasitoid wasps associated with the diamondback moth, *Plutella xylostella*, in the eastern Cape, South Africa. In: Endersby, N.M. and Ridland, P.M. (eds) *Proceedings of the 4th International Workshop on the Management of Diamondback Moth and Other Crucifer Pests*. Department of Natural Resources and Environment, Melbourne, Australia (in press).

Tabashnik, B.E. (1994) Evolution of resistance to *Bacillus thuringiensis*. *Annual Review of Entomology* 85, 1551–1559.

Tabashnik, B.E., Cushing, N.L., Finson, N. and Johnson, M.W. (1990) Field development of resistance to *Bacillus thuringiensis* in diamondback moth (Lepidoptera: Plutellidae). *Journal of Economic Entomology* 83, 1671–1676.

Talekar, N.S. (ed.) (1992) *Diamondback Moth and Other Crucifer Pests. Proceedings of the Second International Workshop*. Asian Vegetable Research and Development Center, Shanhua, Taiwan.

Talekar, N.S. (1996) Biological control of diamondback moth in Taiwan – a review. *Plant Protection Bulletin* 38, 167–189.

Talekar, N.S. and Shelton, A.M. (1993) Biology, ecology, and management of the diamondback moth. *Annual Review of Entomology* 38, 275–301.

Tsunoda, S. (1980) Eco-physiology of wild and cultivated forms in *Brassica* and allied genera. In: Tsunoda, S., Hinata, K. and Gomez-Campo, C. (eds) *Brassica Crops and Wild Allies: Biology and Breeding*. Japan Scientific Societies Press, Tokyo, Japan, pp. 109–120.

Ullyett, G.C. (1947) Mortality factors in populations of *Plutella maculipennis* Curtis (Tineidae: Lep.), and their relation to the problem of control. *Entomology Memoirs, Department of Agriculture and Forestry, Union of South Africa* 2, 77–202.

Vos, H.C.C.A.A. (1953) Introduction to Indonesia of *Angitia cerophaga* Grav., a parasite of *Plutella maculipennis* Curt. *Contributions, Central Agricultural Research Station* 134, 1–32.

Waladde, S.M., Leutle, M.F. and Villet, M.H. (2001) Parasitism of *Plutella xylostella* (Lepidoptera, Plutellidae): field and laboratory observations. *South African Journal of Plant and Soil* 18, 32–37.

Walker, A. and Fitton, M.G. (1992) Records of recent hymenopterous parasitoids attacking diamondback moth in Africa. *Newsletter 1992.* IOBC Global Working Group on Biological Control of *Plutella*, Asian Vegetable Research and Development Center (AVRDC), Tainan, Taiwan, Republic of China..

Waterhouse, D.F. and Sands, D.P.A. (2001) *Classical Biological Control of Arthropods in Australia.* ACIAR Monograph No. 77, Canberra ACT, Australia.

Wilson, F. (1960) A review of the biological control of insects and weeds in Australia and Australian New Guinea. *Technical Communication No. 1*. Commonwealth Institute of Biological Control, Ottawa, Canada.

Yaseen, M. (1978) The establishment of two parasites of the diamond-back moth *Plutella xylostella* (Lep.: Plutellidae) in Trinidad, W.I. *Entomophaga* 23, 111–114.

24

Biological Control for Increased Agricultural Productivity, Poverty Reduction and Environmental Protection in Africa

Peter Neuenschwander,[1] Jürgen Langewald,[1]
Christian Borgemeister[2] and Braima James[1]
[1] *International Institute of Tropical Agriculture, Cotonou, Bénin;*
[2] *Institute of Plant Diseases and Plant Protection, Hannover, Germany*

Introduction

The ecological basis for integrated pest management

In paradise, things were simple: harmony with nature was the order of the day and, up to now, some people think such harmony should be re-established. Let us look briefly at this concept, which supposes that agricultural development is possible without disturbing the ecosystem.

In evolution, humans made it eventually to become the top predator, being responsible for the extinction of quite a number of big animals, as they spread across the globe. Numbers and biomass for such a top predator are small, about ten times lower than its prey, and the numbers of our hunter-gatherer forbears were indeed extremely low. Of course, this harmony with nature was not that idyllic.

Through the agricultural revolution about 11,000 years ago, we set out on a path that eventually made humans also the top herbivore. This allowed vastly increased numbers of our species to survive and to settle every niche on earth (Diamond, 1999). In fact, today we observe, depending on the region, an almost total human domination of the earth's ecosystems (Vitousek *et al.*, 1997).

The resulting agriculture set us on a collision course with our competitors for food, mostly arthropods, microorganisms, nematodes and the like. The more we improved wild plant species through active, as well as inadvertent, selection and by planting them in monoculture, the more attractive they became for these competing organisms, which thereby became pests (the term is used here for insects and diseases, etc.). Out of a total of possibly several million insects and slightly lower numbers of nematodes, for instance, only about

350–500 have ever become serious pests. The other ones have life strategies that do not allow them to exploit these new abundant food sources or are kept under control by their natural enemies.

In intensified agricultural systems, mainly in the industrialized countries, over the last 50 years, a prevailing tactic to compete with these pests has been to destroy them with pesticides, sometimes on a calendar basis, which initially works beautifully, but hence invites more widespread use. This in turn destroys the natural enemies of organisms that were well controlled, leading to secondary pest outbreaks and the 'pesticide treadmill' (van den Bosch, 1978). Today whiteflies, leafminers, diamondback moth and tetranychid mites are considered among the main pests worldwide – and most do not deserve to be labelled pests; they have been made into pests.

At present, the high human population pressure asks for higher productivity on the existing cultivated land. Expansion to farm new areas is only possible if declared protected areas and nature reserves are being encroached upon, and/or if highly marginal areas are taken into cultivation, with even lower returns. In fact, low-yield agriculture can be considered as the biggest danger for the world's natural environment (Avery, 1997). Combating pests is therefore an absolute necessity and a return to subsistence farming in regions where it is now rare, or its promotion in regions where it is still prevalent and where pest problems might be less acute, is not an option.

Today, IPM is the approved method of choice; but in reality what is this IPM? In the early 1970s, realizing that pesticides had created huge environmental and health problems, IPM was devised primarily as a pesticide abatement strategy and a management tool for rational pesticide use. This was later expanded with the quest for a holistic approach (Huffaker *et al.*, 1976). In addition to improving host plant resistance through genetic improvement of crops, many interventions were developed to improve soil fertility and water retention and to preserve natural enemies. These measures allow for a productive agriculture that is compatible with preserving biodiversity, a prerequisite for sustainable agriculture (Musters *et al.*, 2000; McNeely and Scherr, 2001). Today, integration within the entire ecosystem and its human managers through what is known as integrated natural resource management is needed in order to achieve sustainable crop production (Lewis *et al.*, 1997; Verkerk *et al.*, 1998; Sayer and Campbell, 2001).

At present, worldwide, statistics show, however, an increase in pesticide use, which was implemented despite vocal assurances of practising IPM, paralleled by an increase in percentage pest damage (Oerke *et al.*, 1999; Zadoks and Waibel, 2000; Hazell and Wood, 2001). This scenario does not presage sustainability; evidently something is wrong!

The situation in Africa

Despite highly diverse agricultural systems (Raemaekers, 2001), Africa still has low agricultural productivity and a high and increasing percentage of poor and undernourished people. Demographic trends clearly show the need for increased agricultural productivity. In 1970 there were 3.7 ha of agricultural land per rural inhabitant; by 1998 the ratio had declined to 2.2. In West and Central Africa, the corresponding figures are even more alarming, namely a decline from 2.2 ha to 1.5 ha over the same period. In many places, growing population pressure has not led to a widespread intensification of agricultural production as in South-East Asia, but contributed to a decrease in productivity, partly due to soil degradation. Over the period from 1993 to 1998, a slow-down of this decline was observed, which is, however, mostly attributable to rural–urban migration and, in some areas, to the negative effects of the HIV/AIDS pandemic on the life expectancy of rural people (IITA, 2001).

Despite this gloomy picture, several crops, like cassava, maize and yams, have seen a two- to threefold increase in production in the past 20 years. This was achieved mainly through expansion of cropping area and to a lesser degree through intensification. Such production is, however, only sustainable if soil degradation and soil fertility are maintained, which necessitates a larger input of fertilizer (1997: Africa 18 kg ha^{-1} versus about 90 kg ha^{-1} worldwide). On farm, improvements are expected from the introduction of adapted crop varieties, their integration into sustainable production systems with higher labour productivity, and, finally, effective plant health management systems, the part of rural development addressed in this book. Plant health management is to provide sustainable solutions to pest problems, developed together with farmers to ensure that such solutions are adapted to local conditions, accepted and implemented by rural communities. Success is thereby measured not only in production terms, but includes sustainability, income generation, added value from postharvest conservation and transformation, human and animal health, all together then being expressed as human wellbeing (IITA, 2001; Keatinge *et al.*, 2001). All these criteria are directly affected by plant protection measures.

In Africa, pesticides are mainly used in cash crops, like cotton, cut flowers, and the peri-urban horticultural sector. Though insecticide use is often unsophisticated and abusive, the declared strategies do not differ much from those on other continents. There is, however, much less public awareness of the danger of pesticides than in other continents, and end-user protection is unknown and/or inadequate. Unfortunately, insecticide applications, often unjustified and on a calendar basis, are still the norm and insecticides are sometimes used improperly, for instance for meat conservation. Persistent cotton insecticides are used on vegetables, and preharvest intervals often are not respected. Where insecticides are openly available, they are relatively cheap, and sometimes provided free by donors, and rarely any cost:benefit calculations are applied (Schwab *et al.*, 1995). Apart from the health risk and environmental pollution, such misuse eventually leads to serious resistance problems, as for instance documented for whiteflies on cotton in Sudan (Dittrich *et al.*, 1985) or diamondback moth, *Plutella xylostella* (L.) worldwide (Chapter 23). Even when sales of persistent insecticides, such as hydrochlorines like dieldrin, that have been banned in Europe and the USA are stopped, the problem of removing and destroying obsolete stocks of such insecticides remains a large burden.

On most food crops and in most places, however, African subsistence farmers do not apply insecticides. Apart from the fact that pesticides are often too expensive and not available for various reasons, pesticide purchases are not high on the agenda. Confronted by different risks, the farmers' strategy is not to invest in reducing risk, but to spread risk. This is done by diversifying crop types and planting different cultivars on as large an area as labour and land access allow, in the hope of harvesting enough to survive, whatever disaster may befall. In this situation, the only widely applied pest control practice consists of cultural control measures. Those are mainly concerned with the reduction of carry-over of pests from one crop cycle to the next, through crop rotation, removal of plant residues or shortening of the growing season. In some cases this strategy can still be regarded as a 'best bet' solution, as in the case for *Helicoverpa* control (Chapter 21, see also Chapter 22). Hence, on food crops, by comparison with other continents, Africa has rather few induced pest problems, like whiteflies, leafminers or spider mites, for instance. Not all pests are insecticide induced, though, and crops, like cowpea and yams, which are indigenous to Africa, have a vast suite of pests that sometimes cause considerable losses.

Most food in Africa is, however, produced from exotic crops, with their origins in South America or South/South-East Asia. While polyphagous insects from Africa succeeded in accepting these plants as food sources and became pests, many important pest problems on these crops arise when organisms from the original regions of these crops are inadvertently introduced without their natural enemies. In Africa, as on other continents, the risk of such introductions is increasing with increasing trade, travel and transport.

In these different situations with crops and their pests of different origins, manipulating the environment so as to minimize problems evidently calls for differentiated approaches.

Strategy

Biological control-based IPM in all its aspects offers sustainable solutions as documented in this book. These techniques are, however, highly knowledge intensive, which is probably their key constraint to a widespread development of such a strategy. We admit that it is easier for the extension services and other farmer-support groups in Africa to propagate insecticide use than to develop and establish, sometimes ecozone-specific, 'biological' solutions, and it is certainly profitable for the chemical companies to promote insecticides and use them under the banner of IPM. However, experience on other continents demonstrates that most insecticide use does not lead to sustainable solutions. Organic farming has been demonstrated to be both economically and environmentally sustainable, by combining maintenance of soil fertility with relatively high yields achieved almost without insecticides (Mäder *et al.*, 2002 and citations therein). Here in Africa we start from a strong position, where the environment in many situations is still largely intact, with natural enemies and resistance mechanisms in farmers' cultivars still abundantly available. We certainly want to preserve this advantage, which has to be so painfully regained in Europe and the USA, while still working towards higher yields than those presently attained. This can be achieved by using organic and inorganic fertilizer (Lamers *et al.*, 1998; Vanlauwe *et al.*, 2002), low harvest index grain legumes or cover crops (Carsky *et al.*, 2001; Schulz *et al.*, 2001), measures that are totally compatible with environmentally friendly IPM and which do not promote subsistence agriculture.

When equilibrium levels achieved by biological control are not satisfactory because the natural enemies are either not sufficiently efficient or are somehow prevented from developing their full potential, then we arrive at a typical IPM situation where several interventions have to be combined. In this case, the so-called 'judicious use' of insecticides becomes an option. As pointed out by Stonehouse (1995), the use of action thresholds by farmers may be enhanced by information learned from outsiders, but this should complement, rather than supplant, existing knowledge and experience, otherwise action thresholds themselves become the excuse/reason for unsound pesticide regimes.

It has been argued that stimulating the farmer to become an experimenter, and a generator of IPM technologies in the framework of integrated crop management, is the only way to achieve a Green Revolution from 'within' (van Huis and Meerman, 1997), or what has been termed the 'doubly green revolution' (Conway, 2001). In this book, we therefore promote, to the highest degree possible, IPM strategies that take advantage of existing farmers' knowledge and of biological control agents present in most systems. Practical IPM options with mostly no, or very little, reliance on insecticides are being presented. In this holistic view, plant health management has to be applied from the beginning of a cropping cycle, often at pre-planting seasons (e.g. for improved fallows), during planting material selection, and before pest problems become critical. Taking into account tri-trophic interactions between pests and their antagonists, as determined by soil and water that influence the physiological conditions of the host plant, the first requirement therefore is a healthy crop that can withstand pest pressure. In fact, marginal growth conditions are sometimes responsible for the local failure of otherwise successful biological control (Neuenschwander *et al.*, 1990). Thus, good agronomy (i.e. good use of soil, water, fertilizer, labour) (Vanlauwe *et al.*, 2002) is not only the basis for crops to grow, but also for the success of biological control. Sometimes the responsible antagonists can be manipulated and increased in the environment by the presence of plants for shelter and food and as alternative hosts outside the fields, which are crucial for natural enemies (parasitoids, predators, and entomopathogenic nematodes, fungi and bacteria) (Landis *et al.*, 2000), but most often they cannot be managed directly. The other basis of sound IPM consists in fully exploiting and developing host plant resistance. Good choice of seed and planting material (resistant or tolerant varieties) are most important and are in almost all cases totally compatible with the use of natural enemies, i.e. biological control (Thomas and Waage, 1996).

The theoretical basis for biological control and its integration in the farming environment has been developed in the past 50 years and has reached a stage of maturity, where useful generic ecological and economic simulation models are available (May and Hassell, 1988; Gutierrez *et al.*, 1999; Thomas and Kunin, 1999). The present book therefore makes no effort at reviewing this vast literature, but takes these models for granted and ready for use where the situation so demands. The newest challenge in this field consists in linking these methods with data of Geographical Information Systems (GIS), thus achieving adapted spatial-temporal information for prediction (Brewster and Allen, 1997; Chapters 3, 4, 19 and 20).

New IPM technologies are conceived as options for integration into the existing system. It must be noted, however, that none of the options described (including the very important 'do nothing' option) work if the target farmers, and also traders, urban and rural consumers, politicians, communicators, etc., are not involved from the start and give their feedbacks, their contributions and their criticism. A thorough analysis of the process of adoption of innovations demonstrated the importance of this interactive process, with feedback from early adopters, who become the main driving force in further development. Successful technology thereby is the result of a synthesis of the researcher and key stakeholder knowledge sets, achieved through iterations of interactions (Douthwaite *et al.*, 2001a,b) and dependent on matching innovation with potential users (Waller *et al.*, 1998). In the case studies presented in this book, various farmers' participatory techniques have been employed, to ensure that promising contributions to technologies did not end up unused on the shelf (Chapters 5, 15, 17, 19 and 20).

In conclusion, this strategy is not calling for an idyllic agriculture that keeps poor farmers poor, but for an intelligent use of all resources for keeping pests at acceptable levels, now and in the future. This book presents the state of the art of the impact of biological control in African farming systems, including forestry, at the beginning of the third millennium. Biological control was taken in its broadest sense and will now be discussed separately according to different intervention modes.

Case studies

Classical biological control of arthropod pests of exotic origin

Classical biological control, i.e. the release of natural enemies from the place of origin of an exotic pest, has a long and impressive history around the world and also in Africa. Here we concentrate on the tropical areas of sub-Saharan Africa, thereby neglecting some biological control successes that were confined to the temperate or Mediterranean climate areas, particularly in the Republic of South Africa (Chapters 1 and 2).

The highly successful biological control project against cassava mealybug launched a new development at the IITA with repercussions across Africa (Chapter 3). Continent-wide in implementation, it was successful in all affected ecological zones throughout sub-Saharan Africa. This spawned similar projects that provided permanent, sustainable, stable solutions to the respective pests at no costs but vast benefits to the farmers and without affecting non-target organisms. These projects also led to extensive training, at all levels, of national collaborators, who later became available for new projects, and created the needed public awareness of the benefits of biological control.

A more recent success story concerns the biological control of cassava green mite (Chapter 4) by introduced predatory mites, supported by an exotic entomophtoralian fungus, both originating from climatically matched areas in South America. The two control strategies are unique. This is the first example of classical biological control of phytophagous mites by phytoseiid predators executed on a large scale. In the past, such biological control projects were mostly restricted to the glasshouses environment by applying inundative strategies. Apart from recent successes with naturally occurring epizootics of the imported entomophtoralian fungus *Entomophaga maimaiga* Humber, Shimazu, Soper and Hajek against gypsy moth (Hajek, 1999), the establishment of an exotic entomophagous fungus has never been successful before.

In one case classical biological control was also successful in lepidopteran stem borers (Chapter 9). *Chilo partellus* (Swinhoe), which attacks maize and sorghum, was accidentally introduced from Asia to East Africa, and recently the larval parasitoid *Cotesia flavipes* Cameron was collected in Pakistan and India and brought to Kenya. Data from impact assessment studies in Kenya show that *C. partellus* populations have declined by more than 50% following the releases and establishment of the parasitoid.

Another stunning example for the successful use of classical biological control is the story of the control campaigns against potato tuber moth, *Phthorimaea operculella* (Zeller), in southern Africa (Chapter 5). Here two exotic parasitoids were introduced from South America, the area of origin of *P. operculella*, to Africa. Due to their differing requirements concerning relative humidity, the two parasitoids complement each other, providing a seasonal succession. This illustrates how important basic knowledge on the biology and ecology of natural enemies can be for successful introductions.

Other recent cases where classical biological control provided the main solution to a pest problem in tropical Africa concern two exotic whiteflies (Chapter 6). Since the early 20th century, biological control of many exotic pests of introduced forest trees has become a successful routine; the latest success being the control of the black pine aphid (Chapter 7) and an exotic woodwasp (Chapter 8). The history of classical biological control of conifer aphids in Africa (Chapter 7) illustrates the paramount importance of taxonomy for the success of such an approach. The cypress aphid, *Cinara cupressivora* Watson and Voegtlin, was initially described as *C. cupressi* (Buckton), and only after extensive taxonomic research were its correct identity and probable area of origin determined. Consequently, in the beginning the search for biological control agents extended over many regions, reducing the likelihood that an effective agent would be found. By contrast, the black pine aphid, *Cinara cronartii* Tissot and Pepper, was correctly identified immediately after its accidental introduction into South Africa, and a highly efficient and host-specific parasitoid was imported from the USA, the area of origin of *C. cronartii*, and subsequently released in South Africa. Impact assessment studies later revealed that the pest had been brought under control by the parasitoid. Of the three presented case studies of classical biological control against conifer aphids, one was a full success (*C. cronartii*), while the two others were at best partial successes. Assessing the impact of partial successes in biological control is particularly difficult, and very often funds do not suffice and/or donors are not willing to support such 'academic' exercises. However, understanding partial successes and also failures will largely help us to understand how classical biological control works, and improve and facilitate the planning and design of future projects.

A series of highly successful examples of biological control of alien weeds, often confined to South Africa and neighbouring countries, is described in Chapter 2. Many of the targeted plants, like forestry trees or some cacti, had been introduced and became weeds when they spread, displacing unique natural vegetation or encroaching on farm land. This ensuing conflict of interest situation requires careful integration of management practices and choice of biological control agents.

We generally state that biological control restores the population balance. This statement neglects the fact that quarantine has removed upper trophic levels, like hyperparasitoids, which gives the introduced natural enemy a headstart in its new environment, at least until indigenous hyperparasitoids have caught up (Bellows, 2001; Chapter 3). Furthermore, plant attributes, for instance pubescence, have been recognized as important factors influencing the outcome of biological control across trophic levels in many systems (Cortesero *et al.*, 2000), but they have received only scant attention in the reported projects from Africa (Chapters 4, 9 and 21).

New associations, whereby natural enemies of insects related to the invading pest have become successful (Hokkanen and Pimentel, 1989), are described in Chapter 9. Biological control by redistribution of natural enemies within Africa has been attempted against some native and exotic lepidopteran stem borers, but so far only with limited success. For instance, a number of new association parasitoids were introduced into South Africa for control of the sugarcane stem borer *Eldana saccharina* (Walker), but in all but one case the parasitoids did not establish. The maize cob borer *Mussidia nigrivenella* Ragonot occurs in various parts of Africa, but its pest status is limited to West and Central Africa. It is thought that *M. nigrivenella* is under biological control in East Africa and several efficient East African parasitoids have so far been identified. Releases of some of these parasitoids will soon commence in West and Central Africa.

In the case of the exotic leaf-mining fly *Liriomyza trifolii* Burgess, indigenous parasitoids of African *Liriomyza* species achieved total control in Senegal (Neuenschwander *et al.*, 1987), and the same impact is probably achieved wherever *L. trifolii* is spreading. This effect is based on the fact that these leafminer parasitoids gain their clues for oviposition not so much from the host larva itself but from frass and host plant emanations. No such new associations were observed in projects against exposed homopteran pests, for instance, where the parasitoid females usually check their host meticulously for suitability. There exist probably many other cases where indigenous natural enemies brought exotic invaders under control, as is for instance seen by the fact that many introduced plants do not become weeds; but the present book does not bring forth any concrete examples.

The technique of deliberately looking for new associations between natural enemies and hosts for inoculative biological control has recently been heavily criticized because of its potential environmental risks. In the USA, planned releases of a scelionid wasp and a fungus, both from Australia, for the control of grasshoppers were eventually forbidden because of a perceived risk to indigenous grasshopper species (Lockwood, 1993). Some biological control agents against the exotic weed *Chromolaena odorata* (L.) King and Robinson might qualify as new associations (depending on the revision of the taxonomic status of this weed species), but despite clearance and release, they could not establish (Chapter 10). In stem borer control, new associations were studied and some released parasitoids indeed transferred to the new hosts (Chapter 9). Again, host plant and frass cues are responsible for the female wasps laying their eggs in the stem borer tunnels.

In recent years, classical biological control has come under increasing scrutiny for its non-target effects. Several reviews have addressed this sometimes politically potent topic and several international symposia arrived at the general conclusion that biological control practitioners should follow the rules established by weed scientists (Wajnberg *et al.*, 2001). In the context of tropical Africa, where many insect faunas are simply not yet well known, the claim to test for potential effects on (all?) indigenous non-target species is more easily said than done. Lynch and Thomas (2000) put some of these overblown claims into perspective and conclude from a large database that significant community level effects must be suspected mainly from polyphagous predators.

With the exception of Chapters 2, 10, 11 and 14, the influence of biological control on non-target species in the environment was not a prime concern. In fact, no particular risk assessment on indigenous species, beyond useful insects, like honeybees, silk worms, and other insect parasitoids, has been made in the numerous introductions done, for instance, by IITA. Nevertheless, after all documented introductions of successful natural enemies, no negative impact on indigenous organisms was detected in any of the projects (Neuenschwander and Markham, 2001). The authors concluded that almost all introductions would have passed also under the strict observance of the FAO Code of Conduct (FAO, 1996), which came into active effect only after the inception of many of the reported case studies in this book. Today, the introduction of the histerid *Teretrius* (*Teretriosoma*) *nigrescens* Lewis, predator of the larger grain borer, *Prostephanus truncates* (Horn), would probably pose more problems than was the case at the time (Chapter 20). In Africa, it seems that only one case of unwanted transfer to an unplanned host is documented, namely when a lace bug introduced from Mexico into East Africa to control lantana weed switched to attack the sesame crop (Greathead, 1968).

In the more recent biological control projects, non-target effects were tested for, at high costs. A coccinellid of the genus *Nephaspis* was tested in quarantine by CAB *International* beyond what had been done before (Chapter 6). Similarly, for the biological control of weeds, indigenous plants were also screened (Chapter 2). There is no denying that the concerns over non-target effects of biological control have had an enormous impact on the practice of biological control (Messing, 2000; Chapters 2 and 7). However, such concerns, if handled wrongly, can actually stop all classical biological control efforts in their tracks, particularly in an environment of scarcity of information where saying no to a request for importation of natural enemies is the easiest way out for a harassed bureaucrat in a quarantine office. Moreover, in situations where introductions of exotic pests threaten the food security of vast regions, applying extremely tight importation guidelines for natural enemies under the flag of 'protecting the environment against biological pollution' can be a political cover and may not take into account the aspirations of all stakeholders.

All this work would not have been possible without the necessary taxonomic support given by the Natural History Museum in London, the Musée d'Histoire Naturelle of Paris, and many others. Throughout many of the studies reported in this book, taxonomic assistance was provided by the IITA 'biodiversity center' (museum) in Cotonou, Bénin, from where the regional network of insect museums is being coordinated, under the umbrella of BioNET International, a worldwide network of taxonomists and museums. Many stories in this book describe situations where taxonomy of the pest or its natural enemy was of prime importance and where sometimes much time and resources were lost/wasted because of insufficient taxonomic knowledge (Chapters 3, 4, 6, 7, 9, 10, 12, 14 and 22).

Most identifications are still done exclusively on the basis of morphological characters. Increasingly, these are being supplemented with results from techniques that use the PCR to amplify minute amounts of DNA. Today, such techniques can be applied even to the smallest insects (Townson *et al.*, 1999; Manzari *et al.*, 2002; Prinsloo *et al.*, 2002). Eventually, this technique will be using solid-state microarrays, which might also become available for field identifications (Jaccoud *et al.*, 2001). Several chapters describe situations where such opportunities would allow for a quantum improvement in monitoring (Chapters 4, 6, 9 and 17).

Biological control of weeds has a long and successful history, particularly in South Africa (Chapter 2). Recently, the successful biological control of water hyacinth in some southern and western African countries, and particularly in Lake Victoria, has received public acclaim (Chapter 11). The success was built on the network of African, Australian, European and American collaborators to introduce and rear specific weevils against this floating waterweed and to help the national programmes to adapt rearing and monitoring technologies to their conditions. The same approach towards other floating waterweeds has led to similar, though less spectacular, impact. Another weed where control, though only modest, seems to be sufficient to improve the situation is *C. odorata*. As with so many others, this species was introduced as a cover crop, but has become troublesome, despite some attributes that make it a useful species in some farming situations (Chapter 10). This is also an example where small differences in provenance seem to have a great impact on the success of the introduced agents.

Because of limitations in funds, more than out of theoretical considerations (Denoth *et al.*, 2002), projects often stopped after the first good natural enemy had been found. Only a few, particularly older, projects progressed and tried to introduce indiscriminately a large array of natural enemies (Chapters 1, 7 and 8). In recent projects, whenever several exotic species were introduced, they were usually targeted for specific, still empty ecological niches, like the absence of egg parasitoids or the lack of root feeders, for instance. The particularly well-endowed project that led to the complete control of cassava mealybug (Chapter 3) had the financial capacity to investigate the scientific basis of biological control beyond the bare necessities. This paid off handsomely, as described above, in laying the scientific and manpower (and women power!) basis for future projects. The present book's chapters convincingly demonstrate that investment in research for development for biological control leads to high returns. Several projects had to rely on extended foreign exploration before the winning natural enemy had been found (Chapters 3, 4, 7, 8, 10, 11, 14, 19 and others). We would therefore like to argue that any added funding should preferentially be used to step up foreign explorations.

The debate about non-target effects pales, however, in comparison with the whole of humankind's disruptive activities on ecosystem health (Headrick and Goeden, 2001). These authors give a succinct overview of the effect of biological control on entire ecosystems and on how biological control can be used to nudge ecosystems into a desired historical state, for instance. In fact, most projects mentioned in this book concern disturbed, and sometimes highly disturbed, systems, and only a few examples of biological control successes in re-establishing a pristine environment, mostly on remote islands, are mentioned (Chapter 1). In the present chapters, the desired state was usually one of higher sustained agricultural productivity for humans. In this context, biological control must be judged by a comparison with the alternatives, which most often are either doing nothing or spraying pesticides.

To move the discussion about unplanned environmental effects of biological control beyond the actually observed confrontation, one way out consists in the use of quantitative food webs (Memmott, 2000; Schönrogge and Crawley, 2000). In the case studies presented in this book, using only semi-quantitative information, we can already discern a definite gradient from the cassava mealybug food-web, with its one-to-one interactions between a specific exotic parasitoid attacking a practically monophagous exotic herbivore on an exotic host plant (Chapter 3), to the cassava green mite, with several exotic predators that also interact with other mite species and derive sustenance from various other sources, including other plants (Chapter 4), to the maize and sugarcane stem borer systems, with numerous exotic and indigenous parasitoids attacking a variety of stem borers on indigenous grasses and crops as well as on non-gramineous wild plants (Chapter 9). Clearly the potential non-target effects are most likely in the last example, which does, however, not mean that they are necessarily detrimental to the ecosystem.

Biological control projects, despite their potentially huge returns, are initially comparatively expensive. Prevention, at least through checkpoints at international airports for transcontinental flights, of new introductions is overall more effective and cheaper, though less glamorous, than having to develop a biological control project with uncertain success. Governments and institutes therefore created quarantine services that should assure that all incoming and outgoing plant material is free of pests. Such quarantine procedures, which are often under-funded and under-appreciated, prevent introductions of foreign pests into new areas. As most borders of African countries do not follow ecological barriers and often do not have effective quarantine checks, with the possible exception at airports, imposing quarantine regulations is difficult. The authorities in the various countries are assisted by the Inter-African Phytosanitary Council (IAPSC), whose task it is to harmonize the application of quarantine regulations and to educate quarantine officers and the public.

Africa, in line with the origins of its food crops, has received many more insects from other continents, particularly from South America, than it has provided. Relatively few pests in other countries seem to be of African origin and therefore open to biological control by means of natural enemies from Africa. The whitefly *Bemisia afer* (Priesner & Hosny) recently reached outbreak levels in South America (P. Anderson, CIAT, Cali, 2002, personal communication), thus opening up the possibility for a classical biological control project with parasitoids from Africa. Parasitoids against the tephritids *Ceratitis capitata* Wiedemann (Chapter 18) and *Bactrocera* (*Dacus*) *oleae* (Gmelin) (Neuenschwander, 1982) have been searched for in Africa since the early 20th century, and Chapters 1 and 2 give further examples.

Manipulating indigenous natural enemies

In several projects, indigenous natural enemies were shown to be important, though often not sufficient to bring about acceptable population levels of certain pests (Chapters 6, 9, 19 and 21). Can the performance of these natural enemies be influenced by manipulating the cropping system, and would such measures increase yield or yield stability for the farmer? None of the chapters of this book has addressed this question and a review by Lenné and Wood (1999) reveals that vegetational diversity can be quite a mixed blessing for pest management, though varietal mixtures are definitively a viable strategy for sustainable productivity in subsistence agriculture (Smithson and Lenné, 1996; Zhu *et al.*, 2000). The performance of natural enemies might also depend on alternative hosts, food and shelter outside the agricultural fields. From Africa, little is known about the effect of landscape structure on biological control, as demonstrated from ecologically much simpler environments in the north (Thies and Tscharntke, 1999). Furthermore, how possible improvements in landscape structure or natural environments affect agrobiodiversity and how saving biodiversity should be economically justified by society (Moran *et al.*, 1997) is even more uncertain, and beyond the scope of this book. While preserving biodiversity is one of the most important issues facing us today, we can only repeat that IPM and other interventions should be judged in their usefulness not against presumed natural conditions, but against the status quo, taking into account the available alternatives.

Competitive exclusion

Several institutions are investigating the use of fungi to occupy a particular niche in the crop plant to prevent access by pathogens or even to consume invading pathogens. These indigenous endophytic fungi, which live between the root cells of banana, for instance, are being applied experimentally in tissue culture and in the field to prevent damage by nematodes and banana weevil (Chapters 12 and 16). It is conceivable that in vegetatively grown crops, where endophytes are capable of colonizing the entire root system (or in bananas the corm), a permanent change of the planting material, for instance a permanent antagonism against a particular pathogen, could be achieved. To the farmers, plants containing this particular endophyte would be indistinguishable from new resistant varieties.

In maize stores, some *Aspergillus* species produce the carcinogenic and toxic compound aflatoxin, for which the industrialized world is declaring a very low tolerance. Yet, in West Africa, 99% of all children were found to have sometimes extremely high aflatoxin levels in their blood and the resulting growth faltering and high liver cancer rates were recently documented (Gong *et al.*, 2002). Good store keeping and store/food hygiene are being investigated and the results promulgated through awareness campaigns. Based on previous successes in US cotton (Cotty, 1994), a biological control approach is actively pursued in West Africa by experimentally replacing aflatoxin-producing *Aspergillus* strains in the soil with atoxigenic ones that possess a competitive advantage in certain ecozones. Results thus far demonstrate the tremendous potential of biological control of *Aspergillus* spp. in maize in West Africa (Chapter 13). The technology also offers opportunities for stored products like groundnuts (Dorner and Cole, 2002).

Host plant resistance as a main component in an IPM programme

One of the oldest strategies deployed in IPM is host plant resistance. Co-evolution between plants and pests/diseases led to a large variety of host defence strategies. The precursors to our crop plants and the varieties developed from them over thousands of years exhibit a wide range of characters that stop organisms from attacking them. This reservoir of host plant resistance mechanisms of various types should form the basis for modern crop plants, particularly those that are to be deployed in subsistence agriculture. Modern plant breeding has developed new strategies to assure that this 'hardiness' of farmers' varieties does not get lost (Ortiz, 2002). Despite this, breeding efforts, in the name of satisfying consumer preferences, sometimes embark on removing plant compounds, like cyanide in cassava, that are a deterrent at least to non-specialized herbivores (Chapter 3). Host plant resistance plays an important, sometimes crucial, role in plant protection in several systems (Chapters 4, 6, 9, 16, 17, 19–23) and Sharma, H.C. *et al.* (2002) give further examples.

A special case of host plant resistance concerns genetically modified organisms (GMOs). This expression is somewhat misleading since traditional breeding also modifies the genome of crops. The new avenue of genetic engineering is now becoming available also for small-scale farmers, for instance in *Golden*Rice (Kryder *et al.*, 2000). Safety guidelines for implementing the deployment of GMOs are being developed widely (McLean *et al.*, 2002;

ISNAR, 2002; Sharma, K.K. *et al.*, 2002), though in most countries in Africa legislation permitting deployment of GMOs is not yet in place. The recent promulgation of relevant legislation in Nigeria is a promising sign and will allow the eventual testing of GMOs. Taking the lead from the acceptance of synthetic insecticides after World War II, Zadoks and Waibel (2000) warn, however, of an overenthusiastic use of GMOs. At present, the following non-target effects are the topics of newly initiated projects: Testing of the effects of (sterilized) *Bacillus thuringiensis* (Berliner) (*Bt*) transgenic pollen on mite predators and of *Bt*-treated caterpillars on their parasitoids. While such negative effects of GMOs need investigation, there is no doubt that widespread use of *Bt* cotton and maize has been paralleled by a strong reduction in insecticide use, as recently reported from the US (Gianessi *et al.*, 2002) and China (Huang *et al.*, 2002). A recent example concerning cotton in Australia, investigating integration of GMOs with other IPM options, clearly demonstrates the benefits of this new tool (Mensah, 2002a,b) and gives hope that this technology, embedded in an IPM context, will eventually bring environmentally friendly production increases also to small-scale African farmers.

Inundative augmentation of biological control agents: entomopathogens and mycoherbicides

Inundative augmentation strategies require many different approaches compared with classical biological control, particularly when going commercial. In a pioneering effort of international coordination, a commercial mycopesticide, Green Muscle™, was developed against grasshoppers and locusts in the past 13 years (Chapter 14). Assembling a long list of collaborators, searching for pathogens across the drier parts of the Afrotropical region, mass-rearing of a rare indigenous pathogen of locusts and grasshoppers, compiling the registration dossier, filing for use permits for application of the product, entering into licensing agreements for full commercialization, and assisting countries in adapting their legislation to the use of biopesticides were the different steps that led to success, which required talents beyond good entomopathology. Green Muscle™ is now recommended by FAO, commercially produced in the Republic of South Africa and, with the assistance of a new project, will soon be produced and widely used in West Africa. For a sustainable implementation, the product and its commercial producers need further assistance to develop its integration into an IPM scheme. At present, the costs of producing Green Muscle™ in a two-phase process resulting in aerial conidia are not as low as large-scale solid fermentation systems. This type of production assures, however, quality and bridges the gap between research plants and potential future high-volume industrial production (Cherry *et al.*, 1999).

On the basis of this know-how, new projects for the development of entomopathogens against termites (Chapter 15), banana weevil (Chapter 16) and *Maruca* (together with pheromone monitoring) (Chapter 19), but also fruit flies (Chapter 18), diamondback moth (Chapter 23) and mycoherbicides against water hyacinth (Chapter 11) and speargrass, *Imperata cylindrica* (Anderss.)

C.E. Hubbard, are being developed and give some promising first results (F. Beed, IITA, Cotonou, Bénin, 2002, unpublished results).

In other continents, like South America, entomopathogen use against potato tuber moth is now moving into the extension stage (Cisneros and Vera, 2001); but it does not seem that this technology is envisaged yet for use in Africa (Chapter 5). Another system, where mycoparasite mixtures are being developed to replace fungicides, is on the verge of becoming operational in cacao (Krauss and Soberanis, 2001), for instance in several West and Central African countries (Holmes and Flood, 2002).

Evaluation of a large database leads to the conclusion that the safety in inundative augmentative biological control using entomopathogens is justified not by the host range of the agents (often considerable) or even the population effects in the field on non-target organisms (which occur as many times as not), but by the fact that the effect is transient, due to lack of persistence of the agents (Lynch and Thomas, 2000) (characteristics that would also be shared by pyrethrum, for instance, but not by its long-lived synthetic derivatives). Not every entomopathogen lends itself as a useful biopesticide: Gelernter and Lomer (2000) listed the conditions that need to be satisfied for the development of a commercial product. Technical efficacy is a pre-condition for success. This must then be accompanied by the practicability of the product, its commercial viability, the sustainability of production and market, propelled by a measurable public benefit. Similar criteria have been listed by Lacey *et al.* (2001), who come to the conclusion that entomopathogens indeed have a future when they will be broadly appreciated and used particularly as part of IPM schemes. Until now, only *Bt*-based products have been commercially successful. However, with the costs of research for development carried by public institutions, as was the case with the development of Green Muscle™, barriers to commercial involvement can be lowered (Lomer *et al.*, 2001). For the sustainable implementation of such a technology, trained personnel need to be available in the countries to carry on with releases of mass-produced agents after the research for development project has finished (Müller *et al.*, 2002).

In most cases the development of a 'product' is the most sustainable way to implement this technology, which is governed by a regulatory framework that is rather different from the one concerning classical biological control (Chapter 14). Regulations and guidelines for the registration of microbial pesticides have been in place in most industrialized countries for less than 15 years. The FAO guidelines on the registration of biological pest control agents provide an essential basic framework in this direction, and were followed by the FAO Code of Conduct for the Import and Release of Exotic Biological Control Agents in 1996 (FAO, 1996). In West Africa, the CILSS (Comité inter-états de lutte contre la sécheresse dans le Sahel) countries have adopted a regulation framework recently, and similar frameworks are available in South Africa and Madagascar. In East Africa, such regulations do not exist and biopesticides are registered under laws concerning synthetic pesticides. Kenya, especially, has registered a range of biopesticides, mostly *Bt* products. A dramatic increase in interest in microbial control solutions has been observed in countries like Kenya, Senegal and Burkina Faso that produce vegetables for the EU market, since the EU reduced allowed pesticide residue levels for imported crops (Malins and Cox, 1997). Care must, however, be taken that while environmentally safe products are exported, produce not consumed locally is not polluted. Finally, when going commercial, intellectual property rights are becoming an issue, following the Convention of Biological Diversity.

IPM situations and chemical control

Typical IPM situations, where several interventions have to be combined and where the so-called 'judicious use' of insecticides becomes an option and where trade-offs with environmental and human health are being made (Crissman *et al.*, 1998), are common. Experience offered in the present book in this regard provides contrasting situations, which are reviewed here in ascending importance of the insecticide component.

Some IPM situations require a combination of methods to combat various pests but are not centred around the application of synthetic insecticides. Maize stem and cob borers as well as the parasitic weed *Striga* are the key pests on maize in Africa. Most of the species are indigenous. In the case of stem borers, exchange of parasitoids across Africa offers classical biological control solutions (Chapter 9). Packages of options that are being tested combine the manipulation of grasses and legumes, which attract or deter stem borers and which at the same time lead to suicidal germination of *Striga* seeds (Chapter 17).

New recombinations of viruses and a new *Bemisia tabaci* (Gennadius) whitefly vector strain in Uganda have led to an unprecedented outbreak of cassava mosaic disease, which is being combated by the deployment of various resistant cassava varieties. Though various parasitoids are active, most impact is attributable to the deployment of resistant cassava varieties (Chapter 6). This widespread renewal of cassava varieties offers the opportunity to deploy varieties that, through the strong hairiness of their tips, offer shelter to the most effective exotic predator of cassava green mite and therefore improve the impact of biological control on this pest (Chapter 4).

The major insect pest of banana, which is the food crop of Africa with the highest total value, is the banana weevil. This long-lived, low-density insect poses special problems and only a combination of cultural control combined with lures and traps has a chance to reduce its damage to an acceptable level (Chapter 16).

Finally, great challenges are posed by two weeds, among which the exotic weed *Chromolaena* can also be considered a valued component of the fallow vegetation (Chapter 10) and a suppressor of speargrass, *I. cylindrica*. Up to now, biological control has been moderately successful only in Ghana. Speargrass, on the other hand, is the most important agricultural weed in Africa, against which up to now only cultural control practices, like rotation with smothering legumes or long fallows, are effective (Becker and Johnson, 1998). With shortening fallow durations, *I. cylindrica*, similarly to *Striga* spp., has become increasingly difficult to control. Biological control options resulting in the development of a mycoherbicide are being addressed, but it will take years before any such option becomes commercially available, if at all.

Stictococcus vayssierei Richard, an indigenous root scale, seems to have a similar link to intensification of agriculture, this time in the forest zone of Africa. It has suddenly become the most important pest insect on cassava in several Central African countries. Its pest status and solutions for its control are still being investigated, but already it is clear that it is an indicator of newly deforested land (R. Hanna, Cotonou, IITA, 2002, personal communication).

Fruit flies of mangoes (Chapter 18) are a group of highly damaging indigenous and rather polyphagous insects. IPM becomes particularly difficult because none of the numerous parasitoids is on its own providing sufficient control, particularly if the fruits are destined for the export market. Various control options that worked on other fruitflies on other continents are therefore now being tried in Africa. Direct treatment of fruits with synthetic insecticides can be avoided through the application of protein hydrolysate–based insecticide-laced baits. The sterile insect technique, so efficient in Mexico against fruit flies, will probably not find the necessary economic conditions in continental Africa for it to become viable.

Often, integration of insecticide applications with biological control has not worked. It might in fact lead to unexpected consequences within the food web, such as when the added top predator eliminates other predators that had survived the insecticide treatment (Fagan *et al.*, 1998). In some cases, replacement of hard insecticides with neem (*Azadirachta indica* A. Juss) or papaya (*Carica papaya* L.) leaf extracts provided an acceptable solution (Chapter 19). Thus, neem extracts had the highest potential for combining insecticide with releases of *Encarsia* parasitoids against whitefly (Simmonds *et al.*, 2002) and had no negative effect on parasitoids of diamondback moth (Chapter 23). They are generally safe for spiders, adults of numerous beneficial insect species and eggs of many predators (Schmutterer, 1997). Extracts from *Chenopodium* leaves proved similarly successful as postharvest grain protectants against stored product beetles (Tapondjou *et al.*, 2002). It must, however, be stressed that plant extracts must be carefully evaluated and cannot be considered as non-toxic just because they are 'natural' (Trumble, 2002).

A similar situation exists with cowpea, which is attacked by numerous pests that can only be combated effectively through knowledge-intensive IPM, which should replace the often-observed unhealthy use of broad-spectrum insecticides. Most cowpea pests are indigenous, but key pests like thrips, *Maruca vitrata* Fabricius, pod-sucking bugs and aphids have been shown to be open to biological control solutions. Combinations of habitat management, use of botanicals, through farmers' participatory research and farmers' field schools, and the integration with loan projects by the International Fund for Agricultural Development (IFAD), provided a successful framework for improving and stabilizing yields in nine African countries (Chapter 19).

Another situation, where insecticides are often recommended and used in a pre-emptive and un-reflective manner, concerns maize stores (Chapter 6). Though the major pest, the larger grain borer, is sometimes kept under sufficient biological control by the introduced predator *T. nigrescens*, insecticide treatments are still necessary in many situations against this and other store pests. A simulation model was developed that serves as a decision-making tool for extension services, to be translated into specific recommendations for a particular area (Holst *et al.*, 1999). It indicates under which conditions and when to leave the maize store intact, to sell or eat the grain, or to treat the grain (with insecticides, preferably botanicals). The different store options are now being tested in collaboration with national programmes and should replace unnecessary insecticide treatments.

As already discussed, several cash crops have become the entry points for unsustainable and dangerous use of inappropriate insecticides on food crops, like cowpea. In addition, the treatments on, say, cotton, also affect existing biological control on neighbouring food crop fields by drift (Chapter 3). *Helicoverpa armigera* (Hübner) remains the most important cotton pest, its importance being linked in many cases with excessive use of pesticides (Chapter 21). Experiences from organic cotton farming in South Africa seem to support this hypothesis. In South Africa, particularly small-scale farmers increasingly adopt *Bt* cotton, with yield increases between 40% and 300%. The lower the management skills of the farmers were, the higher was the yield increase (Joubert *et al.*, 2001). In the francophone African countries, a large proportion of cotton is still sprayed on a calendar basis (three to six ULV sprays per season). The introduction of very low volume application with variable doses, whereby half of the annual dose is applied according to calendar and half according to pest thresholds, has improved the situation (CIRAD, 1994). At present, because of the economic crisis in cocoa production and consumer pressure from Europe and the USA, cocoa, which attracted a lot of insecticides in the past, is often grown with little or no insecticide input. New varieties, cultural control measures and the use of endophytes offers the perspective of arriving at a high-quality organic cocoa production (Holmes and Flood, 2002). Likewise, falling world market prices, coupled with increasing competition from new producers like Vietnam, has led to a sharp reduction in coffee production in Africa. Because of these economic constraints, African coffee farmers today apply considerably less synthetic pesticide than in the past. Viable IPM options for a more sustainable plant protection are available and need to be implemented for African coffee production to regain its former market share (Chapter 22). In all these situations, research institutions often seek linkages with NGOs that have the comparative advantage. In several situations, synthetic insecticides could eventually be replaced by botanicals prepared on farm.

Vegetables are one of the cropping systems that attract most insecticide abuse in Africa and elsewhere (Chapter 9 for green maize; Chapters 6, 18, 19 and 23). In the case of the diamondback moth, *P. xylostella*, the synthetic sex pheromone is effectively used to monitor the flight activity of the pest to improve the timing of insecticide applications. 'Softer' IPM tactics include the use of Indian mustard (*Brassica juncea* (L.)) as a trap crop and the use of botanical insecticides like neem. The latter has the advantage of being highly toxic to the pest but has no direct negative impact on two important parasitoids of *P. xylostella* (Chapter 23).

New introductions of previously innocuous insects have become important challenges in vegetables. In Senegal in the early 1980s, the agromyzid leafminer *L. trifolii*, a major pest on other continents, had reached outbreak levels on a large variety of vegetable species, before collapsing totally when insecticide sprays were stopped and indigenous parasitoids of other indigenous *Liriomyza* spp. brought them under control (Neuenschwander *et al.*, 1987). In southern Africa, the tetranychid mite *Tetranychus evansi* Baker & Pritchard has become the most serious pest of tomatoes in all smallholder systems. Surveys in Brazil have identified several phytoseiids as potential candidates for biological control (M. Knapp, Nairobi, 2002, personal communication). In Central Africa, new outbreaks of the tephritid *Dacus punctatifrons* (Karsch) are now causing great damage, with the danger that they will attract treatments with broad-spectrum insecticides (R. Hanna, IITA, Cotonou, 2002, personal communication). To replace synthetic insecticides with microbial control is a highly promising approach in peri-urban vegetable production. Some *Bt*-based products are already available on the African market and products based on insect viruses are currently being developed (Langewald and Cherry, 2000). Botanicals such as neem are particularly popular in vegetable production in many African counties, as the repellent effect on defoliating insects protects leafy vegetables for about a week (Schmutterer, 1990). Interestingly, a comprehensive list of botanicals, extraction methods, and target pests in the tropics (Stoll, 2002) has practically no overlap in its reference list with the one of this book. Is this an indication of lack of communication between scientists and practitioners of IPM?

As so often recommended (Schwab *et al.*, 1995), insecticide loads can be limited further if the number of applications is reduced following monitoring of the pest (Chapters 5, 6, 18, 19–21 and 23) or if less persistent and more specific insecticides are used. In line with its goal and definition of IPM, and in agreement with international limitations on the use of pesticides in projects funded by or through the World Bank, FAO strongly advises its partners that IPM research should exclude Persistent Organic Pollutants (POPs) Class I and, where feasible, Class II compounds as components towards IPM implementation and/or as components of IPM strategies and programmes. Promotion of so-called Rational Pesticide Use (Bateman, 2002) is supported by a corresponding database (the DROPDATA base developed by CAB *International*), but particularly in Africa, its implementation is hindered at the promotional efforts by insecticide companies.

Implementation: farmer–scientist–extension interactions

Classical biological control in the past needed mostly interactions with government and inter-governmental agencies, the IAPSC for instance, and the collaborating institutions for coordinating foreign exploration, quarantine, and to overcome the still considerable transport problems that have already affected the earliest practitioners.

For IPM, training and the involvement of extension services and farmers are more important and have long been acknowledged as an essential factor for agricultural development. The problem lies in the use of methods and even more in the attitude of those involved. Farmers' participatory techniques have now become widely recognized as more effective in promoting IPM than the hitherto frequently popular top-down extension approaches. Increasingly now, farmer participatory research (FPR) and participatory learning (PL) approaches make biological control and IPM research more understandable and useful. The potential of FPR and PL to increase impact of IPM hinges largely on improving farmers' capacity to understand and combine knowledge of biological and ecological processes with farming experiences. A pragmatic approach to farmers' participation had already been recommended by the early practitioners (Bentley, 1994). Today, participatory approaches involve diverse types of interactions, such as facilitation of farmers' experiments (Haverkort *et al.*, 1991); farmer collaboration in plant breeding (Witcombe, 1996; Weltzien *et al.*, 2000); farmer testing of 'best-bet' options generated by researchers (Snapp, 1999); action-research and social learning (Röling and Wagemakers, 1998; Ashby *et al.*, 2000; Defoer and Budelman, 2000); and action learning at farmer field schools (van de Fliert, 1993; Hagmann, 1999; Hagmann *et al.*, 1999; Connell, 2000; Ooi, 2000).

Many projects described in this book have experimented with various kinds of participatory approaches, which then also become the means for feedback. It is heartening to see that many projects were carried out in close collaboration with stakeholders, the starting point being already existing technologies, farmers' experience and indigenous traditional knowledge (Chapters 4, 9–11, 16, 17, 19, 20 and 22).

The different approaches led to higher scientific literacy and clearer guidelines for farmers to make informed decisions to solve location-specific problems. There is, however, the need to synthesize the various participatory approaches and specify what would need to be done differently at the level of farmers and community organizers, extensionists, researchers and policy makers to improve their impact. In response to these kinds of challenges, the System-wide Programme on Integrated Pest Management (SP-IPM), a CGIAR-based coalition of plant protection research and development scientists, initiated mentored case study analyses. It established IPM pilot sites in diverse production ecologies to assist participating organizations gain experience in developing effective farmer–scientist–extension partnerships to analyse problems and develop solutions locally (James *et al.*, 2003).

For these pilot site gains to have impact, scaling up technologies to wider areas with similar ecological and socio-economic conditions and scaling out to new domains are essential. This is best achieved where pioneer farmers, who themselves improve on old and new technologies, lead the adaptation process (Douthwaite *et al.*, 2001b). This process, whereby the knowledge about new technologies trickles down through society and is adapted on its way, has, however, often not been documented in a scientific manner that would allow for general conclusions. Typical indicators for impact would include documentation about reductions in harmful pesticide regimes, and the capacity of farmers to increase profitability by responding to market forces.

Many projects have, however, performed formal economic impact assessments and demonstrated sometimes huge economic returns (Chapters 2, 3–5, 7–9, 11, 14, 15 and 19). These projects were shown to benefit all farmers, including the poorest, and therefore to foster broad-based rural growth and improve social well-being. Some of them demonstrated improved human and animal health through reduction or prevention of insecticide misuse or reduction of mycotoxins, like aflatoxins for example, and all reduced environmental degradation and promoted partnerships at all levels.

Vision

This book demonstrates that many sustainable options for environmentally friendly agricultural production exist to increase agricultural productivity decisively and profitability in Africa. They are not the ideas of dreamers, whose recommendations aim to turn the wheel backward and keep poor African farmers poor. While the industrialized nations are producing agricultural surpluses based on heavy subsidies, Africa still can intensify agriculture to raise food quantity and quality with minimal harm to the environment. External inputs like fertilizers are urgently needed to improve soil fertility. Mechanization, particularly for weed control, is needed to free up farmers' time for raising alternative income. We believe that in this drive for intensifying African agriculture, biological control-based IPM will provide the best solution to plant protection problems. This will avoid 'insecticide treadmill systems' as found in many industrialized regions, but also developing countries (Antle *et al.*, 1998).

Successful implementation of IPM often needs the right political decisions and a change in attitude across a broad spectrum of stakeholders, at national and international levels. The ultimate decision for their future development remains with governments in Africa. They will have to develop the necessary infrastructure and institutional arrangement to effectively promote national plant protection capacities and guarantee compliance with the related international agreements, conventions and treaties.

Important agricultural pests for which no sustainable solution has been found anywhere are actually few. For most problems, at least some options that provide higher long-term productivity with maximum sustainable revenue actually exist. The best results are thereby often achieved not by technology baskets that maximize yields, but by those that optimize yields with more moderate inputs and thereby achieve sustainability. We are well aware that in the rush for instant impact, often needed because of a desperate situation, this view is not easily accepted and even more difficult to implement, yet we believe that there are no viable alternatives to such an approach.

Modern IPM does not start with an insecticide overdose followed by a gradual and haphazard reduction in the number of insecticide treatments. IPM offers alternatives based on an ecological approach, but is knowledge intensive. Fortunately, because of the discussed features promoting natural control of many pests, rather few cases of the 'insecticide treadmill' syndrome are yet observed in Africa as compared with other continents. This is, however, sometimes achieved more by default, i.e. the lack of access to insecticides, than by conscious design. Though the evaluation of environmental indicators is controversial (Bailey *et al.*, 1999), it has been shown that a change away from insecticide use towards some sort of ecologically conscious farming, even where short-term economic benefits are marginal, is beneficial in the long term (Lewis *et al.*, 1997; Mäder *et al.*, 2002).

We share the dream of productive agriculture that does not destroy its natural resource base and is compatible with the preservation of biodiversity in field, in fallow or in designated protected areas (Musters *et al.*, 2000; McNeely and Scherr, 2001). In fact, at present, the economic reasons for conserving the remaining wild areas are overwhelming (Balmford *et al.*, 2002; Chapter 2). Together with Hazell (1999), we raise, however, caution against the naïve belief in complete restoration of natural habitats that would also remain agriculturally productive. In the preceding chapters, the reader was shown examples of sustainable plant protection schemes that can be taken as stepping-stones into a more sustainable agriculture than is the case at present. It seems that an example could be made with Cuban agriculture, where many of the practices described or advocated in this book have been put into reality. It is perhaps telling that this was achieved for other reasons than by free decision of the farmers (Oppenheim, 2001). This example – and one could add others – of course illustrates the importance of the political parameters at work, which differ from those found today in most African countries and which differ from the current paradigm of unrestricted freedom, with benefits to the individual and long-term costs to the community. The ultimate question then becomes to what degree personal liberties can be reduced in the name of sustainable development, a question that is clearly beyond the scope of this book.

Even in today's political situation, there are many changes that can be wrought, provided the partial solutions are sound and people concerned are involved in their further development, adaptation and implementation. Given the support by the development investors, we, the researchers in plant protection, hope to continue in the same vein. The teams and instruments across all of Africa are in place and there are problems that cry out for a solution.

References

Antle, J.M., Cole, D.C. and Crissman, C.C. (1998) Further evidence on pesticides, productivity and farmer health: potato production in Ecuador. *Agricultural Economics* 18, 199–208.

Ashby, J., Braun, A.R., Garcia, T., Guerrero, M.P., Hernandez, L.A., Quirós, C.A. and Roa, J.I. (2000) *Investing in Farmers as Researchers. Experience with Local Agricultural Research Committees in Latin America*. CIAT, Cali, Colombia.

Avery, D.T. (1997) Environmentally sustaining agriculture. *Choices* 12, 10–17.

Bailey, A.P., Rehman, T., Park, J., Keatinge, J.D.H. and Tranter, R.B. (1999) Towards a method for the economic evaluation of environmental indicators for UK integrated arable farming systems. *Agriculture, Ecosystems and Environment* 72, 145–158.

Balmford, A., Bruner, A., Cooper, P., Costanza, R., Farber, S., Green, R.E., Jenkins, M., Jefferiss, P., Jessamy, V., Madden, J., Munro, K., Myers, N., Naeem, S., Paavola, J., Rayment, M., Rosendo, S., Roughgarden, J., Trumper, K. and Turner, R.K. (2002) Economic reasons for conserving wild nature. *Science* 297, 950–953.

Bateman, R. (2002) Droplets of wisdom. *Biocontrol News and Information* 23, 45N-46N.

Becker, M. and Johnson, D.E. (1998) Legumes as dry season fallow in upland rice-based systems of West Africa. *Biology and Fertility of Soils* 27, 358–367.

Bellows, T.S. (2001) Restoring population balance through natural enemy introductions. *Biological Control* 21, 199–205.

Bentley, J.W. (1994) Facts, fantasies, and failures of farmer participatory research. *Agriculture and Human Values* spring–summer 1994, 140–150.

Brewster, C.C. and Allen, J.C. (1997) Spatiotemporal model for studying insect dynamics in large-scale cropping systems. *Environmental Entomology* 26, 473–482.

Carsky, R.J., Becker, M. and Hauser, S. (2001) *Mucuna* cover crop fallowsystems: potential and limitations. *Sustaining Soil Fertility in West Africa.* Soil Science Society of America and American Society of Agronomy, Madison, Wisconsin, USA. Special Publication 58, 11–135.

Cherry, A.J., Jenkins, N.E., Heviefo, G., Bateman, R. and Lomer, C.R. (1999) Operational and economic analysis of a West African pilot-scale production plant for aerial conidia of *Metarhizium* spp. for use as a mycoinsecticide against locusts and grasshoppers. *Biocontrol Science and Technology* 9, 35–51.

CIRAD (1994) *Coton Doc, système multimédia sur le cotonnier et ses ennemis en Afrique francophone au sud du Sahara.* CDROM, CIRAD, Montpellier, France.

Cisneros, F. and Vera, A. (2001) Mass-producing *Beauveria brongniartii* inoculum, an economical, farm-level method. In: *The International Potato Center 2001. Scientist and Farmer, Partners in Research for the 21st Century.* Program Report 1999–2000. Lima, Peru, pp. 155–160.

Connell, J.G. (2000) Scaling-up: the roles of participatory technology development and participatory extension approaches. In: Stur, W.W., Horne, P.M., Hacker, J.B. and Kerridge, P.C. (eds) *Working with Farmers: the Key to Adoption of Forage Technologies.* ACIAR Publication PR095, Canberra, Australia pp. 69–82.

Conway, G. (2001) The Doubly Green Revolution: a context for farming systems research and extension in the 21st century. *Journal for Farming Systems Research-Extension.* Special issue 2001, 1–16.

Cortesero, A.M., Stapel, J.O. and Lewis, W.J. (2000) Understanding and manipulating plant attributes to enhance biological control. *Biological Control* 17, 35–49.

Cotty, P.J. (1994) Influence of field application of an atoxigenic strain of *Aspergillus flavus* on the populations of *A. flavus* infecting cotton bolls and on the aflatoxin content of cottonseed. *Phytopathology* 84, 1270–1277.

Crissman, C.C., Antle, J.M. and Capalbo, S.M. (1998) *Economic, Environmental and Health Tradeoffs in Agriculture: Pesticides and the Sustainability of Andean Potato Production.* Kluwer, Dordrecht, The Netherlands.

Defoer, T. and Budelman, A. (eds) (2000) *Managing Soil Fertility in the Tropics: a Resource Guide for Participatory Learning and Action Research.* The Royal Tropical Institute (KIT) in collaboration with IIED, IER, FAO, 5 volume resource guide (4 books plus 1 CD), Amsterdam, The Netherlands.

Denoth, M., Frid, L. and Myers, J.H. (2002) Multiple agents in biological control: improving the odds? *Biological Control* 24, 20–30.

Diamond, J. (1999) *Guns, Germs, and Steel.* W.W. Norton, New York, USA.

Dittrich, V., Hassan, S.O., and Ernst, G.H. (1985) Sudanese cotton and the whitefly: a case study of the emergence of a new primary pest. *Crop Protection* 4, 161–176.

Dorner, J.W. and Cole, R.J. (2002) Effect of application of nontoxigenic strains of *Aspergillus flavus* and *A. parasiticus* on subsequent aflatoxin contamination of groundnuts in storage. *Journal of Stored Products Research* 38, 329–339.

Douthwaite, B., de Haan, N., Manyong, V. and Keatinge, D. (2001a) Blending 'hard' and 'soft' science: the 'follow-the–technology' approach to catalizing and evaluating technology change. *Conservation Ecology* (online) http://www.consecol.org/Journal/ unpub/785738/final_version/main.html

Douthwaite, B., Keatinge, J.D.H. and Park, J.R. (2001b) Why promising technologies fail: the neglected role of user innovation during adoption. *Research Policy* 30, 819–836.

Fagan, W.F., Hakim, A.L., Ariawan, H. and Yuliyantiningsih, S. (1998) Interactions between biological control efforts and insecticide applications in tropical rice agroecosystems: the potential role of intraguild predation. *Biological Control* 13, 121–126.

FAO (1996) *International Standards for Phytosanitary Measures. Part II: Import Regulations. Code of Conduct for the Import and Release of Exotic Biological Control Agents.* Food and Agriculture Organization of the United Nations, Rome, Italy.

Gelernter, W.D. and Lomer, C.J. (2000) Factors in the success and failure of biological control of above-ground insects by pathogens. In: Gurr, G. and Wratten, S. (eds) *Measures of Success in Biological Control.* Kluwer, Dordrecht, The Netherlands, pp. 297 – 322.

Gianessi, L.P., Silvers, C.S., Sankula, S. and Carpenter, J.E. (2002) *Plant Biotechnology: Current and Potential Impact for Improving Pest Management in U.S. Agriculture. An Analysis of 40 Case Studies.* National Center for Food and Agricultural Policy, Washington, DC, USA.

Gong, Y.Y., Cardwell, K., Hounsa, A., Egal, S., Turner, P.C., Hall, A.J. and Wild, C.P. (2002) Dietary aflatoxin exposure and impaired growth in young children from Bénin and Togo: cross sectional study. *British Medical Journal* 325, 20–21.

Greathead, D.J. (1968) Biological control of *Lantana*. A review and discussion of recent developments in East Africa. *Pest Articles and News Summaries* 14, 167–175.

Gutierrez, A.P., Yaninek, J.S., Neuenschwander, P. and Ellis, C.K. (1999) A physiologically-based tritrophic metapopulation model of the African cassava food web. *Ecological Modelling* 123, 225–242.

Hajek, A.E. (1999) Pathology and epizootilogy of *Entomophaga maimaiga* infections in forest lepidoptera. *Microbiology and Molecular Biology Reviews* 63, 818–835.

Hagmann, J. (1999) *Learning Together for Change Through Facilitating Innovation in Natural Resource Management Through Learning Process Approaches in Rural Livelihoods in Zimbabwe.* Margraf Verlag, Weikersheim, Germany.

Hagmann, J., Chuma, E., Murwira, K. and Connolly, M. (1999) *Putting Process into Practice: Operationalising Participatory Extension.* In: ODI Agricultural Research and Extension (AGREN) Network Paper 94. http://www.odi.org.uk/agren/papers/agrenpaper_94.pdf

Haverkort, B., van der Kamp, J. and Waters-Bayer, A. (1991) *Joining Farmers' Experiments.* Intermediate Technology Publications, London, UK.

Hazell, P. (1999) Agricultural growth, poverty alleviation, and environmental sustainability: having it all. *2020 Brief* 59. IFPRI http://www.ifpri.org/2020/briefs/ number59.htm

Hazell, P. and Wood, S. (2001) From science to technology adoption: the role of policy research in improving natural resource management. *Agriculture, Ecosystems and Environment* 82, 385.

Headrick, D.H. and Goeden, R.D. (2001) Biological control as a tool for ecosystem management. *Biological Control* 21, 249–257.

Hokkanen, H. and Pimentel, D. (1989) New associations in biological control: theory and practice. *Canadian Entomologist* 121, 829–840.

Holmes, K. and Flood, J. (2002) Biocontrol of pests and diseases of tropical tree crops with specific reference to cocoa and coffee. In: Vos, J. and Neuenschwander, P. (eds) *Proceeding of the West African Regional Cocoa IPM Workshop.* CABI Bioscience, Egham and IITA, Cotonou, pp.17–24. http://www.cabi-commodities.org/Acc/ACCrc/ACCrc.htm

Holst, N., Markham, R.H. and W.G. Meikle, (1999) Integrated pest management of postharvest maize in developing countries. http://www.purl.dk/net/9906–0408 and http://www.agrsci.dk/plb/nho/vms/

Huang, J., Rozelle, S., Pray, C. and Wang, Q. (2002) Plant biotechnology in China. *Science* 295, 674–677.

Huffaker, C.B., Simmonds, F.J. and Laing, J.E. (1976) The theoretical and empirical basis of biological control. In: Huffaker, C.B. and Messenger, P.S. (eds) *Theory and Practice of Biological Control.* Academic Press, New York, USA, pp. 41–78.

IITA (2001) *IITA Strategic Plan 2001–2010, Supporting Document.* IITA, Ibadan, Nigeria.

ISNAR (2002) *A Conceptual Framework for Implementing Biosafety: Linking Policy Capacity and Regulation.* ISNAR Briefing Paper 47.

Jaccoud, D., Peng, K., Feinstein, D. and Kilian, A. (2001) Diversity arrays: a solid state technology for sequence information independent genotyping. *Nucleic Acids Research* 29, E25.

James, B., Neuenschwander, P., Markham, R.H., Anderson, P., Braun, A., Overholt, W., Khan, K., Makkouk, K. and Emechebe, A. (2003) Bridging the gap with the CGIAR Systemwide Program on Integrated Management. In: Maredia, K., Dakouo, D. and Mota-Sanchez, D. (eds) *Integrated Pest Management in the Global Arena.* CAB International, Wallingford, UK.

Joubert, G.D., Venter, M.J., Theron, G.C., Swanepol, A., Eulitz, E.G., Schröder, H.F. and Macaskill, P. (2001) South African experience with Bt cotton. *Proceedings of the 60th Plenary Meeting of the International Cotton Advisory Committee*, Victoria Falls, Zimbabwe, pp. 28–31.

Keatinge, J.D.H., Breman, H., Manyong, V., Vanlauwe B. and Wendt, J. (2001) Sustaining soil fertility in West Africa in the face of rapidly increasing pressure for agricultural intensification. In: Tian, G., Ishida, F. and Keatinge, J.D.H. (eds) *Sustaining Soil Fertility in West Africa.* American Society of Agronomy, Madison, Wisconsin, USA, Special Publication 58, 1–22.

Krauss, U. and Soberanis, W. (2001) Biocontrol of cocoa pod diseases with mycoparasite mixtures. *Biological Control* 22, 149–158.

Kryder, R.D., Kowalski, S.P. and Krattiger, A.F. (2000) The intellectual and technical property components of pro-Vitamin A rice (*Golden*Rice™): a preliminary freedom-to-operate review. *ISAA Briefs* 20. ISAA, Ithaca.

Lacey, L.A., Frutos, R., Kaya, H.K. and Vail, P. (2001) Insect pathogens as biological control agents: do they have a future? *Biological Control* 21, 230–248.

Lamers, J., Bruentrup, M. and Buerkert, A. (1998) The profitability of traditional and innovative mulching techniques using millet crop residues in the West African Sahel. *Agriculture, Ecosystems and Environment* 67, 23–35.

Landis, D.A., Wratten, S.D. and Gurr, G.M. (2000) Habitat management to conserve natural enemies of arthropod pests in agriculture. *Annual Review of Entomology* 45, 175–201.

Langewald, J. and Cherry, A. (2000) Prospects for microbial control in West Africa. *Biocontrol News and Information* 21, 51N–56N.

Lenné, J.M. and Wood, D. (1999) Vegetational diversity in agroecosystems: a mixed blessing for successful pest management. *BCPC Symposium Proceedings 73: International Crop Protection Achievements and Ambitions,* pp. 75–98.

Lewis, W.J., van Lenteren, J.C., Phatak, S.C. and Tumlinson, J.H. (1997) A total system approach to sustainable pest management. *Proceedings of the National Academy of Sciences USA* 94, 12243–12248.

Lockwood, J.A. (1993) Benefits and costs of controlling rangeland grasshoppers (Orthoptera: Acrididae) with exotic organisms: search for a null hypothesis and regulatory compromise. *Environmental Entomology* 22, 904–914.

Lomer, C., Bateman, R.P., Johnson, D.L., Langewald, J. and Thomas, M.B. (2001) Biological control of locusts and grasshoppers. *Annual Review of Entomology* 46, 667–702.

Lynch, L.D. and Thomas, M.B. (2000) Nontarget effects in the biocontrol of insects with insects, nematodes and microbial agents: the evidence. *Biocontrol News and Information* 21, 117N-130N.

Mäder, P., Fließbach, A., Dubois, D., Gunst, L., Fried, P. and Niggli, U. (2002) Soil fertility and biodiversity in organic farming. *Science* 296, 1694–1697.

Malins, A. and Cox, J.R. (1997) Constraints for developing-country horticultural exporters in meeting the requirements of European markets. *Phytoma la Défense des Végétaux* 498, 27–30.

Manzari, S., Polaszek, A., Belshaw, R. and Quicke, D.L.J. (2002) Morphometric and molecular analysis of the *Encarsia inaron* species-group (Hymenoptera: Aphelinidae), parasitoids of whiteflies (Hemiptera: Aleyrodidae). *Bulletin of Entomological Research* 92, 165–175.

May, R.M. and Hassell, M.P. (1988) Population dynamics and biological control. *Philosophical Transactions of the Royal Society* , London B 318, 129–169.

McLean, M.A., Frederick, R.J., Traynor, P.L., Cohen, J.I. and Komen, J. (2002) A conceptual framework for implementing biosafety: linking policy, capacity, and regulation. *ISNAR Briefing Paper* 47.

McNeely, J.A. and Scherr, S.J. (2001) *Common Ground, Common Future. How Ecoagriculture Can Help Feed the World and Save Wild Biodiversity.* IUCN, Gland; Future Harvest, Washington, D.C.

Memmott, J. (2000) Food webs as a tool for studying nontarget effects in biological control. In: Follett, P.A. and Duan, J.J. (eds) *Nontarget Effects of Biological Control.* Kluwer Academic, Boston, Massachusetts, USA, pp. 147–163.

Mensah, R.K. (2002a) Development of an integrated pest management programme for cotton. Part 1: Establishing and utilizing natural enemies. *International Journal of Pest Management* 48, 87–94.

Mensah, R.K. (2002b) Development of an integrated pest management programme for cotton. Part 2: Integration of a lucerne/cotton interplant system, food supplement sprays with biological and synthetic insecticides. *International Journal of Pest Management* 48, 95–105.

Messing, R.H. (2000) The impact of nontarget concerns on the practice of biological control. In: Follett, P.A. and Duan, J.J. (eds) *Nontarget Effects of Biological Control.* Kluwer Academic, Boston, Massachusetts, USA, pp. 45–55.

Moran, D., Pearce, D. and Wendelaar, A. (1997) Investing in biodiversity: an economic perspective on global priority setting. *Biodiversity and Conservation* 6, 1219–1243.

Müller, D., de Groote, H., Gbongboui, C. and Langewald, J. (2002) Participatory development of a biological control strategy of the variegated grasshopper in the humid tropics in West Africa. *Crop Protection* 21, 265–275.

Musters, C.J.M., de Graaf, H.J. and ter Keurs, W.J. (2000) Can protected areas be expanded in Africa? *Science* 287, 1759–1760.

Neuenschwander, P. (1982) Searching parasitoids of *Dacus oleae* (Gmel.) (Dipt., Tephritidae) in South Africa. *Zeitschrift für angewandte Entomologie* 94, 509–522.

Neuenschwander, P., Hammond, W.N.O., Ajuonu, O., Gado, A., Echendu, N., Bokonon-Ganta, A.H., Allomasso, R. and Okon, I. (1990) Biological control of the cassava mealybug, *Phenacoccus manihoti* (Hom., Pseudococcidae), by *Epidinocarsis lopezi* (Hym., Encyrtidae) in West Africa, as influenced by climate and soil. *Agriculture, Ecosystems and Environment* 32, 39–55.

Neuenschwander, P. and Markham, R. (2001) Biological control in Africa and its possible effects on biodiversity. In: Wajnberg, E., Scott, J.K. and Quimby, P.C. (eds) *Evaluating Indirect Ecological Effects of Biological Control.* CAB International, Wallingford, UK, pp. 127–146.

Neuenschwander, P., Murphy, S.T. and Coly, E.V. (1987) Introduction of exotic parasitic wasps for the control of *Liriomyza trifolii* (Dipt., Agromyzidae) in Senegal. *Tropical Pest Management* 33, 290–297.

Oerke, E.C., Dehne, H.W., Schönbeck, F. and Weber, A. (1999) *Crop Production and Crop Protection: Estimated Losses in Major Food and Cash Crops*, 2nd reprint. Elsevier Science, Amsterdam, The Netherlands.

Ooi, P.A.C. (2000) From passive observer to pest management expert: science education and farmers. In: Guijt, I., Berdegue, J.A. and Loevinsohn, M. (eds) *Deepening the Basis of Rural Resource Management*. ISNAR, The Hague, The Netherlands, pp. 167–177.

Oppenheim, S. (2001) Alternative agriculture in Cuba. *American Entomologist* 47, 216–227.

Ortiz, R. (2002) Germplasm enhancement to sustain genetic gains in crop improvement. In: Engels, J.J.M., Brown, A.H.D. and Jackson, M.T. (eds) *Managing Plant Genetic Resources*. CAB International, Wallingford, UK, pp. 275–290.

Prinsloo, G., Chen, Y., Giles, K.L. and Greenstone, M.H. (2002) Release and recovery in South Africa of the exotic aphid parasitoid *Aphelinus hordei* verified by the polymerase chain reaction. *BioControl* 47, 127–136.

Raemaekers, R.H. (2001) *Agriculture en Afrique Tropical*. Direction Générale de la Coopération Internationale, Brussels, Belgium.

Röling, N. and Wagemakers M.A.E. (eds) (1998) *Facilitating Sustainable Agriculture*. Cambridge University Press, Cambridge, UK.

Sayer, J.A. and Campbell, B. (2001) Research to integrate productivity enhancement, environmental protection, and human development. *Conservation Ecology* 5, 32 (online) http:/www.consecol.org/vol5/iss2/art32

Schmutterer, H. (1990) Properties and potential of natural pesticides from the neem tree *Azadirachta indica*. *Annual Review of Entomology* 35, 271–297.

Schmutterer, H. (1997) Side-effects of neem (*Azadirachta indica*) products on insect pathogens and natural enemies of spider mites and insects. *Journal of Applied Entomology* 121, 121–128.

Schönrogge, K. and Crawley, M.J. (2000) Quantitative webs as a means of assessing the impact of alien species. *Journal of Animal Ecology* 69, 841–868.

Schulz, S., Carsky, R.J. and Tarawali, S.A. (2001) Herbaceous legumes: the panacea for West African soil fertility problems? *Sustaining Soil Fertility in West Africa*. Soil Science Society of America and American Society of Agronomy, Madison, Wisconsin, USA, Special Publication 58, 179–196.

Schwab, A., Jäger, I., Stoll, G., Görgen, R., Prexler-Schwab, S. and Altenburger, R. (1995) *Pesticides in Tropical Agriculture. Hazards and Alternatives*. Margraf Verlag, Weikersheim.

Sharma, K.K., Sharma, H.C., Seetharama, N. and Ortiz, R. (2002) Development and deployment of transgenic plants: biosafety considerations. *In Vitro Cell Development and Biology* 86, 1–11.

Sharma, H.C., Singh, B.U. and Ortiz, R. (2002) Host plant resistance to insects: measurements, mechanisms and insect–plant–environment interactions. In: Ananthakrishnan, T.N. (ed.) *Insects and Plant Defence Dynamics*. Science Publishers, Enfield, pp. 133–159.

Simmonds, M.S.J., Manlove, J.D., Blaney, W.M. and Khambay, B.P.S. (2002) Effects of selected botanical insecticides on the behaviour and mortality of the glasshouse whitefly *Trialeurodes vaporariorum* and the parasitoid *Encarsia formosa*. *Entomologia Experimentalis et Applicata* 102, 39–47.

Smithson, J.B. and Lenné, J.M. (1996) Varietal mixtures: a viable strategy for sustainable productivity in subsistence agriculture. *Annals of Applied Biology* 128, 127–158.

Snapp, S. (1999) Mother and baby trials: a novel farmer participatory trial design and methodology being tried out in Malawi. *Target* 17, 8.

Stoll, G. (2002) *Protection Naturelle des Végétaux en Zones Tropicales.* Margraf, Weikersheim.

Stonehouse, J.M. (1995) Pesticides, thresholds and the smallscale tropical farmer. *Insect Science and its Application* 16, 259–262.

Tapondjou, L.A., Adler, C., Bouda, H. and Fontem, D.A. (2002) Efficacy of powder and essential oil from *Chenopodium ambrosioides* leaves as postharvest grain protectants against six stored product beetles. *Journal of Stored Product Research* 38, 395–402.

Thies, C. and Tscharntke, T. (1999) Landscape structure and biological control in agroecosystems. *Science* 285, 893–895.

Thomas, C.D. and Kunin, W.E. (1999) The spatial structure of populations. *Journal of Animal Ecology* 68, 647–657.

Thomas, M. and Waage, J. (1996) *Integration of Biological Control and Host-Plant Resistance Breeding.* CTA, Wageningen, The Netherlands.

Townson, H., Harbach, R.E. and Callan, T.A. (1999) DNA identification of museum specimens of the *Anopheles gambiae* complex: an evaluation of PCR as a tool for resolving the formal taxonomy of sibling species complexes. *Systematic Entomology* 24, 95–100.

Trumble, J.T. (2002) Caveat emptor: Safety considerations for natural products used in arthropod control. *American Entomologist* 48, 7–13.

van de Fliert, E. (1993) *Integrated Pest Management: Farmer Field Schools Generate Sustainable Practices: a Case Study in Central Java Evaluating IPM Training.* Wageningen Agricultural University Papers 93–3, Wageningen, The Netherlands.

van den Bosch, R. (1978) *The Pesticide Conspiracy.* Doubleday, New York, USA.

van Huis, A. and Meerman, F. (1997) Can we make IPM work for resource-poor farmers in sub-Saharan Africa? *International Journal of Pest Management* 43, 313–320.

Vanlauwe, B., Diels, J., Sanginga, N. and Merckx, R. (eds) (2002) *Integrated Plant Nutrient Management in Sub-Saharan Africa: From Concept to Practice.* CAB International, Wallingford, UK.

Verkerk, R.H.J., Leather, S.R. and Wright, D.J. (1998) The potential for manipulating crop-pest-natural enemy interactions for improved insect pest management. *Bulletin of Entomological Research* 88, 493–501.

Vitousek, P., Mooney, H.A., Lubchenco, J. and Melillo, J.M. (1997) Human domination of earth's ecosystems. *Science* 277, 494–499.

Wajnberg, E., Scott, J.K. and Quimby, P.C. (2001) *Evaluating Indirect Ecological Effects of Biological Control.* CAB International, Wallingford, UK.

Waller, B.E., Hoy, C.W., Henderson, J.L., Stinner, B. and Welty, C. (1998) Matching innovations with potential users, a case study of potato IPM practices. *Agriculture, Ecosystems and Environment* 70, 203–215.

Weltzien, E., Smith, M.E., Meitzner, L.E. and Sperling, L. (2000) *Technical and Institutional Issues in Participatory Plant Breeding from the Perspective of Formal Plant Breeding: a Global Analysis of Issues, Results and Current Experience.* CGIAR Systemwide Program on Participatory Research and Gender Analysis Working Document No. 3. 106pp. http://www.prgaprogram.org/

Witcombe, J.R. (1996) Decentralization versus farmer participation in plant breeding: some methodology issues. *New Frontiers in Participatory Research and Gender Analysis. Proceedings of the International Seminar on Participatory Research and*

Gender Analysis for Technology Development 9–14 September CIAT Publication 294, Cali, Colombia, pp. 135–154.

Zadoks, J.C. and Waibel, H. (2000) From pesticides to genetically modified plants: history, economics and politics. *Netherlands Journal of Agricultural Science* 48, 125–149.

Zhu, Y., Chen, H., Fan, J., Wang, Y., Li, Y., Chen, J., Fan, J., Yang, S., Huß, L., Leungk, H., Mewk, T.W., Tengk, P.S., Wangk, Z. and Mundt, C.C. (2000) Genetic diversity and disease control in rice. *Nature* 406, 718–722.

Index

Page references in *italics* refer to information presented in the form of tables or figures.